PHOTOCHEMISTRY OF SMALL MOLECULES

PHOTOCHEMISTRY OF SMALL MOLECULES

HIDEO OKABE
National Bureau of Standards

A WILEY–INTERSCIENCE PUBLICATION • JOHN WILEY & SONS
NEW YORK • CHICHESTER • BRISBANE • TORONTO

Library of Congress Cataloging in Publication Data

Okabe, Hideo, 1923–

Photochemistry of small molecules.

"A Wiley-Interscience publication."

Bibliography: p.

Includes index.

1. Photochemistry. I. Title.

QD708.2.033 541'.35 78-6704

ISBN 0-471-65304-7

Printed in the United States of America

10 9 8 7 6 5 4 3 2 1

Preface

This book has been written in an attempt to cover the remarkable progress made in recent years in the field of photochemistry of small molecules with up to five atoms in the gas phase.

The advancement of flash photolysis–kinetic spectroscopy, laser technology, and other new techniques has made possible the detection of photochemical primary products and detailed studies of photodissociation dynamics. The secondary processes of atoms and radicals formed in the primary process have also been studied extensively. The reactivities of electronically excited atoms (C, O, S, etc.) and radicals (C_2O, CH_2, etc.) have been found to be very different from those of corresponding ground state atoms and radicals. The results of these studies, together with the traditional end product analysis and quantum yield measurements, have greatly aided in our understanding of the photochemical processes, particularly for small molecules. These recent developments, as well as the underlying principles, are described in some detail in dealing with the photochemistry of about 80 small molecules.

Studies of the photochemical processes of small molecules are not only of intrinsic interest but also are important in understanding the photochemistry of isotope enrichment, of air pollution in the troposphere and stratosphere, and of the atmospheres of other planets.

This book is aimed at the physical chemist, spectroscopist, and atmospheric scientist interested in photochemistry. As a reference book it lists about 1200 papers, including some original classic studies and those of recent years up to July 1977. The space limitation and the tremendous amount of publication in the last decade have prevented inclusion of some important papers. I apologize to those whose work has not been quoted.

Since the photochemical reaction is initiated by absorption of light in the visible, ultraviolet, and vacuum ultraviolet regions, an understanding of atomic and molecular spectroscopy is required. *Chapter I* gives a brief introduction to the electronic states and transitions in atoms and simple molecules.

The primary photochemical process brought about by light absorption is dealt with in *Chapter II*. A great deal of information on photodissociation dynamics (such as the lifetimes of the excited states of reactant molecules and the electronic, translational, and internal energy distribution in photofragments) has recently been obtained, particularly for some diatomic and triatomic molecules. Quantum mechanical theory dealing with dynamics of photodissociation has been proposed by many workers.

Various experimental techniques of photochemistry are described briefly in *Chapter III*.

Further details are given by Calvert and Pitts (4), Noyes and Leighton (22), and McNesby et al. (685a).

Production, detection, and reactivities of various electronically excited atoms are given in *Chapter IV*. The importance of electronically excited atom reactions in photochemistry has been recognized only recently.

Photochemical processes and electronic states of simple molecules with up to five atoms and radicals with up to four atoms in the gas phase are covered in *Chapters V through VII*. The absorption coefficients available for many molecules are shown in figures, as they are important in understanding the quantitative aspect of photochemistry. Bond dissociation energies given are calculated mostly from enthalpies of formation of atoms, radicals, and molecules tabulated in the *Appendix*.

Finally, enrichment of isotopic species has been achieved for a number of atoms and molecules using an appropriate monochromatic light source that preferentially excites an isotopic species of interest in mixtures of other isotopic species. The photochemistry associated with isotopic enrichment is briefly described in *Chapter VIII*. Great efforts have been made recently to obtain information on the detailed photochemical processes involving smog formation, stratospheric pollution, and atmospheres of other planets, and brief discussions of these subjects are also presented in the chapter.

I would like to express my appreciation to Professor W. A. Noyes, Jr., who introduced me to the field of photochemistry and to the late Dr. E. W. R. Steacie, the late Professor W. Groth, and Professor J. R. McNesby, who taught me various aspects of photochemistry. I am particularly grateful to Dr. R. E. Rebbert who has carefully read Chapters I, II, and VII and to Dr. A. H. Laufer for his critical reading of Chapters III through VI and VIII. Their numerous suggestions have greatly improved the manuscript. Thanks are due to Dr. M. D. Scheer for his continuous encouragement and to many friends and colleagues, especially to Drs. D. Garvin, P. J. Ausloos, R. F. Hampson, M. Krauss, and V. H. Dibeler and Professor J. P. Simons for their help and discussion during the preparation of the manuscript. I would like to thank Professor S. A. Rice at the University of Chicago who has given me the opportunity to write this book. I am indebted to Mrs. P. A. Davis who did an excellent job in typing the entire manuscript.

HIDEO OKABE

Gaithersburg, Maryland
March 1978

Contents

PHOTOCHEMISTRY OF SMALL MOLECULES

Introduction

Photochemistry deals with physical and chemical change initiated by the interaction of light with a molecule. It was first reorganized in 1817 on theoretical grounds by Grotthus and later, in 1843, by Draper as a result of experiments on the union of hydrogen and chlorine mixtures that light absorption is intimately connected with chemical change. This relationship, although qualitative, forms the first foundation of the modern photochemistry. According to the law of Grotthus and Draper only light that is absorbed by the molecule can be effective in inducing chemical change in the system. Chemical change occurs when the molecule is raised to an excited state possessing more than sufficient energy to break the weakest bond in the molecule. The bond dissociation energy ranges from about 1 eV in O_3 to more than 11 eV in CO. The corresponding range of light wavelength is from 12,400 to below 1130 Å. Radiation in the microwave (1 to 10 cm) and infrared regions (1 to 10 *μm*) absorbed by the molecule produces the rotationally and vibrationally excited molecule but is not effective in inducing chemical change.

Recently, however, chemical change has been obtained in polyatomic molecules by illuminating them with intense focused CO_2 laser pulses at 10.6 *μm*. In this case more than 25 photons must be absorbed simultaneously by the molecules to initiate chemical change. In most photochemical studies, however, the light intensity is of the order of 10^{14} to 10^{16} quanta sec^{-1}, in which case, following the Stark Einstein law, one photon process per one molecule can reasonably be applied. For example, we assume that a gas at a concentration of 10^{18} molec cm^{-3} absorbs 10% of the light from a source emitting 10^{15} quanta sec^{-1} cm^{-3} and that electronically excited molecules with a lifetime of 100 nsec are produced. The number of excited molecules during illumination is 10^7 cm^{-3}. If we further assume the same absorption coefficient for ground and excited molecules, the probability of second photon absorption is only $10^7/10^{18} = 10^{-11}$. On the other hand, if a laser pulse of 1 J in 10 nsec at 3000 Å is used, the total quanta output is 1.6×10^{18} per pulse. A 10% absorption would produce 1.6×10^{17} excited molecules. Hence, the probability of two-photon absorption increases to about 0.19 at the end of the pulse. If the excited molecules dissociate immediately into radicals, the probability of light absorption by the radicals depends on how rapidly the radicals decay by chemical reactions. If the reactions are slow and the radicals disappear within milliseconds, which is much longer than the typical flash time of 10 to 100 μsec, a substantial fraction of radicals may be dissociated further by the secondary photolysis prior to their disappearance.

Some simple molecules, such as O_3 and NO_2, absorb visible light, but most simple molecules absorb light only in the ultraviolet (2000 to 4000 Å) and vacuum ultraviolet (below 2000 Å) regions. The wavelength region of useful light in photochemistry is further limited by the window materials required for photochemical studies. The transmission of a quartz window is sharply reduced below 1700 Å and that of an LiF window decreases rapidly below 1100 Å. Hence, the wavelength region of interest is from about 1100 Å, the transmission limit of LiF windows, to about 7000 Å, the limit of absorption by simple molecules that leads to physical and chemical change.

In this wavelength range the primary products are in general either electronically excited molecules or their dissociation products, that is, radicals and stable molecules. In a few cases, such as in NO and NO_2, parent ions can be formed. In this respect the photochemical reaction is different from the reaction initiated by X-rays and gamma rays where the production of ions and their reactions with molecules are of primary concern.

Photochemical reactions are also different from thermal reactions. In thermal reactions ground state molecules are raised to higher vibrational levels of the electronic ground state by collisions with other molecules or with the walls, and dissociation occurs when sufficient energies are concentrated in the bond to be broken. In contrast, photochemical reactions involve molecules in electronically excited states. Accordingly, the initial mode of dissociation can be very different in the two cases. Consider, for example, the primary photochemical process in C_2H_6, which is

$$C_2H_6 \xrightarrow{h\nu} C_2H_4 + H_2$$

while the initial thermal reaction of C_2H_6 is

$$C_2H_6 \rightarrow 2CH_3$$

The photochemical primary process can be very specific, depending on the region of absorption. For example, the process

$$OCS \xrightarrow{h\nu} CO + S(^1S)$$

occurs predominantly in the region of absorption 1400 to 1700 Å. Likewise, in N_2O the process

$$N_2O \xrightarrow{h\nu} N_2 + O(^1S)$$

is mainly limited in the absorption region 1200 to 1400 Å. The primary process in ozone photolysis is almost exclusively

$$O_3 \xrightarrow{h\nu} O_2(^1\Delta) + O(^1D)$$

in the 2000 to 3100 Å region. No such specific dependence on energy has been found in thermal reactions.

Various units are used in photochemistry to express wavelength and energy. The wavelengths λ are usually expressed in micrometers (10^{-6} m), nanometers (10^{-9} m), and angstrom units (10^{-10} m). In this book the wavelengths are given in angstrom (Å) units. The wavelength region of interest is from about 1100, the transmission limit of LiF windows, to about 7000 Å, the limit of absorption by simple molecules that leads to chemical change.

The wave number $\bar{\nu}$ in cm^{-1} (number of waves per centimeter) is related to the energy E by the Planck relation, $E = hc\bar{\nu}$, where c is the velocity of light in vacuum ($2.99792458 \times 10^{-10}$ cm sec^{-1}) and h is Planck's constant (6.626176×10^{-27} erg sec). In theoretical work the frequency $\nu = c\bar{\nu}$ is commonly used, but in experimental work the wave number is more useful. The wave number is frequently used as a measure of the energy.

In photochemistry the energy of light corresponding to wavelength λ is usually expressed in eV (electron-volts) and kcal mol^{-1} or kJ mol^{-1}. One electron-volt is the kinetic energy of an electron accelerated under a potential difference of 1 V. This energy corresponds to

$$\frac{4.803242 \times 10^{-10}}{299.7925} = 1.602189 \times 10^{-12} \text{ erg molec}^{-1}$$

where 1/299.7925 is the conversion factor of 1 V into electrostatic units and 4.803242×10^{-10} is an electronic charge in electrostatic units.

The energy of light corresponding to wavelength λ in angstrom units is given by

$$E = \frac{hc}{\lambda} = \frac{6.626176 \times 10^{-27} \times 2.9979246 \times 10^{10}}{\lambda \times 10^{-8}} = \frac{1.986478 \times 10^{-8}}{\lambda} \text{ ergs}$$

or

$$\frac{12398.52}{\lambda} \text{ eV}$$

The energy can also be given in calories. Since one thermochemical calorie is equivalent to 4.184×10^7 ergs = 4.184 J, one obtains

$$1 \text{ eV} = \frac{1.602189 \times 10^{-12} \times 6.022045 \times 10^{23}}{4.184 \times 10^7} = 23.0604 \text{ kcal mol}^{-1}$$

Since one wave number (cm^{-1}) corresponds to 1.986478×10^{-16} erg $molec^{-1}$, 1 eV is equal to 8065.47 cm^{-1}.

These and other energy conversion factors are given in the table below.

Energy Conversion Factors

Quantity	Value	Unit
1 eV	1.6021892	10^{-12} erg $molec^{-1}$
	8065.479	cm^{-1}
	23.0604	kcal mol^{-1}
	96.4847	kJ mol^{-1}
1 cm^{-1}	2.85915	cal mol^{-1}
	11.96268	J mol^{-1}
	1.986478	10^{-16} erg $molec^{-1}$
	1.23985	10^{-4} eV $molec^{-1}$
λ (Å)	285915/λ	kcal mol^{-1}
	12398/λ	eV $molec^{-1}$

Based on Table A-1.

chapter I

Spectroscopy of Atoms and Molecules

Photochemical reactions are initiated by absorption of a light quantum by atoms and molecules.

Incident monochromatic light induces the transition from the ground state to a quantized electronically excited state in the case of atoms and to a quantized rotational and vibrational level of an upper electronic state in the case of small molecules. Hence, a knowledge of atomic and molecular spectroscopy is essential in understanding photochemical reactions.

Most light sources used to initiate photochemical reactions are atomic emission lines generated by an electric or microwave discharge, although laser lines have been increasingly used in recent years in the visible and ultraviolet regions.

The simplest photochemical system may be the reactions by electronically excited atoms produced by resonance absorption.

The reactions by electronically excited metastable atoms have been recognized to be very dependent on the state of excitation. To detect these metastable atoms and follow their reaction rates, time-dependent intensities of absorption and emission lines from these atoms have been studied.

The absorption spectra of molecules, particularly of diatomic molecules, are important in understanding the primary photochemical process. For polyatomic molecules an attempt to correlate the absorption spectrum with photochemistry has been less successful, since the spectrum of interest is in most cases diffuse or continuous and hence, almost no information is available as to the nature of the excited state responsible for photodissociation.

A brief overview of atomic and molecular spectroscopy relevant to photochemistry is given in this chapter. For details the reader is referred to the excellent books by Herzberg (13, 14, 16) on atomic and molecular spectroscopy.

I–1. ELECTRONIC STATES OF ATOMS

I–1.1. Atoms with One Outer Electron

The basic concept of atomic spectroscopy and structure was laid down by Niels Bohr in 1913. According to Bohr, atoms can exist only in certain stationary states. These states may be represented by discrete orbits of an electron around a positive nucleus. Light emission or absorption takes place only when the energy of a light quantum or photon, $h\nu$, corresponds exactly to the energy difference of the two stationary states or two discrete orbits. A discrete orbit is an ellipse whose major axis is proportional to the square of the whole number n, called the principal quantum number, which takes values

$$n = 1, 2, 3, \ldots \qquad \text{(I-1)}$$

The minor axis is proportional to n times $l + 1$ where l is called the azimuthal quantum number and takes values

$$l = 0, 1, 2, \ldots, n - 1 \qquad \text{(I-2)}$$

According to wave mechanics, the azimuthal quantum number l is the quantum number of the angular momentum produced by the electron rotating around the nucleus and has the value $\sqrt{l(l + 1)}$ or nearly l in units of $h/2\pi$. The angular momentum may be represented by a vector whose direction is perpendicular to the plane of orbit with the magnitude $\sqrt{l(l + 1)}(h/2\pi)$. Electrons with $l = 0, 1, 2, 3, \ldots$ are called $s, p, d, f, \ldots$ electrons.

In the presence of an electric or magnetic field the orbital angular momentum vector **l** (a boldface letter signifies a vector) can take only certain orientations to the direction of the field. The components of the angular momentum **l** to the direction of the field are designated m_l and have values (in units $h/2\pi$)

$$m_l = l, l - 1, \ldots, -l \qquad \text{(I-3)}$$

m_l is called the magnetic quantum number. In the absence of the field the $2l + 1$ quantum states of different m_l have exactly the same energy; that is, the states have a $(2l + 1)$-fold degeneracy. It was discovered, however, that the quantum numbers, n, l, and m_l are not sufficient to explain the experimental observations, for example, of a doublet splitting in the spectra of alkali atoms and a splitting of spectral lines in a magnetic field. Uhlenbeck and Goudsmit in 1925 postulated that an electron spins about its own axis and the spinning electron produces a magnetic moment along the axis of rotation. The interaction of the magnetic moment of the spinning electron with that of the orbital motion produces the observed doublet splitting. It is further necessary to assume that the magnitude of the angular momentum of the spinning electron is $\sqrt{s(s + 1)}(h/2\pi)$ where s is called the spin quantum number. It is required that s have the value $\frac{1}{2}$ to explain the observed results. The component m_s of the spin angular momentum along the axis of the field has a magnitude either of $+\frac{1}{2}(h/2\pi)$ or $-\frac{1}{2}(h/2\pi)$.

$$m_s = +\tfrac{1}{2} \text{ or } -\tfrac{1}{2} \qquad \text{(I-4)}$$

We have thus four quantum numbers n, l, m_l, and m_s to specify the state of the electron. In the absence of the field the states with the same n and l but different m_l and m_s have the same energy.

I–1.2. Atoms with More than One Outer Electron

Thus far we have been dealing with only one electron. If an atom has more than one electron it is necessary to consider the mutual interaction of individual electrons. The quantum states resulting from more than one electron may be understood if we consider the orbital angular momentum **l** and the spin angular momentum **s** of each electron as vectors. The angular momenta **l** of the individual electrons are coupled with one another to produce a resultant angular momentum **L**, and the spin angular momenta **s** are coupled together to give a resultant spin angular momentum **S**. (Capital letters **L**, **S** are used for the angular momentum of the atom as a whole and lowercase letters **l**, **s** for the individual electrons). The coupling of individual angular momenta follows the rule of vector addition. Thus, for two electrons, for example, the quantum numbers of the resultant orbital angular momenta are $l_1 + l_2, l_1 + l_2 - 1, \ldots |l_1 - l_2|$.

The resultant S values are given by 1 and 0, because each electron contributes $s = \frac{1}{2}$. The coupling between **L** and **S** is in most cases less strong than the coupling of the angular momenta of individual electrons. The resultant angular momentum **J** takes values

$$L + S, L + S - 1, \ldots, |L - S| \qquad \text{(I-5)}$$

with the magnitude $\sqrt{J(J+1)}(h/2\pi)$, just like the angular momenta of individual electrons. The resultant angular momentum **J** is called the total angular momentum of the atom. The energy difference between the states of different L is large, since the coupling or the interaction of individual angular momenta is strong. However, the coupling between **L** and **S** is in many cases weak. Hence, the energy difference between the states of the same L and S but different J is small. The coupling described above is called Russell-Saunders coupling. The quantum states of atoms corresponding to $L = 0$, 1, 2, 3 ... are called S, P, D, F, ... states. The number of J components is, from (I-5), $2S + 1$ and is called the multiplicity. The quantum state of atoms or the term is usually written as ${}^{2S+1}L_J$. For example, the electron configuration and the term of the ground state for C atoms are

$$1s^2 2s^2 2p^2 ({}^3P_0)$$

where the exponents of the electron configuration indicate the number of electrons in each orbit. [See Herzberg (13) for details.]

From the same electron configuration of C atoms, ${}^1D$ and ${}^1S$ states that are energetically higher than the ${}^3P$ state also result. For the derivation of these states one must use the Pauli principle, which states that no two electrons in the same atom can have the same four quantum numbers n, l, m_l, and m_s. In the case of C atoms $l_1 = 1$ and $l_2 = 1$. Hence, $m_l = 1, 0$, and -1. All possible values of M_L from m_{l_1} and m_{l_2} are given by the following table

	$l_1 = 1$				
$m_{l_1} =$	1	0	-1	m_{l_2}	
	2	1	0	1	$l_2 = 1$
	1	0	-1	0	$l_2 = 1$
	0	-1	-2	-1	$l_2 = 1$

As can be seen the values of M_L are

$$\begin{array}{ccccc} 2 & 1 & 0 & -1 & -2 \\ & 1 & 0 & -1 & \\ & & 0 & & \end{array}$$

The first row gives a term D, the second gives a term P, and the third corresponds to a term S. In a similar manner we have from $s_1 = \frac{1}{2}$ and $s_2 = -\frac{1}{2}$, $M_S = 1, 0, -1$, and 0. According to the Pauli principle, $M_L = 2, 1, 0, -1, -2$ and $M_L = 0$ can combine only with $M_S = 0$, giving the ^1D and ^1S terms. On the other hand each of $M_L = 1, 0, -1$ is associated with $M_S = 1, 0, -1$, resulting in the ^3P term.

I–2. QUANTUM STATES OF DIATOMIC MOLECULES

It is shown in the preceding section that atoms can exist only in certain discrete states. These states may also be represented by various orbits of electrons moving about a nucleus of positive charge. Each electron is designated by four quantum numbers and no two electrons can have the same four quantum numbers. The energy values for various atomic states can in principle be obtained by solving the Schrödinger equation using appropriate potential energies for electrons, although exact calculations have been performed only for a few simple atoms. In the case of a diatomic molecule two additional features, the vibration and rotation of the molecule, have to be considered.

I–2.1. Rotational Energy Levels of Diatomic Molecules

A diatomic molecule can rotate about an axis perpendicular to the internuclear axis and passing through the center of gravity.

We assume that the distance r between the masses m_1 and m_2 does not change with rotation (a rigid rotator model). Then the energy levels of such a rigid rotator can be found by solving the Schrödinger equation of a form

$$\sum_{i=1,2} \frac{1}{m_i}\left(\frac{\partial^2\psi}{\partial x_i^2} + \frac{\partial^2\psi}{\partial y_i^2} + \frac{\partial^2\psi}{\partial z_i^2}\right) + \frac{8\pi^2}{h^2}E\psi = 0$$

where ψ is a wavefunction of the coordinates alone and is called the eigenfunction; x_i, y_i, and z_i are the cartesian coordinates; m_i is the mass of the atom i; and E is the total energy. The results of the calculation show that the energies are given by [Herzberg (14), p. 68]

$$E = \frac{h^2 J(J+1)}{8\pi^2 \mu r^2} = \frac{h^2 J(J+1)}{8\pi^2 I} \tag{I-6}$$

where h is Planck's constant, I is the moment of inertia, μ is the reduced mass given by $\mu = [(m_1 m_2)/(m_1 + m_2)]$ and J is the rotational quantum number, which can take the integral values 0, 1, 2 . . . .

In classical mechanics the energy of the rigid rotator is

$$E = \frac{P^2}{2I} \tag{I-7}$$

where P is the angular momentum. From (I-6) and (I-7) we obtain

$$P = \frac{h\sqrt{J(J+1)}}{2\pi} \tag{I-8}$$

Thus, the magnitude of the angular momentum of the rigid rotator is $\sqrt{J(J+1)}$ in units of $h/2\pi$. The value of the angular momentum is the same given previously for the angular momentum of the electron (see Section I–1). If the wave number is used instead of energy, (I-6) becomes

$$F(J) = \frac{E}{hc} = \frac{h}{8\pi^2 cI} J(J+1) = BJ(J+1) \tag{I-9}$$

where the constant

$$B = \frac{h}{8\pi^2 cI}\ (\text{cm}^{-1}) \tag{I-10}$$

is called the rotational constant and $F(J)$ in cm^{-1} gives the rotational term value. In actual diatomic molecules the internuclear distance r is not constant but changes periodically with time near the equilibrium distance r_e. Furthermore, because of the action of centrifugal force the internuclear distance also increases with increasing rotation of molecules. Hence, the moment of inertia increases when J increases. Consequently, $F(J)$ decreases at larger

J. A detailed calculation shows that the actual rotational term is given by

$$F(J) = BJ(J+1) - DJ^2(J+1)^2 \quad \text{(I-11)}$$

where D is a constant much smaller than B. A deviation from a rigid rotator approximation becomes evident at larger J values.

I–2.2. Vibrational Energy Levels of Diatomic Molecules

We consider first a diatomic molecule as a simple harmonic oscillator, that is, a force F acting between the two atoms is proportional to the displacement x from the equilibrium position of the two nuclei. We obtain the potential energy V

$$V = \tfrac{1}{2}kx^2 \quad \text{(I-12)}$$

where k is called the force constant.

Solving the Schrödinger equation for $V = \tfrac{1}{2}kx^2$, we obtain energy values E

$$E = \frac{h}{2\pi}\sqrt{\frac{k}{\mu}}\left(v + \tfrac{1}{2}\right) = h\nu_{osc}\left(v + \tfrac{1}{2}\right) \quad \text{(I-13a)}$$

where $\nu_{osc} = 1/2\pi\sqrt{k/\mu}$ is the vibrational frequency of the oscillator, v is the vibrational quantum number, and μ is the reduced mass. The vibrational quantum number can take only integral values $v = 0, 1, 2, \ldots$. If we use wave number units (cm^{-1}) instead of energy units (ergs) we obtain

$$G(v) = \omega\left(v + \tfrac{1}{2}\right) \quad \text{(I-13b)}$$

where ω is the vibrational frequency in cm^{-1}.

In actual molecules, the energy difference of the successive vibrational levels is not constant as in the case of a harmonic oscillator, but it decreases as v increases. The vibrational term in the case of an anharmonic oscillator is

$$G(v) = \omega_e\left(v + \tfrac{1}{2}\right) - \omega_e x_e\left(v + \tfrac{1}{2}\right)^2 + \omega_e y_e\left(v + \tfrac{1}{2}\right)^3 + \cdots \quad \text{(I-13c)}$$

where v is the vibrational quantum number and $\omega_e x_e$, $\omega_e y_e$ are the anharmonicity constants ($\omega_e \gg \omega_e x_e \gg \omega_e y_e$). If we substract the following zero-point energy from (I-13c)

$$G(0) = \tfrac{1}{2}\omega_e - \tfrac{1}{4}\omega_e x_e + \frac{1}{8\omega_e y_e} + \cdots \quad \text{(I-14)}$$

we obtain

$$G_0(v) = \omega_0 v - \omega_0 x_0 v^2 + \omega_0 y_0 v^3 + \cdots \quad \text{(I-15)}$$

where

$$\begin{aligned}
\omega_0 &= \omega_e - \omega_e x_e + \tfrac{3}{4}\omega_e y_e + \cdots \\
\omega_0 x_0 &= \omega_e x_e - \tfrac{3}{2}\omega_e y_e + \cdots \\
\omega_0 y_0 &= \omega_e y_e + \cdots
\end{aligned}$$

Equation (I-15) gives the vibrational energy with respect to the zero-point energy.

I–2.3. Electronic States of Diatomic Molecules

In the previous sections we discuss the rotational and vibrational energies of a diatomic molecule. The rotational energy change involved between $J = 0$ and 1 is, from (I-9),

$$\bar{\nu}\,(\text{cm}^{-1}) = 2B$$

where B is the rotational constant. Since B is of the order of 1 cm^{-1}, the energy increase from $J = 0$ to 1 is of the order of 2 cm^{-1}. On the other hand, the change of energy for vibrational transition $v = 0$ to 1 is of the order of 1000 cm^{-1}. The rotational and vibrational energies that can be supplied by infrared absorption are so small compared with the dissociation energy of molecules that they are not important in initiating photochemical reactions in low intensity photolysis. The photochemical reaction is induced by absorption of light in the visible and ultraviolet regions. In these regions the electronic transition plays a major role. However, the vibrational and rotational energies in the ground electronic state may contribute to photodissociation in conjunction with the electronic transition (see Section I–4).

In contrast with atoms where electrons rotate about a spherically symmetrical field, electrons in a diatomic molecule move about a field that is cylindrically symmetric with respect to the internuclear axis. As a result, the orbital angular momentum of the electrons is defined along the internuclear axis. In analogy with the orbital angular momentum of electrons of atoms in a strong electric field, the component of the angular momentum of electrons along the internuclear axis takes the values M_L in units of $h/2\pi$ as

$$M_L = L, L - 1, L - 2, \ldots, -L \tag{I-16}$$

where L is the quantum number of the orbital angular momentum of electrons. The quantum number of the component of the orbital angular momentum of the electrons in the molecule is customarily designated as $\Lambda = |M_L|$. The quantum number Λ takes the values

$$\Lambda = 0, 1, 2, \ldots, L \tag{I-17}$$

The electronic states corresponding to $\Lambda = 0, 1, 2, 3, \ldots$ are represented by $\Sigma, \Pi, \Delta, \Phi \ldots$. The states with different Λ are usually widely separated in energy because of the strong interaction between the angular momentum and the electric field along the axis. The states with $\Lambda > 0$ are doubly degenerate because M_L and $-M_L$ have the same energy, although the direction of electronic rotation is opposite. The Σ state is nondegenerate. In analogy with atoms the spins of the individual electrons in the diatomic molecule form a resultant spin **S**. The resultant **S** has a component $\mathbf{M}_S$ along the internuclear axis in units of $h/2\pi$. M_S in molecules is written as Σ and has values

$$\Sigma = S, S-1, S-2, \ldots, -S \tag{I-18}$$

The quantum number of the total angular momentum given by Ω is the algebraic addition of $\Lambda + \Sigma$ along the internuclear axis

$$\Omega = |\Lambda + \Sigma| \tag{I-19}$$

Because of the interaction between **Λ** and **S** the energy of the state given by Λ is further divided into $2S+1$ substates when $\Lambda > 0$. Thus the electronic energy corresponding to Λ splits into $2S+1$ multiplet components. When $\Lambda = 0$, $2S+1$ states have the same energy, namely, they are degenerate. The molecular states in the diatomic molecule are designated as, just as for atoms,

$$^{2S+1}\Lambda_{\Lambda+\Sigma} \tag{I-20}$$

where a left superscript $2S+1$ is the multiplicity and a subscript $\Lambda + \Sigma$ denotes the multiplet component. If $\Lambda < S$, the number of molecular states is different from the corresponding number of atomic states. For example, for $\Lambda = 1$ and $S = \frac{3}{2}$, the resulting molecular states are $^4\Pi_{5/2}$, $^4\Pi_{3/2}$, $^4\Pi_{1/2}$, and $^4\Pi_{-1/2}$. On the other hand, the atomic states for $L = 1$, $S = \frac{3}{2}$ are $^4P_{5/2}$, $^4P_{3/2}$, and $^4P_{1/2}$, since J can take only positive values in the absence of a field in the atomic system. Hence, $\Lambda + \Sigma$ rather than Ω should be used to label the multiplet components for the molecular system. (If Ω is used the resulting molecular states are $^4\Pi_{5/2}$, $^4\Pi_{3/2}$, and $^4\Pi_{1/2}$, and $^4\Pi_{-1/2}$ is missing.)

I–2.4. Coupling of Rotation and Electronic Motion in Diatomic Molecules; Hund's Coupling Cases

In the diatomic molecule the total angular momentum designated by **J** consists of the electron spin, the orbital angular momentum of electrons, and the angular momentum of nuclear rotation. The electronic energies of the states depend primarily on the extent of coupling (or interaction) of the spin (**Σ**) and orbital angular momentum (**Λ**) of electrons about the internuclear axis. For $\Lambda \neq 0$ the coupling between **Λ** and **Σ** is strong, and accordingly the states with the same Λ but different Ω have considerably different energies. This is called Hund's case a (see Fig. I–1*a*). The total angular momentum **J**, a resultant produced from weak coupling between **Ω** and the angular momentum of nuclear rotation, takes the values

$$J = \Omega, \Omega + 1, \Omega + 2, \ldots \tag{I-21}$$

If $\Lambda = 0$ or for light molecules coupling between **Λ** and **Σ** is weak and the states with the same Λ but different $\Lambda + \Sigma$ have only a small energy difference.

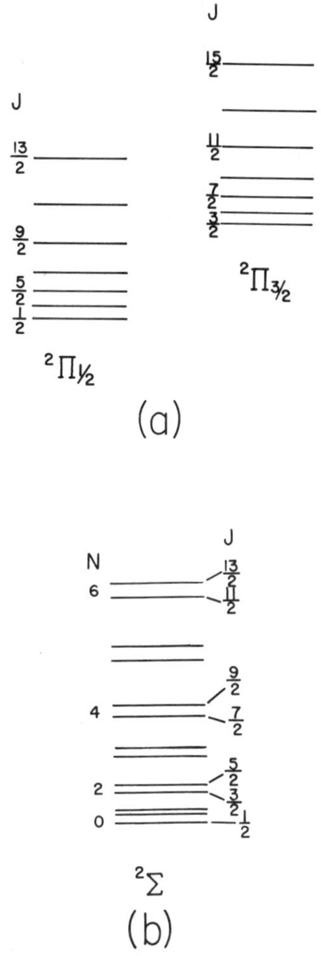

Fig. I–1. **(*a*) Hund's coupling case for the $^2\Pi$ state. The energy separation between $^2\Pi_{1/2}$ and $^2\Pi_{3/2}$ is large because of the large Λ, Σ interaction. *J* in $^2\Pi_{3/2}$ starts from $\frac{3}{2}$. (*b*) Hund's coupling case for the $^2\Sigma$ state. The lowest level ($N = 0$) is a singlet. For higher *N* each level is split into two because of the weak interaction between *N* and *S*. From G. Herzberg, *Molecular Spectra and Molecular Structure. Spectra of Diatomic Molecules*, 2nd. ed. pp. 220 and 222. Copyright 1950 by Litton Educational Publishing, Inc. Reprinted by permission of Van Nostrand Reinhold Company.**

In this case $\mathbf{\Lambda}$ and the nuclear rotation are weakly coupled to form a resultant $\mathbf{N}$. The quantum number N of the total angular momentum apart from spin can take values

$$N = \Lambda, \Lambda + 1, \Lambda + 2, \ldots \tag{I-22}$$

Weak coupling between $\mathbf{N}$ and $\mathbf{S}$ forms a resultant $\mathbf{J}$ with the values

$$J = N + S, N + S - 1, \ldots, |N - S| \tag{I-23}$$

(See Fig. I–1*b*.) Thus, Λ splits into $2S + 1$ components. This mode of coupling is called Hund's case b. Cases a and b are the most important among the five coupling cases discussed by Hund.

I–3. QUANTUM STATES OF POLYATOMIC MOLECULES

I–3.1. Rotational Levels of Polyatomic Molecules

In the case of diatomic molecules rotational energy levels are given in terms of the moment of inertia about the axis perpendicular to the internuclear axis and going through the center of mass (see Section I–2.1). In polyatomic molecules three mutually perpendicular axes, called the principal axes, must be considered about which the moment of inertia (the principal moment of inertia) is maximum or minimum. Depending on the magnitude of the moment of inertia four types of molecules result—spherical top, symmetric top, asymmetric top, and linear polyatomic molecules. If the three principal moments of inertia are equal, a molecule is called a spherical top. If two of the three principal moments of inertia are equal, a molecule is a symmetric top. If three moments of inertia are all different, a molecule is an asymmetric top. A linear polyatomic molecule is the special case of a symmetric top in which two principal moments of inertia are equal and the third moment of inertia is zero or nearly zero. The term values for the linear polyatomic molecule are the same as those for diatomic molecules and are given by (I-9). A general discussion of rotational levels for polyatomic molecules is given by Herzberg [(16), p. 82].

According to international nomenclature, the rotational constants are designated as

$$A > B > C \tag{I-24}$$

and the corresponding axes are called a axis, b axis, and c axis. If $B = C$, the molecule is called a prolate symmetric top and the a axis about which the moment of inertia is smallest is called the top axis or figure axis. On the other hand, if $A = B$, the molecule is called an oblate symmetric top and the top axis about which the moment of inertia is largest is the c axis. For the prolate symmetric top, the rotational term values are given by

$$F(J, K) = BJ(J + 1) + (A - B)K^2 \tag{I-25}$$

where $A = h/8\pi^2 cI_A$, $B = h/8\pi^2 cI_B$, and K is the quantum number of the component of the angular momentum along the top axis. For the oblate symmetric top, the rotational term values are

$$F(J, K) = BJ(J + 1) - (B - C)K^2 \tag{I-26}$$

Energy levels with the same J decrease with an increase of K in this case, while in the prolate symmetric top energy levels with the same J increase with increasing K. The component $\mathbf{K}$ of the total angular momentum $\mathbf{J}$ along the top axis, has the magnitude $K(h/2\pi)$. Since $\mathbf{K}$ is a component of

J, K cannot be greater than J. For each K, J takes a series of values

$$J = K, K + 1, K + 2, \ldots \tag{I-27}$$

Both formaldehyde and nitrogen dioxide have the rotational constants $A > B \simeq C$. Hence, they are nearly symmetric tops. As shown in Fig. I–2, the top axis in formaldehyde is along the z axis, while in nitrogen dioxide, the top axis is along the y axis.

I–3.2. Vibrational Levels of Polyatomic Molecules

In the case of polyatomic molecules there are $3N - 6$ vibrational degrees of freedom for nonlinear molecules and $3N - 5$ vibrational degrees of freedom for linear molecules, where N is the number of atoms in the molecule. The vibrational term values without degenerate vibrations are [Herzberg (15), p. 208]

$$G(v, v_2 \ldots) = \sum_i \omega_i(v_i + \tfrac{1}{2}) + \sum_i \sum_{k>i} x_{ik}(v_i + \tfrac{1}{2})(v_k + \tfrac{1}{2}) + \cdots \tag{I-28}$$

instead of (I-13c) for diatomic molecules. Here ω_i is the frequency (cm^{-1}) of the ith normal vibration corresponding to ω_e of diatomic molecules, v_i

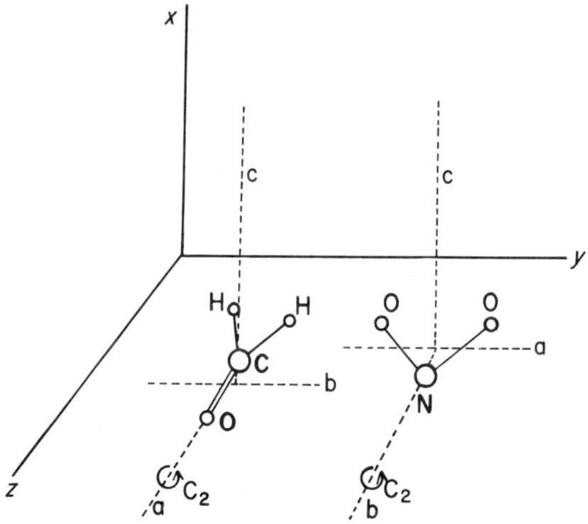

Fig. I–2. The ground state structure of formaldehyde and nitrogen dioxide; both belong to the point group C_{2v}; C_2 is a twofold axis of symmetry which is the z axis; the x axis is perpendicular to the molecular plane: H—C—H angle = 116°, O—N—O angle = 134°. The axes corresponding to the rotational constants A, B, C (cm^{-1}) ($A > B > C$) are designated as a, b, and c (723, 831). $A_0 = 9.4053$, $B_0 = 1.2953$, $C_0 = 1.1342$ in cm^{-1} for formaldehyde and $A_0 = 8.0012$, $B_0 = 0.43364$, $C_0 = 0.41040$ in cm^{-1} for nitrogen dioxide; a is the top axis. Both molecules belong to a near symmetric top.

is the vibrational quantum number of the ith normal vibration or harmonic motion, and x_{ik} is the anharmonicity constant corresponding to $\omega_e x_e$ for diatomic molecules.

I–3.3. Electronic States of Polyatomic Molecules

The electronic states of a polyatomic molecule are generally classified according to the symmetry properties of the molecule in the equilibrium position. The electronic eigenfunctions either change sign or remain unchanged by the symmetry operations allowed for the point group to which the molecule belongs. The symmetry property of the electronic eigenfunction is important to determine whether a given transition is allowed and to obtain the direction of the transition moment. Conversely, if the direction of the transition moment is known experimentally, it is possible to assign the symmetry of the excited state from which photodissociation occurs (see Section II–5).

Another expression of the electronic states of polyatomic molecules is the so-called molecular orbital notation. In this notation each electron is associated with the eigenfunction (or orbital) corresponding to the motion of the electron in the field of the fixed nuclei and the average field of other electrons. The eigenfunction has the same symmetry properties as one of the symmetry species of the point group to which the molecule belongs. For example, for the point group C_{2v} molecular orbitals are represented by a_1, a_2, b_1, and b_2. Electronic energies of the molecular orbitals can be estimated from the correlation between the united atom and the separated atoms.

The ground electronic state corresponds to the electron configuration that is made up by filling each orbital with two electrons of opposite spin starting from the lowest orbital. An excited electronic state can be obtained by moving one outer electron into higher orbitals. For example, the ground state electron configuration of H_2O is in order of increasing energy

$$(1a_1)^2(2a_1)^2(1b_2)^2(3a_1)^2(1b_1)^2$$

where the coefficients 1, 2, 3 represent the order of increasing energy and superscripts are the number of electrons. The molecular orbital representation of the electronic states is useful to estimate electronic energies and configurations of various electronic states. However, a detailed discussion is beyond the scope of this book and the reader is referred to Herzberg (16).

Following Herzberg (16), an example is given in this section to explain the group theoretical notation of the electronic state of a molecule belonging to the point group C_{2v}, such as formaldehyde, nitrogen dioxide, and water. Four symmetric types (or species) of electronic state are given for the

C_{2v} molecules. Figure I–2 shows the structures of the ground state H_2CO (planar) and NO_2. They have a twofold axis of symmetry (the z axis of the cartesian coordinates), that is, the same configuration is obtained by 180° rotation of the molecule about the z axis. A twofold symmetry axis is designated by C_2. In addition ground state H_2CO and NO_2 have two symmetry planes yz and xz through the symmetry axis. These planes are denoted by σ_v. C_2 and σ_v are called symmetry elements. The electronic eigenfunction either changes sign or remains unchanged by the symmetry operation (that is, 180° rotation about the z axis or reflections at the symmetry planes). When the eigenfunction changes sign by the operation it is called antisymmetric and is written as a minus sign, and if the eigenfunction remains unchanged, it is called symmetric and is designated by a plus sign. Depending on the plus or minus sign as determined by the two symmetry operations, that is, either by 180° rotation about the z axis or reflection at the symmetry plane yz, four types (species) of electronic states are obtained. These four states may be designated as $++$, $+-$, $--$, and $-+$, corresponding to $A_1, A_2, B_1,$ and B_2 symmetry species (see Table I–1). According to the IUPAC recommendation, the x axis is taken perpendicular to the molecular plane and the z axis along the C_2 symmetry axis [Moule and Walsh (723)]. The species of the translation in the x, y, and z directions (T_x, T_y, T_z) and rotation about the x, y, and z axes (R_x, R_y, R_z) are also indicated in the last column of Table I–1. These species are important for selection rules given in Section I-10. The ground states of formaldehyde and nitrogen dioxide belong to the species A_1.

Table I–1. Symmetry Species and Characters for the C_{2v} Point Group

C_{2v}	$C_2(z)$	$\sigma_v(xz)$	$\sigma_v(yz)$	
A_1	$+1$	$+1$	$+1$	T_z
A_2	$+1$	-1	-1	R_z
B_1	-1	$+1$	-1	T_x, R_y
B_2	-1	-1	$+1$	T_y, R_x

Note: A molecule belonging to the point group C_{2v} has four symmetry species, A_1, A_2, B_1, and B_2. The electronic eigenfunctions of these species change sign $(-)$ or remain unchanged $(+)$ as indicated by the symmetry operations C_2 (180° rotation about the symmetry axis z) and σ_v (reflection at the symmetry planes). T_x, T_y, T_z are translations along x, y, z directions and R_x, R_y, R_z are rotations about the x, y, and z axes. T_x, T_y, and T_z belong to the species B_1, B_2, and A_1, respectively.

I–4. THERMAL CONTRIBUTION TO PHOTODISSOCIATION

The dissociation of a molecule may occur when the photon energy $h\nu$ exceeds the bond dissociation energy D_0, where D_0 is referred to the lowest vibrational and rotational levels of the potential energy curve of the ground state. Most photochemical experiments are performed near room temperature and not at 0°K. Consequently, a fraction of molecules would be in vibrationally and rotationally excited levels. The molecules in those excited levels of the ground state may undergo dissociation by photon absorption even below the photon energy corresponding to D_0.

Because of the experimental difficulty in obtaining sufficiently intense monochromatic light of varying wavelengths near threshold, only a few studies have been done to measure the extent of photodissociation as a function of incident wavelength. Since the vibrational quanta are normally larger than the rotational quanta, it is expected that the rotational contribution near room temperature is much larger than the vibrational contribution for photodissociation in view of the Maxwell-Boltzmann law described in the following section.

I–4.1. Vibrational Population in Diatomic Molecules

According to the Maxwell-Boltzmann distribution law, in thermal equilibrium the number of molecules possessing an energy E at a given absolute temperature T is proportional to $e^{-E/kT}$ where k is Boltzmann's constant equal to 0.6952 cm^{-1} K^{-1}. The relative population of molecules at 300°K as a function of the vibrational energy (cm^{-1}) is given in Fig. I–3. If the vibrational energy is not quantized, the distribution of vibrational energies among molecules would be a continuous function of energy with a maximum at $E = 0$. Since the vibrational energy is quantized, most diatomic molecules

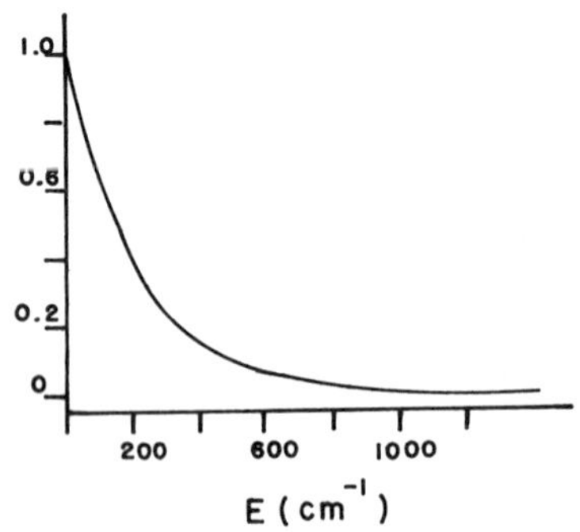

Fig. I–3. Thermal distribution of molecules as a function of the vibrational energy E in cm^{-1} at $T = 300$°K. The ordinate is $e^{-E/kT}$. The vibrational energies are in reality not continuous as shown, but quantized. From G. Herzberg, *Molecular Spectra and Molecular Structure I, Spectra of Diatomic Molecules*, 2nd ed., p. 122. Copyright 1950 by Litton Educational Publishing, Inc. Reprinted by permission of Van Nostrand Reinhold Company.

are in the lowest vibrational levels at room temperature. However, heavy molecules such as Cl_2 and I_2 have relatively small vibrational quanta (557 and 213 cm^{-1}, respectively) and, as a result, appreciable fractions, namely, 7 and 36%, respectively, for Cl_2 and I_2, are in the first vibrational levels. The sum of the Boltzmann factors $e(-G_v/kT)$ over all states (0 to v), which is called the vibrational partition function, is

$$Q_v = 1 + e^{-G_1/kT} + e^{-G_2/kT} + \cdots e^{-G_v/kT} \tag{I-29}$$

Therefore, the number of molecules at a level $v(N_v)$ is

$$N_v = \frac{N_0}{Q_v} e^{-G_v/kT} \tag{I-30}$$

where N_0 is the total number of molecules. For most cases the second term of (I-29) is much smaller than one and $N_v \simeq N_0 e^{-G_v/kT}$.

I–4.2. Rotational Population in Diatomic Molecules

The rotational population in diatomic molecules at the absolute temperature T, unlike the vibrational population, is not simply given by the Boltzmann factor $e^{-E/kT}$, because each rotational level has a $(2J + 1)$-fold degeneracy. The number of molecules at a given rotational level J and at T is

$$N_J \propto (2J + 1)e^{-F_J/kT} \tag{I-31}$$

In a rigid rotator approximation, F_J is from (I-9)

$$F_J = BJ(J + 1) \tag{I-32}$$

The rotational partition function (or state sum) Q_r is

$$Q_r = 1 + 3e^{-2B/kT} + 5e^{-6B/kT} + \cdots \tag{I-33}$$

where B is the rotational constant. Thus, the number of molecules in J is

$$N_J = \frac{N_0}{Q_r}(2J + 1)e^{-BJ(J+1)/kT} \tag{I-34}$$

where N_0 is the total number of molecules. The function $(2J + 1)e^{-BJ(J+1)/kT}$, which is proportional to the number of molecules at the rotational level J and at T, is plotted as a function of J in Fig. I–4 with $B = 10.44$ cm^{-1} and $T = 300$°K (HCl, the ground state). Since $2J + 1$ increases linearly with an increase of J, while the exponential term decreases with an increase of J, the population goes through a maximum. In the case of HCl, $J_{max} = 2.66$ [Herzberg (14), p. 124] and $Q_r = 20$ at 300°K. The fractional number of molecules at a given J is obtained by dividing the corresponding ordinate in Fig. I–4 by 20.

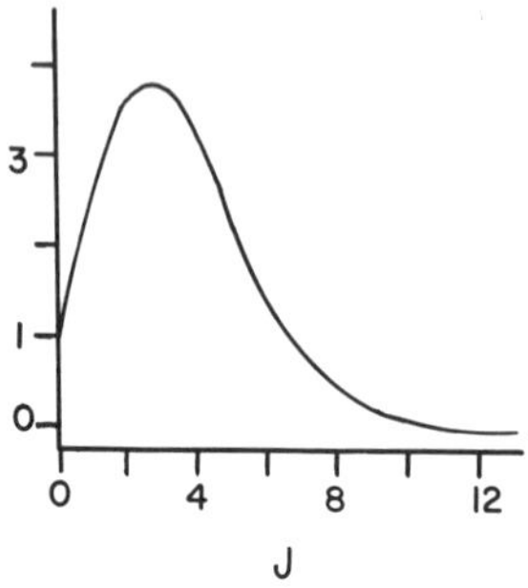

Fig. I–4. Rotational population at 300°K as a function of the total angular momentum *J*. The population of rotational levels is given by $(2J+1)\,e[-BJ(J+1)/kT]$, where the rotational constant $B = 10.44$ cm^{-1}, $k = 0.6952$ cm^{-1} K^{-1} (HCl in the ground state). From G. Herzberg, *Molecular Spectra and Molecular Structure I, Spectra of Diatomic Molecules*, 2nd ed., p. 122. Copyright 1950 by Litton Educational Publishing, Inc. Reprinted by permission of Van Nostrand Reinhold Co.

I–4.3. Thermal Contribution to Photolysis and Fluorescence

Few examples have been found in the literature to demonstrate the effect of internal energy on the photochemical primary process.

The bond dissociation energy D_0(NO—O) of NO_2 is 71.89 kcal mol^{-1}, corresponding to the wavelength 3977 Å. Therefore, photodissociation is energetically possible only below 3977 Å if the experiment is performed at 0°K. However, it has been observed that even at 4047 Å, the dissociation takes place to an extent of 36% (810) at room temperature (see Fig. I–5, curve 1). It must be assumed that the internal energy has contributed to dissociation in addition to the photon energy at 4047 Å.

The yield of NO_2 fluorescence, on the other hand, starts to decrease at 4150 Å and disappears completely below 3977 Å (see curve 2 of Fig. I–5). The three vibrational frequencies of NO_2 are $\nu_1 = 1322$ cm^{-1}, $\nu_2 = 751$ cm^{-1}, $\nu_3 = 1616$ cm^{-1}. Of these the bending vibration ν_2, the lowest frequency, would contribute mainly to the vibrational energy at room temperature. From (I-30) the population of molecules in the first vibrational level is about 3%. Since the photon energy at 4047 Å is 435 cm^{-1} less than the dissociation energy of NO_2 at 0°K the vibrationally excited molecules (with 751 cm^{-1} energy) can dissociate by absorption of light at 4047 Å. However, the vibrationally excited molecules amount to only 3% of the ground state molecules. A substantial fraction of molecules must then be rotationally excited.

The rotational term values of NO_2 may be given approximately by those of the symmetric top, that is, (I-25)

$$F(J, K) = \bar{B}J(J+1) + (A - \bar{B})K^2$$

where A is the rotational constant with respect to the top axis and $\bar{B}$ is an average of the other two nearly equal constants. The rotational constants A and $\bar{B}$ are 8.0012 and 0.4220 cm^{-1}, respectively (16) for NO_2. Figure I–6 shows the rotational population of NO_2 at 300°K as a function of rotational energy in cm^{-1}. Curve 1 is the population of molecules when $K = 0$ is

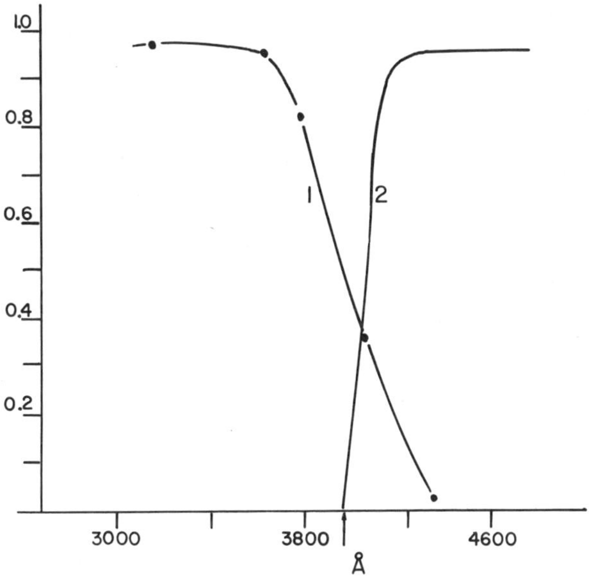

Fig. I–5. Photodissociation and fluorescence yields of NO_2 as a function of incident wavelength at 300°K; the arrow indicates the threshold wavelength (3977 Å) for dissociation at 0°K. Curve 1: primary yield of photodissociation; curve 2: relative fluorescence yield. From Pitts et al. (810) and Lee and Uselman (619). Reprinted by permission of the American Institute of Physics and by The Chemical Society. Copyright 1964 by the American Institute of Physics.

assumed, that is, if NO_2 is a linear molecule. Since NO_2 is the near symmetric top, component K must also be taken into consideration. A series of curves corresponding to various K values are obtained according to (I-25) and (I-31). In these curves the rotational energies corresponding to $J < K$ are missing. The superposition of these curves is given as curve 2 in Fig. I–6. Curve 2 starts at the same population as in curve 1, but its maximum shifts towards higher rotational energy compared to curve 1. The shaded area indicates the fractional population of molecules with rotational energies above 450 cm^{-1} and is 33% of the total number of molecules, which is nearly equal to the percent of decomposition. Thus, almost all rotationally excited molecules with energies above 435 cm^{-1} (the difference of D_0 and the photon energy at 4047 Å) must have contributed to dissociation by absorption of the 4047 Å light. Although the detailed processes involving light absorption and the resulting photodissociation are not well understood, it is reasonable to assume that rotational energies are converted at least partially upon light absorption to the antisymmetric vibration ν_3 of NO_2 before dissociation

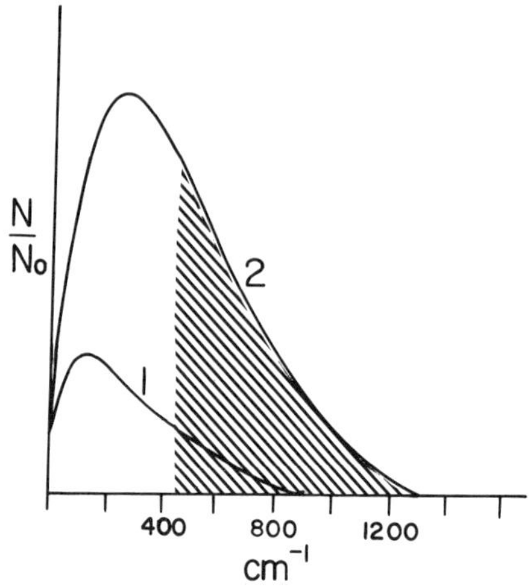

Fig. I–6. The fractional rotational population (relative) of NO_2 at 300°K as a function of rotational energy $E_{J,K}$ in cm^{-1}; rotational energy $E_{J,K} = \bar{B}J(J+1) + (A - \bar{B})K^2$, $\bar{B} = 0.4220\ cm^{-1}$, $A = 8.0012\ cm^{-1}$. From Pitts et al. (810) N/N_0 is proportional to $(2J+1)e^{-E_{J,K}/kT}$. Curve 1: the rotational population assuming a linear configuration ($K = 0$); curve 2: the population of the actual molecule (symmetric top). The shaded area shows the molecules with energies above 450 cm^{-1}.

takes place. It is, however, unlikely that all rotational energies can contribute to the vibration leading to dissociation. It is more likely that the fraction of molecules with more than sufficient rotational energy for dissociation is larger than 33%. The similar temperature dependence on the photolysis of ozone near 3100 Å has recently been found.

The thermochemical threshold energy at 0°K for the process

$$O_3 \xrightarrow{h\nu} O(^1D) + O_2(^1\Delta)$$

is 92.2 ± 0.4 kcal mol^{-1}, which corresponds to the incident wavelength 3100 ± 15 Å.

Kajimoto and Cvetanović (555) have found that the quantum yield of $O(^1D)$ production in the region 2300 to 2800 Å is temperature independent, but at 3130 Å $\phi_{O(^1D)}$ decreases from 0.53 to 0.23 as the temperature decreases from 40° to −75°C. Similar temperature dependence has been found by Kuis et al. (591). They have attributed the decrease of $\phi_{O(^1D)}$ to the loss of rotational energy.

I–5. ELECTRONIC TRANSITION IN ATOMS

To correlate the experimentally measurable absorption intensities and emission lifetimes by atoms with theoretically derived quantities such as the oscillator strength and the transition moment, it is necessary to derive equations of transition probabilities of absorption and spontaneous emission. The selection rules in atoms are discussed in Section I–10.1.

I–5.1. Einstein Transition Probabilities

The absorption and emission of radiation by atoms may be treated by perturbation of an electric dipole of the system by an external electric field of radiation. The perturbation energy by the oscillating electric field of radiation acting on the electric dipole $\boldsymbol{\mu}$ in the x direction is $E_x\mu_x$, where E_x and μ_x are the x components of the electric vector $\mathbf{E}$ and the dipole moment vector $\boldsymbol{\mu}$, respectively. The dipole moment $\boldsymbol{\mu}_x$ is equal to ex, where e is the electric charge and x is the cartesian coordinate of a particle in the system. According to wave mechanics, when the perturbation energy $\mathbf{E}_x\boldsymbol{\mu}_x$ is introduced into the wave equation, it can be shown [Pauling and Wilson (23), p. 302] that the transition probability per second from a state m to a state n induced by light is

$$\frac{8\pi^3}{3h^2c}|\mathbf{R}_{xnm}|^2\rho_{\bar{\nu}} \tag{I-35}$$

where c is the velocity of light, $\rho_{\bar{\nu}}$ is the radiation density in units of erg cm^{-3} per wave number and $\mathbf{R}_{xnm}$ is

$$\mathbf{R}_{xnm} = \int \psi_n^* \boldsymbol{\mu}_x \psi_m \, dx \tag{I-36}$$

(ψ^* is the complex conjugate of the eigenfunction ψ).

Similar expressions hold for the y and z directions. Thus, we obtain for isotropic radiation the transition probability from m to n per second

$$\frac{8\pi^3}{3h^2c}(|\mathbf{R}_{xnm}|^2 + |\mathbf{R}_{ynm}|^2 + |\mathbf{R}_{znm}|^2)\rho_{\bar{\nu}} = \frac{8\pi^3}{3h^2c}|\mathbf{R}_{nm}|^2\rho_{\bar{\nu}} \tag{I-37}$$

The quantity $\mathbf{R}_{nm}$ is a vector with components $\mathbf{R}_{xnm}$, $\mathbf{R}_{ynm}$, $\mathbf{R}_{znm}$ in the x, y, z directions and is called the transition moment. The transition probability per second from the lower state m to the upper state n is also expressed by the Einstein transition probability of absorption B_{nm}:

$$B_{nm}\rho_{\bar{\nu}} \tag{I-38}$$

From (I-37) and (I-38) we obtain

$$B_{nm} = \frac{8\pi^3}{3h^2c}|\mathbf{R}_{nm}|^2 \ (\text{erg}^{-1}\ \text{cm}^2\ \text{sec}^{-1}) \tag{I-39a}$$

If the lower state is degenerate

$$B_{nm} = \frac{8\pi^3}{3h^2cg_m}|\mathbf{R}_{nm}|^2 \tag{I-39b}$$

where g_m is the degeneracy.

Einstein provides an important relationship between absorption and emission of radiation by atoms. The number of systems undergoing transitions from m to n per second is

$$N_m B_{nm} \rho_{\bar{\nu}}$$

The number of systems undergoing the reverse transitions is

$$N_n \{A_{mn} + B_{mn} \rho_{\bar{\nu}}\}$$

where A_{mn} is the Einstein transition probability of spontaneous emission and B_{mn} is the Einstein transition probability of induced emission. At equilibrium with radiation the number of transitions from m to n must be equal to the number from n to m

$$\frac{N_m}{N_n} = \frac{A_{mn} + B_{mn}\rho_{\bar{\nu}}}{B_{nm}\rho_{\bar{\nu}}}$$

From this relationship and Planck's radiation expression of energy density,

$$\rho_{\bar{\nu}} = \frac{8\pi hc\bar{\nu}^3}{e^{h\nu/kT} - 1}$$

Einstein has shown

$$B_{mn} = B_{nm} \tag{I-40}$$

$$A_{mn} = 8\pi hc\bar{\nu}^3 B_{nm}\ (\text{sec}^{-1}) \tag{I-41}$$

where c is the velocity of light and $\bar{\nu}$ is the wave number of the transition. From (I-39a) and (I-41) we obtain

$$A_{mn} = \frac{64\pi^4\bar{\nu}^3}{3h} |\mathbf{R}_{nm}|^2 \tag{I-42a}$$

If the upper level is degenerate

$$A_{mn} = \frac{64\pi^4\bar{\nu}^3}{3hg_n} |\mathbf{R}_{nm}|^2 \tag{I-42b}$$

where g_n is the degeneracy of the upper level.

I–5.2. Absorption Intensity of Atoms

In this section the correlation between the Einstein transition probability and the integrated absorption of atoms is derived.

Consider a parallel beam of light of the wave number range from $\bar{\nu}$ to $\bar{\nu} + d\bar{\nu}$ and of intensity $I_{\bar{\nu}}$ passing through a layer of atoms between l and $l + dl$. The decrease of intensity by absorption of light of wave numbers between $\bar{\nu}$ and $\bar{\nu} + d\bar{\nu}$ is

$$-dI_{\bar{\nu}}\, d\bar{\nu} = dl hc\bar{\nu} B_{nm} \rho_{\bar{\nu}}\, \delta N_{\nu} \tag{I-43}$$

The intensity $I_{\bar{\nu}}$ is given in units of erg cm^{-2} sec^{-1} per wave number and N is the number of atoms per cm^3 of which $\delta N_{\bar{\nu}}$ is capable of absorbing light of wave number between $\bar{\nu}$ and $\bar{\nu} + d\bar{\nu}$. The intensity $I_{\bar{\nu}}$ is related to the energy density $\rho_{\bar{\nu}}$ by

$$I_{\bar{\nu}} = c\rho_{\bar{\nu}} \tag{I-44}$$

From (I-43) and (I-44) we have

$$-dI_{\bar{\nu}}\, d\bar{\nu} = dl h\bar{\nu} B_{nm} I_{\bar{\nu}}\, \delta N_{\bar{\nu}} \tag{I-45}$$

The absorption coefficient, $\alpha_{\bar{\nu}}$ of N atoms per cm^3, is defined by

$$I_{\bar{\nu}} = I_0^{-\alpha_{\bar{\nu}} l}$$

where I_0 and $I_{\bar{\nu}}$ are incident and transmitted intensities at a wave number $\bar{\nu}$ and l is the path length in centimeters.

Differentiating, we find for the intensity decrease over the wave number range from $\bar{\nu}$ to $\bar{\nu} + d\bar{\nu}$ and the path length between l and $l + dl$

$$-dI_{\bar{\nu}}\, d\bar{\nu} = dl I_{\bar{\nu}} \alpha_{\bar{\nu}}\, d\bar{\nu} \tag{I-46}$$

From (I-45) and (I-46), we obtain $\alpha_{\bar{\nu}}\, d\bar{\nu} = h\bar{\nu} B_{nm}\, \delta N_{\bar{\nu}}$, and integration over the whole line [with (I-39a)] leads to

$$\int_0^{\infty} \alpha_{\bar{\nu}}\, d\bar{\nu} = h\bar{\nu} N B_{nm} = \frac{8\pi^3 \bar{\nu} N}{3hc} |\mathbf{R}_{nm}|^2 \tag{I-47}$$

Using the relationship $B_{nm} = A_{mn}/8\pi hc\bar{\nu}^3$ from (I-41) and $A_{mn} = 1/\tau$ (τ is the mean life of the state n), we find

$$\int_0^{\infty} \alpha_{\bar{\nu}}\, d\bar{\nu} = \frac{A_{mn} N}{8\pi c\bar{\nu}^2} = \frac{N}{8\pi c\bar{\nu}^2 \tau} \tag{I-48a}$$

where c is the velocity of light, $\bar{\nu}$ is the wave number of the absorption line, and τ is the radiative life of the upper state. If the upper and lower states are degenerate

$$\int_0^{\infty} \alpha_{\bar{\nu}}\, dv = \frac{N}{8\pi c\bar{\nu}^2 \tau} \frac{g_n}{g_m} \tag{I-48b}$$

where g_n, g_m are the degeneracies of the upper and lower states, respectively.

I–5.3. Oscillator Strength [Barrow (2), p. 81]

The intensity of the absorption line is sometimes expressed as a dimensionless quantity called oscillator strength f. The oscillator strength is the ratio of the observed integrated absorption coefficient $\int \alpha_{\bar{\nu}}\, d\bar{\nu}$ to that of N electrons bound to N atoms with a Hooke's law type of force. Treating N bound electrons as N harmonic oscillators, we obtain as the integrated absorption coefficient of N electrons the quantity

$$\frac{N\pi e^2}{mc^2} \tag{I-49}$$

Hence, the oscillator strength is

$$f = \frac{mc^2}{N\pi e^2} \int \alpha_{\bar{\nu}}\, d\bar{\nu} \tag{I-50}$$

where m and e are the electronic mass and charge, respectively, and c is the velocity of light. $\int \alpha_{\bar{\nu}}\, d\bar{\nu}$ is in units of cm^{-2}. If the absorption coefficient $k_{\bar{\nu}}$ of atoms under 1 atm at 0°C is used instead of $\alpha_{\bar{\nu}}$ of N atoms, (I-50) becomes

$$f = 4.20 \times 10^{-8} \int k_{\bar{\nu}}\, d\bar{\nu} \tag{I-51}$$

From (I-50) and (I-48b) it follows that the relationship between the oscillator strength and the lifetime is

$$f = \frac{mc}{8\pi^2 e^2 \bar{\nu}^2} \frac{1}{\tau} \frac{g_n}{g_m} = \frac{1.50}{\bar{\nu}^2} \frac{1}{\tau} \frac{g_n}{g_m} \tag{I-52}$$

where g_n and g_m are degeneracies of the upper and lower states, respectively, $\bar{\nu}$ is the wave number of the absorption line in cm^{-1}, and τ is the lifetime in seconds.

If the absorption line profile is determined by Doppler broadening alone, the integrated absorption coefficient given by (I-50) can be expressed in terms of α_0, the absorption coefficient at the peak, and $\Delta\nu_D$, the Doppler width at a half maximum in wave numbers.

For the Doppler broadened absorption line (see Section I–6.1)

$$\alpha_{\bar{\nu}} = \alpha_0 e^{-\{2(\bar{\nu}-\bar{\nu}_0)/\Delta\bar{\nu}_D \sqrt{\ln 2}\}^2}$$

we obtain [Mitchell and Zemansky (21), p. 99]

$$\int \alpha_{\bar{\nu}}\, d\nu = \frac{1}{2}\sqrt{\frac{\pi}{\ln 2}}\, \alpha_0\, \Delta\nu_D \tag{I-53}$$

From (I-50) and (I-53) one obtains

$$\alpha_0 = \frac{2}{\Delta\bar{\nu}_D} \sqrt{\frac{\ln 2}{\pi}} \frac{\pi e^2}{mc^2} Nf \tag{I-54a}$$

$$= \frac{0.832 \times 10^{-12}}{\Delta\bar{\nu}_D} Nf \tag{I-54b}$$

where $\Delta\bar{\nu}_D$ is the Doppler width in cm^{-1}. By measuring α_0 at a given concentration of atoms, the oscillator strength f can be calculated. Clyne and Townsend (222) have obtained α_0 of Br and I atom absorption in the vacuum ultraviolet using a Doppler broadened resonance lamp. The known concentrations of halogen atoms are generated from fast chemical reactions such as

$$O + Br_2 \rightarrow BrO + Br$$

$$Cl + ICl \rightarrow Cl_2 + I$$

The authors calculated the oscillator strengths for various halogen atom transitions using (I-54b). Using a similar technique, Parkes et al. (797) have measured the oscillator strength of the O resonance triplet (^3P_J) near 1300 Å. The oscillator strength of the H atom transition at 1216 Å has been evaluated by Michael and Weston (701), Barker and Michael (60), and Morse and Kaufman (718).

I–6. RESONANCE ABSORPTION AND EMISSION BY ATOMS

Emission lines from atoms have been used as the light sources for various purposes: (*1*) to initiate photochemical reactions of molecules, (*2*) to follow the concentration of atoms as a function of time, (*3*) to measure the oscillator strength of the atomic absorption, and (*4*) to produce electronically excited atoms as in atom-sensitized reactions. In case *1* the intensity of the light source is the main concern, while in cases *2* through *4* the resonance radiation, that is, the emission due to the transition from the electronically excited state initially formed to the ground state, is almost exclusively used and the shape of the emission line is important.

In the following section the contour of the resonance absorption and emission lines are described.

I–6.1. Line Profile in Resonance Absorption; Natural, Doppler, and Pressure Broadening

Although the atomic absorption occurs between well-defined quantum states, the absorption line width is not vanishingly small because of the finite lifetime of excited atoms, thermal motions, and collisions with atoms of the same kind and with foreign gas atoms. If the pressure of absorbing atoms is kept low (below 0.01 torr) and the foreign gas pressure is below a few torr, the line width of the atoms is governed mainly by the lifetime of the excited atoms and by Doppler broadening. If a continuous light source is used in conjunction with a spectrograph of extremely high resolution, in principle the absorption line profile shown in Fig. I–7 would be obtained.

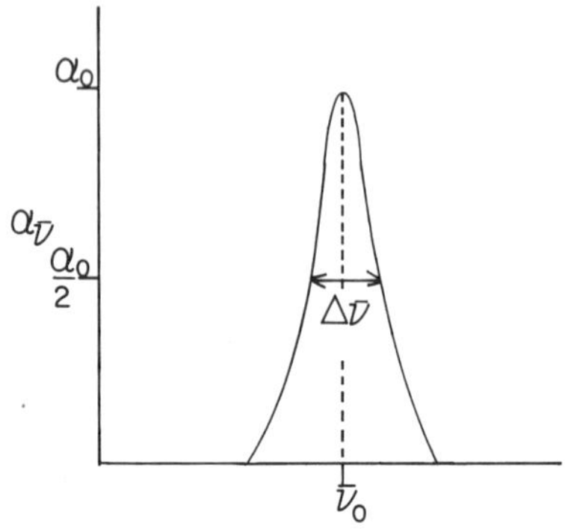

Fig. I–7. Absorption coefficient $\alpha_{\bar{\nu}}$ (cm^{-1}) of an atomic line; α_0 is the absorption coefficient at the line center $\bar{\nu}_0$, $\Delta\bar{\nu}$ is the line width at a half maximum. $\Delta\bar{\nu}$ is a function of the lifetime of the excited state, thermal motions of atoms, and pressure of a foreign gas.

The absorption coefficient $\alpha_{\bar{\nu}}$ (cm^{-1}) of N atoms per cm^3 at a wave number $\bar{\nu}$ is defined by

$$I_{\bar{\nu}} = I_0 e^{-\alpha_{\bar{\nu}} l} \tag{I-55}$$

where l is the path length in centimeters and $I_{\bar{\nu}}$ and I_0 are, respectively, the transmitted and incident light intensities at a wave number $\bar{\nu}$. In practice, however, since the atomic line width is so narrow it is not possible to obtain $\alpha_{\bar{\nu}}$ experimentally even with a very high resolution spectrograph. The absorption coefficient has a maximum value α_0 at $\bar{\nu} = \bar{\nu}_0$. The half width of the absorption line $\Delta\bar{\nu}$ is defined as the width of the line at $\alpha = \alpha_0/2$.

The quantity $\alpha_0 l$ is called opacity, optical depth, or optical thickness in astrophysics. The absorption cross section σ (cm^2) is defined as

$$\sigma = \frac{\alpha}{N} \tag{I-56}$$

where N is the number of atoms per cm^3.

Although collisions with other atoms become important when the pressure is high, we first consider the case where the gas pressure is sufficiently low so that only the natural and Doppler broadening need to be considered.

Natural Line Width. According to Heisenberg's uncertainty principle, a spectral line produced by a transition from an upper to a lower state is broadened because of the finite lifetime of the excited state. This natural line broadening is usually very small (of the order of 0.001 cm^{-1}). The relationship between the mean lifetime τ of the excited state and the line broadening $\Delta\bar{\nu}_N$ is

$$\Delta\bar{\nu}_N\ (cm^{-1}) = \frac{1}{2\pi c\tau} = \frac{5.30 \times 10^{-12}}{\tau\ (sec)} \tag{I-57}$$

where c is the velocity of light in cm sec^{-1} and τ is in seconds. The values of $\Delta\bar{\nu}_N$ for the Xe 1296 and 1470 Å lines are 1.89×10^{-3} and 1.42×10^{-3} cm^{-1}, respectively (1043). Even with a spectrograph of very high resolution, the instrumental broadening (0.2 cm^{-1}) amounts to 100 times as much as the natural line width.

Doppler Broadening. If an electronically excited atom emits light while it is moving with a velocity v whose component along the line of sight is v_x, the emitted wavelength appears shifted by $\lambda_0 v_x/c$, where λ_0 is the emitted wavelength from the atoms at rest. This is called the Doppler effect. Because of the Doppler effect, spectral lines emitted or absorbed by atoms are broadened according to the Maxwell-Boltzmann velocity distribution of atoms. This line broadening, $\Delta\bar{\nu}_D$, is given by the formula

$$\Delta\bar{\nu}_D\,(\text{cm}^{-1}) = \frac{2\sqrt{2R\ln 2}}{c}\,\bar{\nu}_0\sqrt{\frac{T}{M}} \quad \text{(I-58a)}$$

$$= 7.16 \times 10^{-7}\,\bar{\nu}_0\sqrt{\frac{T}{M}} \quad \text{(I-58b)}$$

where R is the gas constant equal to 8.314×10^7 erg deg^{-1} mol^{-1}, $\bar{\nu}_0$ is the wave number at the center of the absorption line, T is the gas temperature in °K, and M is the atomic weight. The Doppler width $\Delta\bar{\nu}_D$ for the Xe 1296 Å line is 83.2×10^{-3} cm^{-1} and for the Xe 1470 Å line it is 73.4×10^{-3} cm^{-1} at 300°K. Doppler broadening is of comparative magnitude with instrumental broadening at very high resolution. If natural broadening is neglected and only Doppler broadening is considered, the absorption coefficient $\alpha_{\bar{\nu}}$ is [Mitchel and Zemansky (21) p. 99]

$$\alpha_{\bar{\nu}} = \alpha_0 e^{-\varpi^2} \quad \text{(I-59a)}$$

where ϖ is

$$\varpi = \frac{2(\bar{\nu} - \bar{\nu}_0)}{\Delta\bar{\nu}_D}\sqrt{\ln 2} \quad \text{(I-60)}$$

Integration of (I-59a) gives

$$\int_0^\infty \alpha_{\bar{\nu}}\,d\bar{\nu} = \frac{1}{2}\sqrt{\frac{\pi}{\ln 2}}\,\alpha_0\,\Delta\bar{\nu}_D \quad \text{(I-59b)}$$

Pressure and Stark Broadening. When the pressure of a gas is low, that is, below 10 mtorr, the line width is mainly determined by natural and Doppler broadening. However, when the gas pressure is high or a foreign gas is introduced, the absorption line is further broadened. The broadening due to collisions with foreign gas atoms is called Lorentz broadening and the broadening by the absorbing gas of the same kind is called Holtsmark broadening. The broadening due to collisions with electrons and ions is called Stark broadening.

Lorentz broadening. The Lorentz broadening $\Delta\bar{\nu}_L$ is given by [Mitchell and Zemansky (21), p. 160]

$$\Delta\bar{\nu}_L\ (\text{cm}^{-1}) = \frac{Z_L}{c\pi} \tag{I-61}$$

where Z_L is the number of collisions with foreign gas atoms per second per absorbing atom and c is the velocity of light. According to (I-61), Lorentz broadening increases linearly with an increase of the foreign gas pressure. This relationship has been verified experimentally for a number of gases. The Lorentz broadening for the Hg 2537 Å line becomes almost equal to the Doppler broadening at 300°K at about 100 torr of He or Ne. Lorentz broadening is characterized by asymmetry and the shift of the peak of the line. According to Jablonski Lorentz broadening is due to the temporary formation of a quasi-molecule between an absorbing atom A and a foreign gas molecule F. The transition between the ground state atom A and an excited state atom A* without foreign gas is now represented, when a foreign gas F is introduced, by the transition between the ground state quasi-molecule A—F and the excited state quasi-molecule A*—F. The line broadening is explained by transitions that occur between the two potential energy curves of the quasi-molecule [see Mitchell and Zemansky (21), p. 175]

Holtsmark broadening. The line broadening due to collisions with atoms of the same kind is called Holtsmark broadening. The half width of self-broadening $\Delta\bar{\nu}_{SB}$ that is, the width of the absorption line at half maximum (see Fig. I–7) is proportional to the concentration of the gas. According to Hutcherson and Griffin (507)

$$\Delta\bar{\nu}_{SB}\ (\text{cm}^{-1}) = K\,\frac{e^2 f N}{c^2 m \bar{\nu}_0} \tag{I-62}$$

where K is a constant, e and m are the charge and mass of an electron, c is the velocity of light, f is the oscillator strength, and $\bar{\nu}_0$ is the wave number at the line center. According to Hutcherson and Griffin (507) $\Delta\bar{\nu}_{SB}$ for the Kr 1236 Å line is only about 0.01 cm^{-1} at 0.8 torr which is only one-tenth of the Doppler width (0.11 cm^{-1}). Above 1 torr self-broadening is appreciable and the line shape becomes asymmetric as a result of the molecular formation (Kr_2) [(Wilkinson (1041)].

Stark broadening. The absorption line is also broadened by the interaction of atoms with moving ions and electrons. If atoms are produced under high current density, Stark broadening may be important.

I–6.2. Line Profile in Resonance Emission; Four Types of Resonance Lamps

The profile of the resonance emission is important in some fields of photochemistry, such as atom photosensitization, isotope separation, measurement of oscillator strength, and kinetic spectroscopy, where time dependent atom concentrations have to be followed.

Four types of resonance lamps may be considered depending on the design and running conditions of the lamp. In every case it is assumed that broadening is entirely due to the Doppler effect because of the low pressures of gases commonly used for the resonance lamp.

Type A: Doppler Broadened Resonance Lamp Operated near Room Temperature. If atom concentrations in the resonance lamp are sufficiently low so that the secondary emission by absorbing atoms can be neglected and the primary emission comes out without being absorbed, the emission line shape may be determined by [Mitchell and Zemansky (21), p. 109]

$$I_{\bar{\nu}} = \mathscr{C}e^{-\bar{\omega}^2} \tag{I-63}$$

where $\mathscr{C}$ is a constant and $\bar{\omega}$ is given by (I-60). Expression (I-63) is similar to the Doppler broadened absorption line profile given in (I-59a). The line profile is shown in Fig. I–8*a*. Expression (I-63) may be used if an emitting layer is thin and the temperature inside the lamp is the same as that outside.

Clyne and Townsend (222) have used the resonance radiation from excited atoms with a Doppler line profile given by (I-63) to measure the oscillator strengths of atomic transitions. The light source is the resonance radiation from a lamp containing low concentrations of flowing atoms ($<10^{13}$ atoms cm^{-3}) which are excited by a second resonance lamp operated by a microwave generator. The resonance line from the second lamp is strongly self-reversed.

Type B: Doppler Broadened Resonance Lamp Operated at Higher Temperatures. The conditions are the same as for the type A lamp, that is, the atom concentrations are low and the emitting layer is thin except that the lamp temperature is high. The expression in this case is

$$I_{\bar{\nu}} = \mathscr{C}e^{-(\bar{\omega}/\beta)^2} \tag{I-64}$$

and

$$\beta = \frac{\text{width of the emission line}}{\text{width of the absorption line}} = \sqrt{\frac{T_2}{T_1}}$$

where T_2 is the temperature of the lamp and T_1 is that of the outside in °K and $T_2 > T_1$. Figure I–8*b* shows the line contour of the emission.

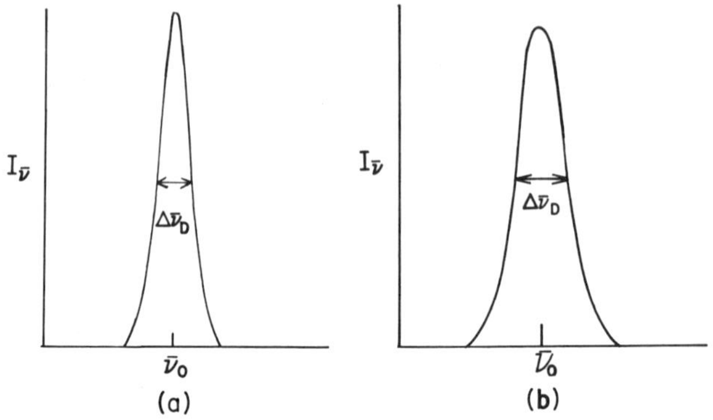

Fig. I–8. From Mitchel and Zemansky (21), p. 109, reprinted by permission of Cambridge University Press. (*a*) Line contour of the Doppler broadened resonance emission near room temperature. $\Delta\bar{\nu}_D$ is the Doppler half width and I_ν is the intensity at a wave number $\bar{\nu}$

$$I_\nu = \mathscr{C}e^{-\varpi^2}$$

where

$$\varpi = \frac{2(\bar{\nu} - \bar{\nu}_0)}{\Delta\bar{\nu}_D}\sqrt{\ln 2}$$

(*b*) Line contour of the Doppler broadened resonance lamp at higher temperatures

$$I_{\bar{\nu}} = \mathscr{C}e^{-(\varpi/\beta)^2}$$

where

$$\beta = \sqrt{\frac{T_2}{T_1}}, \; T_2 > T_1$$

T_2, inside temperature of the lamp; T_1, outside temperature of the lamp.

Parkes et al. (797) have used the resonance lines from flowing O atoms near 1300 Å excited by a microwave generator. The emitting layer is thin ($\sim$2 mm). The lamp temperature T_2 is assumed to be 600°K. The authors have measured the oscillator strength of O atoms near 1300 Å using this lamp.

Type C: Doppler Broadened Resonance Lamp Modified by Self-Absorption. The atom concentration is so low that only the primary radiation is considered. The resonance radiation is partially absorbed by the layer of ground state atoms on its way out. The form of the emission line is [Mitchell and Zemansky (21), p. 109]

$$I_{\bar{\nu}} = \mathscr{C}(1 - e^{-\alpha_{\bar{\nu}} l}) \qquad \text{(I-65a)}$$

where $\mathscr{C}$ is a constant, $\alpha_{\bar{\nu}} = \alpha_0 e^{-\varpi^2}$, and l is the thickness of the emitting layer. The line profile depends on the optical thickness $\alpha_0 l$. If $\alpha_0 l < 1$, the line shape approaches that of the Doppler broadened emission line. If $\alpha_0 l > 1$, the profile shows a flat top due to self absorption. An ideal resonance lamp, that is, a lamp emitting a Doppler broadened line, is usually too weak to use as a photochemical light source. If $\alpha_{\bar{\nu}} l < 1$ equation (I-65a) becomes

$$I_{\bar{\nu}} = \mathscr{C} \alpha_{\bar{\nu}} l = \mathscr{C}' e^{-\varpi^2} \tag{I-65b}$$

This equation is equivalent to (I-63).

Osborn et al. (783) have studied the emission profile of the Hg 2537 Å line. The lamp contains ^{202}Hg atoms and is water-cooled to 26°C. It emits the 2537 Å resonance line when activated by a microwave generator. The lamp can be used to selectively excite the ^{202}Hg isotope in natural Hg, which contains several Hg isotopes. The emission profile of the 2537 Å line may be given by (I-65a) with $\alpha_0 l = 16.9$. The profile is shown in Fig. I–9. The emission

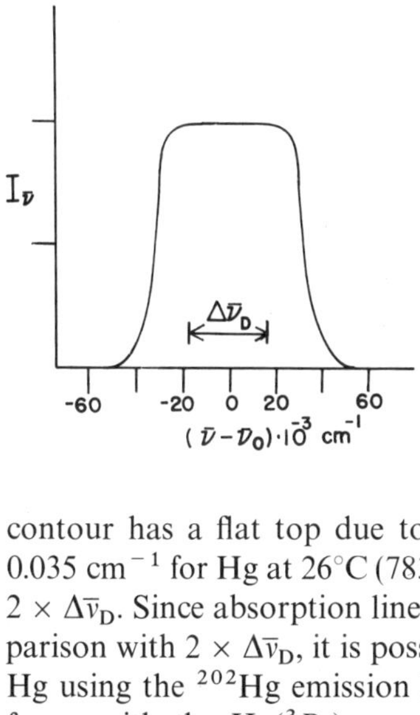

Fig. I–9. Type C resonance lamp; Doppler broadened resonance line of Hg 2537 Å modified by self-absorption

$$I_{\bar{\nu}} = \mathscr{C}(1 - e^{-\alpha_{\bar{\nu}} l}) \qquad \alpha_{\bar{\nu}} = \alpha_0 e^{-\varpi^2}$$

$$\varpi = \frac{2(\bar{\nu} - \bar{\nu}_0)}{\Delta\bar{\nu}_D}\sqrt{\ln 2} \qquad \alpha_0 l = 16.9$$

α_0 is the absorption coefficient at the center of the line; $\Delta\bar{\nu}_D$ is the Doppler width. After Osborn et al. (783).

contour has a flat top due to self-absorption. The Doppler width $\Delta\bar{\nu}_D$ is 0.035 cm^{-1} for Hg at 26°C (783). The half width of the emission line is about $2 \times \Delta\bar{\nu}_D$. Since absorption lines of other isotopes are well separated in comparison with $2 \times \Delta\bar{\nu}_D$, it is possible to selectively excite the ^{202}Hg in natural Hg using the ^{202}Hg emission lamp. If natural Hg is mixed with a gas that forms with the $Hg(^3P_1)$ a stable compound removable from the system, isotope separation can be achieved (783). (See Section VIII–1 for further discussion.) Table I–2 shows $\alpha_0 l$ values for some resonance lamps calculated from (I–54b) and (I–58b) using f in the table. Atom concentrations of $N = 3.2 \times 10^{13}$ atoms cm^{-3} (1 mtorr at 25°C) and $l = 1$ cm are assumed. Most resonance lamps of low intensity show line profiles characterized by a flat top, since $\alpha_0 l \gg 1$

Type D: Resonance Lamp with Two Layers. In some atomic resonance lamps, such as hollow cathode and photochemical rare gas resonance lamps,

Table I–2. Absorption Cross Section, $\sigma_0 = \alpha_0/N$ (cm^2), at the peak of the Doppler Broadened Resonance Line and the Optical Thickness $\alpha_0 l$

Atom	λ (Å)	σ_0	$\alpha_0 l$ [a]	f [b]
Ar	1067	2.75×10^{-13}	8.8	0.061
Kr	1236	1.4×10^{-12}	45	0.158
Xe	1470	2.9×10^{-12}	93	0.260
Hg	2537	6.1×10^{-13}	19.5	0.021
Hg	1849	3.5×10^{-12}	112	1.0
Cd	2288	1.64×10^{-11}	525	0.92
O	1302	7.3×10^{-14}	2.3	0.012
H	1215	3.0×10^{-13}	11	0.416
N	1200	2.8×10^{-13}	9	0.094

[a] $N = 3.2 \times 10^{13}$ atoms cm^{-3} and $l = 1$ cm are assumed; α_0 is calculated from (I-54b) using f given in the table. See text.
[b] From Refs. 21, 32, 33.

it is necessary to consider another absorption layer in the lamp near the exit window. Figure I–10 shows a two layer model applicable to this type of lamp. The lamp has two layers, the emitting and absorbing layers of length l_1 and l_2, respectively. Since the temperature of the emitting layer may be somewhat higher than that of the absorbing layer, $\alpha_1 < \alpha_2$ is assumed. The atom concentrations are also assumed to be sufficiently low so that reemission along the beam direction can be neglected. The second layer produces self reversal of the emission line, that is, the central portion is missing as shown in Fig. I–11 (d_2). The line profile is given by Braun and Carrington, (140) Bruce and Hannaford (154)

$$I_{\bar{v}} = \mathscr{C}(1 - e^{-\alpha_1 l_1}) \cdot e^{-\alpha_2 l_2} \tag{I-66}$$

where $\mathscr{C}$ is a constant and α_1 and α_2 are the absorption coefficients at a wavenumber $\bar{v}$ for the layers of thickness l_1 and l_2, respectively. Many high

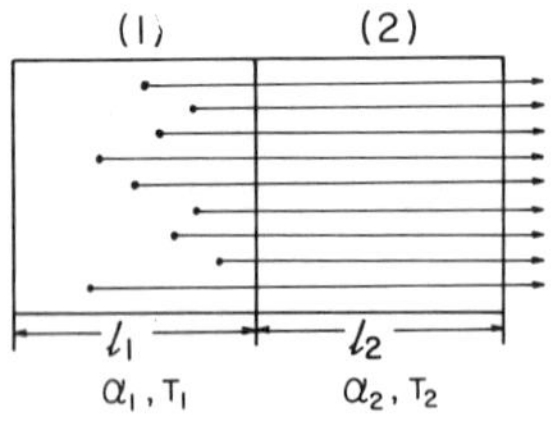

Fig. I–10. Two layer model of resonance lamp (1), an emitting–absorbing layer; the primary radiation is partially absorbed by a layer l_1 on its way out; (2) an absorbing layer; the resonance line is transmitted through a second absorbing but nonemitting layer of length l_2. The second layer produces self-reversal because of the large optical thickness ($\alpha_2 l_2 \gg 1$). Because of the temperature difference of the two layers ($T_1 > T_2$), the absorption coefficient α_1 is smaller than α_2. From Bruce and Hannaford (154), reprinted by permission of Pergamon Press.

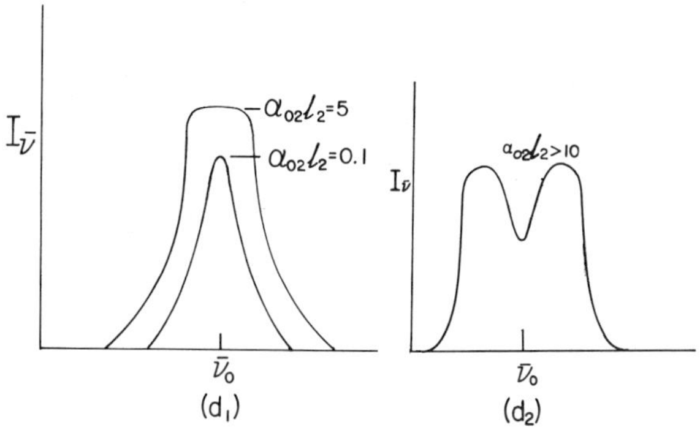

Fig. I–11. Type D resonance lamp; intensity distribution of resonance lamp according to a two layer model

$$I_{\bar{\nu}} = \mathscr{C}(1 - e^{-\alpha_1 l_1}) \cdot e^{-\alpha_2 l_2}$$

where $\mathscr{C}$ is a constant, α_1 and α_2 are the absorption coefficients at a wave number $\bar{\nu}$, respectively, for the emitting and absorbing layers of thickness l_1 and l_2; α_0 is the absorption coefficient at the center of the line. (d_1) $\alpha_{01}l_1 \approx 0.1$ is assumed; (d_2) $\alpha_{01}l_1 \approx 10$ is assumed. In most photochemical lamps both $\alpha_{01}l_1$ and $\alpha_{02}l_2$ are >10; namely, type d_2 and the center portion is missing. The intensity is in arbitrary units.

intensity resonance lamps are probably provided with emission profiles as shown in Fig. I–11 (d_2), or the central portion may be completely missing, and are not suitable for quantitative measurements of absorption line strength. Slanger (898) has estimated the width of the Xe 1470 Å resonance line to be about 15 cm^{-1}, which is about 200 times $\Delta\bar{\nu}_D$ (0.073 cm^{-1}).

On the other hand, the Ca 4227 Å line in a hollow cathode discharge has a width of 50×10^{-3} cm^{-1} [Bruce and Hannaford (154)], which is close to the Doppler width (46×10^{-3} cm^{-1}). The lack of self-absorption may be understood from a small value of $\alpha_0 l = 0.01$.

High lamp output is the primary requirement for photochemical studies. Since the intensity is proportional to the number of excited atoms per second, higher intensities can be achieved at higher pressures of atoms. On the other hand, self-absorption (particularly by the absorbing layer) and self-quenching are more pronounced at higher pressures. Hence, the output shows a maximum at a pressure P_{max} as shown in Fig. I–12. This pressure is about 500 mtorr in Xe and Kr light sources.

I–6.3. Measurement of the Absorption Intensity Using a Resonance Lamp

In principle it is possible to obtain α_0 if the width of the incident monochromatic light is sufficiently small compared with the Doppler width of the absorption line. However, in practice the light intensity becomes so weak

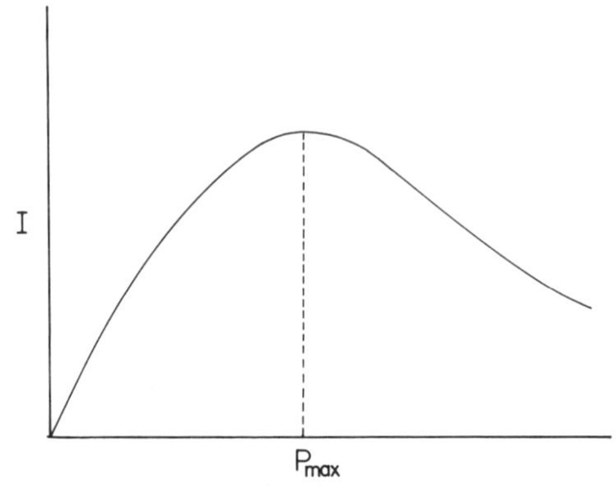

Fig. I–12. Output of a resonance lamp as a function of pressure. *I* is the lamp output of a resonance line. P_{max} is a pressure of a gas being excited to give a maximum output of the resonance line. *I* initially increases with an increase of pressure but decreases with a further increase of pressure because of self-absorption and self-quenching. P_{max} is about 0.5 torr for Kr and Xe lamps.

that it would not be possible to measure the absorption intensity with a combination of a continuous emission source and a high resolution spectrograph. Instead, a resonance lamp of type A or B has been used to measure the absorption intensity. The absorption A by N atoms, in a cell (length l) using a type A or B resonance lamp, that is, a lamp with the Doppler emission profile, is expressed by [Mitchell and Zemansky (21), p. 122]

$$A = \frac{\int_{-\infty}^{+\infty} I(\varpi)(1 - e^{-\alpha_{\bar{\nu}} l})\, d\varpi}{\int_{-\infty}^{+\infty} I(\varpi)\, d\varpi} \tag{I-67}$$

where $I(\varpi)$ is the intensity distribution of the emission line, $\alpha_{\bar{\nu}}$ is the absorption coefficient at a wave number $\bar{\nu}$, and ϖ is given by $[2(\bar{\nu} - \bar{\nu}_0)/\Delta\bar{\nu}_D]\sqrt{\ln 2}$. $I(\varpi)$ takes the form (I-64) for the Doppler broadened emission line

$$I_{\bar{\nu}}(\varpi) = \mathscr{C} e^{-(\varpi/\beta)^2} \tag{I-68}$$

where $\mathscr{C}$ is a constant and $\beta > 1$. The absorption line profile is also assumed to be Doppler broadened, that is, $\alpha_{\bar{\nu}} = \alpha_0 e^{-\varpi^2}$.

Equation (I-67) may be expressed as a power series [Mitchell and Zemansky (21), p. 323]

$$A = a(\alpha_0 l) - b(\alpha_0 l)^2 + c(\alpha_0 l)^3 - \cdots \tag{I-69}$$

where a, b, c are constants. The atomic absorption A for various β and $\alpha_0 l$ has been calculated by Mitchell and Zemansky [(21), p. 323]. Hence, for a given β, $\alpha_0 l$ can be obtained by measuring A. If $\alpha_0 l$ is small A is linearly proportional to α_0 according to (I-69). For the Doppler broadened absorption line (I-54b) can be applied

$$\alpha_0 = \frac{0.832 \times 10^{-12}}{\Delta\bar{\nu}_D} Nf \tag{I-70}$$

The oscillator strength f can be obtained from α_0 and N. Bemand and Clyne (95) have computed A for various types of resonance lamps (type A, B, and C lamps). They have concluded that the Beer-Lambert law holds up to $A \simeq 0.7$ for type A, an ideal resonance lamp. However, for type C, the Doppler profile with self-absorption ($\alpha_0 l = 10$), the Beer-Lambert law is obeyed only up to $A \simeq 0.27$. If a type D lamp with strongly self-reversed profile is used, the Beer-Lambert law does not hold. A modified form

$$I_t = I_0 e^{-(\alpha_0 l)^x} \qquad \text{with } 1 > x > 0 \tag{I-71}$$

where I_t and I_0 are transmitted and incident light intensities, respectively, are often used empirically to correlate the atom concentrations with absorption [Donovan et al. (309)]. This relationship is valid only for the limited concentration range [Bemand and Clyne (95)].

To obtain absolute concentrations of atoms using (I-70), it is necessary to know α_0. This value can be determined by measuring absorption of the system containing known concentrations of atoms, and the linear relationship between α_0 and N may be obtained in the limited concentration range. From this linear relationship, unknown atom concentrations can be obtained by measuring absorption.

I–7. BAND INTENSITIES IN THE MOLECULAR SYSTEM

In Section I–5 it is shown that the transition probability from a lower state designated by the eigenfunction ψ'' to an upper state ψ' is proportional to the square of the transition moment $\mathbf{R}$ [see (I-36)] defined as

$$\mathbf{R} = \int \psi'^* \boldsymbol{\mu} \psi'' \, dv \tag{I-72}$$

where dv is the volume element of the coordinates. (Throughout this book a single prime refers to the upper state and a double prime signifies the lower state. For transitions between the two states the upper state is always written first regardless of whether the process is absorption or emission.) Based on the Born-Oppenheimer approximation, we assume the total eigenfunction ψ of molecules is the product of the electronic and vibrational eigenfunctions ψ_e and ψ_v, neglecting to a good approximation the rotation of the molecule [see Herzberg (14), p. 200]

$$\psi = \psi_e \psi_v \tag{I-73}$$

Then (I-72) becomes

$$\mathbf{R} = \int \psi_e'^* \psi_v' \boldsymbol{\mu} \psi_e'' \psi_v'' \, dv \tag{I-74}$$

We further assume that the electronic transition moment defined as

$$\mathbf{R}_e = \int \psi_e'^* \boldsymbol{\mu} \psi_e'' \, dv_e \tag{I-75}$$

changes slowly with the internuclear distance and may be replaced by an average value $\bar{\mathbf{R}}_e$ for transitions between various v' and v'' levels. Equation (I-74) becomes

$$\mathbf{R} = \bar{\mathbf{R}}_e \int \psi_v' \psi_v'' \, dv \tag{I-76}$$

The integral $\int \psi_v' \psi_v'' \, dv$ is called the overlap integral and $q(v', v'')$, the square of the overlap integral, is called the Franck-Condon factor.

$$q(v', v'') = \left[\int \psi_v' \psi_v'' \, dv \right]^2 \tag{I-77}$$

The Franck-Condon factor summed over all v' or v'' including a continuum is unity

$$\sum_{v'} q(v', v'') = \sum_{v''} q(v', v'') = 1 \tag{I-78}$$

From (I-76) and (I-77)

$$\mathbf{R} = \bar{\mathbf{R}}_e q^{1/2}(v', v'') \tag{I-79}$$

The Einstein transition probability of spontaneous emission from a level v' to all v'' levels is given by

$$A(v') = \sum_{v''} A(v', v'') = \frac{1}{\tau(v')} \tag{I-80}$$

where $\tau(v')$ is the radiative life of a level v' and can be measured experimentally. The individual $A(v', v'')$ may be obtained experimentally from the observed emission intensities $I(v', v'')$ of the transitions v' to v''.

$$\frac{A(v', v'')}{A(v')} = \frac{I(v', v'')}{I(v')} \tag{I-81}$$

where I is in units of quanta $\sec^{-1}$.

The absorption intensities of bands are often expressed in terms of the oscillator strength $f(v', v'')$ or the electronic transition moment $\mathbf{R}_e$. According to (I-52) the oscillator strength in the molecular system is

$$f(v', v'') = \frac{mc}{8\pi^2 e^2} \cdot \frac{A(v', v'')}{\bar{\nu}^2(v', v'')} \frac{g(v')}{g(v'')} = 1.50 \frac{A(v', v'')}{\bar{\nu}^2(v', v'')} \frac{g(v')}{g(v'')} \tag{I-82a}$$

where $\bar{\nu}$ is in cm^{-1}, A is in sec^{-1}, and $g(v')$, $g(v'')$ are degeneracies of the upper and lower levels, respectively. The average electronic transition moment for various v' to v'' transitions is obtained from (I-42a) and (I-79)

$$|\bar{\mathbf{R}}_e|^2 = \frac{3h}{64\pi^4} \frac{A(v', v'')}{\bar{\nu}^3(v', v'')q(v', v'')} \tag{I-83a}$$

$$= 3.19 \times 10^{-30} \cdot \frac{A(v', v'')}{\bar{\nu}^3(v', v'')q(v', v'')} \tag{I-83b}$$

$|\bar{\mathbf{R}}_e|^2$ is usually expressed in terms of atomic units, $a_0^2 e^2 = 6.459 \times 10^{-36}$ cm^2 esu^2, where a_0 is the Bohr radius and e is the electronic charge. Hence, $|\bar{\mathbf{R}}_e|^2$ in atomic units is

$$|\bar{\mathbf{R}}_e|^2 = 4.94 \times 10^5 \cdot \frac{A(v', v'')}{\bar{\nu}^3(v', v'')q(v', v'')} \tag{I-83c}$$

where $\bar{\nu}$ is in cm^{-1} and A is in sec^{-1}.

From (I-82a) and (I-83a), the oscillator strength is given in terms of the transition moment

$$f(v', v'') = \frac{8mc\pi^2}{3he^2} \bar{\nu} |\bar{\mathbf{R}}_e|^2 q(v', v'') \frac{g(v')}{g(v'')} \tag{I-82b}$$

For the rotational transition from J'' to J' in the (v', v'') vibrational band, the oscillator strength is given by

$$f(J', J'') = f(v', v'') \frac{S(J', J'')}{2J'' + 1} \tag{I-82c}$$

where $S_{J'J''}$ is called the rotational line strength or the Hönl-London factor [see Herzberg (16) p. 226].

In some cases $A(v')$ can be measured, but $A(v', v'')$ cannot be measured by experiment. If the calculated values of Franck-Condon factors for transitions from a given v' to various v'' levels are available, we obtain $A(v', v'')$ from $A(v')$ by

$$\frac{A(v', v'')}{A(v')} = \frac{\bar{\nu}^3(v', v'')q(v', v'')}{\sum_{v''} \bar{\nu}^3(v', v'')q(v', v'')} \tag{I-84}$$

(I-84) is derived from (I-83a) and $A(v') = \sum_{v''} A(v', v'')$ assuming $|\bar{\mathbf{R}}_e|^2$ is the same for all transitions.

From (I-82a) and (I-84) we obtain

$$f(v', v'') = 1.50 \frac{A(v')\bar{\nu}(v', v'')q(v', v'')}{\sum_{v''} \bar{\nu}^3(v', v'')q(v', v'')} \frac{g(v')}{g(v'')} \tag{I-85}$$

The oscillator strength can also be obtained experimentally from the integrated absorption coefficient.

From (I-51)

$$f(v', v'') = 4.20 \times 10^{-8} \int k_{\bar{\nu}}\, d\bar{\nu} \tag{I-86}$$

where $k_{\bar{\nu}}$ is the absorption coefficient in units of atm^{-1} cm^{-1} and $\bar{\nu}$, the wave number is in cm^{-1}.

The oscillator strength summed over all v' levels from the common v'' is called the electronic absorption oscillator strength $f_{el}(v'')$

$$f_{el}(v'') = \sum_{v'} f(v', v'') \tag{I-87}$$

From (I-82b) and (I-87) the electronic absorption oscillator strength can be expressed as

$$f_{el}(v'') = \frac{\sum_{v'} \bar{\nu}(v', v'')q(v', v'')}{\bar{\nu}(v', v'')q(v', v'')} f(v', v'') \tag{I-88}$$

The relationship between the fluorescence lifetime and the integrated absorption coefficient in the molecular system has been derived by Strickler and Berg (939) in the case of a strongly allowed transition.

The fluorescence lifetime τ from the v' level to all v'' levels of the lower state is from (I-80) and (I-83a)

$$\frac{1}{\tau} = A(v') = \sum_{v''} A(v', v'') = \frac{64\pi^4}{3h} \sum_{v''} \bar{\nu}^3(v', v'')q(v', v'')|\bar{\mathbf{R}}_e|^2 \tag{I-89}$$

The integrated absorption coefficient for v'' to all v' levels of the upper state is from (I-47) and (I-78)

$$\int \frac{\alpha_{\bar{\nu}}}{\bar{\nu}} d\bar{\nu} = \frac{8\pi^3 N}{3hc} |\bar{\mathbf{R}}_e|^2 \tag{I-90}$$

where $\alpha_{\bar{\nu}}$ is the absorption coefficient in cm^{-1}, and $\bar{\nu}$ is the wave number in cm^{-1}. From (I-90) and (I-89) we obtain

$$A(v') = \frac{8\pi c}{N} \sum_{v''} \nu^3(v', v'')q(v', v'') \int \frac{\alpha_{\bar{\nu}}}{\bar{\nu}} d\bar{\nu} \tag{I-91a}$$

If the absorption coefficient $k_{\bar{\nu}}$ (atm^{-1} cm^{-1}) defined by

$$I_t^{\bar{\nu}} = I_0^{\bar{\nu}} e^{-k_\nu p l}$$

is used instead of $\alpha_{\bar{\nu}}$ (cm^{-1}) (p-pressure in atmospheres at 0°C; l-path length in centimeters; see Section I–8).

$$A(v') = 2.80 \times 10^{-8} \sum_{v''} \bar{\nu}^3(v', v'')q(v', v'') \int \frac{k_{\bar{\nu}}}{\bar{\nu}} d\bar{\nu} \qquad \text{(I-91b)}$$

and if the upper and lower levels are degenerate,

$$A(v') = 2.80 \times 10^{-8} \sum_{v''} \bar{\nu}^3(v', v'')q(v', v'') \frac{g(v'')}{g(v')} \int \frac{k_{\bar{\nu}}}{\bar{\nu}} d\bar{\nu} \qquad \text{(I-91c)}$$

where $g(v')$ and $g(v'')$ are degeneracies of the upper and lower levels, respectively. The quantity $\sum_{v''} \bar{\nu}^3(v', v'')q(v', v'')$ may be obtained experimentally from the intensities of the fluorescence spectrum, $I_{\bar{\nu}}(v', v'')$, as a function of wave number in the following way. From $I_{\bar{\nu}}(v', v'') = N(v')A(v', v'')$ (N is the number of molecules in the level v') and remembering $\sum_{v'} q(v', v'') = 1$, we obtain for the continuous fluorescence spectrum

$$\sum_{v''} \bar{\nu}^3(v', v'')q(v', v'') = \frac{\int \bar{\nu}^3(v', v'')q(v', v'')d\bar{\nu}}{\int q(v', v'')d\bar{\nu}} = \frac{\int I_{\bar{\nu}}(v', v'')\,d\bar{\nu}}{\int \frac{I_{\bar{\nu}}(v', v'')}{\bar{\nu}^3} d\bar{\nu}}$$

I–8. ABSORPTION COEFFICIENT IN THE MOLECULAR SYSTEM

The knowledge of the extent of absorption (absorption coefficient) in the molecular system is important in photochemistry since the absorption coefficient must be known to obtain the quantum yield, the oscillator strength of the bands, and the lifetime of the electronically excited state.

In this section the Beer-Lambert law and the measurement of the absorption coefficient in the molecular system are described. The deviation from the Beer-Lambert law for unresolved molecular bands is also discussed.

I–8.1. The Beer-Lambert Law in the Molecular System

If a parallel beam of purely monochromatic light with the intensity $I_0^{\bar{\nu}}$ at a wave number $\bar{\nu}$ passes through a vessel of length l containing a gas at a given pressure or concentration b, the transmitted light intensity $I_t^{\bar{\nu}}$ is

$$I_t^{\bar{\nu}} = I_0^{\bar{\nu}} e^{-abl} \qquad \text{(I-92a)}$$

or

$$I_t^{\bar{\nu}} - I_0^{\bar{\nu}} 10^{-abl} \qquad \text{(I-92b)}$$

where a is a proportionality constant called the absorption coefficient. Equation (I-92) is generally known as the Beer-Lambert law.

Various units are used to express the absorption coefficient. b is given either in pressure units (mm Hg, atm) or in concentration units (molec cm^{-3}, mol dm^{-3}). l is usually expressed in centimeters. If the pressure unit is used the temperature to which the pressure is referred must be provided. The pressure unit is commonly used with base e and sometimes with base 10. The concentration unit, molec cm^{-3}, is commonly used with base e and is called the absorption cross section. If the concentration is expressed in mol dm^{-3}, base 10 is usually used and the absorption coefficient is called the molar absorption coefficient. In this book absorption coefficient is denoted by k when the pressure unit (atm) is used, by σ when the concentration unit molec cm^{-3} is used, and by ϵ when the concentration unit of mol dm^{-3} is used.

Thus, we have three expressions for absorption coefficients k, σ, and ϵ given by

$$I_t^{\bar{\nu}} = I_0^{\bar{\nu}} e^{-kpl} \qquad (0 \text{ or } 25°\text{C}) \tag{I-92c}$$

$$I_t^{\bar{\nu}} = I_0^{\bar{\nu}} e^{-\sigma Nl} \tag{I-93}$$

$$I_t^{\bar{\nu}} = I_0^{\bar{\nu}} 10^{-\epsilon cl} \tag{I-94}$$

where p is in atm at a reference temperature T, N is the number of molecules per cm^3, c is in moles dm^{-3}, and l is in centimeters. The absorption coefficient as given by k, σ, and ϵ has units of atm^{-1} cm^{-1}, cm^2 molec^{-1}, and dm^3 mol^{-1} cm^{-1}, respectively. The value of k that refers to 0°C is 1.09 k at 25°C. The absorption cross section σ is related to k at 0°C by

$$\sigma = 3.72 \times 10^{-20} k$$

The relationship between ϵ and k at 0°C is

$$\epsilon = 9.73k$$

Equation (I-92) is equivalent to (I-55) in the atomic system. The Beer-Lambert law holds in the atomic system only when the spectral width of monochromatic light is infinitesimal or when both the emission and absorption lines follow ideal Doppler profile.

The Beer-Lambert law holds in the molecular system when the absorption bands are continuous or diffuse and the molecular interaction is negligible. Conversion factors among various units are given in the Appendix in Table A–3.

I–8.2. Deviation from the Beer-Lambert Law

When the absorption bands consist of fine rotational lines, the Beer-Lambert law is not obeyed, that is, the absorption coefficient is not constant when the pressure changes. Figure I–13 shows a plot of log I_0/I_t against pressure.

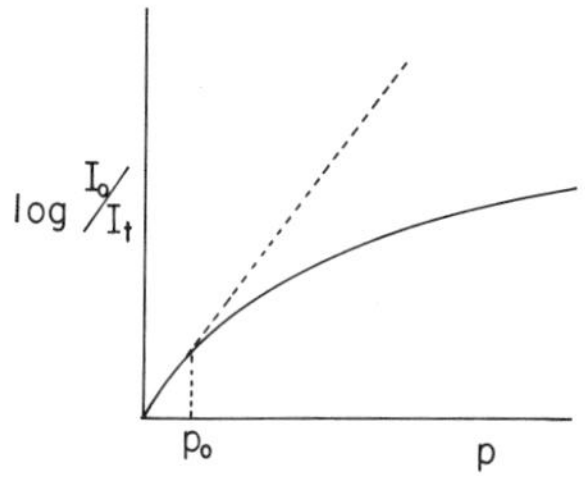

Fig. I–13. Deviation from the Beer-Lambert law. At $P < P_0$ the absorption follows the relationship $I_t = I_0 e^{-kPl}$ while at $P > P_0$ the absorption deviates from the law. The widths of the absorption lines are much narrower than the spectral width of the light source.

The slope is linear at low pressures ($P < P_0$), but it deviates from linearity at high pressures.

The deviation from the Beer-Lambert law results in part from the lack of instrumental resolution and is discussed in detail by Nielsen et al. (741).

Qualitative explanation may be given using Fig. I–14, which shows three rotational absorption lines with widths much narrower than the width

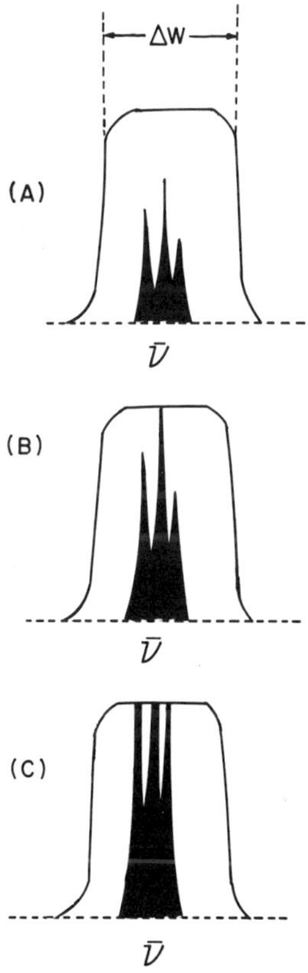

Fig. I–14. Deviation from the Beer-Lambert law due to the lack of spectral resolution: the widths of three absorption bands are much narrower than the width (Δw) of the monochromatic radiation. When the absorption by a molecule is small (*A*) it obeys the law. As the pressure is increased, the absorption at the line center is complete (*B*). Further increase (*C*) causes deviation from the law.

(Δw) of a monochromatic line emerging from a slit of a monochromator. The Beer-Lambert law holds up to a pressure shown in Fig. I–14*B*. When the pressure is increased further the deviation occurs, since the absorption at the line center is complete and only the off-center portion of the absorption lines contributes to an increase of absorption. At a fixed pressure and path length the apparent absorption coefficient sometimes decreases as the slit width increases. Figure I–14*A* explains that the area taken by the absorption lines becomes proportionately smaller as the slit width is increased. Meinel (690) has shown that the apparent absorption coefficient k of the NO (0, 0) band does in fact decrease with an increase of the slit width.

I–8.3. Measurement of the Integrated Absorption Coefficient

When the absorption bands consist of many fine rotational lines (the cases for many diatomic molecules) it is not possible to obtain an individual absorption coefficient because of the lack of instrumental resolution. To circumvent this difficulty, Bethke (103) has added sufficiently high pressures of Ar to broaden the absorption lines until the widths become comparable to the instrumental resolution. He has obtained the oscillator strength of NO (103) and O_2 (104) in the far ultraviolet from the integrated absorption coefficient using (I-86). On the other hand, Farmer et al. (341) have used a fixed slit width corresponding to 0.3 Å and have obtained the integrated absorption coefficient of the O_2 Schumann-Runge system from the linear portion of a plot similar to Fig. I–13.

I–8.4. Temperature Dependence of the Continuous Absorption Spectrum

The wave mechanical expression for the absorption coefficient for the transition $v' \leftarrow v''$ is from (I-77), (I-82b), and (I-86)

$$\int k_{\bar{\nu}}\, d\bar{\nu} = \mathscr{C}\bar{\nu}\left[\int \psi'_{v'}\psi''_{v''}\, dv\right]^2 \tag{I-94a}$$

where $\bar{\nu}$ is the wave number corresponding to a transition from v'' (vibrational levels of a lower state) to v' (vibrational levels of an upper state), $\mathscr{C}$ is a constant, and $\psi'_{v'}$ and $\psi''_{v''}$ are the nuclear eigenfunctions of vibrational levels of an upper and a lower electronic state, respectively. In this section we consider the temperature dependence of the absorption coefficient for the transition from the ground to the upper repulsive state in the case of diatomic or quasi diatomic molecules. A marked temperature dependence of absorption coefficient has been found for Cl_2 and a number of halogenated methanes near the long wavelength limit of continuous absorption. The temperature dependence may be explained with the aid of Fig. I–15 in which

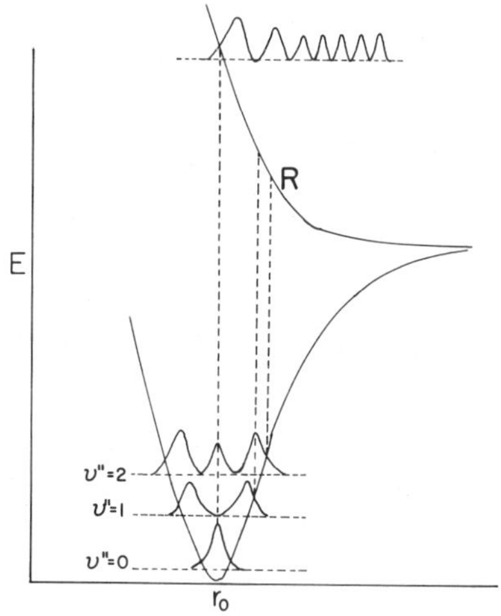

Fig. I–15. The transition from the ground to the repulsive upper state for diatomic or quasidiatomic molecules. The probability density distributions are shown for the ground $v'' = 0$, 1, and 2 levels and for the upper state. The absorption coefficient shows a maximum for the vertical transition where the overlap integral is largest.

is shown the probability density distributions $|\psi_v|^2$ of the upper (R) and the ground (G) state vibrational levels near the equilibrium internuclear distance r_0. The absorption coefficient shows a maximum where the square of the overlap integral $\int \psi'_v \psi''_{v''}\, dv$ is largest (actually the absorption coefficient shows a maximum at slightly shorter wavelength than that of the square of the overlap integral due to $\bar{\nu}$ in (I-94a)).

The largest overlap integral arises when the internuclear distance remains unchanged during the transition according to the Franck-Condon principle [see Section I-9.2 (**I-9**)]. The increase in absorption with an increase of temperature can be understood from the increased fraction of the molecules in $v'' = 1, 2, \ldots$ etc according to (I-30)

$$\frac{N_v}{N_0} = \frac{e^{-G_v/kT}}{Q_v} \tag{I-94b}$$

where N_0 and N_v are the total number of molecules and the number of molecules in a given vibrational level v, respectively, Q_v is the vibrational partition function, and G_v is the vibrational term value. The absorption coefficients from higher vibrational levels of the ground state have maxima towards longer wavelengths than that from the lowest level of the ground state as can be seen from Fig. I–15. If the vibrational term value is small, the absorption coefficients at 300°K show an increase compared with those at low temperatures towards the long wavelength side of an absorption maximum, while for large term values the differences of absorption coefficients between 300°K and low temperatures are small. Fink and Goodeve (353a) have measured the molar absorption coefficients of CH_3Br at 20°C near the long wavelength limit of absorption. Their results are shown in Fig. I–16 by the full line. They have also calculated the absorption coefficients on the basis of wave mechanics assuming that the repulsive curve is due to the dissociation of the quasidiatomic molecule CH_3Br into CH_3 + Br, and using a CH_3—Br vibrational term of 610 cm^{-1} only. Their calculated results for $v'' = 0$ are given in Fig. I–16 by the broken line. The two curves show a marked difference in absorption for wavelengths above 2200 Å. With a vibrational term of 610 cm^{-1} 5.5% of the molecules are in $v'' = 1$ at 300°K. A number of halogenated methanes show similar temperature dependence of absorption coefficient near the long wavelength end of absorption [Hubrich et al. (486b), Chou et al. (210a)].

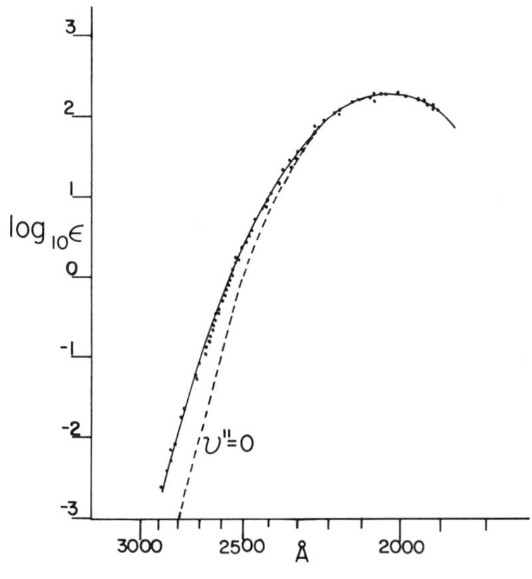

Fig. I–16. The molar absorption coefficients of CH_3Br near the long wavelength limit of absorption. (———) The experimental values at 20°C; (----) wave mechanical calculations for the transitions from $v'' = 0$ to the upper repulsive curve assuming dissociation into CH_3 + Br. Only a CH_3–Br vibrational term value of 610 cm^{-1} is used. From Fink and Goodeve (353a), Reprinted by permission of The Royal Society.

I–9. ELECTRONIC TRANSITIONS IN DIATOMIC MOLECULES

Absorption and emission of light in the molecular system occur, just like in the atomic system, between the two quantum states, although these states consist not only of electronic but also of vibrational and rotational states.

Most photochemical reactions are brought about by light absorption in the visible and ultraviolet regions. In these regions the total energy T' of the upper state and the total energy T'' of the lower state are the sum of the electronic (T'_e, T''_e), vibrational (G', G''), and rotational (F', F'') energies.

The energies (in wave numbers) of photons absorbed or emitted between the two states are [Herzberg (14), p. 151] given by

$$\begin{aligned}\bar{\nu}\,(\text{cm}^{-1}) &= T' - T'' = (T'_e - T''_e) + (G' - G'') + (F' - F'') \\ &= \bar{\nu}_e + \bar{\nu}_v + \bar{\nu}_r \end{aligned} \tag{I-95}$$

For a given electronic transition $\bar{\nu}_e$ is a constant. The vibrational and rotational term values are given by (I-13c) and (I-11). The vibrational and rotational constants of the upper state are usually different from those of the lower state.

I–9.1. Vibrational Structure in the Electronic Transition

Since the energy separations between successive rotational levels are much smaller than those between vibrational levels, we can observe only the vibrational structure in the electronic bands with a low resolution spectrograph. At room temperature most molecules are in the lowest vibrational level of the ground electronic state, that is, $v'' = 0$ (see Section I–4.1). The absorption bands consist of the transitions from $v'' = 0$ to various vibrational levels of the upper electronic state $v' = 0, 1, 2, \ldots$. A series of these bands is called a progression and is shown on the right of Fig. I–17. The band corresponding to the transition from $v'' = 0$ to $v' = 0$ is called the 0–0 band.

I–9.2. The Franck-Condon Principle

It has been observed that the intensity distribution of the progression $v'' = 0$ to $v' = 0, 1, 2 \ldots$ is not uniform but usually has a maximum. Such an observation is explained by the Franck-Condon principle. This principle assumes that the electronic transition occurs so fast compared with the nuclear motion that the relative position and velocity of the nuclei remain approximately the same during the electronic transition. Hence, the maximum intensity appears for the v'' to v' transition in which the internuclear distance remains unchanged. In terms of potential energy curves shown in Fig. I–18, the transition to give the maximum intensity corresponds to a vertical transition from $v'' = 0$ to $v' = 2$. In the language of wave-mechanics the probability of the transition is largest if the square of the overlap integral

$$\int \psi'_v \psi''_v \, dr$$

is greatest, where ψ'_v and ψ''_v are the vibrational eigenfunctions of the upper and lower states, respectively, and r is the internuclear distance; see Herzberg

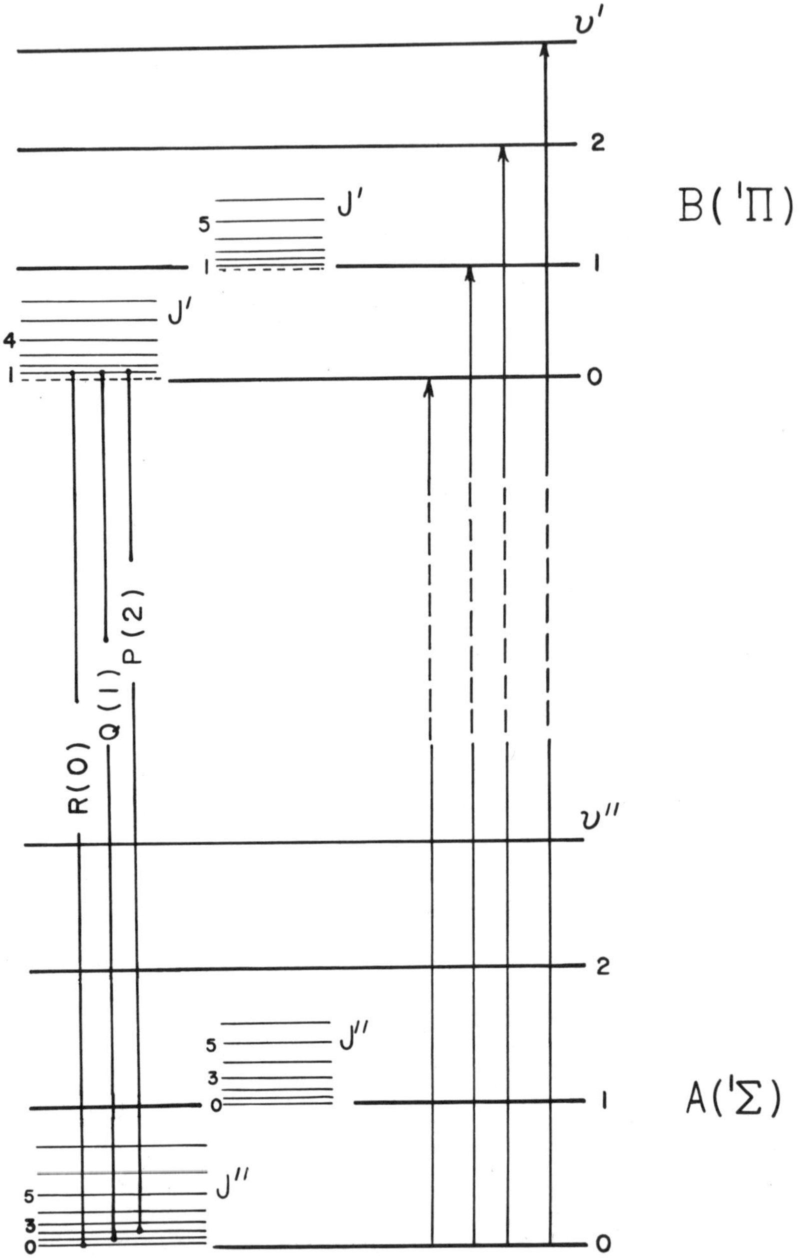

Fig. I–17. Vibrational and rotational structure in electronic transition between $B(^1\Pi)$ and $A(^1\Sigma)$. The vibrational progression from $v'' = 0$ to $v' = 0, 1, 2, 3\ldots$ is shown to the right. The rotational transitions are given to the left; $J'' \rightarrow J' = J'' + 1$ forms the R branch, $J'' \rightarrow J' = J''$ forms the Q branch and $J'' \rightarrow J' = J'' - 1$ forms the P branch of a band. $J' = 0$ is missing for the $B(^1\Pi)$ state and hence, the Q branch starts from $J'' = 1$.

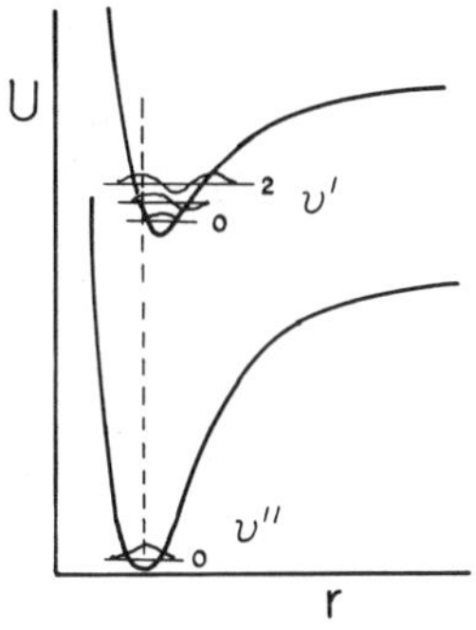

Fig. I–18. **Franck-Condon principle in a diatomic molecule; the intensity is maximum for the transition $v'' = 0$ to $v' = 2$, where the overlap of the vibrational eigenfunctions $\psi''(v'' = 0)$ and $\psi'(v' = 2)$ is the largest. The dashed line indicates electron jump to the upper state. From G. Herzberg, *Molecular Spectra and Molecular Structure I, Spectra of Diatomic Molecules*, 2nd ed., p. 199. Copyright by Litton Educational Publishing, Inc. Reprinted by permission of Van Nostrand Reinhold Company.**

(14), p. 199. Figure I–18 shows that the maximum of ψ_v'' at $v'' = 0$ coincides with that of ψ_v' at $v' = 2$ in electron jump. Hence, the square of the integral has a maximum value for the transition $v'' = 0$ to $v' = 2$.

I–9.3. Rotational Structure in the Electronic Transition

With a spectrograph of high resolution we can observe fine structure in vibrational bands. For a given vibrational transition we obtain transitions between various rotational levels of the upper electronic state and those of the lower electronic state with energies in wave numbers

$$\bar{\nu} = \bar{\nu}_0 + F_v'(J') - F_v''(J'') \tag{I-96}$$

where $F'(J')$ and $F''(J'')$ are the rotational term values of given upper and lower states and given vibrational levels, respectively. The quantity $\bar{\nu}_0$, called the band origin, is the difference of the term values for the transition $J' = 0$ to $J'' = 0$. Transitions between various J' and J'' are possible only for $\Delta J = J' - J'' = 0, \pm 1$ (see selection rules Section I–10.2) if one of the states is $\Lambda \neq 0$. These transitions are shown on the left of Fig. I–17 for $B(^1\Pi) \leftarrow A(^1\Sigma)$. If the transitions are from J'' to $J' = J'' + 1$ ($\Delta J = +1$) we obtain a series of lines for $J'' = 0, 1, 2, 3 \ldots$ that is called the R branch. For the transitions from J'' to $J' = J''$ ($\Delta J = 0$) we obtain the Q branch and from J'' to $J' = J'' - 1$ ($\Delta J = -1$) we obtain the P branch. These transitions are given by the formulae

$$R \text{ branch:} \quad \bar{\nu} = \bar{\nu}_0 + F_v'(J+1) - F_v''(J) \tag{I-97}$$

$$Q \text{ branch:} \quad \bar{\nu} = \bar{\nu}_0 + F_v'(J) - F_v''(J) \tag{I-98}$$

$$P \text{ branch:} \quad \bar{\nu} = \bar{\nu}_0 + F_v'(J-1) - F_v''(J) \tag{I-99}$$

The band origin ($J'' = 0$ to $J' = 0$ transition) is missing, since J' in the upper state $B(^1\Pi)$ starts from $J' = 1$. The Q branch is entirely absent for a $^1\Sigma - {}^1\Sigma$ transition, since $\Delta J = 0$ is forbidden.

I–10. SELECTION RULES IN ATOMS AND MOLECULES

Selection rules play an important role in interpreting fine structure of atomic and molecular transitions and in understanding the symmetry species of two states between which a transition occurs.

Usually rotational analysis of an electronic band is required to obtain symmetry properties of an excited state. Recently, however, the symmetry of an excited state of some triatomic molecules has been determined by using polarized monochromatic light and measuring the direction of dissociation products with respect to the direction of polarization (see Section II–5 for further details).

I–10.1. Selection Rules in Atoms

In Section I–5 it is shown that the transition probability from a state m to a state n induced by radiation is proportional to $|\mathbf{R}_{nm}|^2$ where $\mathbf{R}_{nm} = \int \psi_n^* \boldsymbol{\mu}_{nm} \psi_m \, dv$ (the transition moment). The transition from the m to the n state by absorption of radiation is allowed only when the transition moment is nonzero. The transition is forbidden if the transition moment is zero. Even if the transition is not allowed as electric dipole transition, it can be induced by the interaction of radiation with the magnetic dipole moment or the electric quadrupole moment with much less probability. The rules to determine whether two states can combine with each other under the influence of radiation are called the selection rules.

Wave mechanical calculation of the transition moment in atoms shows that the integral is nonzero only when the total angular momentum **J** (the vector addition of the orbital and spin angular momenta of electrons) remains the same or changes by one when the transition occurs by absorption of radiation. The selection rule for electric dipole radiation is

$$\Delta J = 0, \pm 1 \tag{I-100}$$

except that $J = 0 \rightarrow J = 0$ is forbidden. The rule is obeyed regardless of the type of coupling.

For Russell-Saunders coupling, which is valid for light atoms, we have in addition

$$\Delta S = 0 \tag{I 101}$$

that is, a transition between different multiplicities is forbidden. For the orbital angular momentum **L**, the selection rule is

$$\Delta L = 0, + 1 \tag{I-102}$$

Laporte Rule. Without making a detailed calculation of the transition moment $\int \psi_n^* \boldsymbol{\mu}_{nm} \psi_m \, dv$ it is possible to show whether a transition between the two states is forbidden or allowed if we know the symmetry properties of the two eigenfunctions. If the eigenfunction changes sign upon inversion at the origin, that is, when the coordinates x, y, z are replaced by $-x$, $-y$, $-z$, it is called odd and if it does not change sign, it is called even. Since $\boldsymbol{\mu}_{nm}$ changes sign by such an inversion, $\psi_n^* \boldsymbol{\mu}_{nm} \psi_m$ changes sign if both ψ_n and ψ_m are even or both odd. If $\psi_n^* \boldsymbol{\mu}_{nm} \psi_m$ is odd, the integral $\int \psi_n^* \boldsymbol{\mu}_{nm} \psi_m \, dv$ must be zero, since the integral over the positive coordinates is exactly counterbalanced by that over the negative coordinates. Accordingly, the selection rule is that only the transitions between even and odd states are allowed. The rule is called the Laporte rule. According to the rule, no transition is allowed, for example, between $2p^3(^2D^o)$ and $2p^3(^2P^o)$ states of N atoms (superscript o signifies odd terms), since they are derived from the same electronic configuration $1s^2 2s^2 2p^3$ and both states are odd [see Herzberg (13), p. 154].

I–10.2. Electronic Transitions in Diatomic Molecules

In the molecular system the total eigenfunction ψ may be expressed to a first approximation as the product of the electronic (ψ_e), vibrational (ψ_v), and rotational (ψ_r) eigenfunctions

$$\psi = \psi_e \psi_v \psi_r \tag{I-103a}$$

The selection rules for diatomic molecules are determined as for atoms by the transition moment

$$\mathbf{R} = \int \psi'^* \boldsymbol{\mu} \psi'' \, dv \tag{I-103b}$$

where ψ' and ψ'' are the total eigenfunctions for the upper and lower states, respectively, and $\boldsymbol{\mu}$ is the dipole moment vector with components $\boldsymbol{\mu}_x$, $\boldsymbol{\mu}_y$, and $\boldsymbol{\mu}_z$ along x, y, and z directions (z is along the internuclear axis). For allowed transitions the transition moment must be different from zero, and for forbidden transitions the transition moment must be zero. To a first approximation, neglecting the interaction between electronic and nuclear motions, the transition moment $\mathbf{R}^X$ along the space fixed X axis may be written

$$\begin{aligned}\mathbf{R}^X = &\int \psi_e'^* \boldsymbol{\mu}_x \psi_e'' \, dv_e \int \psi_v' \psi_v'' \, dr \int \psi_r'^* \cos \alpha_x \psi_r'' \, dv_r \\ &+ \int \psi_e'^* \boldsymbol{\mu}_y \psi_e'' \, dv_e \int \psi_v' \psi_v'' \, dr \int \psi_r'^* \cos \alpha_y \psi_r'' \, dv_r \\ &+ \int \psi_e'^* \boldsymbol{\mu}_z \psi_e'' \, dv_e \int \psi_v' \psi_v'' \, dr \int \psi_r'^* \cos \alpha_z \psi_r'' \, dv_r\end{aligned} \tag{I-103c}$$

where α_x, α_y, α_z are the angles between the molecule fixed x, y, z axes and the space fixed X axis, respectively. Similar equations are obtained for R^Y and R^Z. Hence the selection rules for the overall (electronic, vibrational, rotational) transition are that each integral $\int \psi_e'^* \boldsymbol{\mu}_x \psi_e'' \, dv_e$, $\int \psi_v' \psi_v'' \, dr$, $\int \psi_r'^* \cos \alpha_x \psi_r'' \, dv_r$, and so on must be different from zero for allowed transitions.

In diatomic molecules the vibrational eigenfunction is totally symmetric and the integral $\int \psi'_v \psi''_v \, dr$ is always different from zero. Hence, there is no selection rule for the vibrational transition and we need to consider the properties of electronic and rotational eigenfunctions only.

Selection Rules for Electronic Transitions. For allowed electronic transitions $\psi_e'^* \boldsymbol{\mu}_e \psi_e''$ must be symmetric with respect to a reflection at the plane through the internuclear axis. The electronic eigenfunction ψ_e of a Σ state either changes sign upon reflection at the plane (Σ^- state) or remains unchanged (Σ^+ state). On the other hand, $\boldsymbol{\mu}_z$, the component of the dipole moment in the direction of the internuclear axis z, does not change sign by such a reflection. Therefore, $\psi_e'^* \boldsymbol{\mu}_z \psi_e''$ (likewise $\psi_e'^* \boldsymbol{\mu}_x \psi_e''$ and $\psi_e'^* \boldsymbol{\mu}_y \psi_e''$) changes sign by a reflection at the plane if $\psi_e'^*$ is the eigenfunction of a Σ^+ state and ψ_e'' is that of a Σ^- state, or vice versa, that is, a transition between Σ^+ and Σ^- is forbidden. A transition Σ^+–Σ^+ is allowed only in the z direction, since $\psi_e'^* \boldsymbol{\mu}_z \psi_e''$ does not change sign by a reflection at the plane through the internuclear axis, while $\psi_e'^* \boldsymbol{\mu}_x \psi_e''$ and $\psi_e'^* \boldsymbol{\mu}_y \psi_e''$ change sign by reflections at the yz and xz planes, respectively, that is, the transition moment lies in the internuclear axis. The allowed transitions are

$$\Sigma^+ - \Sigma^+, \ \Sigma^- - \Sigma^- \tag{I-104}$$

If two nuclei in diatomic molecules are identical, the electronic eigenfunctions have an additional symmetry, namely, a center of symmetry. If the eigenfunction does not change sign when the electronic coordinates x, y, z are replaced by $-x$, $-y$, $-z$, it is called even and a state to which the eigenfunction belongs is called an even state. A state whose eigenfunction changes sign is called an odd state. The even state is designated by a subscript g and the odd state by u. Since the components of the dipole moment change sign by an inversion at the origin, we have the following allowed transitions

$$\Sigma_g^+ - \Sigma_u^+, \ \Sigma_g^- - \Sigma_u^- \tag{I-105}$$

Additional selection rules for the quantum numbers of the resultant orbital angular momentum of the electrons about the internuclear axis are [Herzberg (14), p. 241] in Hund's cases a and b (see Section I–2.4)

$$\Delta\Lambda = 0, \ \pm 1 \tag{I-106}$$

and for quantum number of the component of the resultant spin about the axis [in Hund's case a]

$$\Delta\Sigma = 0 \tag{I-107}$$

Hence, transitions ${}^2\Pi_{1/2}$–${}^2\Pi_{1/2}$, ${}^2\Pi_{1/2}$–${}^2\Delta_{3/2}$—but not ${}^2\Pi_{1/2}$–${}^2\Pi_{3/2}$, ${}^2\Pi_{1/2}$–${}^2\Delta_{5/2}$—are allowed. The selection rule for the quantum number of the total electronic angular momentum about the internuclear axis is

$$\Delta\Omega = 0, \ \pm 1 \tag{I-108}$$

but $\Delta\Omega = 0$ is forbidden for $\Omega = 0 \to \Omega = 0$. The quantum number of the total angular momentum apart from spin follows the rule [Hund's case b]

$$\Delta N = 0, \pm 1 \tag{I-109}$$

but $\Delta N = 0$ is forbidden for a Σ–Σ transition.

Selection Rules for Rotational Levels in Electronic Transition. A detailed calculation shows that the integrals, $\int \psi_r'^* \cos\alpha_x \psi_r''\, dv_r$, $\int \psi_r'^* \cos\alpha_y \psi_r''\, dv_r$, $\int \psi_r'^* \cos\alpha_z \psi_r''\, dv_r$, in (I-103c) are different from zero only when $\Delta J = 0, \pm 1$. Hence, the selection rule for the rotational transition between the two electronic states is

$$\Delta J = J' - J'' = 0, \pm 1 \tag{I-110a}$$

where J is the quantum number of the total angular momentum (electronic orbital angular momentum, electron spin, and angular momentum of nuclear rotation). The transition $\Delta J = 0$ is, however, forbidden for a Σ–Σ transition.

Rotational levels in diatomic molecules have different symmetry properties for an inversion of the nuclear coordinates at the origin depending on whether J is even or odd. The rotational eigenfunctions for even J do not change sign by such an inversion, while they do change sign for odd J. Thus, the rotational levels are designated either positive $(+)$ or negative $(-)$, depending on whether the total (electronic and rotational) eigenfunction remains unchanged or changes sign by an inversion at the origin. Since the components of the transition dipole moment change sign by the inversion the transition is allowed only between $+$ and $-$ levels. Hence the selection rule is

$$+ \to - \quad , \quad - \to + \tag{I-110b}$$

For homonuclear molecules we have an additional symmetry property for rotational levels. That is, the total eigenfunction changes its sign or remains unaltered by an exchange of the two nuclei. If the total eigenfunction does not change its sign the level is called symmetric and if it does change sign the level is antisymmetric. Transitions between the two symmetric (or two antisymmetric) levels are allowed by electric dipole, since the components of the dipole moment remain unchanged by an exchange of the nuclei. The selection rule is then

$$\text{sym} \leftrightarrow \text{sym} \quad , \quad \text{antisym} \leftrightarrow \text{antisym} \tag{I-110c}$$

I–10.3. Electronic Transitions in Polyatomic Molecules

The selection rules for electronic, vibrational, and rotational transitions in polyatomic molecules can be derived just as for atoms and for diatomic molecules from the properties of the transition moment.

For allowed transitions the transition moment

$$\mathbf{R} = \int \psi'^{*} \boldsymbol{\mu} \psi'' \, dv \tag{I-111}$$

must be different from zero, where ψ' and ψ'' are the total eigenfunctions of the upper and lower states, respectively, and $\boldsymbol{\mu}$ is the dipole moment vector and has components $\boldsymbol{\mu}_x$, $\boldsymbol{\mu}_y$, and $\boldsymbol{\mu}_z$.

To a first approximation the total eigenfunction may be expressed as the product of the electronic ψ_e, vibrational ψ_v, and rotational ψ_r eigenfunctions neglecting the interaction between electronic, vibrational, and rotational motions [see Herzberg (16), p. 128]

$$\psi = \psi_e \psi_v \psi_r$$

and the transition moment along the space fixed X axis may be given by

$$\begin{aligned} \mathbf{R}^X = {} & \int \psi_e'^{*} \boldsymbol{\mu}_x \psi_e'' \, dv_e \int \psi_v'^{*} \psi_v'' \, dv_v \int \psi_r'^{*} \cos \alpha_x \psi_r'' \, dv_r \\ & + \int \psi_e'^{*} \boldsymbol{\mu}_y \psi_e'' \, dv_e \int \psi_v'^{*} \psi_v'' \, dv_v \int \psi_r'^{*} \cos \alpha_y \psi_r'' \, dv_r \\ & + \int \psi_e'^{*} \boldsymbol{\mu}_z \psi_e'' \, dv_e \int \psi_v'^{*} \psi_v'' \, dv_v \int \psi_r'^{*} \cos \alpha_z \psi_r'' \, dv_r \end{aligned} \tag{I-112}$$

where α_x, α_y, α_z are the angles between the molecule fixed x, y, z axes and the space fixed X axis, respectively. Similar expressions for R^Y and R^Z can be derived. Hence the selection rules for the overall (electronic, vibrational, rotational) transition are that each integral, $\int \psi_e'^{*} \boldsymbol{\mu}_x \psi_e'' \, dv_e$, $\int \psi_v'^{*} \psi_v'' \, dv_v$, $\int \psi_r'^{*} \cos \alpha_x \psi_r'' \, dv_r$ and so on, in (I-112) must be different from zero for allowed transitions.

Selection Rules for Electronic Transitions. For allowed electronic transitions the transition moment

$$\int \psi_e'^{*} \boldsymbol{\mu} \psi_e'' \, dv_e \tag{I-113}$$

must be different from zero. This means $\psi_e'^{*} \psi_e''$ must belong to the same symmetry species as $\boldsymbol{\mu}_x$, $\boldsymbol{\mu}_y$, or $\boldsymbol{\mu}_z$, so that the integrand $\psi_e'^{*} \boldsymbol{\mu}_x \psi_e''$, $\psi_e'^{*} \boldsymbol{\mu}_y \psi_e''$, or $\psi_e'^{*} \boldsymbol{\mu}_z \psi_e''$ is totally symmetric, namely, $\psi_e'^{*} \boldsymbol{\mu}_x \psi_e''$ and so on must be symmetric for all symmetry operations allowed for the point group to which a molecule belongs. As an example, if the molecule belongs to the point group C_{2v}, such as formaldehyde and water, $\boldsymbol{\mu}_x$, $\boldsymbol{\mu}_y$, $\boldsymbol{\mu}_z$ behave like the coordinates x, y, z for symmetry operations as shown in Table I-1. That is, $\boldsymbol{\mu}_x$, $\boldsymbol{\mu}_y$, and $\boldsymbol{\mu}_z$, respectively, belong to the species B_1, B_2, and A_1. Hence, for allowed transitions in the point group C_{2v}, the product $\psi_e' \psi_e''$ must belong to B_1, B_2, or A_1. If the ground state is A_1 (usually the case for stable C_{2v} molecules) only transitions to the B_1, B_2, and A_1 states are allowed. If the transition moment is parallel to the top axis, a transition is called a parallel (||) transition. If the transition moment is perpendicular to the top axis, a transition is called a perpendicular ($\perp$) transition. In the case of formaldehyde the top axis is the z axis (see Fig. I–2.) Hence, transitions to B_1 and B_2 are $\perp$ transitions and that to A_1 is a || transition. The ${}^1A_2$–${}^1A_1$ transition in formaldehyde is forbidden, since there is no component of the dipole moment corresponding to the ${}^1A_2$ symmetry species. Although the transition is electronically forbidden, there are some cases where the transition becomes weakly allowed because of the vibronic interaction. The selection rule is that if the transition moment

$$\int \psi'^{*}_{ev} \boldsymbol{\mu} \psi''_{ev}\, dv \tag{I-114}$$

is different from zero, the transition is allowed. For example, the ${}^1A_2$–${}^1A_1$ transition of formaldehyde, forbidden by the electronic transition becomes allowed if a b_1 or b_2 type of vibration is excited (A_2 and b_1 give B_2 vibronic symmetry. A_2 and b_2 result in B_1 symmetry). The vibronic interaction is strong in some cases, such as the bending vibration in linear triatomic molecules. The vibronic interaction in this case is called the Renner-Teller effect [see Herzberg (16), p. 26].

Selection Rules for Vibrational Levels in Electronic Transitions. If the interaction between the electronic and vibrational motion is neglected, the selection rule for allowed vibrational transitions of a symmetric molecule follows the rule that the product $\psi'^{*}_v \psi''_v$ must be symmetric with respect to all symmetry operations of the point group to which the molecule belongs. That is, ψ'_v and ψ''_v belong to the same symmetry species, assuming the electronic transition is also allowed. There is no vibrational selection rule for diatomic and unsymmetrical molecules.

Selection Rules for Rotational Levels in Electronic Transitions. General discussion of rotational selection rules in the electronic transition involving linear, symmetric top, spherical top, and asymmetric top molecules are given by Herzberg (16).

In this section we consider the cases of photochemically interesting molecules such as formaldehyde and nitrogen dioxide in which both the ground and upper states belong to the point group C_{2v} and may be represented by symmetric tops, that is, $A > B \simeq C$ (a is the top axis). See Section I–3.1 and Fig. I–2.

The selection rules for the rotational quantum number K (the quantum number of the component of the total angular momentum $\mathbf{J}$ about the top axis) and for J are, when the transition moment is parallel to the figure axis (|| bands),

$$\Delta K = 0,\ \Delta J = 0,\ \pm 1 \text{ if } K \neq 0 \tag{I-115}$$

$$\Delta K = 0,\ \Delta J = \pm 1 \text{ if } K = 0$$

Hence, the Q branch is missing for $K = 0$ in $\|$ bands. On the other hand, when the transition moment is perpendicular to the figure axis ($\perp$ bands) the rules are

$$\Delta K = \pm 1, \qquad \Delta J = 0, \pm 1 \tag{I-116}$$

The most prominent feature of $\perp$ bands is that the group of lines formed by Q branches stand out while the lines of the P and R branches form an unresolved background [see Herzberg (15), p. 424]. Assuming that z is the figure axis, these rules are derived from calculations that the integral $\int \psi_r'^* \cos \alpha_z \psi_r'' \, dv_r$ in (I-112) is different from zero only for $\Delta K = 0, \Delta J = 0, \pm 1$ and the integrals, $\int \psi_r'^* \cos \alpha_x \psi_r'' \, dv_r$, $\int \psi_r'^* \cos \alpha_y \psi_r'' \, dv_r$, are different from zero only for $\Delta K = \pm 1, \Delta J = 0, \pm 1$.

In the case of formaldehyde, transitions to B_1 and B_2 states from the ground state give rise to $\perp$ bands and a transition to A_1 yields $\|$ bands, since B_1 and B_2 states have symmetries like x and y, respectively, while A_1 has symmetry like z, which is the top axis (see Fig. I–2). On the other hand, in nitrogen dioxide, transitions to A_1 and B_1 yield $\perp$ bands and a B_2–A_1 transition gives $\|$ bands, since the top axis is now the y direction, which coincides with the symmetry of the B_2 state. Ritchie et al. (831) have found from the rotational analysis that the 2491 Å bands of NO_2 show characteristic feature of the $\|$ bands and have been assigned to a transition $^2B_2$–$^2A_1$.

chapter II

Primary Photochemical Processes in Simple Molecules

The photochemical reaction comprises a series of events starting from the initial act of light absorption and ending in the production of stable molecules generally different from the reactant molecules.

A series of photochemical processes may be divided into the primary and secondary processes.

The primary process may be defined, following Noyes and Leighton [(22), p. 153], as the initial act of light absorption and the immediately following processes of the initially formed electronically excited state.

The latter processes may include unimolecular processes such as fluorescence, dissociation, and isomerization, as well as bimolecular processes exemplified by deactivation and reactions of the excited state by collisions with the same or different kind of molecules.

The reactions of the excited molecule may be important for molecules such as SO_2 and NO in the regions of absorption where the bond energies are larger than the energies of incident light. For some molecules the electronically excited states neither dissociate nor fluoresce immediately and the reactions of the excited states become of importance even in the vacuum ultraviolet absorption. Such cases are found in acetylene and carbon dioxide in the vacuum ultraviolet and in formaldehyde in the near ultraviolet photolysis.

The secondary process includes the reactions of atoms and radicals produced in the primary process either directly from the excited molecule or by collisions of the excited molecule with the same or different kind of molecules. The secondary reactions are the reactions of atoms and radicals with each other and the reactions of atoms and radicals with molecules.

In the case of a diatomic molecule the study of absorption and emission spectrum is important in understanding the primary process. On the other hand, in a polyatomic molecule spectroscopy generally contributes little in understanding the primary process.

II–1. THE PRIMARY PROCESSES IN DIATOMIC MOLECULES

The primary processes following light absorption by diatomic molecules should illustrate the simplest case of photochemical processes since there is only one bond to be broken and the resulting products are two atoms. The molecule may, upon light absorption, undergo immediate dissociation or may perform many vibrations and rotations before it dissociates into atoms. Sometimes the molecule is raised to a stablc excited state from which it fluoresces if it suffers no collisions. The electronic state that fluoresces or dissociates into atoms can be assigned from fine structure analysis of the absorption or emission spectra. The dissociation products corresponding to a given electronic state can be derived from the correlation rules, the convergence limits of the band spectra, and the relevant thermochemical data. If the absorption spectra are continuous no information is obtained from spectroscopy on the electronic states and atomic products produced. In these cases one must identify the products directly by optical absorption, mass spectrometry, and other techniques or indirectly by state-and-energy-dependent chemical reactivities of atoms produced.

II–1.1. Spectroscopic Studies of Diatomic Molecules

The absorption spectra of diatomic molecules are in some cases continuous, without any fine structure, while in other cases they show discrete structure followed by a continuum. The spectra may be explained with the aid of the potential energy curves.

Continuous Absorption Spectra. Figure II–1 shows the potential energy curves for direct dissociation by absorption of a light quantum $h\nu$. The upper curve has no potential minimum and a dissociation results from light absorption at all wavelengths. The absorption spectra are continuous in these cases. Generally, the photon energy corresponding to the long wavelength limit of absorption far exceeds the bond dissociation energy, since,

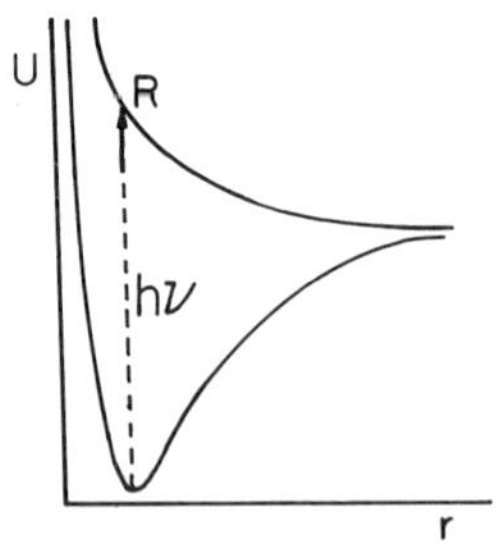

Fig. II–1. Potential energy curves showing absorption continuum; the upper curve is repulsive and an immediate dissociation results by light absorption at all wavelengths.

according to the Franck-Condon principle (Section I–9.2), the transition probability by absorption of light energy corresponding to the bond energy is very small. The absorption spectra of F_2 and HI are such examples.

Absorption Spectra Showing Band Convergence. Figure II–2*a* shows schematically the convergence of the progression $v' \leftarrow v'' = 0$ (v' signifies the vibrational levels of the upper state and v'' means those of the ground state; the upper state is always written first) followed by a continuum towards higher wave numbers. The vertical arrow indicates the convergence limit. Figure II–2*b* presents the potential energy curves showing that a molecule dissociates into one ground and one excited atom above energies corresponding to the convergence limit. The spectrum of Br_2 shown in Fig. V–17 is an example. The convergence limit is at 5108 Å, below which the molecule dissociates into one ground state atom $^2P_{3/2}$ and one excited atom $^2P_{1/2}$. The bond dissociation energy can be obtained from the energy corresponding to 5108 Å (2.427 eV) less the electronic energy of Br ($^2P_{1/2}$), 0.456 eV, which gives 1.970 eV. Similarly, the convergence limit of the I_2 spectrum is at 4995 Å and the bond energy is 1.539 eV.

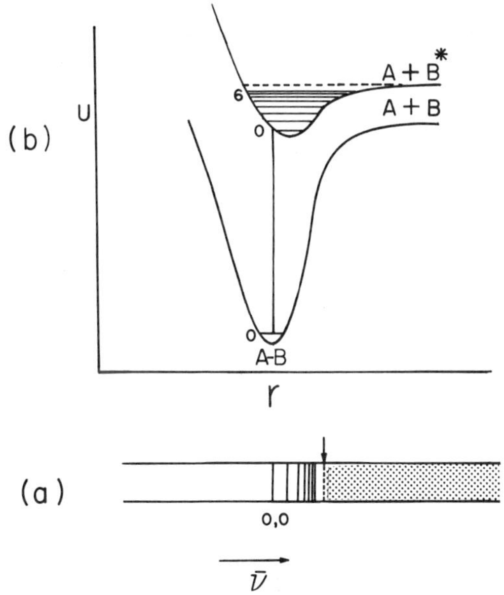

Fig. II–2. Potential energy curves showing the convergence limit in diatomic molecules. (*a*) Absorption spectrum showing the vibrational progression and the convergence limit (indicated by the arrow). A continuum follows the convergence limit towards higher wave number. (*b*) Potential energy curves showing the convergence limit. If photon energies exceed the convergence limit, one ground atom and one excited atom are formed.

Predissociation. If we suppose the vibrational level $v' = 1$ in the electronically excited state B in Fig. II–3 lies above the dissociation energy of a molecule, it is possible that the molecule dissociates by absorption of a light quantum corresponding to the transition to $v' = 1$ or above. The dissociation may occur via the continuum of the low-lying stable state A or a repulsive state. If the dissociation occurs within one vibration (10^{-13} sec) rotational lines disappear while vibrational lines $v' = 1$ and above broaden as shown on the extreme left.

On the other hand, if the dissociation occurs within one rotation (10^{-11} sec) rotational lines become diffuse while vibrational lines remain sharp. The interaction of the discrete levels in the state B with the continuum belonging to the state A results in the diffuseness of the absorption lines. The width of the absorption line depends on the extent of the interaction between the A and B states. If the interaction is stronger the width of the line is larger and the lifetime of the excited state is shorter. The above type of dissociation, namely, the dissociation occurring sometime after light absorption, is called predissociation. Predissociation is characterized by a sudden broadening of vibrational or rotational lines.

The predissociation can in principle be recognized by line broadening in absorption. However, it is not always possible to find line broadening in absorption, since line broadening by predissociation must exceed that by the Doppler effect. The Doppler line width is typically of the order of 0.05 cm^{-1}. From (I-57) the corresponding lifetime is 10^{-10} sec. If the lifetime is less than 10^{-10} sec, it may be possible to detect diffuseness of absorption lines by a high resolution absorption spectrograph. (A typical high resolution spectrograph has a resolution of 500,000 corresponding to 0.1 cm^{-1} in the near ultraviolet region.) If, however, the lifetime is longer than 10^{-9} sec, the

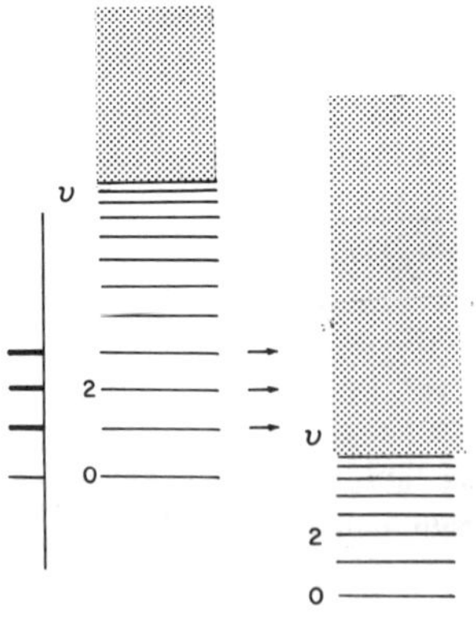

Fig. II–3. Energy level diagram for predissociation process. The horizontal arrows indicate predissociation of levels $v' = 1$ and above. The widths of these levels are broadened by predissociation as shown to the extreme left. From G. Herzberg, *Molecular Spectra and Molecular Structure I, Spectra of Diatomic Molecules*, 2nd ed., p. 410. Copyright by Litton Educational Publishing, Inc. Reprinted by permission of Van Nostrand Reinhold Company.

diffuseness by predissociation is obscured by the Doppler width. Consequently, the predissociation is not detectable even with the high resolution spectrograph. On the other hand, if the predissociation occurs above a certain level in an emission spectrum, the emission lines originating above this level are weak or completely missing (breaking-off of bands). The typical lifetime of the excited state in the visible or ultraviolet is about 10^{-8} sec. Hence, emission becomes very weak even for a vibrational level with a lifetime of 10^{-9} sec. Thus, the breaking-off of emission bands is a much more sensitive criterion of the occurrence of predissociation than the diffuseness in absorption. The β and δ absorption bands of NO, for example, do not show any diffuseness in rotational structure beyond the dissociation limit at 1910 Å [Herzberg et al. (462)] while no emission bands have been seen above $v' = 7$ for the β bands and $v' = 1$ for the δ bands. These vibrational levels are just above the dissociation energy.

II–1.2. Photochemical Studies of Diatomic Molecules

Fluorescence and Quenching. If an electronically excited state is formed by absorption of light with energies below that required to break a bond, the only processes that occur are the transitions to various levels of a lower state with emission of light (fluorescence).

In the case of diatomic molecules, the crossover to the ground state from the initially formed excited state does not occur, since the level separation of the ground state at the point of the crossover is much larger than the radiative width. Hence, the quantum yield of fluorescence is always unity [see Bixon and Jortner (110)].

Fluorescence from the excited molecule is quenched by ground state molecules as well as by foreign gas molecules. The following Stern-Volmer type quenching equation can be derived (167, 729) assuming the steady state conditions and a weak absorption. Consider the following quenching reactions

$$\mathrm{A} \xrightarrow{h\nu} \mathrm{A}^* \qquad I_a$$

$$\mathrm{A}^* \rightarrow A + h\nu \qquad k_f$$

$$\mathrm{A}^* + \mathrm{A} \rightarrow 2A \qquad k_q$$

$$\mathrm{A}^* + \mathrm{M} \rightarrow \mathrm{A} + \mathrm{M} \qquad k_{\mathrm{M}}$$

Here A* is an electronically excited molecule formed by light absorption and k_f is the fluorescence decay rate in seconds k_q and k_{M} are the self-quenching rate constant, and quenching rate constant by a nonabsorbing foreign gas M respectively in units of $(\text{concentration})^{-1}\ \text{sec}^{-1}$ and I_a is the number of quanta absorbed per sec per cm^3 by the molecules A.

Assuming steady state conditions for the concentration of excited molecules (A^*) we have

$$I_a = k_f(A^*) + k_q(A^*)(A) + k_M(A^*)(M)$$

We define quenching Q by

$$Q = \frac{I_f}{I_0}$$

where I_f is the fluorescence intensity and I_0 is the intensity of incident light.

Since $I_f = k_f(A^*)$, it follows that

$$\frac{I_a}{Q} = I_0\left[1 + \frac{k_q}{k_f}(A) + \frac{k_M}{k_f}(M)\right]$$

For weak absorption $I_a \simeq I_0\epsilon(A)l$ where ϵ is the absorption coefficient and l is the path length. Hence, it follows that

$$\frac{(A)}{Q} = \mathscr{B}[1 + a_A(A) + a_M(M)] \qquad \text{(II-1)}$$

where $\mathscr{B}$ is a constant involving absorption coefficient, path length, and other proportionality factors. The quantities, $a_A = k_q/k_f$, $a_M = k_M/k_f$, are called quenching constants and are in units of (concentration)$^{-1}$. The fluorescence decay rate k_f (sec^{-1}) and the self-quenching cross section σ_q^2 (cm^2) may be obtained from the observed intensity of fluorescence $I_f(t)$ as a function of decay time t (sec) after a pulsed excitation

$$I_f(t) = I_f(0)e^{-[k_f + \sigma_q^2(16\pi RT/M_A)^{1/2}(A)]t} \qquad \text{(II-1a)}$$

where M_A is the molecular weight of a diatomic molecule A, R is the gas constant equal to 8.314×10^7 erg mol^{-1} K^{-1}, T is in °K and (A) is the number of molecules per cm^3. The quantity $\sigma_q^2(16\pi RT/M_A)^{1/2}(A)$ signifies the number of quenching collisions per second as is further discussed in Section IV–1.

Using (II-1a) lifetimes and quenching cross sections for $Br_2(B^3\Pi)$ and I_2 $(B^3\Pi)$ states have been measured by many workers (153, 186, 187, 213a, 781, 852) as a function of vibrational and rotational levels.

Dissociation and Predissociation. If the incident photon energy exceeds the bond energy, the primary process is either the complete dissociation or the simultaneous occurrence of dissociation and fluorescence. The former case is represented in Fig. II–4, where the two upper states, R_1 and R_2 are both repulsive. The R_1 state is correlated with ground state atoms A and B, while the R_2 state is associated with one ground and one excited atom, A and B*. A typical example is the absorption spectrum of HI (see Fig. V–6)

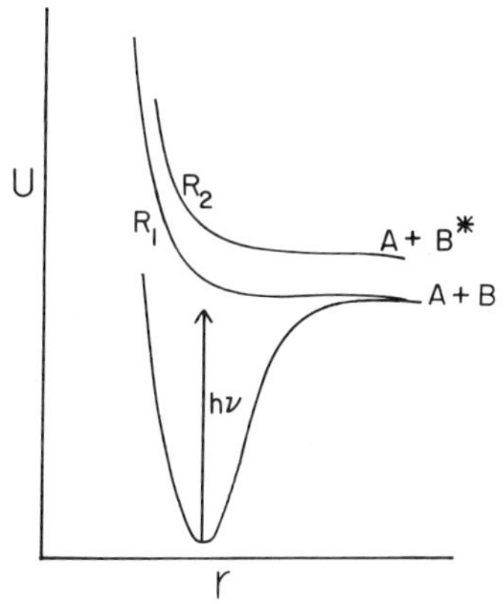

Fig. II–4. Potential energy curves showing continuous absorption. Two upper repulsive curves R_1, R_2, are associated, respectively, with ground state atoms, A + B, and ground and excited atoms, A + B*.

where R_1 is the $^3\Pi_1$ correlated with $H(^2S) + I(^2P_{3/2})$ and R_2 is the $^3\Pi_{0^+}$ associated with $H(^2S) + I(^2P_{1/2})$ (219). The production of $I(^2P_{1/2})$ is energetically possible below 3100 Å. The fraction of $I(^2P_{1/2})$ produced below 3100 Å photolysis has been obtained experimentally by measuring the translational energy of H atoms (219) following the pulsed laser photolysis of HI or by optical absorption of $I(^2P_{3/2}, {}^2P_{1/2})$ resulting from the flash photolysis of HI.

The latter case, that is, the simultaneous occurrence of dissociation and fluorescence, is shown in Fig. II–5. It shows the stable upper state E correlated with one ground and one excited atom, A and B*, and the repulsive state R arising from ground state atoms, A and B. If the photon energy slightly exceeds the bond energy, the molecule may be raised to the repulsive state R and dissociates immediately or it may be brought into the stable state E by light absorption from which the molecule may either fluoresce or predissociate into A + B by a radiationless transition to the repulsive state R.

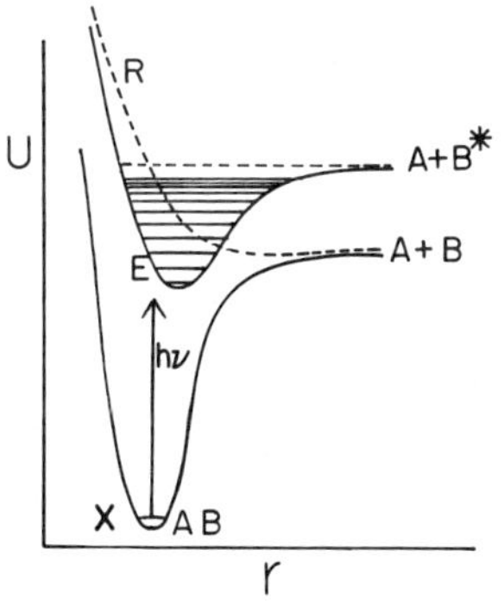

Fig. II–5. Potential energy curves showing predissociation. The upper stable state E is crossed by the repulsive curve R. The predissociation into ground state atoms A + B occurs above the intersection by crossover from the E to R state. The molecule in the state E dissociates into A + B* above energies exceeding the convergence limit. Simultaneous occurrence of fluorescence and predissociation may be found in the region between the crossing point and the convergence limit.

Capelle and Broida (187) have measured the variation of the fluorescence lifetimes from the $I_2(B^3\Pi_{0_u^+})$ state (corresponding to the E state in Fig. II–5) as a function of the vibrational energy using (II-1a). The lifetimes have two minima at about $v' = 3$ and $v' = 24$. The results indicate a competing predissociation process to the two ground state atoms via the repulsive $^1\Pi_{1_u}$ state (the R state in Fig. (II–5)), which crosses the $B^3\Pi_{0_u^+}$ state at $v' = 3$ and $v' = 24$. The predissociation is so weak that its manifestation in the absorption spectra is highly unlikely.

If the incident photon has more than sufficient energy required for the production of A + B*, the excited molecule may be formed in the R state and dissociates immediately into ground state atoms. Or the molecule may be raised to the E state, from which it dissociates into A + B*. The fraction of B* produced is thus dependent on the fraction of light absorbed by the transition to the E state. The primary processes of Cl_2, Br_2, and I_2 may be represented by this scheme. The fraction of B* formed may be a function of incident wavelength, and it can be determined only by a direct observation of the products formed. This is illustrated below by the photolysis of Cl_2 in the ultraviolet.

The absorption spectrum of Cl_2 shows a continuum in the 2500 to 4500 Å region (see Fig. V–15). Busch et al. (159) have shown that the photolysis of Cl_2 at 3471 Å produces $Cl(^2P_{3/2}) + Cl(^2P_{3/2})$ rather than $Cl(^2P_{1/2}) + Cl(^2P_{3/2})$. Hence, the transition is entirely to the $R(^1\Pi_{1_u})$ state rather than to the E $(^3\Pi_{0_u^+})$ state in this region.

II–2. THE PRIMARY PROCESSES IN SIMPLE POLYATOMIC MOLECULES

II–2.1. Fluorescence in Simple Polyatomic Molecules

As described in Section II–1.2, if the incident photon energy is below that of the weakest bond, an electronically excited molecule formed by light absorption may give rise to fluorescence. The lifetime of the radiative transition to the ground state is related to the integrated absorption coefficient [see Section I–7, (I-91a)].

Anomalously Long Radiative Lives. It has been found for some simple molecules, such as CS_2, SO_2, and NO_2, that the observed fluorescence lifetimes are much longer than those expected from the integrated absorption coefficient. Common to these molecules are the extensive perturbation of rotational structure observed in the ultraviolet and visible regions. Douglas was the first who suggested (324) that the anomalous lifetime can be explained on the basis of an extensive interaction between the initially formed excited state by light absorption and another state that does not combine with the ground state. Figure II–6 shows a schematic energy level diagram that illustrates the Douglas mechanism. The molecule is brought into a certain rotational level below the dissociation energy D_0 of the excited state E_1 by light absorption.

The rotational level interacts with many adjacent levels belonging to another excited state E_2. The E_2 state may be a metastable state located between the ground state and E_1, or it can be the ground state. If the E_2 state is the ground state, the level density at the energy corresponding to $h\nu$ may be much larger than that of the E_1 state. Accordingly, the level in the E_1 state may interact with many levels of the ground state. In other words, the level in the E_1 state is diluted many times by the levels of the ground state. Since the transitions from these levels are forbidden, the radiative life is many times greater than that expected from the integrated absorption coefficient. It may also be stated that the molecule, initially in the E_1 state, goes over to the E_2 state where it spends most of its time before it comes back to the E_1 state and fluoresces. Bixon and Jortner (110) have applied to small molecules a theory of electronic relaxation in polyatomic molecules. They have concluded that in the case of small molecules such as NO_2, CS_2, and SO_2 no intramolecular electronic relaxation processes occur. Hence, the quantum yield of fluorescence must be unity in the collision-free limit. The unexpectedly long collision-free lifetimes of SO_2 fluorescence found by Brus and McDonald (156) appear to verify unit quantum efficiency of fluorescence. Since the excited states responsible for the long radiative life are easily quenched by wall collisions, the partial loss of the excited molecules by collisions accounts for the measured quantum efficiency of less than unity. On the other hand, large polyatomic molecules such as naphthalene and anthracene have less than unit quantum efficiency since the intramolecular relaxation process occurs rapidly (110). The measured lifetime is also

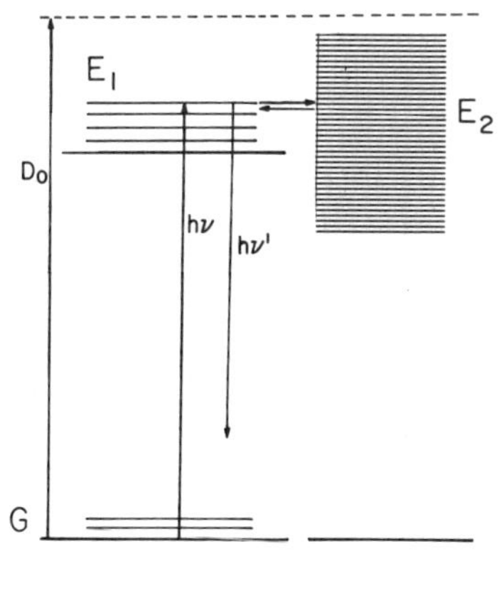

Fig. II–6. The anomalously long radiative life. According to Douglas (324) a molecule is brought by light absorption to a rotational level or levels in the upper state E_1 below the dissociation energy D_0. The rotational level in the E_1 state is perturbed by many adjacent levels in the E_2 state, which does not combine with the ground state *G*. As a result the fluorescence lifetime is lengthened. The extent of lengthening depends on the ratio of rotational level density in the E_1 state to that in the E_2 state at the point of interaction.

much less than that expected from the oscillator strength, since the intramolecular relaxation process is much faster than the radiative process.

II–2.2. Photodissociation in Simple Polyatomic Molecules

The photodissociation commonly occurs by absorption of light with energies exceeding the weakest bond. Several types of dissociation may be considered in simple polyatomic molecules.

Types of Dissociation in Triatomic Molecules. The most common type of photodissociation in triatomic molecules XYZ is a dissociation at the weaker of the two bonds

$$XYZ \xrightarrow{h\nu} X + YZ \qquad \text{(II-2)}$$

where the X—Y bond is weaker than the Y—Z bond. Typical examples are the dissociations of OCS and NOCl

$$OCS \xrightarrow{h\nu} CO + S$$
$$NOCl \xrightarrow{h\nu} Cl + NO$$

If a dissociation takes place from a repulsive upper state or a stable state with a shallow potential minimum at a larger X—Y distance than that of the ground state, the molecule dissociates within one vibration, that is, within 10^{-13} sec at all wavelengths of absorption. The absorption spectrum is continuous. The excess energy of photons beyond that required to break the X—Y bond appears as the translational energy of the dissociation products. Such cases have been found in the photolysis of H_2O in the vacuum ultraviolet [Welge and Stuhl (1033)] and in the photolysis of NOCl at 3471 Å [Busch and Wilson (164)].

Another type of dissociation is the simultaneous scission of the two bonds and the formation of a new bond

$$XYZ \xrightarrow{h\nu} XZ + Y \qquad \text{(II-3)}$$

This type of dissociation appears to be common in the vacuum ultraviolet photolysis but less common in the near ultraviolet, indicating a large potential barrier for such a process. The photolysis of H_2O [McNesby et al. (683)] at 1236 Å leads partially to the formation of H_2 molecules

$$H_2O \xrightarrow{h\nu} H_2 + O$$

An upper state responsible for the dissociation must have an equilibrium configuration with an H—O—H angle much smaller than that in the ground state. The molecular product, H_2, is likely to be formed with some vibrational energy, although it has not been observed experimentally.

A still less common type of dissociation is the production of three atoms

$$XYZ \xrightarrow{h\nu} X + Y + Z \qquad \text{(II-4)}$$

This type of dissociation is energetically possible in the vacuum ultraviolet photolysis. However, no such examples have clearly been demonstrated. The nonoccurrence of this process may be due to a much easier conversion of the photon energy into the translational energy of the fragments rather than into the internal energy of the diatomic product which subsequently dissociates into atoms.

Types of Dissociation in Polyatomic Molecules. If a polyatomic molecule contains several bonds of various energies, a common dissociation mode is the scission along the weakest bond R—X by photon absorption similar to (II-2) in triatomic molecules

$$RX \xrightarrow{h\nu} R + X \qquad \text{(II-5)}$$

In a series of alkyl iodide photolyses at 2662 Å Riley and Wilson (834) have shown that the dissociation occurs within one rotation. The energy beyond that required to break the CH_3—I bond appears predominantly as the translational energy of CH_3 just as in the triatomic molecules. Similar results have been obtained for halomethyl bromide photolysis by Simons and Yarwood (892). The extent of internal energy of alkyl radicals increases from 12% for the methyl radical to 50% for the propyl radical. The energy partitioning in photodissociation is discussed further in Section II–4. There are, however, many exceptions to this rule in polyatomic molecules. For example, the dissociation of ethane into two methyl radicals (the CH_3—CH_3 bond is the weakest in ethane) is only a minor process in photolysis [Calvert and Pitts (4), Ausloos and Lias (49)].

Molecular Detachment Processes. The molecular detachment processes similar to (II-3) in the triatomic molecules, that is, the rupture of the two bonds followed by the formation of a new bond, appear to be more common in polyatomic molecules. Typical examples are the photolysis of formaldehyde [DeGraff and Calvert (271)] in the near ultraviolet

$$H_2CO \xrightarrow{h\nu} H_2 + CO$$

the photolysis of methane in the vacuum ultraviolet [Mahan and Mandal (658)]

$$CH_4 \xrightarrow{h\nu} CH_2 + H_2$$

and the photolysis of ethane in the vacuum ultraviolet [Okabe and McNesby (759)]

$$C_2H_6 \xrightarrow{h\nu} C_2H_4 + H_2$$

If the molecule in the upper state dissociates within one vibration, the radical-atom production of the type (II-5) seems a more favorable process than the molecular detachment process, since the latter process would require the simultaneous excitation of the bending and stretching vibrations. Molecular detachment processes appear to require incident photon energies far above the thermochemical energies because of the presence of large potential barriers for these dissociation paths (see Section II–3.2).

II–2.3. Predissociation in Simple Polyatomic Molecules

When a discrete upper state is above the dissociation limit, a molecule in the discrete upper state raised by light absorption may dissociate into products through the interaction with an overlapping dissociation continuum and at the same time give rise to fluorescence or it may cross over to another metastable state. The situation is similar to the predissociation in diatomic molecules discussed in Section II–1.2, although few clear-cut cases have been found in absorption spectra of polyatomic molecules.

The predissociation is manifest in sudden weakening of fluorescence above the dissociation limit shown in SO_2 fluorescence near 2200 Å by Okabe (767) (see Fig. VI–18) and in the sudden breaking-off of the formaldehyde emission near the 0–0 band of the $^1A_2$–$^1A_1$ transition [Brand and Reed (126)] at 3500 Å. Evans and Rice (337) have observed that in electronically excited chloro- and bromoacetylene above the dissociation limit the fluorescence and dissociation are competitive. They attribute the competition to a slow transfer of internal energy to the ν_3 C—Cl or C—Br stretching vibration necessary for dissociation to take place. A striking example of predissociation has been found in the photolysis of acetylene by Stief et al. (928). The electronically excited acetylene formed by light absorption in the vacuum ultraviolet has a lifetime of the order of microseconds. Since no fluorescence has been observed from excited acetylene, it is likely that the excited molecule undergoes rapid internal conversion to the ground state from which it dissociates into $C_2 + H_2$ or $C_2H + H$. Slow decomposition may be attributed to an inefficient transfer of internal energy to vibrations leading to dissociation [see Okabe (773)].

The predissociation of NO_2 near 3979 Å illustrates a rare example of that in the polyatomic molecules observed both in absorption and emission. Douglas and Huber (322) have seen an onset of diffuseness of rotational bands at 3979 Å, although vibrational structure is not affected at the onset. The fluorescence of NO_2 gradually weakens below 4200 Å of incident wavelength and it completely disappears at 3979 Å in agreement with the onset in the absorption spectrum [Lee and Uselman (619)]. The process corresponds to

$$NO_2(^2B_1) \rightarrow NO(X^2\Pi) + O(^3P)$$

Another interesting example of predissociation is the photolysis of thiophosgene [Okabe (774)].

The photodissociation process

$$Cl_2CS \xrightarrow{h\nu} ClCS + Cl$$

occurs below 4500 Å. Above 4500 Å the primary process is the fluorescence from the excited state $^1A_2$ in the collision-free limit region.

$$Cl_2CS(^1A_2) \rightarrow Cl_2CS + h\nu'$$

The fluorescence disappears almost completely below 4500 Å, although the absorption spectrum still shows fine structure in the region 4000 to 6000 Å. The predissociation manifests itself in the dependence of the primary quantum yield of dissociation ϕ_{Cl} on pressure as shown in Fig. II–7. This pressure dependence may be explained on the basis of the competition of the excited state between dissociation

$$Cl_2CS(^1A_2) \rightarrow ClCS + Cl$$

and quenching by thiophosgene.

$$Cl_2CS(^1A_2) + Cl_2CS \rightarrow 2Cl_2CS$$

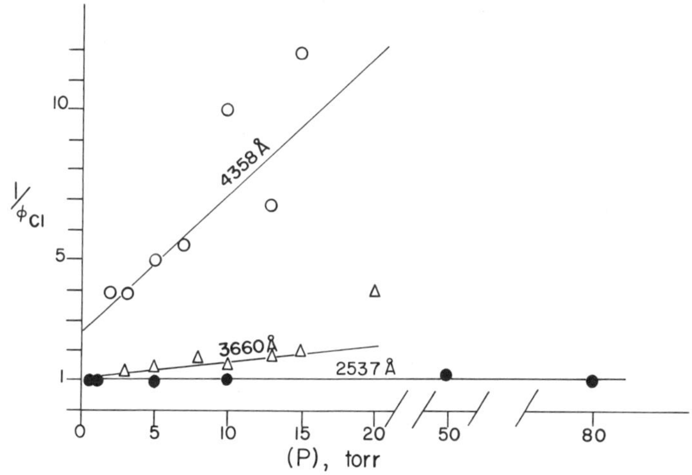

Fig. II–7. Predissociation of thiophosgene. The primary quantum yield of photodissociation, ϕ_{Cl}, of thiophosgene, $Cl_2CS \rightarrow ClCS + Cl$, decreases as the pressure increases at 3660 and 4358 Å, indicating the finite lifetime of the excited state $^1A_2$. The lifetimes of the $^1A_2$ state at 3660 and 4358 Å are about 6 and 55 nsec, respectively. At 2537 Å ϕ_{Cl}, is unity and has no pressure dependence, suggesting direct dissociation. From Okabe (774), reprinted by permission. Copyright 1977 by the American Institute of Physics.

From the pressure dependence of $1/\phi_{Cl}$ the lifetimes of $Cl_2CS(^1A_2)$ formed at 4358 and 3660 Å are about 55 and 6 nsec, respectively.

On the other hand, the ϕ_{Cl} near 2537 Å is unity and has no pressure dependence, indicating the occurrence of direct dissociation in accordance with the observation that the absorption spectrum shows no structure in this region.

The photolysis of carbon disulfide in the vacuum ultraviolet yields strong fluorescence of $CS(A^1\Pi)$ [Okabe (769)].

$$CS_2 \xrightarrow{h\nu} CS(A^1\Pi) + S$$
$$CS(A^1\Pi) \rightarrow CS(X^1\Sigma) + h\nu'$$

The fluorescence excitation spectrum in the region 1200 to 1330 Å shown in Fig. II–8 clearly demonstrates the occurrence of predissociation from Rydberg states. A similar example is the photolysis of HCN in the vacuum ultraviolet [Davis and Okabe (264)]

$$HCN \xrightarrow{h\nu} CN(B^2\Sigma^+) + H$$
$$CN(B^2\Sigma^+) \rightarrow CN(X^2\Sigma^+) + h\nu'$$

The fluorescence excitation spectrum shows the v_2' vibrational progression observed in the absorption spectrum of HCN in the 1330 to 1470 Å region.

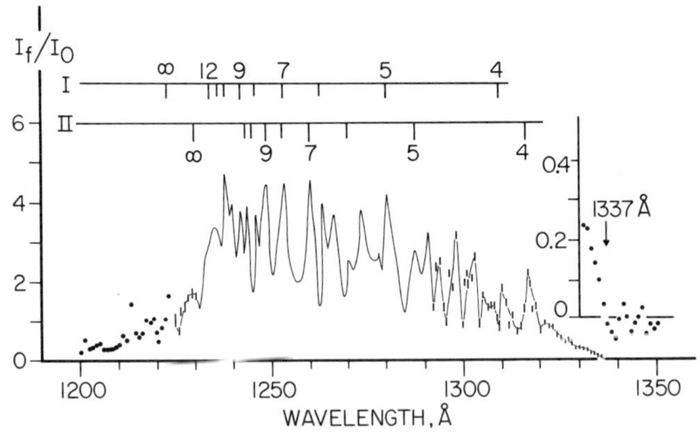

Fig. II–8. Fluorescence $CS(A^1\Pi \rightarrow X^1\Sigma)$ excitation spectrum of CS_2 (pressure ~0.1 torr). I and II designate Rydberg series. The vertical arrow indicates the threshold wavelength of incident light to excite the fluorescence. The excitation spectrum follows the absorption spectrum, indicating $CS(A^1\Pi)$ is predissociated from Rydberg states of CS_2. From Okabe (769), reprinted by permission. Copyright 1972 by the American Institute of Physics.

II–3. CORRELATION RULES IN PHOTODISSOCIATION

Useful rules limiting the number of possibilities for products dissociated from a certain electronically excited state are the correlation rules. In many cases absorption spectra of polyatomic molecules are diffuse or continuous in the region of photochemical interest. Hence, in such cases it is not possible to assign from the analysis of the spectra the symmetry species of the excited state undergoing dissociation. However, the multiplicity, that is, whether the state is a singlet or triplet, can be deduced, for example, from the integrated absorption coefficient. If an absorption is an allowed singlet to singlet transition, the integrated absorption coefficient is of the order of 10^7 atm^{-1} cm^{-2}, corresponding to an oscillator strength of 0.5 (see (I-51)]. If the absorption spectrum extends over the range of 10,000 cm^{-1}, the corresponding maximum absorption coefficient will be about $k_{max} =$ 1000 atm^{-1} cm^{-1}. If a transition is from a singlet to a triplet state, the expected maximum absorption coefficient is much less than 1, corresponding to an oscillator strength much less than 10^{-3}. It is of course not possible to state definitely whether the transition is spin forbidden from the absorption coefficient alone, since the transition may be symmetry forbidden. The analysis of the rotational bands would be required for the definitive assignment of the transition. If the multiplicity of the excited state is known, it is possible to apply the spin correlation rules (or spin conservation rules) between the excited molecule and the dissociation products. If the symmetry of the excited state is known from spectroscopic analysis of fine structure, or from theory, we can further apply the symmetry correlation rules between the excited molecule and its dissociation products. The primary photochemical processes of molecules cited below are discussed in more detail in Chapters VI and VII.

II–3.1. Examples of Spin Correlation Rules

The spin correlation rules have been developed by Wigner and Witmer [Herzberg (14), p. 315]. We consider a simple case where a molecule dissociates into two fragments that have the spin vectors $\mathbf{S}_1$ and $\mathbf{S}_2$. The resultant spin vector $\mathbf{S}$ of the molecule can have the quantum numbers

$$S = (S_1 + S_2), (S_1 + S_2 - 1), \ldots, |S_1 - S_2|$$

For example, if $S_1 = S_2 = \frac{1}{2}$, the resultant spin S is 1 or 0 (triplet or singlet). On the other hand if $S_1 = \frac{1}{2}$, $S_2 = 1$, $S = \frac{3}{2}$ and $\frac{1}{2}$ (quartet and doublet) result. These and other cases are shown in Table II–1. The spin correlation rules may hold for the molecules containing light atoms where the spin orbit interaction is small. In molecules composed of heavy atoms the rule may not be strictly obeyed.

Table II–1. Multiplicities of Dissociation Products and Those of the Molecule

Product	Molecule
Singlet + singlet	Singlet
Singlet + doublet	Doublet
Doublet + doublet	Singlet, triplet
Doublet + triplet	Doublet, quartet
Doublet + quartet	Triplet, quintet
Triplet + triplet	Singlet, triplet, quintet

Source: G. Herzberg, *Molecular Spectra and Molecular Structure I, Spectra of Diatomic Molecules*, 2nd ed., p. 319. Copyright 1950 by Litton Educational Publishing, Inc. Reprinted by permission of Van Nostrand Reinhold Company.

Carbon Dioxide. In the region 1200 to 1672 Å the primary process is entirely the production of $O(^1D)$

$$CO_2 \xrightarrow{h\nu} CO(^1\Sigma^+) + O(^1D)$$

Since the excited state of CO_2 is a singlet (16) in this region the process follows the spin correlation rules, although the production of $O(^3P)$ is energetically possible.

Carbonyl Sulfide. The primary processes of OCS in the region 1900 to 2550 Å are

$$OCS \xrightarrow{h\nu} CO(^1\Sigma^+) + S(^1D)$$

$$OCS \xrightarrow{h\nu} CO(^1\Sigma^+) + S(^3P)$$

The $S(^1D)$ production is 67% and the $S(^3P)$ production is 24%. The upper OCS state is most likely the repulsive $^1A'$ state [Ferro and Reuben (347)]. Because of the presence of the heavy S atom the correlation rules are partially broken.

Carbon Disulfide. The primary process of CS_2 in the region 1900 to 2250 Å is entirely the production of $S(^3P)$, although the production of $S(^1D)$ is energetically possible below 1930 Å. The upper state must be the $^1B_2$ [(16), p. 600]

$$CS_2 \xrightarrow{h\nu} CS(^1\Sigma) + S(^3P)$$

In the vacuum ultraviolet photolysis the main primary process is also the production of $S(^3P)$ [Okabe (769)]

$$CS_2 \xrightarrow{h\nu} CS(A^1\Pi) + S(^3P)$$

Because of the strong absorption of CS_2 in the vacuum ultraviolet, the excited state must be a singlet. In the case of CS_2 photolysis the spin correlation rules are entirely disobeyed.

Nitrous Oxide. The primary process of N_2O in the region 1850 to 2300 Å is

$$N_2O \xrightarrow{h\nu} N_2(^1\Sigma^+) + O(^1D)$$

in accordance with the spin conservation rule, since the excited N_2O is most likely a singlet. [(16), p. 596] The production of NO has not been observed, since the process

$$N_2O \xrightarrow{h\nu} N(^4S) + NO(X^2\Pi)$$

violates the spin correlation rules, although the process is energetically possible below 2520 Å.

Ozone. The ozone photolysis presents a striking example of the spin conservation rules. The primary process at 2537 Å is [Schiff (857)]

$$O_3 \xrightarrow{h\nu} O_2(^1\Delta) + O(^1D)$$

The excited state of O_3 corresponding to the Hartley band is probably $^1B_2$ [Hay et al. (450)].

The spin correlation rules alone do not predict whether the products are $O_2(^1\Delta) + O(^1D)$ or $O_2(^3\Sigma_g^-) + O(^3P)$, since both processes are spin allowed. If the symmetry species of the excited ozone is definitely established, for example, by photofragment spectroscopy (see Section II–5), the symmetry correlation rules described in the following section can be applied to predict which process is more likely to occur.

II–3.2. Examples of Symmetry Correlation Rules

If the symmetry of an excited state is known, we may be able to apply the symmetry correlation rules between the excited state and its dissociation products.

Water. The structure of ground state water is shown in Fig. II–9. Ground state water belongs to the point group C_{2v}. The x axis is chosen to be perpendicular to the plane of the molecule, which is the yz plane. The z axis is a twofold (C_2) symmetry axis. The correlation diagram of water and its dissociation products is shown in Fig. II–10.

A broad continuum observed in the region 1450 to 1850 Å corresponds to a transition to the unstable $\tilde{A}(^1B_1)$ state [Horsley and Fink (485)]. (A tilde signifies an electronic state for nonlinear polyatomic molecules, to make a distinction from symmetry species.) Welge and Stuhl (1033) have found the dissociation products $H + OH(X^2\Pi)$ by flash photolysis in this region. The

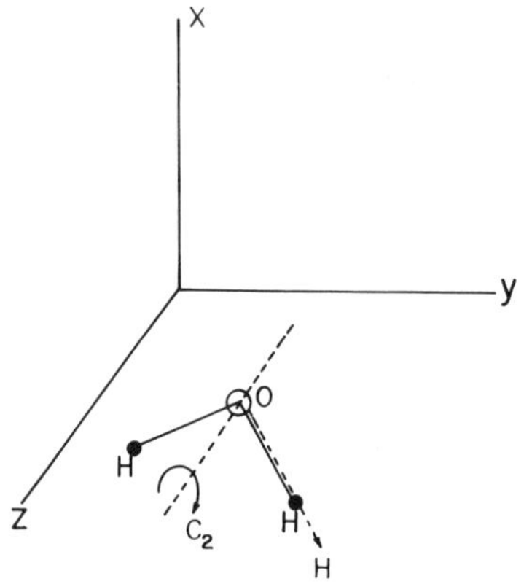

Fig. II–9. The ground state configuration of water. Water belongs to the point group C_{2v}. The x axis is perpendicular to the molecular plane. The z axis is the symmetry axis (C_2). The H—O—H angle is 105°.

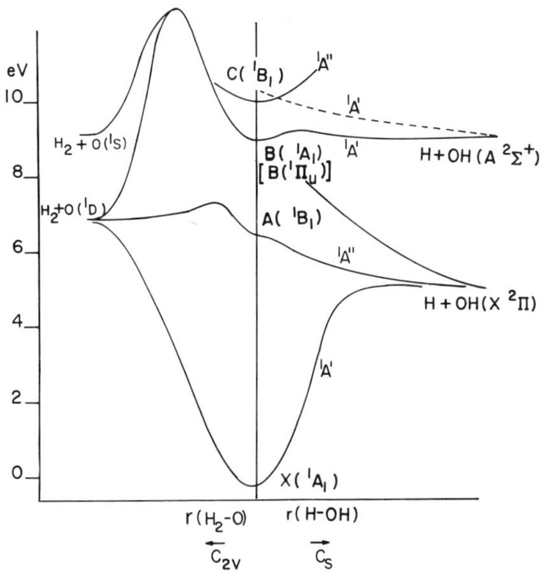

Fig. II–10. Correlation diagram of water and its dissociation products. Dissociation into H + OH is given to the right and dissociation into H_2 + O is given to the left. Dissociation is assumed to take place in the molecular plane. The former dissociation involves a change of symmetry from C_{2v} to C_s and the latter involves no change of symmetry. The $\tilde{A}(^1B_1)$ dissociates mainly into H + OH($X^2\Pi$) with little change of the H—O—H angle, producing nonrotating OH. The $\tilde{B}(^1A_1)$ state dissociates mainly into H + OH($A^2\Sigma^+$) with a large change in bond angle, producing rotating OH($A^2\Sigma^+$). The $\tilde{C}(^1B_1)$ may dissociate into H + OH($A^2\Sigma^+$) through the repulsive $^1A'$ state. The products H_2 + O(^1D) comes mainly from the $\tilde{A}(^1B_1)$, state since the dissociation path from $\tilde{B}(^1A_1)$ requires a large potential barrier [after Tsurubuchi (980)].

Table II–2. Correlation of Species of Point Groups C_{2v} and D_{3h} with Those of Lower Symmetry

C_{2v}	$C_s[\sigma(yz) \to \sigma]^a$	D_{3h}	$C_{2v}[\sigma_h \to \sigma_v(yz)]^a$
A_1	A'	A_1'	A_1
A_2	A''	A_1''	A_2
B_1	A''	A_2'	B_2
B_2	A'	A_2''	B_1
		E'	$A_1 + B_2$
		E''	$A_2 + B_1$

Source: G. Herzberg, *Molecular Spectra and Molecular Structure III, Electronic Spectra and Electronic Structure of Polyatomic Molecules*, p. 577. Copyright 1966 by Litton Educational Publishing, Inc. Reprinted by permission of Van Nostrand Reinhold Company.

[a] Dissociation is assumed to take place in the molecular plane (yz) (see Fig. II–9)

dissociation from the $\tilde{A}(^1B_1)$ state apparently is brought about without much change in geometry since no rotation has been found in the OH radicals (see Section II–6.5). Figure II–10 shows that C_{2v} symmetry changes to C_s as the H atom moves away along the original OH bond in the molecular plane. The $^1B_1$ state then becomes $^1A''$ in C_s (see Table II–2). The $^1A''$ state correlates with H + OH($X^2\Pi$) (see Table II–3). The second mode of dissociation in this region is the production of H_2 and O(^1D). This process has been shown to occur to an extent of several percent in the vacuum ultraviolet [Stief et al. (935), McNesby et al. (683)]. The dissociation products can be correlated with $\tilde{A}(^1B_1)$ (see Table II–4). According to Tsurubuchi (980), the dissociation from $H_2O(\tilde{A}\,^1B_1)$ requires a small potential barrier. Therefore, the $\tilde{A}(^1B_1)$ state mainly dissociates into H + OH($X^2\Pi$) rather than into H_2 + O(^1D).

The second absorption region (1250 to 1400 Å) corresponds to a transition to the $\tilde{B}(^1A_1)$ state (485), which is almost linear in the equilibrium conformation. The $\tilde{B}(^1A_1)$ state becomes $^1A'$ in C_s conformation, which in turn correlates with the products H + OH($A^2\Sigma^+$). Since the $\tilde{B}(^1A_1)$ is nearly linear, the H—O—H angle would increase by light absorption. The simultaneous excitation of bending and stretching vibrations exerts a torque to the departing OH($A^2\Sigma^+$). The results by Carrington (191), who has observed OH($A^2\Sigma^+$) in a high degree of rotation, are in accordance with this hypothesis. The $\tilde{B}(^1A_1)$ state becomes $^1\Pi_u$ in a linear conformation [Gangi and Bader (385a)] and the $^1\Pi_u$ state should correlate with OH($X^2\Pi$) + H rather than OH($A^2\Sigma^+$) + H as shown in Fig. II–10. Hence, if the water molecule

Table II–3. Correlation of Species of Linear Molecules with Those of Molecules of Lower Symmetry

Linear Molecule	$C_s(\sigma_v \to \sigma)^a$	$C_{2v}(z \to z)^b$
Σ^+	A'	A_1
Σ^-	A''	A_2
Π	$A' + A''$	$B_1 + B_2$
Δ	$A' + A''$	$A_1 + A_2$

Source: G. Herzberg, *Molecular Spectra and Molecular Structure III, Electronic Spectra and Electronic Structure of Polyatomic Molecules*, p. 576. Copyright 1966 by Litton Educational Publishing, Inc. Reprinted by permission of Van Nostrand Reinhold Company.

[a] Dissociation is assumed to occur in the plane containing the internuclear axis (the z axis).

[b] The linear axis (the z axis) corresponds to the main axis (the z axis) of C_{2v} molecule.

dissociates from the linear conformation, that is, from the $\tilde{B}(^1\Pi_u)$ state the products must be $OH(X^2\Pi) + H$. Since the production of $OH(A^2\Sigma^+)$ is only about 5% of the primary processes, the major process must be the production of $OH(X^2\Pi) + H$. It is likely, therefore, that the dissociation into OH + H occurs mainly from the linear conformation $\tilde{B}(^1\Pi_u)$ and to a smaller extent from the bent $\tilde{B}(^1A_1)$ conformation.

Table II–4. Correlation of Species of Atoms with Molecules

Atom	C_{2v}	C_s	$D_{\infty h}(C_{\infty v})^a$
S_g	A_1	A'	Σ_g^+
P_g	$A_2 + B_1 + B_2$	$A' + 2A''$	$\Sigma_g^- + \Pi_g$
D_g	$2A_1 + A_2 + B_1 + B_2$	$3A' + 2A''$	$\Sigma_g^+ + \Pi_g + \Delta_g$

Source: G. Herzberg, *Molecular Spectra and Molecular Structure III, Electronic Spectra and Electronic Structure of Polyatomic Molecules*, p. 575 Copyright 1966 by Litton Educational Publishing, Inc. Reprinted by permission of Van Nostrand Reinhold Company

[a] For $C_{\infty v}$ disregard the subscripts g and u.

The dissociation from the $\tilde{B}(^1\Pi_u)$ state, however, would yield highly excited $OH(X^2\Pi)$ in rotation since the transition $\tilde{B}(^1\Pi_u) \leftarrow \tilde{X}(^1A_1)$ involves a bent to linear change of configuration, although the rotational distribution of ground state OH in the second region of absorption has not been shown to support this hypothesis. According to Tsurubuchi (980) the productions of $H_2 + O(^1D)$ and $H_2 + O(^1S)$ from the $\tilde{B}(^1A_1)$ state require large potential barriers as shown in Fig. II–10. The transition to the third excited singlet state $\tilde{C}(^1B_1)$ occurs in the region 1194 to 1241 Å. The 1240 Å bands of water studied by Johns (533), show diffuseness that depends on the rotational quantum number; that is, the lines are sharpest at the lowest rotational level and become diffuse at higher rotational quantum numbers. The predissociation then must be caused by a state of different symmetry, in which case rotational excitation is required for these states to interact with each other. Since the state corresponding to the 1240 Å bands is $\tilde{C}(^1B_1)$, it is reasonable to assume that the predissociation is caused by the adjacent $\tilde{B}(^1A_1)$ state or a $^1A'$ repulsive state.

Formaldehyde. The ground state of formaldehyde is $\tilde{X}^1A_1$ in C_{2v} symmetry (see Fig. I-2) and it becomes $^1A'$ in C_s as the H atom is dissociated from formaldehyde in the molecular plane. The $^1A'$ state correlates with the products $H + HCO(\tilde{X}^2A')$. Therefore, the ground state formaldehyde correlates with the products $H + HCO(\tilde{X}^2A')$ (see Fig. II–11). The results of the photolysis (686) suggest two primary processes, H + HCO and $H_2 + CO$ in the ultraviolet region. These products must be correlated with the excited state $^1A_2$ initially brought about by light absorption in the near ultraviolet. Neither $H_2 + CO$ nor H + HCO correlates with $^1A_2$. Therefore, the dissociation processes must involve transitions $^1A_2 \rightarrow {}^1A'$ (C_s conformation) or $^1A_2 \rightarrow {}^1A_1$ (C_{2v} conformation), that is, heterogeneous predissociation [(16), p. 458] (the two interacting electronic states have different symmetry). The dissociation of ground state H_2CO to $H_2 + CO$ requires a large potential barrier, 4 to 7 eV, depending on the configuration of the transition state [Jaffe et al. (521)]. This large potential barrier for the dissociation arises from the required energy to bring the two H atoms close together to form a new bond. To achieve such a configuration the totally symmetric stretching and H–C–H bending vibrations may have to be excited simultaneously to a high degree. It seems reasonable to assume that the excited $^1A_2$ state crosses over to the ground state where the molecule may achieve such a configuration after many complex Lissajous motions of the nuclei resulting in dissociation into $H_2 + CO$. On the other hand, the dissociation into H + $HCO(^2A')$ may occur from a repulsive $^1A'$ state within one vibration since only the ν_5 type assymmetric C—H stretching vibration has to be strongly excited. The dissociation into H + H + CO must involve a large potential

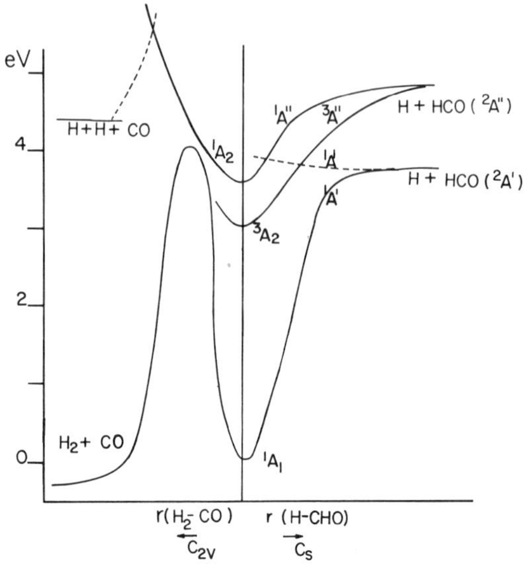

Fig. II–11. Correlation diagram of formaldehyde and its dissociation products. Dissociation into H + HCO is given to the right and dissociation into H_2 + CO (or H + H + CO) is given to the left. It is assumed the dissociation occurs in the molecular plane. The former dissociation involves a change from C_{2v} to C_s, and the latter involves no change of symmetry. The $^1A_2$ state is assumed to dissociate into H + HCO($^2A'$) and H_2 + CO through the repulsive $^1A'$ and the ground state $^1A_1$, respectively. The dissociation into CO + H + H from the $^1A_2$ state requires a large potential barrier.

barrier since even at 2537 Å there is no evidence of this process [DeGraff and Calvert (271)]. The production of the excited HCO($^2A''$) with which the $^1A_2$ correlates is not energetically possible below about 5 eV. Miller and Lee (706) emphasize the importance of the $^3A_2$ in the photodissociation process in analogy with other aldehyde compounds.

Ammonia. The excited state of ammonia corresponding to the absorption in the region 1700 to 2200 Å is $^1A_2''$ in D_{3h} symmetry [Douglas (319)]. The main primary process in this region is the production of $NH_2(\tilde{X}^2B_1)$ + H [Groth et al. (429)]. The $\tilde{A}(^1A_2'')$ state correlates with $NH_2(\tilde{X}^2B_1)$ + H assuming the occurrence of dissociation in the molecular plane, that is, the molecular plane of ammonia, σ_h, becomes that of NH_2, σ_v ($\sigma_h \rightarrow \sigma_v$). On the other hand, the ground state ammonia dissociates into $NH_2(\tilde{A}^2A_1)$ + H. See Fig. II–12 and Tables II–2 and II–3 for the correlations of ammonia with its dissociation products. Figure II–12 is constructed on the basis that curves of the same symmetry species do not cross each other. The dissociation of

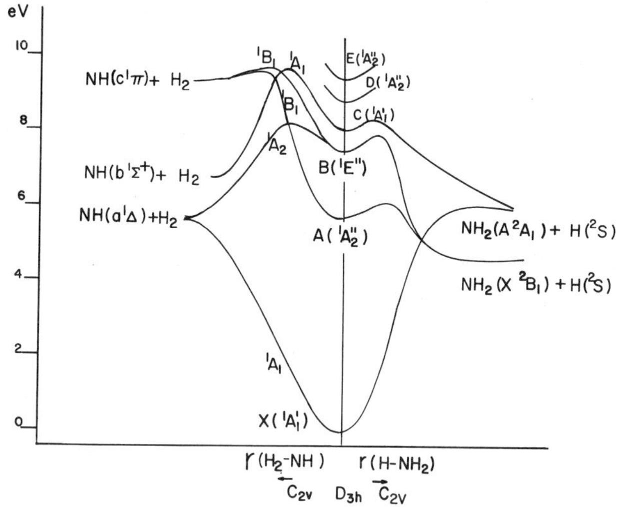

Fig. II–12. Correlation diagram of ammonia and its dissociation products. Dissociation into NH_2 + H is given to the right and dissociation into NH + H_2 is given to the left. The dissociation is assumed to occur in the molecular plane. The former dissociation involves a symmetry change from D_{3h} to C_{2v} and the latter involves changes $D_{3h} \rightarrow C_{2v} \rightarrow C_{\infty v}$. The $\tilde{A}(^1A_2'')$ state dissociates mainly into $NH_2(\tilde{X}\,^2B_1) + H(^2S)$. The $\tilde{B}(^1E'')$ state dissociates mainly into $NH_2(\tilde{X}\,^2B_1) + H(^2S)$ and partially into $NH(a^1\Delta) + H_2$ and $NH(c^1\Pi) + H_2$. The dissociation path $NH_3[\tilde{C}(^1A_1')] \rightarrow NH(b^1\Sigma^+) + H_2$ requires a large potential barrier and is seen with incident photon energies above 9.3 eV. The products $NH_2(\tilde{A}\,^2A_1) + H(^2S)$ are formed probably from the $\tilde{C}(^1A_1')$ state.

ammonia into NH + H_2 is likewise assumed to occur in the molecular plane via C_{2v} symmetry as an intermediate, that is, $\sigma_h \rightarrow \sigma_v(yz)$.

$$NH_3 \rightarrow HN\!\left\langle{}^{H}_{H}\right. \rightarrow NH + H_2$$

$$D_{3h} \qquad C_{2v} \qquad\qquad C_{\infty v}$$

Hence, from Table II–2 we have the correspondence for $D_{3h} \rightarrow C_{2v}$

$$A_1' \rightarrow A_1,\ A_2'' \rightarrow B_1,\ E'' \rightarrow A_2 + B_1$$

Further, the symmetry axis of C_{2v}(z axis) is assumed to correspond to the figure axis of the departing NH molecule, that is, $z \rightarrow z$. From Table II–3 we obtain for $C_{2v} \rightarrow C_{\infty v}$

$$A_1 \rightarrow \Sigma^+, \qquad B_1 \rightarrow \Pi, \qquad A_2 \rightarrow \Sigma^-, \Delta$$

Hence, the dissociation products of the type $NH + H_2$ from ammonia are

$$\tilde{A}(^1A_2'') \rightarrow NH(c^1\Pi) + H_2$$
$$\tilde{B}(^1E'') \rightarrow NH(a^1\Delta) + H_2$$
$$\tilde{B}(^1E'') \rightarrow NH(c^1\Pi) + H_2$$
$$\tilde{C}(^1A_1') \rightarrow NH(b^1\Sigma^+) + H_2$$

Accordingly, the dissociation of $\tilde{A}(^1A_2'')$ into $NH(a^1\Delta) + H_2$ is not likely to occur, although the reaction is energetically possible. Experimentally, only 4% of the primary process accounts for this dissociation with light of wavelengths above 1700 Å (see the photolysis of NH_3, Section VII–1). The second excited state $\tilde{B}(^1E'')$ correlates with $NH_2(\tilde{X}^2B_1) + H$, $NH(a^1\Delta) + H_2$, and $NH(c^1\Pi) + H_2$. The fact that the production of $ND + D_2$ from ND_3 is still 3% at 1470 Å (634) indicates the presence of a potential barrier for the process $ND_3(\tilde{B}^1E'') \rightarrow ND(a^1\Delta) + D_2$ as shown in Fig. II–12. At 1236 Å the D_2 yield increases to 25%, indicating an increased importance of the processes $ND_3(\tilde{B}^1E'') \rightarrow ND(a^1\Delta) + D_2$, $ND(c^1\Pi) + D_2$, and $ND_3(\tilde{C}^1A_1') \rightarrow ND(b^1\Sigma^+) + D_2$. The $C(^1A_1')$ state correlates with $NH_2(\tilde{A}^2A_1) + H$ and $NH(b^1\Sigma^+) + H_2$. Okabe and Lenzi (762) have found that the NH_2 emission from ammonia photolysis is observable only above 7.56 eV of incident photon energy, although the threshold energy for this process is much less (about 6 eV), since the $\tilde{A}(^1A_2'')$ state does not correlate with $NH_2(\tilde{A}^2A_1) + H$ as shown in Fig. II–12. The dissociation path to form $NH(b^1\Sigma^+) + H_2$ from $C(^1A_1')$ apparently requires a large potential barrier [Gilles et al. (394), Gelernt et al. (389)], since $NH(b^1\Sigma^+)$ is formed only above about 9.2 eV of photon energy. The production of $NH(c^1\Pi)$ has been seen with photon energies above 9.36 eV [Okabe and Lenzi (762)]. The products $NH(c^1\Pi) + H_2$ can be correlated with $\tilde{A}(^1A_2'')$ and $\tilde{B}(^1E'')$.

Nitrous oxide, carbon dioxide, carbonyl sulfide. The quantum yield of $O(^1S)$ production from N_2O photolysis shows a maximum near 1300 Å of incident wavelength [McEwan et al. (680), Black et al. (113)] that coincides with the 1290 Å peak of absorption. This peak is assigned by Winter (1053) to the $^1\Sigma^+$ state of N_2O. From symmetry correlation rules it is reasonable to assume

$$N_2O \xrightarrow{h\nu} N_2O(^1\Sigma^+) \rightarrow N_2(^1\Sigma_g^+) + O(^1S)$$

to be predominantly responsible for the production of $O(^1S)$ in agreement with the results (see Fig. II–13.) The 1810 and 1445 Å bands, corresponding to the transitions to the $^1\Delta$ and $^1\Pi$, correlate with $O(^1D)$, in accordance with experimental observations. The products $NO(^2\Pi) + N(^4S)$ do not correlate with any of the singlet states and have not been observed experimentally (816) while $NO(^2\Pi) + N(^2P, {}^2D)$ have been found in the vacuum ultraviolet photolysis.

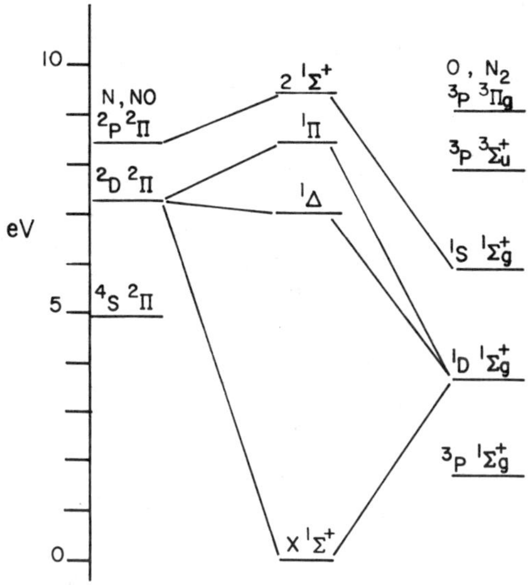

Fig. II–13. Adiabatic correlation diagram for N_2O; energy relative to the ground state $X^1\Sigma^+$. Dissociation of N_2O into N_2 + O is given to the right and dissociation into N + NO is given to the left. N_2O is $C_{\infty v}$ symmetry. Transitions to $^1\Delta$, $^1\Pi$, and $2^1\Sigma^+$ correspond respectively to 1810, 1445, and 1290 Å bands (from M. Krauss, private communication). Photolysis in the 1810 and 1445 Å bands produces $O(^1D)$ while the 1290 Å band yields $O(^1S)$ mainly. The products $NO(^2\Pi)$ + $N(^4S)$ do not correlate with any of the singlet excited states and have not been observed experimentally, while $NO(^2\Pi)$ + $N(^2D, {}^2P)$ production has been observed in the vacuum ultraviolet photolysis.

The $O(^1S)$ production from CO_2 photolysis is also most efficient near 1120 Å of incident wavelength corresponding to the $^1\Sigma_u^+$ state of CO_2 [Lawrence (616) Koyano et al. (584)].

$$CO_2 \xrightarrow{h\nu} CO_2(^1\Sigma_u^+) \rightarrow CO(^1\Sigma^+) + O(^1S)$$

Black et al. (114) have found that the $S(^1S)$ production from carbonyl sulfide is most efficient near 1500 Å, indicating the process

$$OCS \rightarrow OCS(^1\Sigma^+) \rightarrow CO(^1\Sigma^+) + S(^1S)$$

II–4. DISTRIBUTION OF THE EXCESS ENERGY IN PHOTOFRAGMENTS

The energy of incident photons above that required for a given process appears as the translational, rotational, vibrational and electronic energies of fragments. If the process involves a single bond scission, the excess energy is the difference between the photon energy and the dissociation energy of the bond to be broken. The fragments with excess energy are in general more reactive than those without excess energy. Particularly when the product atoms are electronically excited, the reactivities are very different from those in their ground state. The knowledge of energy distribution in fragments is, apart from its intrinsic interest, important in understanding the overall photochemistry.

Recently Simons (894a) has reviewed the dynamics of photodissociation, including the theoretical and experimental developments on energy partitioning in photofragments formed in the primary process and the angular dependence of the fragments with respect to the direction of polarization.

II–4.1. Measurement of the Translational Energy of Photofragments

Two entirely different methods have been used to assess the translational energy of fragments. The first method is based on the chemical reactivities of the atoms produced. The second method is the measurement of time-of-flight of photofragments from the initial zone of formation to a detector.

The Translational Energy of Hydrogen Atoms. The H atoms generated by the photolysis of a molecule HR contain some excess energy of $E_t(\mathrm{H})$ in the form of translation when the incident photon energy $h\nu$ exceeds the bond dissociation energy $D_0(H\text{—}R)$. If R is a halide atom, X, then the energy in excess of that required to break the bond $D_0(\mathrm{H{-}X})$ may be expressed as

$$E_t(\mathrm{H}) = \frac{m_\mathrm{X}}{m_\mathrm{HX}}(h\nu - D_0(\mathrm{H{-}X}) + E_{int}) \qquad \text{(II-6)}$$

where E_{int} is the internal energy of hydrogen halides and m_X and m_HX are masses of X and HX, respectively. It is apparent from (II-6) that H atoms carry almost all the excess energy. If H atoms have some excess energy, the ratio of two reaction rates

$$\mathrm{H + HX \rightarrow H_2 + X} \qquad k_7 \qquad \text{(II-7)}$$

$$\mathrm{H + X_2 \rightarrow HX + X} \qquad k_8 \qquad \text{(II-8)}$$

where k_7 and k_8 are the rate constants for (II-7) and (II-8), respectively, may vary depending on the extent of excess energy, since (II-7) requires more activation energy than (II-8).

Fass (344) has found the ratio k_8/k_7 for X = Br to be 5.3 ± 0.4 at 1850 and 2480 Å independent of temperature in the range 300 to 523°K. On the other hand, the same ratio for thermal H atoms is 26 at 300°K. The H atoms produced from HBr at 1850 and 2480 Å have 2.9 and 1.2 eV translational energies, respectively. The activation energy difference $E_7 - E_8 = 0.8$ kcal mol^{-1} for (II-7) and (II-8) found for thermal H atoms is not observed for the translationally excited (hot) H atoms. In a similar experiment with the DCl—Cl_2 system Wood and White (1056) have found the rate constant ratio, k_8/k_7 for X = Cl, to be 6.5 for D atoms produced at 1849 Å while the same ratio is about 300 for thermal atoms at 300°K. The hot hydrogen atom reaction has been studied for the HCl–Cl_2 system at 1849 Å by Jardine et al. (528) and for the HI–I_2 system by Penzhorn and Darwent (803) and by Holmes and Rodgers (480).

Some workers have used systems similar to (II-7) and (II-8) with different pairs of molecules M_1 and M_2, where M_2 is a dominant absorber of photons. These pairs of molecules are chosen in such a way as to give an activation energy difference of several kcal mol^{-1} for thermal H atoms. Oldershaw and Porter (776) have used M_1 = OCS and M_2 = HI. In this system HI was irradiated by the 3000 Å line to produce H* (translationally excited) atoms

$$HI \xrightarrow{h\nu} H^* + I \tag{II-9}$$

The H* atoms react with OCS to produce CO + HS or are deactivated to thermal H atoms

$$H^* + OCS \rightarrow CO + HS \qquad k_{10} \tag{II-10}$$

$$H^* + OCS \rightarrow H + OCS \qquad k_{11} \tag{II-11}$$

Likewise the H* atoms react with HI to form H_2 + I or are deactivated to thermal H atoms

$$H^* + HI \rightarrow H_2 + I \qquad k_{12} \tag{II-12}$$

$$H^* + HI \rightarrow H + HI \qquad k_{13} \tag{II-13}$$

The thermal H atoms are scavenged by HI

$$H + HI \rightarrow H_2 + I \tag{II-14}$$

From (II-10) to (II-14) we obtain

$$\frac{[H_2]}{[CO]} = \frac{k_{11}}{k_{10}} + \frac{k_{12} + k_{13}}{k_{10}} \frac{[HI]}{[OCS]} \tag{II-15}$$

A plot of $[H_2]/[CO]$ versus $[HI]/[OCS]$ gives the positive intercept $r_0 = [H_2]/[CO]$ as $[HI]/[OCS] \rightarrow 0$. The probability of reaction, *P*, of H atoms with OCS while they are hot is defined as

$$P = \frac{[CO]_0}{[CO]_0 + [H_2]_0} = [1 + r_0]^{-1} \tag{II-16}$$

where $[CO]_0$ and $[H_2]_0$ indicate the concentrations of CO and H_2 at the limit $[HI]/[OCS] \rightarrow 0$. The linear relationship between $[H_2]/[CO]$ and

[HI]/[OCS] has been experimentally observed (776). The probability was found to be larger when the H atoms had more translational energy. Oldershaw et al. (777) also have shown that P increases with the translational energy of H atoms for the N_2O–HBr and N_2O–HI systems.

In these systems $P = [N_2]_0/([N_2]_0 + [H_2]_0)$, where N_2 is produced from the reaction

$$H^* + N_2O \rightarrow N_2 + OH \qquad \text{(II-17)}$$

In another system N_2O–H_2S mixtures were illuminated at 2483 Å. Oldershaw et al. (777) have concluded that in this system 90% of available energy appears in translation of H and SH. Similar results are obtained by Compton et al. (230) for the C_4D_{10}–H_2S system at 1850 Å and for C_4H_{10}–D_2S at 2138 and 2288 Å (232). It is somewhat unexpected that H atoms dissociated from CH_3SH and C_2H_5SH at 2537 and 2288 Å carry almost all excess energy as translation [White et al. (1038)]. However, at shorter wavelengths the internal energies of CH_3S and C_2H_5S appear to increase (1039).

In the C_4D_{10}–HI system illuminated at 2537, 2288, and 1850 Å, Compton and Martin (231) have concluded that the production of I($^2P_{1/2}$) is 7, 20, and 0%, respectively, assuming that in the C_4D_{10}–HBr system no Br($^2P_{1/2}$) is produced.

Effects of Moderators on Hot Hydrogen Atoms. The effects of moderators on hot hydrogen atoms have been studied by many workers by measuring the probability of reaction (P) defined by (II-16) in the presence of moderators. It can be shown [(11), p. 43] that the average loss of kinetic energy ($\overline{\Delta E}/E$) per collision between two elastic spheres with masses M_1 and M_2 and the initial kinetic energies E and 0 is

$$\frac{\overline{\Delta E}}{E} = \frac{2M_1M_2}{(M_1 + M_2)^2} \qquad \text{(11-18)}$$

$\overline{\Delta E}/E$ takes the maximum value when $M_1 = M_2$, that is, the energy loss per collision is largest (50%) when the masses of two spheres are identical. Oldershaw and Porter (776) have observed the moderating effects of hot H atoms by inert gases in the following order

$$\text{He} > \text{Ar} > \text{Kr} > \text{Xe}$$

The results agree qualitatively with the hard sphere model. However, OCS has been shown to be a better moderator than that expected on the basis of the hard sphere model, suggesting that the collisions are inelastic. See a recent review by Vermeil (991) for further details.

Time-of-Flight Measurement of Photofragments. The translational energy of a fragment can be obtained directly by measuring the flight time from the initial zone of production to a detector. In this method a molecular beam of a reactant gas is illuminated by pulsed monochromatic laser light and the dissociated fragments are detected by a mass spectrometer–detector system a few centimeters away [Busch et al. (161)]. By this technique Wilson and his coworkers have measured the translational energy of fragments dissociated from Cl_2 (159), Br_2 (779), IBr and I_2 (160), ICN (638), NOCl (164), NO_2 (162), and alkyl iodide (834). Diesen et al. (290, 291) have also measured the translational energy of Cl atoms dissociated from Cl_2. Figure II–14 shows schematically the flight time distribution of I atoms dissociated from CH_3I at 2662 Å (834). The distribution curve has two peaks, a large one corresponding to slow I atoms and a small one associated with fast I atoms. The large peak represents excited $(^2P_{1/2})$ I atoms and the small peak refers to ground $(^2P_{3/2})$ state I atoms. The $I(^2P_{1/2})$ atoms have less kinetic energy than $I(^2P_{3/2})$ atoms since the available energy after dissociation is converted partially to the electronic energy of I atoms.

According to Busch and Wilson (162) when a parent molecule M dissociates into fragments A + B at right angles both to the laser and molecular beam

$$M \xrightarrow{h\nu} A + B$$

the center-of-mass translational energy of both fragments E_t can be obtained from the flight time of the fragment A by

$$E_t = \frac{m_P}{m_B}\left[\left(\frac{m_A}{2}\right)\left(\frac{d}{t_A}\right)^2\right] + \left(\frac{m_A}{m_B}\right)E_t^P \qquad \text{(II-19)}$$

on the basis of the conservation of energy and momentum. Here m_P, m_A, and m_B are masses of the parent and the fragments A and B, respectively, d is the distance between the production zone and the detection, and t_A is the flight time of fragment A: E_t^P is the translational energy of the parent molecule. The total available energy E_{avl}, which is the sum of the translational energy E_t and the internal energy E_{int} of both fragments, is

$$E_{avl} = E_{int} + E_t = E_{int}^P + h\nu - D_0 \qquad \text{(II-20)}$$

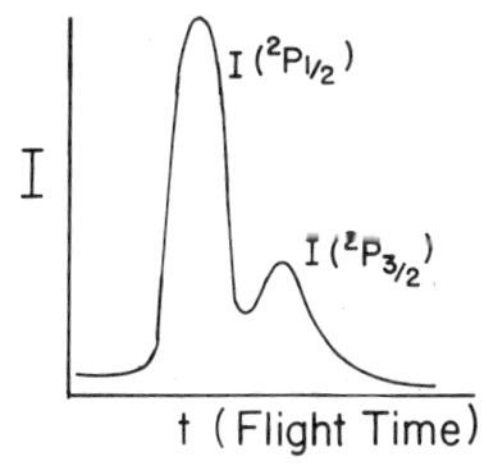

Fig. II–14. The flight time distribution curve of I atoms dissociated from CH_3I at 2662 Å. The curve has two peaks corresponding to slow I atoms (large peak) and fast I atoms (small peak). The slow peak represents excited $(^2P_{1/2})$ I atoms and the fast peak refers to ground state $(^2P_{3/2})$ I atoms. From Riley and Wilson (834), reprinted by permission of The Chemical Society.

where E_{int}^{P} is the internal energy of the parent molecule, $h\nu$ is the incident photon energy, and D_0 is the bond dissociation energy of the parent molecule.

In the case of Cl_2 illuminated by the 3471 Å laser beam, the total translational energy E_t has a maximum corresponding to the production of ground state $^2P_{3/2}$ Cl atoms (159). On the other hand, I_2, when dissociated at 2662 Å, yields one ground state ($^2P_{3/2}$) and one excited ($^2P_{1/2}$) I atom (218).

In the photolysis of NO_2 at 3471 Å the translational energy distribution curve of the fragments shows two distinct peaks (162). Since the difference of these two peaks corresponds approximately to that of the vibrational energy between $v'' = 1$ and $v'' = 0$ of NO it is reasonable to assume that the peak corresponding to the larger translational energy (the faster peak) is associated with the production of NO ($v'' = 0$) and O atoms both in the ground state. The second peak (the slower peak) would correspond to the NO ($v'' = 1$) and O atom production. In the case of NOCl photolysis at 3471 Å (164) only one peak has been observed for the translational energy distribution curve. About 70% of the excess energy beyond that required to break the Cl—NO bond goes into translation and the remaining energy must appear in the form of NO vibration (v'' up to 2), rotation, or electronic excitation of NO and Cl. The photolysis of ICN at 2662 Å (638) yields two peaks in the translational energy distribution. The faster peak corresponds to the production of the electronic ground state CN and I atoms with considerable internal excitation of CN. The slower peak (60%) may correspond to the formation of the ground state I and electronically excited CN($A^2\Pi$). The photolysis of alkyl iodides has been studied at 2662 Å (834). The results of the flight time measurement for I atoms and the alkyl radicals indicate that both ground ($^2P_{3/2}$) and excited ($^2P_{1/2}$) I atoms are clearly produced from methyl and ethyl iodide, and the production of I($^2P_{1/2}$) appears more important than that of I($^2P_{3/2}$). For *n*-C_3H_7I the I($^2P_{1/2}$) atoms are also produced, but the ratio of the $^2P_{1/2}$ to $^2P_{3/2}$ state is less clear and for *iso*-C_3H_7I it is not clear whether excited I atoms are produced at all. The internal energy excitation in methyl radicals is small (12% of the available energy), while in propyl radicals it is large (50% of the available energy).

II–4.2. Measurement of the Internal Energy of Photofragments

In the foregoing section the measurement of the translational energy of photofragments has been described. The internal energy of photofragments can be inferred from the translational energy measurement. However, there are various ways of distributing the internal energy among photofragments. The exact knowledge of such a distribution can be obtained only by directly observing the internal energy of the fragments spectroscopically. Three such techniques are presented below.

Absorption Spectroscopy following Flash Photolysis. In this method radicals produced by flash photolysis are detected spectroscopically before they disappear by reactions. A few examples are presented.

The flash photolysis of water in the region 1400 to 1860 Å has produced OH in the ground state ($X^2\Pi$) [Welge and Stuhl (1033)]. In this wavelength region the excess energy beyond that required to dissociate H_2O into H + OH($^2\Pi$) is 35 to 86 kcal mol^{-1}. Welge and Stuhl have found that the rotational distribution of OH($^2\Pi$) is practically equal to that of room temperature and only a few percent of OH is excited to $v'' = 1$. The experiment has been performed at pressures low enough to avoid collisional perturbation within a detection time of a few microseconds which indicates that the excess energy appears mostly as the translational energy of the fragments, H and OH. Basco et al. (64) have found that CN($X^2\Sigma$) produced from the near ultraviolet flash photolysis of cyanogen halides contains little or no vibrational energy. Donovan and Konstantatos (315) have found that I atoms dissociated from ICN in the near ultraviolet flash photolysis (>2000 Å) are predominantly in the ground state ($^2P_{3/2}$) and I($^2P_{1/2}$) atoms are less than 5%.

Other examples are given in references by Mele and Okabe (692) and by Mitchell and Simons (709).

Flash Photolysis followed by Laser Fluorescence. Radicals produced by flash photolysis are pumped from the ground to a fluorescing excited state by a tunable laser. By measuring the fluorescence intensity as a function of laser wavelength it is possible to map the internal energy distribution of the radicals produced. Jackson and Cody (519) and Cody and Sabety-Dzvonik (224) have measured the rotational and vibrational population of CN($X^2\Sigma$) radicals produced by the flash photolysis of C_2N_2 in the region of absorption 1500 to 1700 Å. They concluded that the primary process is

$$C_2N_2 \rightarrow CN(X^2\Sigma) + CN(A^2\Sigma)$$

The CN($X^2\Sigma$) radicals formed are predominantly in $v'' = 0$ and $v'' = 1$ levels. The authors concluded that about 28% of the excess energy (which is 1 eV) goes into vibration, 22% into rotation, and 40% into translation of the fragments. The observed Boltzmann distribution of rotational energy in CN($X^2\Sigma^+$) suggests that the dissociation occurs only after many vibrations in the excited state formed by light absorption. As a result the rotational, and most probably the translational, energy follows the Maxwell-Boltzmann distribution.

Fluorescence Spectra of Photofragments. If the incident photon energy greatly exceeds the energy of the bond to be broken, the resulting fragments may be produced in a fluorescing excited state. The internal energy distribution of the excited fragments can be found by measuring fluorescence spectra from the excited fragments. Carrington (191) has observed that when a water molecule is excited by the 1216, 1236, and 1302 Å lines the excess energy beyond that required to produce $OH(^2\Sigma)$ goes mostly into rotation of $OH(^2\Sigma)$. Mele and Okabe (692) have photodissociated cyanogen halides and hydrogen cyanide in the vacuum ultraviolet. The vibrational and rotational distributions of the $CN(B^2\Sigma^+)$ produced have been measured from the fluorescence spectra. The vibrational population is much less than that expected from the equipartition of energy in all vibrational degrees of freedom before dissociation. The rotational energy accounts for up to 20% of the excess energy. More than half of the excess energy goes into the translational energy of the fragments.

Mukamel and Jortner (724) have developed a quantum mechanical model to explain the vibrational distribution of $CN(B^2\Sigma^+)$ photodissociated from various halogen cyanides and hydrogen cyanide. More extensive vibrational excitation at shorter incident wavelengths is interpreted as being due to stronger intercontinua coupling between the initially formed excited state and a set of continua, each corresponding to a different internal vibrational state of the fragment $CN(B^2\Sigma^+)$.

II–5. ANGULAR DISTRIBUTION OF PHOTOFRAGMENTS

A new technique to measure the angular distribution of photofragments using polarized monochromatic light has been developed recently by Wilson et al. (161), Bersohn et al. (331), and Diesen et al. (290)

Figure II–15 is a schematic diagram showing how to measure the angular distribution of photodissociated fragments used by Wilson et al. The polarized and pulsed monochromatic light along the y axis with an electric vector **E** illuminates the molecular beam directed along the x axis, which is 90° to the direction of the light beam y. A detector (mass spectrometer) is placed at 90° to both the molecular and light beams (the z direction). The direction of the electric vector **E** can be varied with respect to the z axis, that is, the direction of the photofragments. The angle between the direction of the electric vector **E** and that of the recoil fragments is Θ. It can be shown [Busch and Wilson (163)] that if the direction of the electric vector **E** of the polarized light coincides with the direction of the recoil fragment, the angular distribution of the fragment has approximately a form $\cos^2 \Theta$ and has a maximum when $\Theta = 0$. On the other hand, if the direction of the recoil fragment is 90° to the **E** vector, the angular distribution has a form $\sin^2\Theta$ and is minimum at $\Theta = 0$. For more accurate representation of the angular distribution of recoil fragment it is necessary to consider the velocity of parent molecules. The correct angle θ is given by (163)

$$\theta = \Theta - \sin^{-1} \frac{c}{u} \qquad \text{(II-21)}$$

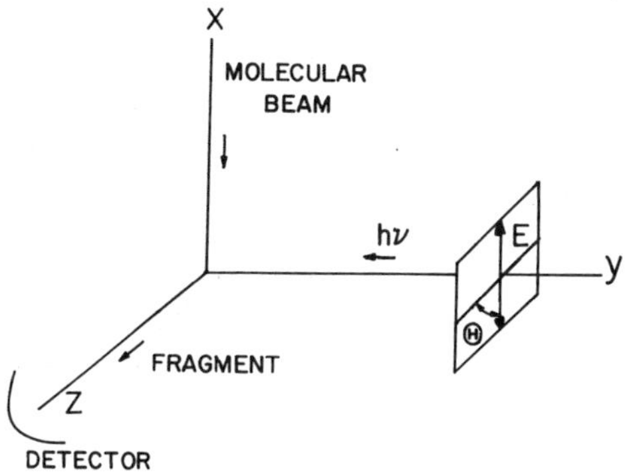

Fig. II–15. Measurement of angular distribution of photofragments [Busch et al. (161)]. The polarized and pulsed monochromatic light with an electric vector E is directed along the *y* axis. The light illuminates the molecular beam of a gas introduced along the *x* axis. The photofragments produced are detected perpendicular to the *xy* plane with a mass spectrometer. The direction of E can be varied with respect to the *z* direction. Θ is an angle between the electric vector and the *z* direction.

where c is the velocity of the parent molecules and u is that of the recoil fragments. The angular distribution of the photofragment $W(\theta)$ is given by (163)

$$W(\theta) = (4\pi)^{-1}[1 + 2bP_2(\theta)] \qquad \text{(II-21a)}$$

where $P_2(\theta)$ is the second degree Legendre polynomial in $\cos\theta$ and is given by $P_2(\theta) = (3\cos^2\theta - 1)/2$. A constant b has values between $-\frac{1}{2}$ and 1 depending on the symmetry, configuration, and lifetime of the excited parent molecule. For positive values of b, $W(\theta)$ has a peak at $\theta = 0°$, while with negative values of b, $W(\theta)$ peaks at $\theta = \pm 90°$ and when $b = 0$, no peak appears. Thus, b determines the shape of $W(\theta)$ and is called the shape factor.

The angular distributions of fragments from NOCl and NO_2 photodissociation at 3471 Å obtained by Busch and Wilson (163, 164) are shown in Fig. II-16 where $W(\theta)$, the probability of finding the fragments (Cl or O atoms), is potted as a function of θ, an angle between the direction of the fragment and that of the electric vector corrected for the velocity of the parent molecule. Both Cl and O atoms have a peak at $\theta = 0$, that is, dissociation is most probable when the direction of the electric vector coincides with that of the recoil fragment. This is shown in Figure II–17 for NOCl. The symmetry species of the excited state may be derived from the results as follows. The ground state NOCl is $^1A'$ in C_S symmetry, that is, the wave function is symmetric to the reflection in the molecular plane yz. The direction of the transition moment vector **R** (see Section I–5), which coincides with the direction of the electric vector **E**, must be symmetric to the reflection in the molecular plane since the dissociation is brought about by the electric vector parallel to

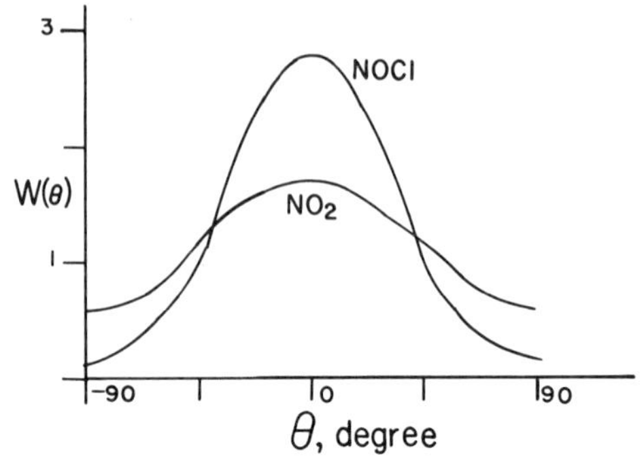

Fig. II–16. The angular distribution curves of photofragments from NOCl and NO_2 at 3471 Å. $W(\theta)$, the probability of finding the fragment (Cl or O atoms), is plotted as a function of θ, the angle between the electric vector and the direction of the recoiling fragment corrected for the velocity of the parent molecule. The probability of finding the recoil atoms is maximum at $\theta = 0$. A narrow peak for the Cl fragment curve indicates that the dissociation of NOCl is instantaneous. From Wilson et al. (163, 164), reprinted by permission. Copyright 1972 by the American Institute of Physics.

the z axis. Therefore, the excited state must also be $^1A'$ (see selection rules in Section I–10). It is assumed here that the direction of recoil nearly coincides with that of the N—Cl bond. In the case of NO_2 photodissociation, a peak occurs at $\theta = 0$, but the angular dependency of the intensity distribution of O atoms is not as pronounced as that of the Cl atoms from NOCl (see Fig. II–16). According to Busch and Wilson (163) the shape of the angular dependence curve of the photofragment is determined by three factors, assuming a dissociation from a single excited state: (1) direction of the transition dipole moment and that of the recoil fragment; (2) rotation of the parent

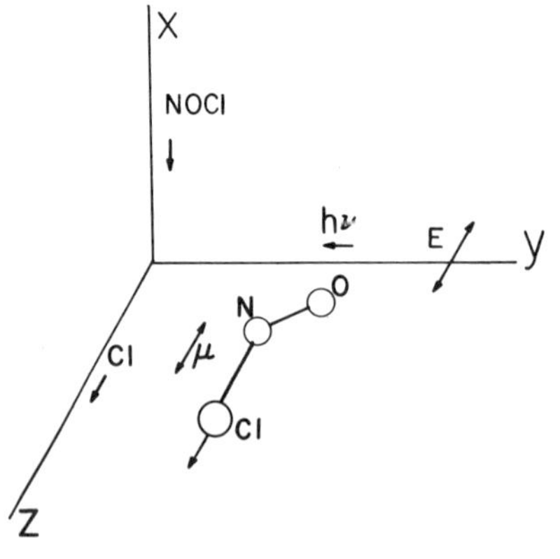

Fig. II–17. The photodissociation of NOCl at 3471 Å. The dissociated Cl atoms recoil parallel to the electric vector. The excited state dissociating into Cl + NO must be $^1A'$. From Busch and Wilson (164), reprinted by permission. Copyright 1972 by the American Institute of Physics.

molecule; (3) the lifetime of the dissociative excited state. The angular distribution curve has the sharpest peak when the transition moment direction coincides with the recoil direction, the rotation of the parent molecule is negligible, and the dissociation is immediate. All three factors may contribute to the less pronounced angular dependency of the distribution curve for NO_2.

From the observed results it is most likely that the direction of the recoil is within 40° to the transition moment direction or approximately the direction of the electric vector. This leaves $^2B_2$ as the only dissociative excited state, since for the two other possible states, $^2B_1$, $^2A_1$, the transition moment directions should be 90 and 67° respectively, to the recoil direction (see Fig. II–18). It has been shown by Diesen et al. (290) and by Busch et al. (159) that the angular distribution curve of Cl atoms dissociated from Cl_2 at 3471 Å peaks at 90° to the direction of the electric vector, which indicates that the dissociative excited state must be $^1\Pi(1_u)$ and not $^3\Pi(0_u^+)$, which should yield a peak when the electric vector is parallel to the recoil direction (the z axis).

Solomon et al. (921) have shown that the photodissociation of formaldehyde into H + CHO in the near ultraviolet occurs perpendicular to the transition moment which is in the molecular plane and perpendicular to the C—O axis. They suggest that the H atom split from formaldehyde is perpendicular to the molecular plane.

A similar technique has been used by Chamberlain and Simons (202, 204) for a dissociation of a triatomic molecule to produce a fluorescing diatomic product BC*

$$ABC \rightarrow A + BC^*$$

$$BC^* \rightarrow BC + h\nu'$$

They have used the plane polarized monochromatic beam to dissociate a molecule and have measured the intensities of the resulting fluorescence

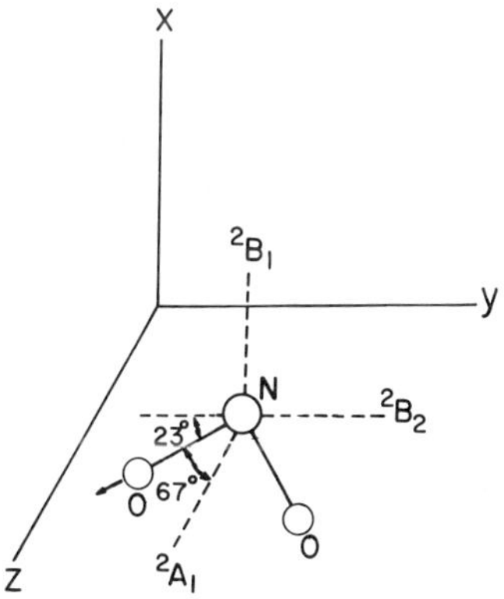

Fig. II–18. The ground state configuration of NO_2. The x axis is perpendicular to the molecular plane. The O—N—O angle is 134°. Three symmetry species $^2B_1$, $^2B_2$, $^2A_1$ that behave like x, y, z, respectively, under the symmetry operations of the point group C_{2v} are shown. The dipole moments for the $^2B_1$–$^2A_1$, $^2B_2$–$^2A_1$, $^2A_1$–$^2A_1$ transitions are respectively along the x, y, z directions. The direction of recoiling O atoms at the 3471 Å photolysis is nearly parallel to the electric vector, indicating that the $^2B_2$ state dissociates into NO + O. From Busch and Wilson (163), reprinted by permission. Copyright 1972 by the American Institute of Physics.

of the fragment both parallel and perpendicular to the direction of the electric vector of the initial beam. If an absorption occurs parallel to the molecular plane, and the resulting fluorescence arises along the internuclear axis of BC*, the fluorescence is predominantly polarized along the direction of the electric vector of the incident beam, if the dissociation occurs immediately. It is then possible to determined the direction of the transition moment of the absorbing molecule dissociating into BC*. The authors have measured the symmetry species of H_2O excited near 1300 Å and those of HCN and BrCN excited above 1250 Å.

II–6. MODELS FOR ENERGY PARTITIONING IN PHOTODISSOCIATION

Recent advancement of kinetic spectroscopy and laser technology has provided rich information on the lifetimes and symmetry properties of the dissociative excited states, as well as the energy partitioning among fragments in photodissociation. Several models have been proposed to explain these results, although it does not appear possible at present to apply a single model even for dissociation of triatomic molecules. This is primarily due to the lack of information on the correctness of assumptions on which these models must depend. Broadly, these models may be divided into three categories: (*a*) statistical, (*b*) impulsive, and (*c*) equilibrium geometry.

II–6.1. Statistical Model

Consider an electronically excited molecule formed by an absorption of monochromatic light of energy beyond that necessary for a bond scission. If a molecule is diatomic, dissociation is often instantaneous. In the case of a polyatomic molecule it may survive for a period of several vibrations before it undergoes dissociation during which time the energy acquired by the molecule may be equally partitioned among all vibrational degrees of freedom. For the dissociation to occur the vibrational energy must flow into the bond to be broken after many Lissajous motions of the molecule. The dissociation may occur from the upper electronic state formed by light absorption or from the ground state after crossing over from the initially formed upper electronic state by internal conversion. The former case may be represented by the photolysis of C_2N_2 at 1600 Å where the products $CN(X^2\Sigma) + CN(A^2\Pi)$ may arise from the excited $C^1\Pi_u$ state of C_2N_2 after many vibrations, as indicated by the Boltzmann distribution of rotational energy in $CN(X^2\Sigma)$ [Cody et al. (224)]. The latter case, that is, the dissociation of the ground electronic state by vibration is known as unimolecular decomposition. Both types of predissociation are referred to as type II predissociation by Herzberg [(16), p. 458], that is, the dissociation brought about by vibration. If the photodissociation occurs via the ground state molecule, the mode of decomposition must be the same as that induced by heat and the same as those of products formed by exothermic addition of radicals and by intense infrared laser photolysis. The model is used with some success to explain energy partitioning in the photolysis of large molecules such as di-*tert*-butyl peroxide [Dorer and Johnson (317)]. Quack and Troe (820) have used their statistical model to calculate the translational distribution of products in NO_2 photodissociation assuming dissociation occurs from the ground state. The calculated results agree well with the results obtained experimentally by Busch and Wilson (162). However, in the case of NOCl photodissociation the calculated results do not agree with the results found by experiment. The statistical model applied to the alkyl iodide photodissociation predicts a much larger fraction of internal energy residing in the alkyl fragment than that found experimentally [Riley and Wilson (834)].

II–6.2. Impulsive Model

This model has been developed by Wilson et al. (162, 459, 478). It may be applicable to the case of direct dissociation. According to this model it is assumed, as shown in Fig. II–19, that upon photon absorption a repulsive potential V

$$V = V_0 e^{-Z/L} \qquad \text{(II-22)}$$

is suddenly created between A and B with a distance Z, where L is a characteristic range parameter. The repulsive energy is initially converted to the translational energy of atoms A and B. The kinetic energy of B subsequently changes into the translational and internal energies of BC when B collides with its partner C. The total translational energy E_t of both fragments A and BC is (162)

$$E_t = \frac{\mu_{\text{A–B}}}{\mu_{\text{A–BC}}} E_{avl} \qquad \text{(II-23)}$$

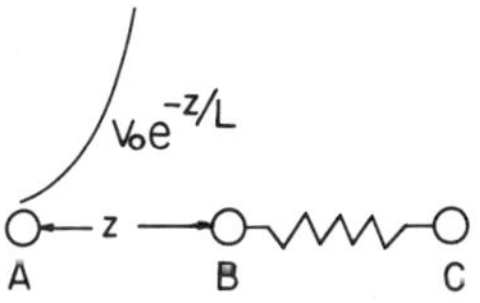

Fig. II–19. Impulsive model for photodissociation of a triatomic molecule. A repulsive potential $V = V_0 e^{-Z/L}$ is suddenly created between A and B upon light absorption where L is a characteristic range parameter. The repulsive energy is initially converted to the translational energies of A and B atoms. The translational energy of B is transformed subsequently into the translational and internal energies of BC. From Wilson et al. (162, 459, 478), reprinted by permission. Copyright 1971 by the American Institute of Physics.

where μ_{A-B} is the reduced mass of atoms A and B, and μ_{A-BC} is the reduced mass of A and the fragment BC. E_{avl} is the energy available initially as the translational energy of A and B. The impulsive model predicts a high fraction of E_{avl} going into E_t, the total translational energy of O and NO in the case of NO_2. In the NO_2 photolysis, according to (II–23) 72% of E_{avl} is expected to go into E_t, while experimentally only 60% E_t is observed (162). On the other hand, the impulsive model somewhat underestimates the fraction of E_{avl} converted into the translational energy of both fragments in alkyl iodide photodissociation (834).

II–6.3. Equilibrium Geometry Model

Mitchell and Simons (709) have proposed that in photodissociation of a triatomic molecule ABC into A + BC

$$\mathrm{ABC} \xrightarrow{h\nu} \mathrm{ABC}^* \rightarrow \mathrm{A} + \mathrm{BC}$$

the equilibrium bond length B—C in an electronically excited state ABC* may be larger than that in the ground state ABC or that in a fragment BC. In this circumstance the fragment BC appears in the vibrationally excited state. Unfortunately, the hypothesis cannot be tested experimentally in the case where dissociation is direct, since only a continuum is observed. Nevertheless, Mitchell and Simons estimate the change in bond order on the basis of molecular orbitals involved in the transition. The transition in the photolysis of NOCl near 2000 Å, for example, probably involves the promotion of an electron from the $1a''$ to the $3a''$ orbital in which case the NO bond order in NOCl decreases; that is, the equilibrium N—O bond length increases from 1.14 to 1.33 Å in the upper state. This increase is consistent with the results of Basco and Norrish (63) which indicate that NO is vibrationally excited up to $v'' = 11$ when produced in the flash photolysis of NOCl near 2000 Å. On the other hand, Basco et al. (64) have found that CN radicals produced from BrCN above 2400 Å photolysis are mainly in the lowest vibrational levels. The change in the CN bond order in BrCN is small if the promotion of an electron from the 2π orbital to the 5σ orbital is involved in the transition. The equilibrium geometry model may also explain the formation of $NCO(A^2\Sigma)$ excited in ν_2' bending vibration in the vacuum ultraviolet photolysis of isocyanic acid. The photolysis of isocyanic acid in the vacuum ultraviolet yields electronically excited NCO radicals [Okabe (765)]

$$\mathrm{HNCO} \xrightarrow{h\nu} \mathrm{H} + \mathrm{NCO}(A^2\Sigma)$$

The fluorescence spectrum of $NCO(A^2\Sigma)$ indicates the formation of $NCO(A^2\Sigma)$ excited in bending vibration up to $v_2' = 4$. Since the ground state isocyanic acid has a linear NCO structure and the equilibrium geometry of $NCO(A^2\Sigma)$ is also linear, it is reasonable to assume that the excited HNCO molecule formed by light absorption is in a bent NCO configuration prior to dissociation. The photodissociated $NCO(A^2\Sigma)$ becomes excited in bending vibration as it returns from the bent to the linear equilibrium structure. While no analysis of the vacuum ultraviolet absorption bands for HNCO has been made to support this model, Dixon and Kirby (287) have observed the formation of the excited HNCO with a bent NCO structure corresponding to the transition in the 2000 to 2800 Å region.

II–6.4. Other Models

The distribution of vibrational and rotational energy in the photofragment has been treated recently by a quantum mechanical model in the case of the collinear dissociation of triatomic molecules [Florida and Rice (365), Band and Freed (59), Ashfold and Simons (47)]. The dissociation occurs by coupling of the initially formed bound excited state and a repulsive final state. Ashfold and Simons have explained, on the basis of a simplified quantum mechanical model, the observed Boltzmann distribution of $CN(B^2\Sigma^+)$ rotational energy produced from the photolysis of BrCN at 1580 Å.

Simons and Tasker (893, 894) have combined the impulsive model and the equilibrium geometry model to explain the vibrational and rotational distributions of photodissociation fragments in halogen cyanides and water. Assuming an exponential repulsive potential given in (II-22) between atoms A and B, and assuming BC to be a harmonic oscillator, the probability of forming the fragment BC in various vibrational levels after dissociation has been calculated (893). As an example, the calculation of Simons and Tasker shows that the probability of vibrational excitation in the ground state dissociated from ICN at 2500 Å shows a maximum at $v'' = 0$. As the range parameter L decreases, that is, as the repulsive curve becomes much steeper, higher vibrational levels will proportionately be more excited. In the case of halogen cyanide dissociation in the vacuum ultraviolet the vibrational distribution of $CN(B^2\Sigma)$ experimentally found by Mele and Okabe (692) is in good agreement with the calculation based on the impulsive model (894). On the other hand, the observed vibrational distribution of $CN(B^2\Sigma)$ dissociated from HCN does not agree with the calculated results unless the change in C—N bond length in the excited HCN molecule is also introduced, since the H atom is too light to exert sufficient recoil. The change in bond length reproduces the observed vibrational distribution. This is shown in Fig. II–20.

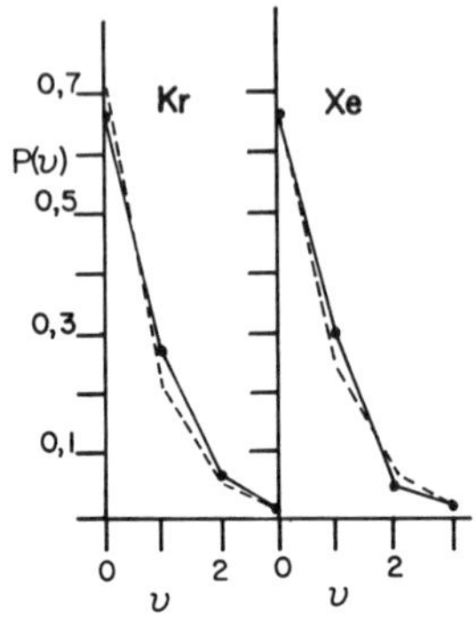

Fig. II–20. Calculated and observed vibrational distributions in CN($B^2\Sigma$) radicals formed by photodissociation of hydrogen cyanide at the Kr and Xe resonance lines. (----) Calculated results; (———) experimental results. For calculated results, $L = 0.11$ Å, $\Delta r = 0.04$ Å for Kr, and $L = 0.13$ Å, $\Delta r = 0.04$ Å for Xe lines are assumed. L is a range parameter and Δr is an increase in C–N bond length of the excited state HCN. From Simons and Tasker (894), reprinted by permission of Taylor and Francis.

II–6.5. Rotational Excitation

Rotational excitation of the photofragment has been observed in water by Carrington (191) and in cyanogen halides by Mele and Okabe (692). A model for rotational excitation of OH($A^2\Sigma$) from water is shown in Fig. II–21.

The orbital angular momentum P of an atom A in photodissociation of a triatomic molecule ABC into A + BC is given by (191, 894)

$$P = \mu_{\text{A–BC}} g b \tag{II-24}$$

where $\mu_{\text{A–BC}}$ is the reduced mass for the separating fragments A and BC, g is the terminal velocity of the atom A, and b, the impact parameter, is the perpendicular distance from the center of mass to the line extrapolated from a distant linear portion of the A atom trajectory as shown in Fig. II–21.

To produce a large amount of rotation in the fragment BC, the angular momentum of the atom A, which is in the oposite direction, must be equally

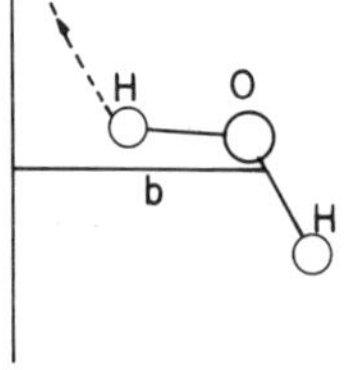

Fig. II–21. Rotational excitation of OH($A^2\Sigma$) by photodissociation of water in the vacuum ultraviolet. Rotational excitation is exerted by H atoms as they dissociate from water with simultaneous excitation of bending and stretching vibrations upon light absorption. The impact parameter b is large [see Simons and Tasker (894)]. Reprinted by permission of Taylor and Francis.

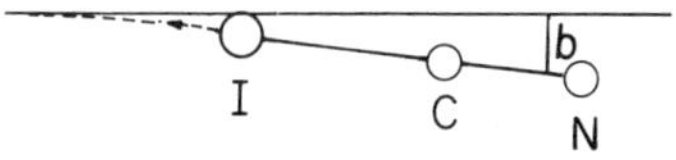

Fig. II–22. A model to explain the rotational excitation of $CN(B^2\Sigma)$ formed in photodissociation of ICN in the vacuum ultraviolet [Simons and Tasker (894)]. *b* is the impact parameter. The I atom dissociates from ICN with little change of linear configuration upon absorption of light, that is, *b* is small. Reprinted by permission of Taylor and Francis.

large. The large rotational excitation of $OH(^2\Sigma)$ observed in the vacuum ultraviolet photolysis of H_2O can be explained on the basis of the large impact parameter *b*. The large *b* value indicates that the dissociation of H atoms is a result of the simultaneous excitation of antisymmetric stretching vibration ν_3 and bending vibration ν_2. When the bending vibration is excited, the otherwise linear trajectory of H atoms along the OH bond deviates from linearity, thus exerting a torque on $OH(^2\Sigma)$ radicals. The bending excitation must be brought about by the change in the H—O—H bond angle associated with the transition from the bent ground state to the almost linear excited state. In halogen cyanide photodissociation the calculated *b* is much smaller than that for water, indicating that the configuration of the excited state is close to the linear configuration of the ground state (see Fig. II–22). However, a small deviation from linearity ($\sim 5°$) is sufficient to explain the observed CN rotation.

II–7. DETERMINATION OF BOND DISSOCIATION ENERGIES

Knowledge of the bond dissociation energy (or simply the bond energy) is important in photochemical studies. In the majority of simple molecules a photodissociation occurs at the weakest bond when the incident photon energy exceeds the bond energy, and occasionally two bond scissions are involved (see Section II–2).

The bond dissociation energy D_0 of a molecule AB may be defined as the energy required to form the products, A + B, in their respective lowest levels of the ground electronic states, from AB in its lowest level of the ground electronic state.

The bond dissociation energy is in general equivalent to the dissociation energy of the ground electronic state of the molecule, although in some cases the ground state molecule may dissociate into excited species by increasing the vibrational energy without altering the electronic state. For example, the ground state N_2O dissociates into $N_2(X^1\Sigma^+) + O(^1D)$ unless the singlet ground state crosses a triplet state with which the products

$N_2(X^1\Sigma^+) + O(^3P)$ correlate. In this case the dissociation energy of the ground state N_2O is larger than the bond dissociation energy. In thermochemistry, $D_0(A-B)$ corresponds to ΔH_0°, the enthalpy increment for the gas phase reaction in the ideal state at 0°K

$$AB(g) \rightarrow A(g) + B(g)$$

$$D_0(A-B) = \Delta H_0^\circ = \Delta Hf_0^\circ(A, g) + \Delta Hf_0^\circ(B, g) - \Delta Hf_0^\circ(AB, g) \quad \text{(II-25)}$$

where $\Delta Hf_0^\circ(A, g)$, $\Delta Hf_0^\circ(B, g)$, and $\Delta Hf_0^\circ(AB, g)$, respectively, signify the standard enthalpies of formation (or standard heats of formation) of A, B, and AB in the gaseous state at 0°K (the degree sign indicates the thermodynamic standard state of 1 atm pressure and the subscript zero denotes the temperature in °K). The standard enthalpy of formation is the increment of enthalpy when a given compound is formed from its elements in their respective standard states at a given temperature. If ΔHf° is given at 298°K instead of 0°K, appropriate corrections have to be applied to each ΔHf_{298}° to obtain ΔHf_0°. Usually corrections amount to less than a few kcal mol^{-1}

II–7.1. Determination of Bond Dissociation Energies from Thermochemical Data

The standard enthalpies of formation, ΔHf_0° at 0°K, have been obtained for many molecules from the measurements of the heats of reaction, the heats of oxidation, and various other methods. See Lewis et al. (19) and Stull et al. (27) for details. The data on ΔHf_0° are compiled in the tables by Wagman et al. (29) and by Stull and Prophet (28). Some of them are listed in Tables A–6 through A–13 of the Appendix. Thus, the bond dissociation energies can be obtained using (II-25) from ΔHf_0° of AB, A, and B.

The bond energies of some diatomic molecules have been obtained by measuring equilibrium constants K_P of the dissociation reactions at high temperatures

$$AB(g) \rightarrow A(g) + B(g)$$

$$K_P = \frac{P_A \cdot P_B}{P_{AB}}$$

where P is the pressure of each species in atmospheres at a given temperature T.

To obtain $D_0(A-B)$, which is equal to ΔH_0° for the dissociation, it is convenient to use the free energy function defined by

$$\text{free energy function} = \frac{F_T^\circ - H_0^\circ}{T} \quad \text{(II-26a)}$$

where F_T° is the Gibbs free energy in the standard state at a temperature T.

The enthalpy change for the reaction at 0°K, ΔH_0° is

$$\frac{\Delta H_0^\circ}{T} = -R \ln K_P - \Delta\left(\frac{F_T^\circ - H_0^\circ}{T}\right) \tag{II-26b}$$

where R is the gas constant equal to 1.987165 cal deg^{-1} mol^{-1} and $\Delta[(F_T^\circ - H_0^\circ)/T]$ is the difference between the sum of the free energy functions of the products and the free energy function of the reactant at the temperature T°K. The free energy functions have been calculated from spectroscopic data for many atoms and molecules and are found for example, in JANAF Thermochemical Tables (28). The method of obtaining ΔH_0° from (II-26b) is known as the third law method.

If the equilibrium constants at various temperatures are accurately known, it is possible to obtain ΔH_T° from

$$\Delta H_T^\circ = -R\frac{d(\ln K_P)}{d(1/T)} \tag{II-27}$$

This method is known as the second law method.

Since it is often difficult to measure K_P over a wide range of temperatures to obtain ΔH_0°, the third law method is preferred whenever the free energy functions are available.

Examples. Perlman and Rollefson (804) have measured the equilibrium constant K_P for the reaction at 1270°K

$$I_2 \leftrightarrow 2I \qquad K_P = \frac{(P_I)^2}{P_{I_2}} = 0.169 \text{ atm}$$

Hence, $-R\ln K_P = 3.53$ cal mol^{-1} deg^{-1}. From JANAF Thermochemical Tables (28)

$$-\Delta\left(\frac{F_T^\circ - H_{298}^\circ}{T}\right) = 24.8 \text{ cal mol}^{-1}\text{ deg}^{-1}$$

From (II-26b) and $T = 1270$°K

$$\frac{\Delta H_{298}^\circ}{T} = 28.33 \text{ cal mol}^{-1}\text{ deg}^{-1}$$

Therefore, $\Delta H_{298}^\circ = 35979$ cal mol^{-1}. Since $\Delta H_0^\circ - \Delta H_{298}^\circ = -544$ cal mol^{-1} we obtain $\Delta H_0^\circ = 35{,}435$ cal mol^{-1}, corresponding to 12,394 cm^{-1}. The convergence limit of I_2 for the transition $B^3\Pi \leftarrow X^1\Sigma$ at 4995 Å (20,016 cm^{-1}) should correspond to the formation $I(^2P_{1/2}) + I(^2P_{3/2})$, because the difference of the convergence limit (20,016 cm^{-1}) and $D_0 = 12394$ cm^{-1} is 7622 cm^{-1}, nearly equal to the energy of $I(^2P_{1/2})$ above the ground state $I(^2P_{3/2})$.

Berkowitz (99) has measured the equilibrium constant K_p for the reaction over the temperature range 2200 to 2500°K

$$C(g) + \tfrac{1}{2}N_2(g) \leftrightarrow CN(g)$$

$$K_P = \frac{P_{CN}}{P_C \cdot P_{N_2}^{1/2}}$$

From $K_P = 330\ \text{atm}^{-1/2}$ at 2340°K and $\Delta[(F_T^\circ - H_{298}^\circ)/T] = 13.60$ cal mol^{-1} deg^{-1}, we obtain by use of (II-26b) $\Delta H_{298}^\circ = -58.78$ kcal mol^{-1}. Since $\Delta H_0^\circ - \Delta H_{298}^\circ = 0.525$ kcal mol^{-1}, the enthalpy change for the reaction at 0°K is $\Delta H_0^\circ = -58.26$ kcal mol^{-1}. Using $\Delta Hf_0^\circ(C, g) = 170$ kcal mol^{-1}, the standard enthalpy of formation for CN(g) at 0°K, $\Delta Hf_0^\circ(CN, g)$, is $\Delta H_0^\circ + \Delta Hf_0^\circ(C, g) = 112$ kcal mol^{-1}. This value is about 11 kcal mol^{-1} higher than that obtained more recently by a photodissociation method (264).

II–7.2. Determination of the Bond Dissociation Energies in Diatomic Molecules

The bond dissociation energy of a diatomic molecule can be determined with the highest precision from the convergence limit of a band progression, predissociation, and breaking-off of emission bands (see Section II–1.1). The bond energy of O_2 (5.115 eV) has been obtained from the convergence limit of weak transition $A^3\Sigma^+ \leftarrow X^3\Sigma^-$ at 41,259 cm^{-1} (5.115 eV) and that of the strong transition $B^3\Sigma^- – X^3\Sigma^-$ at 57,128 cm^{-1} (7.083 eV). The difference of the two convergence limits, 1.967 eV, corresponds to the electronic energy of O(^1D) [see Fig. V–14, the potential energy curves of O_2, and Gaydon (81, p. 70)].

The D_0(S—O) = 5.357 eV has been derived from the predissociation of the bands SO($B^3\Sigma^- \leftarrow X^3\Sigma^-$), corresponding to a transition similar to the E–X in Fig. II–5, assuming the products S(^3P) + O(^3P). In this case it is necessary to confirm from other data that the products are not S(^1D) + O(^3P) [see Gaydon (8), p. 245, Colin (226)]. The breaking-off of the emission spectrum of NO ($C^2\Pi$) above $v' > 0$ has led Callear and Smith (169) to conclude that D_0(N—O) = 6.493 eV.

In some cases only several vibrational bands of a progression are observed and it is necessary to extrapolate these bands to the convergence limit. Birge and Sponer [see Herzberg (14), p. 438] have suggested that if a vibrational level of an electronic state is expressed as (I-15)

$$G_0(v) = \omega_0 v - \omega_0 x_0 v^2 \qquad \text{(II-28a)}$$

(v is the vibrational quantum number, ω_0 and $\omega_0 x_0$ are constants), the separation of successive levels is

$$\Delta G = G_0(v + 1) - G_0(v) = \omega_0 - \omega_0 x_0 - 2\omega_0 x_0 v \qquad \text{(II-28b)}$$

The dissociation energy D_0 of the electronic state is equal to $G_0(v)$ for the level v at $\Delta G = 0$, that is,

$$v = \frac{\omega_0 - \omega_0 x_0}{2\omega_0 x_0} \tag{II-28c}$$

From (II-28c) and (II-28a)

$$D_0 \simeq \frac{\omega_0^2}{4\omega_0 x_0} \tag{II-29}$$

The dissociation energies of the ground state diatomic molecules obtained from (II-29) are usually approximate values since the vibrational energy levels often deviate from (II-28a) at higher vibrational levels. To remedy the situation, a graphical extrapolation that is a plot of the vibrational energy intervals ΔG as a function of v is sometimes used. ΔG deviates at higher v from a linear relationship with v given in (II-28b). In this case the dissociation energy is given by the graphical sum of ΔG over the range from $v = 0$ to v corresponding to $\Delta G = 0$. The extrapolation of the curve until it cuts the v axis appears justified for nonpolar molecules only.

The determination of bond energies in polyatomic molecules is by no means simple, as is described in the following section.

II–7.3. Determination of the Bond Dissociation Energies in Simple Polyatomic Molecules

The bond energy has been frequently determined from the band convergence limit in diatomic molecules. However, in polyatomic molecules no clear-cut example of the band convergence has been found [see Herzberg, (16) p. 482]. Few examples have been found in the literature to show that the observed predissociation limit corresponds to the bond dissociation energy. One such example is the onset of predissociation at 3979 Å in the absorption spectrum of NO_2 [Douglas and Huber (322)] corresponding to the dissociation into NO + O both in the ground states. Another predissociation limit of NO_2 at 2491 Å [Herzberg (16), p. 483] does not correspond to the products $NO(X^2\Pi) + O(^1D)$ which should be formed below 2439 Å. Uselman and Lee (985) have observed the production of $O(^1D)$ by the photolysis of NO_2 below this wavelength and thus confirmed the occurrence of the process. The long wavelength limit of absorption seldom corresponds to the bond dissociation energy, as exemplified by Herzberg [(16), p. 484]. It gives usually only the upper bound of the dissociation energy.

Breaking-Off of the Emission Bands. It has been discussed previously in Section II–1.1 that the breaking-off of the emission spectrum above a certain level is a more sensitive criterion for predissociation. The author has observed a sudden decrease of the fluorescence intensity of SO_2 below the incident wavelength 2189 Å, corresponding to the dissociation energy,

5.66 eV [Okabe (767)]. Brand and Reed (126) have observed no fluorescence of H_2CO above $v'_4 > 0$, indicating the dissociation energy $D_0(\text{H—CHO}) <$ 28,736 $\text{cm}^{-1} = 3.56$ eV. Lee and Uselman (619) have measured the fluorescence intensity of NO_2 as a function of incident wavelength. The fluorescence yield decreases sharply below the incident wavelength 4150 Å, well above the dissociation threshold at 3979 Å. This discrepancy is ascribed to the contribution of rotational energy of the ground state NO_2 to the incident photon energy (see Section I-4.3).

Determination of the Bond Dissociation Energies and Related Thermochemical Data by Photon Impact. In principle it is possible to determine the bond energy directly from the onset of photodissociation yielding the ground or metastable products. In practice only in a few cases have such studies been performed. The main reasons for the lack of study are (1) the difficulty in obtaining monochromatic light sources of sufficient intensity in the desired wavelength region, (2) many molecules do not absorb light in the spectral region corresponding to the formation of ground state products; (3) the difficulty in determining whether the products are formed in primary processes or from the reactions of an excited state. For example, McGee and Heicklen (681) have found the products N_2 and NO_2 in the photolysis of NO in the region where only the excited state is formed.

Pitts et al. (810) and Jones and Bayes (550) have performed the photolysis of NO_2 as a function of incident wavelength as described earlier in Section I–4.3. The O_3 photolysis has also been studied as a function of wavelength by Lin and Demore (637). The threshold wavelength for the production of $O(^1D)$ is about 3200 Å, which is somewhat longer than the thermochemical threshold at 3100 Å for the process $O_3 \rightarrow O_2(^1\Delta) + O(^1D)$. The contribution of internal energy to photodissociation has been suggested by Kajimoto and Cvetanovic (555).

The better method to determine the bond energy is to measure the primary photodissociation product directly as a function of incident wavelength in the region where the molecule absorbs strongly, that is, in the vacuum ultraviolet region. If the primary product B* fluoresces

$$\text{AB} \xrightarrow{h\nu} \text{A} + \text{B}^*$$
$$\text{B}^* \xrightarrow{h\nu} \text{B} + h\nu$$

the bond dissociation energy, $D_0(\text{A—B})$, may be obtained from

$$D_0(\text{A—B}) = h\nu_E - E_0(\text{B}^*) \qquad \text{(II-30)}$$

where $h\nu_E$ is the minimum photon energy required to form fluorescing species B* and $E_0(\text{B}^*)$ is the electronic energy of B*. It is assumed that at threshold the fragments A and B* and the molecule AB have no internal and kinetic energies. If the primary product is an ion, A^+,

$$AB \xrightarrow{h\nu} A^+ + B + e$$

the bond energy is given by

$$D_0(A\text{—}B) = h\nu_I - \text{I.P.}(A) \qquad \text{(II-31)}$$

where $h\nu_1$ is the minimum photon energy required to produce A^+ and I.P.(A) is the ionization potential of the fragment A.

The dissociation process may involve a small potential barrier, in which case D_0(A—B) is an upper limit. In some cases the internal energy of the molecule AB contributes to photodissociation (see Section I–4.3). In such cases bond energies obtained from (II–30) at room temperature must be smaller than D_0(A—B) defined by (II-25), although no clear-cut examples have been found.

Neuimin and Terenin (734) were the first to observe fluorescence from electronically excited radicals such as OH($A^2\Sigma$) and CN($B^2\Sigma$) by irradiating water and acetonitrile in the vacuum ultraviolet. More recently, Dyne and Style (329) have observed fluorescence from the photolysis of HCHO and HCOOH in the vacuum ultraviolet.

The experimental arrangement for measuring the fluorescence threshold is given in Fig. II–23. The light source is a hydrogen discharge lamp operated on a high voltage AC power supply. The lamp is provided with an LiF window of 1 mm thickness. The beam of monochromatic light of desired wavelength in the region 1050 to 2000 Å can be obtained at the exit slit of

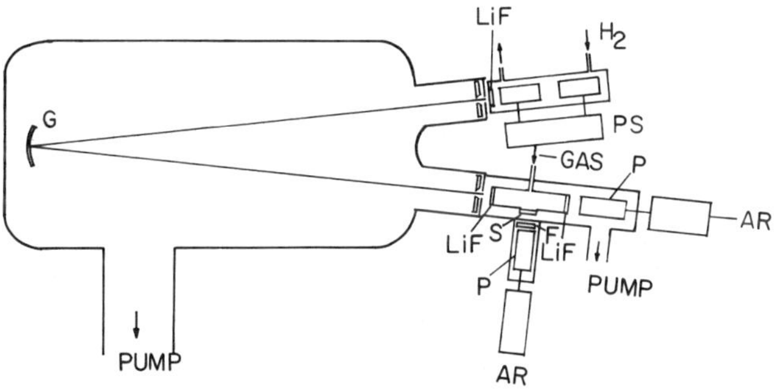

Fig. II–23. Schematic diagram for measuring radical fluorescence intensity as a function of incident wavelength. The light source is a hydrogen lamp with an LiF window. A beam of monochromatic light enters the reaction cell equipped with two LiF windows. The intensity of fluorescence from an electronically excited radical is measured at right angles to the incident beam with the broad band filter–photomultiplier system. *PS*, power supply; *P*, photomultipler; *S*, Suprasil; *AR*, amplifier–recorder; *G*, a concave grating; *F*, a broad band filter.

the monochromator by a rotation of a concave grating G. The fluorescence cell, made of Monel, is provided with three windows. The fluorescence light emerging at right angles to the incident beam passes through a Suprasil window and is detected by an appropriate photomultiplier. An appropriate filter is placed in front of the multiplier to select emission in the desired wavelength region. The intensity of the incident light over the region 1100 to 2000 Å is monitored by another photomultiplier coated with sodium salicylate. For identifying the fluorescence spectrum, a Xe or Kr resonance lamp is used to irradiate a sample directly to yield fluorescence, and its spectrum is taken by scanning the monochromator over the desired wavelength region [see a recent review by Okabe (772)]. To obtain D_0(A—B) by photodissociation the threshold energy $h\nu_E$ and $E_0(\mathrm{B}^*)$ in (II-30) have to be known. In the case of photoionization the threshold energy $h\nu_I$ and the ionization potential of a fragment A in (II-31) must be known to obtain the bond energy. The electronic energies and ionization potential of many radicals have been measured by absorption spectroscopy and by photoionization. Frequently, however, the radical ionization potential is not available although the electronic energy is known.

If both photodissociation and photoionization data are available, the ionization potential can be calculated from

$$\mathrm{I.P.(A)} = h\nu_I - h\nu_E + E_0(\mathrm{B}^*) \qquad \text{(II-32)}$$

Okabe and Dibeler (771) have studied photodissociation and photoionization processes of cyanoacetylene

$$C_2HCN \xrightarrow{h\nu_E} C_2H + CN(B^2\Sigma)$$
$$C_2HCN \xrightarrow{h\nu_I} C_2H^+ + CN + e$$

The threshold photon energy $h\nu_E$ for the production of $CN(B^2\Sigma)$ is 9.41 ± 0.04 eV, corresponding to 1318 Å. The photoionization threshold energy $h\nu_I$ for the production of C_2H^+ is 18.19 ± 0.04 eV, corresponding to 681.6 Å. Figure II–24 shows (*a*) the $CN(B^2\Sigma)$ fluorescence efficiencies and (*b*) the ionization yields of C_2HCN^+ and C_2H^+ as a function of incident wavelength. Both the fluorescence efficiency and C_2HCN^+ ion yield curves show several peaks corresponding to the Rydberg series, indicating that $CN(B^2\Sigma)$ radicals are predissociated from a Rydberg state and C_2HCN^+ ions are also autoionized from Rydberg states. Using $h\nu_I = 18.19 \pm 0.04$ eV, $h\nu_E = 9.41 \pm 0.04$ eV, and $E_0(CN\,B^2\Sigma) = 3.198$ eV, Okabe and Dibeler have obtained $\mathrm{I.P.}(C_2H) = 11.98 \pm 0.06$ eV from (II-32).

In combination with $\Delta Hf_0^\circ(C_2H^+) = 17.47 \pm 0.01$ eV, $\Delta Hf_0^\circ(C_2H)$ is 5.50 ± 0.04 eV or 127 ± 1 kcal mol^{-1}. This value is in excellent agreement with $\Delta Hf_0^\circ(C_2H) = 129 \pm 3$ kcal mol^{-1} obtained from high temperature

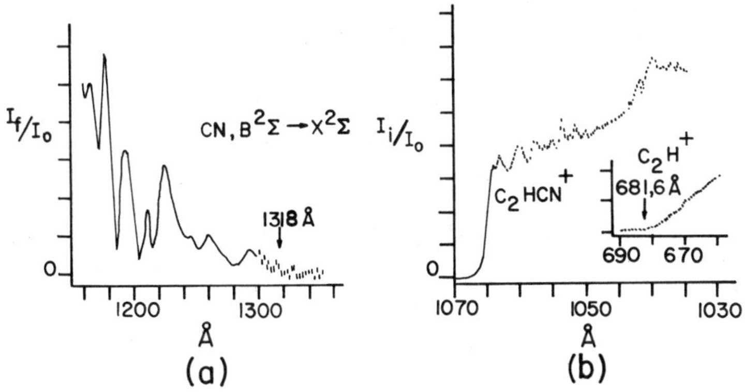

Fig. II–24. Fluorescence and photoionization efficiency curves of cyanoacetylene as a function of incident wavelength. (*a*) Fluorescence efficiency curves for the process $C_2HCN \xrightarrow{h\nu} C_2H + CN(B^2\Sigma)$. The incident threshold wavelength for the production of $CN(B^2\Sigma)$ is 1318 Å. (*b*) Photoionization yield curves for the parent and C_2H^+ ions. The threshold wavelength for the production of C_2H^+ ions is 681.6 Å. From Okabe and Dibeler (771), reprinted by permission. Copyright 1973 by the American Institute of Physics.

Table II–5. Comparison of Bond Dissociation Energies of Cyanogen Compounds Obtained by Photodissociation Photoionization and Electron-Impact Methods

	Bond Dissociation Energies (eV) D_0 (R—CN)		
Molecule	Photodissociation	Photoionization	Electron Impact
HCN	5.20 ± 0.05	5.20 ± 0.02	5.1 ± 0.2
FCN	4.82 ± 0.05	5.01 ± 0.02	—
ClCN	4.33 ± 0.05	4.31 ± 0.02	4.9 ± 0.1
BrCN	3.77 ± 0.05	3.68 ± 0.02	4.4 ± 0.1
ICN	3.11 ± 0.04	3.17 ± 0.02	3.9 ± 0.1
C_2N_2	5.62 ± 0.05	6.22 ± 0.02	6.2 ± 0.3
CH_3CN	5.32 ± 0.03	—	4.8 ± 0.1
C_2HCN	6.21 ± 0.04	—	—

Note: The bond energies obtained by photoionization are somewhat higher than those obtained by photodissociation for FCN, ICN, and C_2N_2, indicating small potential barriers for the ionization processes. On the other hand, D_0(Br–CN) by photodissociation is somewhat higher than that by photoionization. The values by electron impact are in many cases higher than those by photon impact. From Okabe 772).

reactions of graphite with hydrocarbons by a technique given in Section II–7.1. $D_0(C_2H—H) = 124 \pm 1$ kcal mol^{-1} is calculated from ΔHf_0° of C_2H, H, and C_2H_2.

Table II–5 shows the bond dissociation energies of some cyanogen compounds obtained by photon and electron impact methods [Davis and Okabe (264), Okabe and Dibeler (771)]. For C_2N_2, FCN, and ICN photoionization provides bond energies larger than those obtained by photodissociation, indicating the presence of some energy barriers for photoionization processes. On the other hand, for BrCN the photodissociation appears to have a small energy barrier. Bond energies obtained from an electron impact method are in many cases higher than those from photon impact. Various thermochemical data have been obtained from a photodissociation method [Okabe (772)]. Table II–6 gives some recent results.

Standard enthalpies of formation for atoms, radicals, and molecules of photochemical interest are given in the Appendix A–6 through A–13. Most of the bond energies given in the text are computed from these data.

Table II–6. Thermochemical Data Obtained from Photodissociation

Quantity Measured	eV	kcal mol^{-1}
$D_0(SC—S)$	4.463 ± 0.014	102.92 ± 0.32
$D_0(OS—O)$	5.64 ± 0.02	130 ± 0.4
$D_0(NC—N_3)$	4.2 ± 0.1	96 ± 2
$D_0(NCN—N_2)$	0.3 ± 0.1	7 ± 2
$D_0(H—NCO)$	4.90 ± 0.01	112.9 ± 0.2
$D_0(N—CO)$	2.14 ± 0.15	49 ± 3
$D_0(N—CN)$	4.3 ± 0.2	99 ± 5
$E_0(NH\ a^1\Delta)$	1.6 ± 0.1	36 ± 2
$\Delta Hf_0^\circ(C_2H)$	5.50 ± 0.04	127 ± 1
$\Delta Hf_0^\circ(CS)$	2.82 ± 0.02	64.96 ± 0.4
$\Delta Hf_0^\circ(SO)$	0.056 ± 0.03	1.3 ± 0.7
$\Delta Hf_0^\circ(CH_2N_2)$	2.22 ± 0.05	51.3 ± 1

[a] From Okabe (772).

chapter III

Experimental Techniques in Photochemistry

A brief description is given in this chapter of various light sources and materials commonly encountered in conventional photochemical studies. The measurement of quantum yields of photochemical products is important in understanding overall photochemical processes and it is desirable to obtain quantum yields whenever possible. Various chemical actinometers that are frequently used in measuring quantum yields are given. For details see Calvert and Pitts (4).

Light sources, windows, and actinometers for vacuum ultraviolet photochemical studies have been described recently by McNesby et al. (685a). Flash photolysis in conjunction with absorption spectroscopy has been extremely useful in understanding the primary process by directly identifying primary products. Time dependent studies of atom and radical concentrations following flash photolysis have been made by measuring the decay of absorption intensity or resonance fluorescence intensity. Resonance fluorescence is obtained by exciting ground state atoms and radicals to fluorescing states with appropriate resonance lamps. Many important rate constants of atoms and radicals have been measured by these techniques, as is discussed briefly in Section III–5. For details the reader is referred to Ref. 685a. Radical trapping agents have been used frequently in probing the photochemical primary process. Section III–6 deals with this technique.

Samson (25) describes gratings, spectrographs, light sources, window materials, and so forth required for studies in vacuum ultraviolet spectroscopy. Light sources, filters, and detectors for the ultraviolet region of radiation are treated by Koller (17b).

III–1. LIGHT SOURCES

The light sources used in photochemical studies may be classified into three categories depending on its applications: (*a*) atomic line sources, (*b*) molecular band sources, and (*c*) lasers. Atomic line sources are most frequently used for initiating photochemical reactions in the vacuum, near ultraviolet, and visible regions. On the other hand, weak resonance lines with more or less an ideal Doppler shape have been used for atom sensitized reactions and for probing atomic concentrations (see Sections I–6.2 and I–6.3). Molecular band sources in conjunction with a monochromator or spectrograph have been used for measuring absorption spectra of molecules and radical species. Flash discharge lamps of short duration have been used to produce atoms and radicals in large concentrations. Lasers have been especially useful for studies of primary photochemical and photophysial processes and for isotope separations.

III–1.1. Atomic Line Sources in the Vacuum Ultraviolet Region

Light sources must have intensities sufficient to produce measurable amounts of products for analysis within a reasonable time limit, that is, typically above 10^{13} photons sec^{-1}. Stability of output, simplicity of construction, and spectral purity are also required.

Sealed Rare Gas Resonance Lamps. The detailed description of lamp construction and the emission spectra for Xe and Kr has been given by McNesby and Okabe (684), and more recently by Gorden et al. (413). Various other designs of low pressure rare gas resonance lamps have been proposed [Stief and Mataloni, (927), Lane and Kuppermann (598), Loewenstein et al. (645), Bass (71), Gleason and Pertel (401)]. Figure III–1 shows an electrodeless discharge lamp of simple construction operated by a 2450 MHz microwave generator. A getter is used to remove impurity lines due to trace water. [Slanger and Black (899) have found recently that these impurity lines are entirely the CO fourth positive emission.] About 0.7 torr of Xe or Kr is introduced and the lamp is sealed. The lamp has a life of about 50 hr at intensities of 10^{14} to 10^{15} quanta sec^{-1}.

Figure III–2 shows the emission spectra of Ar, Kr, and Xe lamps in the absence of impurity lines.

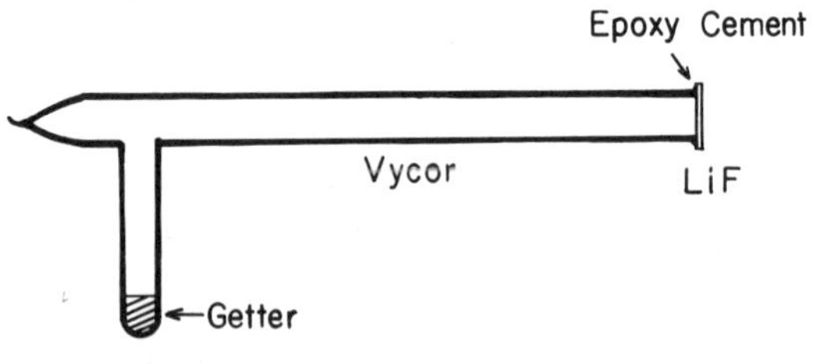

Fig. III–1. Design of the sealed rare gas resonance lamp in the vacuum ultraviolet. The tube is heated and degassed to 10^{-7} torr prior to firing a getter. A pressure of about 0.7 torr of rare gas is introduced and the tube is sealed at the end. The lamp is operated by a 2450 MHz microwave generator. For details see McNesby and Okabe (684).

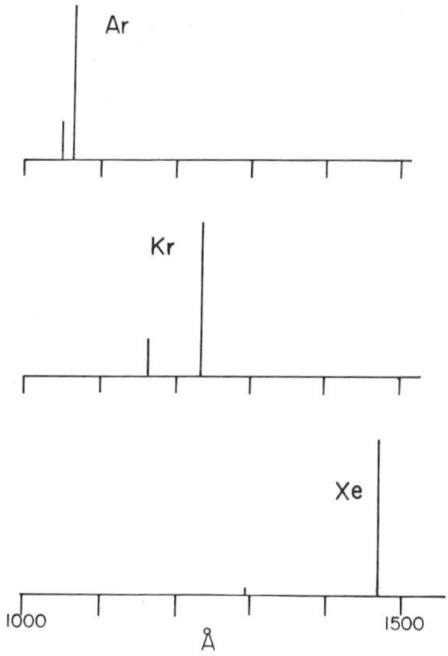

Fig. III-2. Schematic emission spectra of Ar, Kr, and Xe resonance lamps. The intensity ratios of the resonance lines are 1:2 for the Ar (1048, 1067 Å) lines, 1:4 for the Kr (1165, 1236 Å) lines, and 1:50 for the Xe (1295, 1470 Å) lines. Impurity lines are not included in the spectra. See McNesby and Okabe (684), and Gorden et al. (413).

The Ar resonance lamp emits two lines at 1048 and 1067 Å with a ratio of about 1:2 depending on the window thickness. The total output is on the order of 10^{12} to 10^{13} quanta sec^{-1} [Gorden et al. (413), Lane and Kuppermann (598), Laufer and McNesby (602a)]. To remove impurity lines, the Ar lamp must be baked at 350°C for at least 12 hr (620a). Hence, a thin LiF window is sealed to the tube with temperature resistant Ag–AgCl cement. However, the lamp output decreases substantially in a few hours and has to be restored by reheating the lamp. The Kr resonance lamp gives mainly two lines at 1165 and 1236 Å with a ratio of 1:4. Total output is about 10^{14} to 10^{15} quanta sec^{-1} [Okabe (760)].

The Xe resonance lamp produces two main lines at 1295 and 1470 Å with a ratio of 1:50. The total output is about 10^{14} to 10^{15} quanta sec^{-1} [Okabe (760)].

Mercury Resonance Lamp. A low pressure (<0.1 torr) mercury lamp emits strong resonance lines mainly at 1849 and 2537Å. The lamp, containing Hg and a few torr of Ar, is excited either in an electric discharge or an electrodeless microwave discharge. For details see Calvert and Pitts [(4), p. 689]. Beckey et al. (87) have developed a high intensity low pressure Hg lamp that emits 10^{17} quanta sec^{-1} in a 2 cm^3 volume at 1849 Å. The intensity at 2537 Å is about 4 to 10 times that at 1849 Å in an electric discharge [Beckey et al. (87), Glasgow and Willard (400)] (see Fig. III-3). The intensity ratio depends on window thickness and lamp temperatures [Johnson (536)]. A

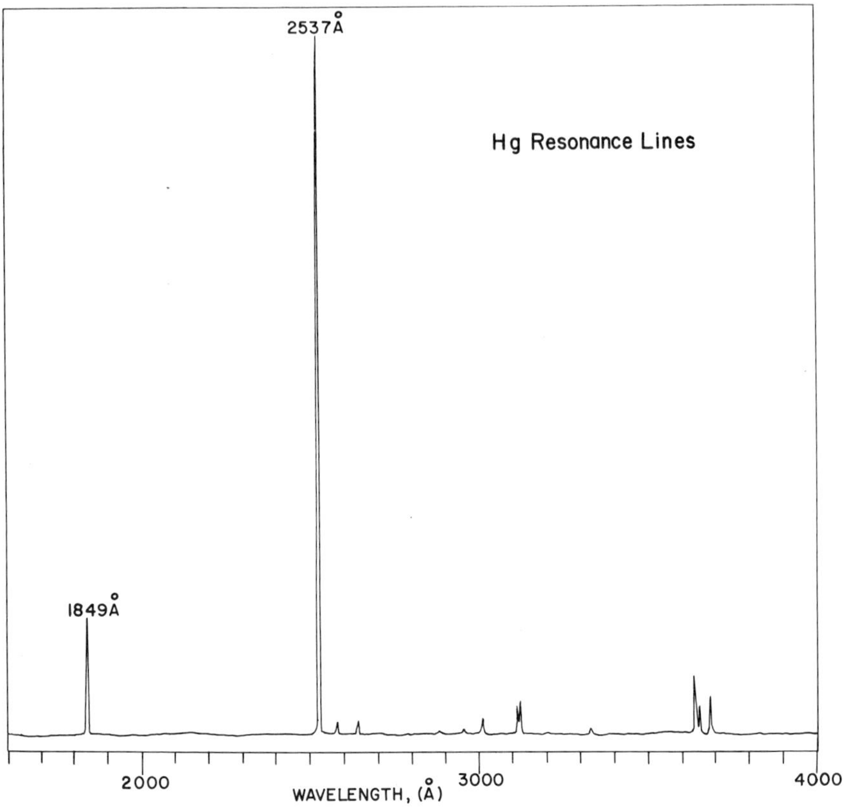

Fig. III–3. The emission spectrum of a low pressure mercury lamp operated in an electric discharge. The intensity at 2437 Å is typically several times that at 1849 Å. Beckey et al. (87) reprinted by permission of *Zeitschrift für Naturforschung*.

low pressure Hg lamp is especially useful for initiating sensitized reactions since the self-absorption of the resonance lines is much less than that of a high pressure Hg lamp.

Sealed Atomic Lamps. Loucks and Cvetanović (649) have constructed a sealed atomic bromine lamp with a Suprasil window. A mixture of 0.5 ml of Br_2 and 0.3 torr of Ar at $-78°C$ is excited by a microwave discharge producing atomic Br lines at 1633, 1582, 1577, and 1575 Å [see also Thompson et al. (968)]. Washida et al. (1010) have used a mixture of CBr_4–Ne at $-40°C$ to obtain the atomic carbon lines (1657, 1931, and 2479 Å). Whittaker and Kintner (1039a) have designed a hollow-cathode carbon resonance lamp at 1657 Å.

Flow Discharge Lamps. Davis and Braun (263) have used low pressure flow discharge lamps operated by a 2450 MHz microwave generator to obtain atomic lines of H, O, N, Cl, Br, C, and S [see also Kikuchi et al. (564, 565)]. These atomic lines are produced by a microwave discharge through respective mixtures of H_2, O_2, N_2, Cl_2, Br_2, CH_4, H_2S with He. The total output is on the order of 10^{14} to 10^{15} quanta sec^{-1}. Table III–1 gives the emission lines of H, O, N, Cl, Br, C, and S lamps and their relative intensities. These lamps are useful for photochemical studies where a large output is required rather than for measuring atomic concentrations.

For measuring the oscillator strength of ground state atoms a resonance lamp with little self-reversal is required (see Section I–6.2). Morse and Kaufman (718) have constructed flow discharge resonance lamps of O, N, and H atoms with extremely small optical depth [see also Clyne et al. (95, 222)], that is, with little self-reversal of the resonance lines as described in Section I–6.1. The lamp intensity is, however, extremely small.

III–1.2. Atomic Line Sources in the Ultraviolet and Visible Regions

Iodine Lamp. Harteck et al. (443) have developed a lamp emitting the 2062 Å line of iodine atoms. A flowing mixture of I_2–Ar at room temperature was excited either by an electric discharge or a microwave discharge. With a quartz window the emission lines are mainly 1876 and 2062 Å with a ratio of 1:30.

Metal Discharge Lamps. Various metal discharge lamps are commercially available. The lamps contain a mixture of metal vapor and a rare gas (Ar for example). Figure III–4 shows the emission spectra of Cd and Zn discharge lamps made by Philips. They are operated on AC at a current of 0.9 Å. The light intensity is much lower than the medium pressure Hg arcs.

Low and Medium Pressure Mercury Lamps. A low pressure (<0.1 torr) Hg lamp emitting the 2537 Å line has been described. Ninety percent of the total radiation is at the 1849 and 2537 Å lines. The lamp is used mainly for initiating Hg sensitized reactions because of small self-absorption. Osborn et al. (783) have designed a microwave discharge lamp of small optical depth. The width of the 2537 Å line is about 0.05 cm^{-1}, while the separation of the ^{202}Hg 2537 Å line from the adjacent isotopic lines is about 0.15 cm^{-1}. Therefore, it is possible to selectively excite the ^{202}Hg in isotopic mixtures (see Section VIII–1.1). Medium pressure (a few hundred torr) Hg lamps of various designs are commercially available. They emit intense lines near 2380, 2537 (reversed), 2652 to 2655, 2804, 2967 to 3022, 3126 to 3132, 3341, 3650 to 3663, 4045 to 4078, 4358, 5461, 5770 to 5790 Å as described by Calvert and Pitts [(4), p. 695]. The lamp is the most suitable light source for

Table III–1.[a] **Atomic Lines Produced by Microwave Excitation in the Vacuum Ultraviolet. Total Pressure is about 1 torr, Intensity is $>10^{14}$ quanta sec^{-1}**

Atom	Gas Mixture	Emission Line (A)[b]	Relative Intensity
H	2% H_2 in He	1216	
O	10% O_2 in He	1302	0.61
		1305	0.97
		1306	1.00
N	1% N_2 in He	1745	0.78
		1743	1.00
		1495	0.15
		1493	0.30
		1412	0.06
Cl	0.1% Cl_2 in He	1397	0.35
		1390	1.00
		1380	0.13
		1364	0.17
		1352	0.05
		1347	0.03
Br	0.1% Br_2 in He	1634	1.00
		1582	0.50
		1577	0.92
		1741	0.25
		1575	0.46
		1532	0.17
		1489	0.15
		1450	0.03
		1385	0.01
C	1% CH_4 in He	1931	1.00
		1658, 1657, 1656	0.52
		1560, 1561	0.58
S	0.2% H_2S in He	1915	0.16
		1900	0.52
		1826	1.00
		1820	0.64
		1807	0.18
		1667	0.13

[a] From Davis and Braun (263).
[b] Corresponding transitions are given in the Appendix, Table A–2.

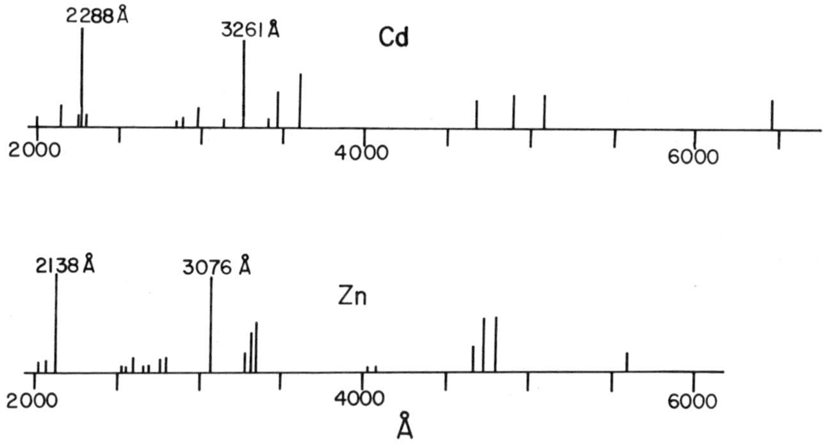

Fig. III–4. The emission spectra of commercial cadmium and zinc electric discharge lamps, 16 W (schematic). From Okabe et al. (770), reprinted by permission of the Air Pollution Control Association.

photochemical studies in the ultraviolet and visible regions, since various emission lines are well separated and of sufficient intensity for photochemical use. They can be made nearly monochromatic by using various filters. Figure III–5 shows the emission spectra in the region 2400 to 3100 Å of a medium pressure Hg lamp made by Philips. The 2537 Å line is pressure-broadened and is almost missing because of self-absorption by Hg atoms near the walls. The lamp is therefore not suitable as a light source for the Hg sensitized reactions.

High Pressure Arcs. Arcs are operated at low voltage high current DC and at 50 to 70 atms. Near point source Xe and Hg arcs are commercially available. A high pressure Xe lamp emits a near continuum in the visible and ultraviolet regions. A high pressure Hg lamp emits continuous spectra superimposed on many diffuse Hg lines. Emission spectra of these lamps are shown in Fig. III–6. They are useful light sources for photochemical studies because of their large output in the near ultraviolet and visible regions. For quantitative work they have to be used in conjunction with filters or a monochromator. Calvert and Pitts [(4), p. 700] give a detailed description on these light sources.

III–1.3. Molecular Band Sources

A hydrogen lamp excited in an electric (1015) or microwave discharge (1008) gives rise to many-lined spectra in the region 1000 to 1800 Å. Figure III–7

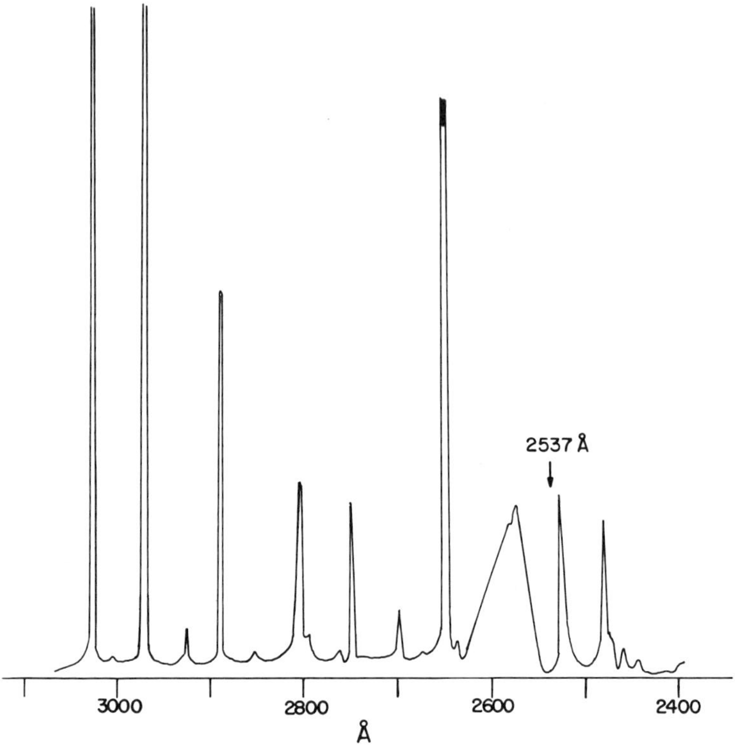

Fig. III–5. The emission spectra of a medium pressure mercury lamp made by Philips, 90 W. Only the 2400 to 3100 Å region is shown. The spectra are taken by the author with a monochromator at a resolution of about 0.3 Å. The 2537 Å line is pressure-broadened and is almost absent because of self-absorption by Hg atoms near the walls, as indicated by the arrow.

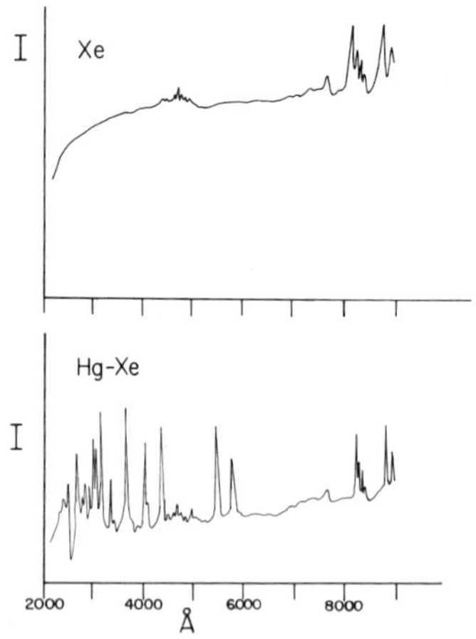

Fig. III–6. Emission spectra of high pressure Xe and Hg–Xe arcs. The arcs operate at low (10–60) voltages and high DC currents. The Xe lamp emits a near continuum in the ultraviolet and visible regions. The Hg–Xe lamp produces intense line spectra superimposed on a continuum. The intensity is in arbitrary units. From a catalog of Oriel Corporation, reprinted by permission of Oriel Corporation.

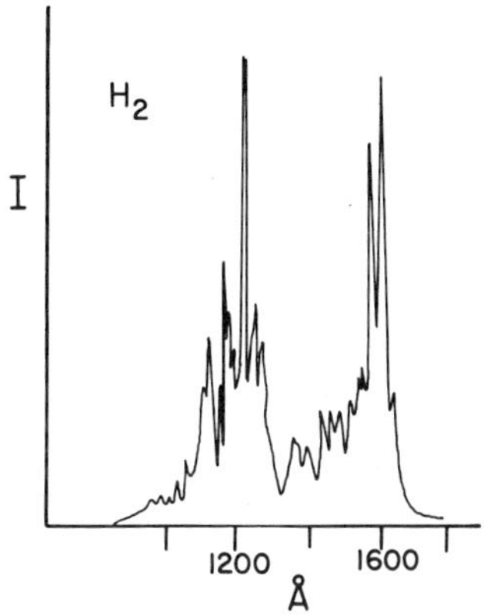

Fig. III-7. The emission spectra of a hydrogen discharge lamp. H_2 pressure is about 2 torr. The light output is about 10^{15} to 10^{16} quanta sec^{-1}. From Brehm and Siegert (145a), reprinted by permission of Springer-Verlag.

shows the emission bands of H_2 at a pressure of about 2 torr. The integrated lamp intensity is on the order of 10^{15} to 10^{16} quanta sec^{-1} [Warneck (1008)].

Wilkinson and Byram (1042) have obtained the emission continua in the regions 1060 to 1500, 1260 to 1700 and 1500 to 2000 Å, respectively, for Ar, Kr, and Xe gases. The gases are excited by a 2450 MHz microwave generator at a pressure of a few hundred torr. The emission continua are attributed to the transitions from electronically excited to ground state diatomic molecules. The emission spectra are shown in Fig. III-8. These continuous sources are too weak for photochemical studies and have mainly been used for measuring the absorption spectrum of gases. A commercial deuterium discharge lamp with a quartz window is suitable for measuring the ultraviolet absorption in the region 1800 to 4000 Å, since the emission spectrum is continuous. The lamp intensity may be too weak for photochemical studies.

Flash Discharge Lamps. Flash discharge lamps are used for producing large amounts of photofragments in a period of a few microseconds. Welge

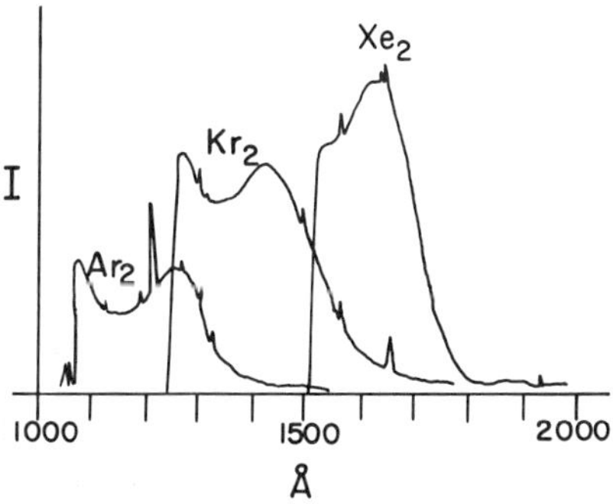

Fig. III-8. The emission continua of a microwave excited discharge of Ar, Kr, and Xe. The gas pressure is about 200 torr. From Wilkinson and Byram (1042), reprinted by permission of the optical society of America.

et al. (1032) have described a high pressure (a few hundred torr) N_2 flash discharge lamp with an intensity of 10^{15} to 10^{16} quanta cm^{-3} per flash in the vacuum ultraviolet region [see also Braun et al. (138)]. Daby et al. (258) have developed an electrodeless (θ-pinch) flash lamp that emits light at an intensity of 10^{17} quanta cm^{-2} per flash in less than 10 μsec.

III–1.4. Lasers

Various lasers have been developed for use in photochemical studies [see Moore's recent review on lasers (714)]. Because of their production of intense monochromatic light and of short pulse time (<1 μsec) lasers have been particularly useful in studying photodissociation dynamics. Dye lasers have a capability of changing the wavelength of light in the region 3400 Å to near infrared. They have been applied to excite one isotopic species from mixtures to achieve isotopic enrichment, as is exemplified in Section VIII–1.

Dewey (285a) has recently succeeded in second harmonic generation between 2171 and 3150 Å using potassium pentaborate ($KB_5O_8 \cdot 4H_2O$) with a conversion efficiency of 9.2%.

Table III–2 gives the wavelengths and approximate power outputs of some commonly used lasers in photochemistry. Solid state lasers have a very high output per pulse. For example 1 J per pulse ruby laser corresponds to 3.5×10^{18} photons. For more general descriptions and applications of lasers, a recent book by Beesley (20) should be consulted. A brief description of each laser is given below.

A ruby laser consists of Al_2O_3 to which a small amount (0.035%) of chromium is added. Laser action takes place between various levels of Cr^{3+} in host Al_2O_3. The transition is from the E level at about 14,000 cm^{-1} to the ground giving rise to a stimulated emission at 6943 Å.

In neodymium laser the transitions occur between levels of impurity ion Nd^{3+} in glass or YAG (yttrium aluminum garnate) as the host material. Lasing wavelengths are 0.914, 1.06, 1.317, 1.336, and 1.35 μm. However, 1.06 μm with YAG is almost always used. Since the lower lasing level is about 2000 cm^{-1} above the ground, that is, the level is empty at room temperature, the overall efficiency is high (1 to 2%).

Lasing action of a nitrogen laser occurs between $C^3\Pi_u$ and $B^3\Pi_g$ states of N_2 with the strongest emission line near 3371 Å. The power output ranges from 100 KW to 2.5 MW in 4 nsec (1056a). The laser is powered by an electric discharge through 10 to 80 torr of N_2. The transition responsible for the 6328 Å line in He–Ne laser is from the Ne $3s_2$ at 166,658 cm^{-1} to $2p_4$ at 150,860 cm^{-1} (Paschen notation). The strong 4416 and 3250 Å lines of a He–Cd laser arise from the transitions of the $Cd^+ 4d^9 5s^2 (^2D_{5/2}) \rightarrow 4d^{10} 5p (^2P_{3/2})$ and $4d^9 5s^2 (^2D_{3/2}) \rightarrow 4d^{10} 5p (^2P_{1/2})$, respectively. The Ar ion

Table III–2. Wavelengths and Power Outputs of Some Commonly Used Lasers in Photochemistry

Type	Wavelength	Power	Remarks
		solid state lasers	
Ruby	6943 Å,[a] 6934 Å	1–100 J per pulse, 1 msec	Xe flash pumping, 0.1% efficiency
Neodymium	0.914, 1.06[a] 1.317, 1.336, 1.35 μm	mean power output 1–20 W 1 J per pulse (100 MW peak power) 20 pulses sec^{-1}	1–2% efficiency
		Gas lasers[b]	
N_2	3371 Å	13 mJ per pulse, 12 nsec	(1056a)
He–Ne	6328 Å	100 mW (CW)[d]	0.05% efficiency
He–Cd	4416 Å, 3250 Å	50 mW (CW)[d], 5 mW (CW)[d]	
Ar^+	4880 Å, 5145 Å	3 W (CW)[d]	0.05% efficiency
CO_2	10.6 μm 9.4 μm	Several kW (CW)[d]	Low total pressure, 20% efficiency
TEA[c]	10.6 μm 9.40 μm	>100 J per pulse, 50–100 nsec	Atmospheric pressure, 10% efficiency

[a] Commonly used wavelength
[b] Only the main laser lines are listed.
[c] Transversely excited atmospheric laser.
[d] Continuous wave.

laser transitions are the $Ar^+ 4p(^4D^\circ_{1/2}) \rightarrow 4s(^2P_{3/2})$ and $4p(^4D^\circ_{5/2}) \rightarrow 4s(^2P_{3/2})$, giving rise to the 4880 and 5145 Å lines, respectively.

A carbon dioxide laser has the main emission lines at 9.4 and 10.6 μm, corresponding to the vibrational transitions (001) → (020) and (001) → (100) as shown in Fig. III–9*a*. The vibrational levels (001) and so forth indicate the quantum numbers of symmetric, bending, and antisymmetric vibrations in that order. The dominant rotational transitions at 10.6 μm are *P*(18), *P*(20), and *P*(22) as shown in Fig. III–9*b*. The population of the upper levels corresponding to the *P*(18), *P*(20), and *P*(22) transitions, that is, $J' = 17$, 19, and 21 is higher than that corresponding to the *R*(18), *R*(20), and *R*(22), that is, $J' = 19$, 21, and 23 because the system is in Boltzmann equilibrium. Hence, the stimulated emission is dominant in the *P* branch transitions, as the population of the *P* branch levels increases by depletion of the population of the *R* branch levels above. The CO_2–N_2 and N_2–He–CO_2 mixtures are commonly used for CO_2 lasers and they are excited by an electric discharge. Laser action occurs as the (001) level of CO_2 is preferentially excited by collisions with vibrationally excited N_2. The addition of He increases the output power by decreasing the gas temperature and at the same time depleting the population of the lower laser levels. The most powerful CO_2

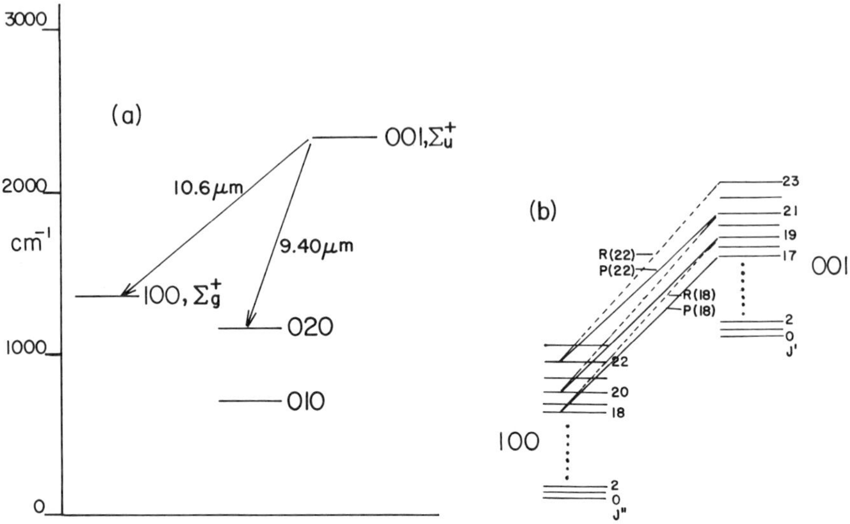

Fig. III–9. The laser transitions in CO_2. (*a*) Energy level diagram in CO_2. The two main laser transitions are (001) → (100) and (001) → (020), giving rise to the laser lines at 10.6 and 9.40 μm, respectively, where (001) and so on means vibrational quantum numbers of symmetric, bending, and antisymmetric vibrations in that order. (*b*) The main rotational lasing transitions between the (001) and (100) levels; the *P*(18), *P*(20), and *P*(22) rotational transitions are the dominant lasing transitions. The populations of the upper levels corresponding to the *R*(18), *R*(20), and *R*(22) transitions, that is, $J' = 19$, 21, and 23, is less than that corresponding to the *P*(18), *P*(20), and *P*(22) transitions, that is, $J' = 17$, 19, and 21, because the system will have a Boltzmann distribution by depolutation of *R* branch levels to the corresponding *P* branch levels. Hence, stimulated emission occurs mainly from the *P* branch levels.

laser has been obtained by using N_2–He–CO_2 mixtures at atmospheric pressure. Instead of two electrodes at each end of a tube many electrodes are placed along the side of the tube to produce transverse discharges. This type of CO_2 laser is operated in a pulsed mode and is called the "transverse excited atmospheric" (TEA) laser. TEA lasers have typically a peak power of 100 MW in 50 to 100 nsec with 10% efficiency.

Ultraviolet and Visible Lasers. Useful light sources in the ultraviolet region are: the fourth harmonic of a neodymium laser at 2662 Å, a pulsed nitrogen laser at 3371 Å, and the second harmonic of a ruby laser at 3471 Å.

Wilson and his coworkers have studied the photodissociation dynamics of NOCl (164), NO_2 (163), and alkyl iodides (834) using ruby and neodymium lasers. A pulsed nitrogen laser has been used to measure the fluorescence lifetime of formaldehyde by Aoki et al. (44).

Tunable dye lasers are convenient sources in the region 3400 Å to near infrared. Yeung and Moore (1074) have developed the tunable ultraviolet source (3000 to 3500 Å) by mixing the 6943 Å ruby laser with an organic dye laser. Using a frequency doubled dye laser, Clark et al. (217) have been able to achieve CO isotopic enrichment in the photolysis of isotopic mixtures of formaldehyde. Various noble gas ion lasers in CW (continuous wave) mode are available in the visible region [Bridges et al. (148)].

Some of the useful ion lasers are an Ar ion laser at 4880 and 5145 Å, a Kr ion laser at 6471 Å, and a Cd ion laser at 3250 and 4416 Å. A pulsed high intensity CO_2 laser at 10.6 μm (943 cm^{-1}) or 9.6 μm (1042 cm^{-1}) has been used for isotope enrichment of SF_6, BCl_3, CF_2Cl_2, and SiF_4 [Lyman and Rockwood (655)].

Vacuum Ultraviolet Lasers. Various lasers excited by an electron beam have been developed. Hodgson and Dreyfus (476) have observed the stimulated emission of the Lyman bands ($B^1\Sigma_u^+ - X^1\Sigma_g^+$) of H_2 near 1600 Å using a pulsed high density electron beam through H_2 at pressures of 10 to 100 torr. The estimated peak power output is about 100 W. Waynant et al. (1024a) also have observed lasing of Lyman bands of H_2 by a pulsed electric discharge through H_2 at a pressure of a few tenths of a torr. The peak power is about 100 KW. A CO laser emits radiation in the 1800 to 2000 Å region (475). The transitions involved are between the $A^1\Pi$ and $X^1\Sigma^+$ states (CO fourth positive bands). Laser pulses of 6 W in 1.5 nsec were generated by pulsed high current electric discharges through 60 torr CO. In both H_2 and CO lasers, the transitions are to high vibrational levels of the ground electronic state. The overall efficiency is still very low.

Hughes et al. (489a) have found laser action at 1261 Å from molecular argon at pressures up to 67 atm by electron beam excitation. The transition is from the Σ_u^+ excited state to the repulsive $^1\Sigma_g^+$ ground state. These lasers are still in the developmental stage and have not been applied to photochemical studies. See a recent review by Bradley (124). Very recently several rare gas halide lasers have been developed [see for example Waynant (1024b)]. These excimer lasers operate in a pulsed mode near atmospheric pressures and emit light of wavelengths from 1750 to 2480 Å depending on the type of gas mixtures.

III–2. MATERIALS FOR PHOTOCHEMICAL STUDIES

III–2.1. Window Materials

Depending on the wavelength region of interest, a proper choice of window material has to be made. Figure III–10 shows transmission curves of various window materials. An LiF window is most frequently used for photochemical

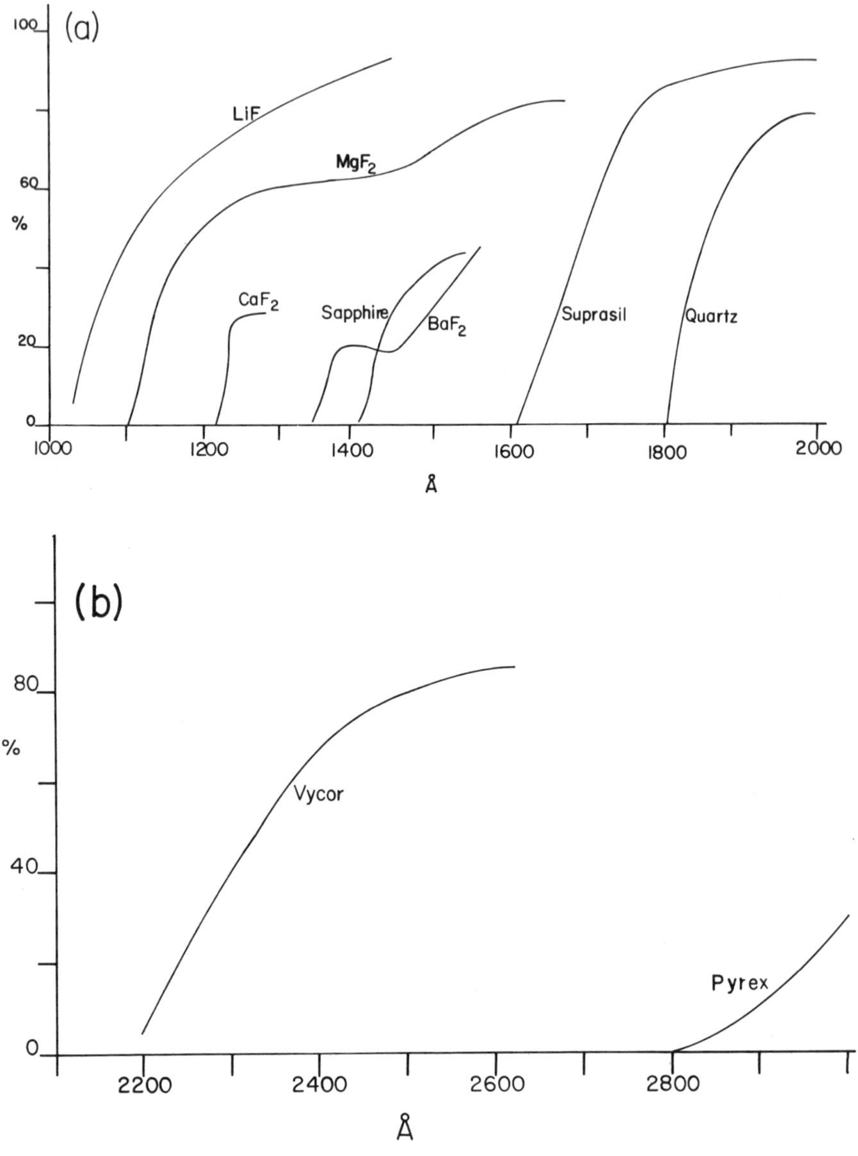

Fig. III–10. Percentage transmission of window materials at room temperature. (*a*) 1.95 mm thick lithium fluoride [Lanfer et al. (602)], 1 mm thick magnesium fluoride [Samson (25)], 0.25 mm thick calcium fluoride [Knudson and Kupperian (579)], 0.32 mm thick sapphire [Laufer et al. (602)], 1.95 mm thick barium fluoride [Laufer et al. (602)], 10 mm thick Suprasil (Amersil, Inc. catalog), 2 mm thick Quartz (Corning Color Filter Glasses), (*b*) 2 mm thick Vycor (96% silica), (Corning Color Filter Glasses), 2 mm thick Pyrex (Corning Color Filter Glasses). Reprinted by permission of The American Institute of Physics, John Wiley & Sons, Heraeus-Amersil Inc. and Corning. Copyright 1965 by The American Institute of Physics.

studies in the vacuum ultraviolet since it has a high transmission down to about 1050 Å. The transmission limit of MgF_2 is at about 1120 Å. However, MgF_2 is less reactive with moisture and is more stable against exposure to intense vacuum ultraviolet radiation than LiF. For the work in the near ultraviolet pure silica (Suprasil) or quartz has been used most frequently.

III–2.2. Filters

To isolate light of desired wavelength from a polychromatic light source various filters have been developed. One can use solution filters to isolate various Hg lines. These solution filters are described by Calvert and Pitts (4). Interference filters of good transmission are commercially available in the ultraviolet and visible regions. Broad band color filters are also commercially available.

1849 Å Filters. Filters to selectively transmit the 1849 Å line from a low pressure Hg arc have been developed by various workers. Toby and Pritchard (974) have used a hexafluoroacetone imine filter to isolate the 1849 Å line. At 100 torr in a 5 cm cell the filter transmits 79% of the 1849 Å line, while it only transmits 0.5% at 2537 Å. Glasgow and Willard (400) have used a flowing O_2–O_3 filter to obtain 50% of the 1849 Å line and only 0.1% of the 2537 Å line.

Vacuum Ultraviolet Interference Filters. Vacuum ultraviolet interference filters have been developed recently by Fairchild (339) and are commercially available. Alternate layers of Al and MgF_2 are vacuum deposited on the polished surface of various substrates (LiF, CaF_2, SiO_2) depending on the region of transmission desired.

III–3. QUANTUM YIELDS

III–3.1. Definition

According to the Stark-Einstein law one quantum of monochromatic light is absorbed per molecule, that is, one photochemically active molecule is formed for one photon absorbed. The law holds for absorption of photons of moderate intensity (10^{13} to 10^{15} quanta sec^{-1}), since the probability of the second photon being absorbed by an electronically excited molecule of a typical lifetime of 10^{-8} sec is negligibly small (see Introduction). If the light intensity is extremely high, multiphoton absorption processes are possible. Many such examples have been found in laser excitation [Moore (714)]. For example, Lyman and Rockwood (655) have found the occurrence of photodissociation in SF_6, BCl_3, and so forth by an intense CO_2 laser at 10.6

μm where more than 30 photons are required for dissociation. The quantum yields of products may depend on various experimental conditions, such as temperature, pressure, and added inert gas. For example, the quantum yield of a product may be temperature independent, if the product is formed in the primary process or by reactions involving no activation energy, although the primary photochemical yield of some molecules may be temperature dependent as shown in the photolysis of NO_2 at 4047 Å or of O_3 at 3130 Å (Section I–4.3), where the contribution of internal energy is required for bond rupture. On the other hand, the quantum yield of CO production in the photolysis of acetone at 3130 Å increases from about 0.1 to 1 as the temperature increase from 25 to over 100°C, which is ascribed to the decomposition of acetyl radicals at higher temperatures

$$CH_3CO \rightarrow CH_3 + CO$$

If the overall quantum yield of a product is one, the product is formed only in the primary process with a quantum yield of unity or the product arises from the combination of radicals produced in the primary process with unit quantum yield. In the photolysis of acetone at 3130 Å and 100°C the quantum yields of products, CO and C_2H_6, are unity and the reactions are represented by

$$(CH_3)_2CO \xrightarrow{h\nu} CO + 2CH_3$$
$$2CH_3 \rightarrow C_2H_6$$

The quantum yield much larger than unity indicates a chain mechanism. As an example, the quantum yield of O_3 disappearance at 2537 Å is 6. A mechanism proposed involves a chain decomposition of O_3 by the energetic primary products, O_2 and O (see Section VI–11).

$$O_3 \xrightarrow{h\nu} O_2^* + O^*$$
$$O_2^* + O_3 \rightarrow 2O_2 + O$$
$$O^* + O_3 \rightarrow O_2 + 2O$$
$$O + O_3 \rightarrow 2O_2$$

where the asterisk indicates electronically excited species. Other examples are given by Calvert and Pitts [(4) p. 645], including the effects of light intensity, wavelength, and pressure. The information on quantum yield is of fundamental importance in understanding photochemical reaction mechanisms.

The quantum yield of a product, Φ, may be defined as

$$\Phi = \frac{\text{number of molecules formed by the overall process}}{\text{number of quanta absorbed by a reactant}} \qquad \text{(III-1)}$$

We may also define the quantum yield of a primary process, ϕ, such as dissociation:

$$\mathrm{A} \xrightarrow{h\nu} \mathrm{A}^* \rightarrow \mathrm{D}_1 + \mathrm{D}_2$$

$$\phi = \frac{\text{number of } A \text{ molecules dissociated by the primary process}}{\text{number of quanta absorbed by } A} \qquad \text{(III-2)}$$

The quantum yields of the stable product, Φ, can be measured experimentally, while the primary quantum yield ϕ may not be obtained directly if products D_1 and D_2 are radicals. In principle, it is possible to obtain the radical quantum yield, using an intense pulsed monochromatic light source, from the ratio of concentrations of radicals produced to light quanta absorbed per pulse. However, this technique has not been used because of the technical difficulties in measuring absolute concentrations of radicals produced and in obtaining a stable intense monochromatic light source at a desired wavelength, and so on. If more than one primary process is involved, we must define the primary quantum yield in terms of each primary process, such as ϕ_1 and ϕ_2 by the primary processes 1 and 2, respectively. We can also define the quantum yield of disappearance of the reactant A, $\Phi_{-\mathrm{A}}$, as the number of reactant molecules that have disappeared during the overall process divided by the number of quanta absorbed by the reactant. $\Phi_{-\mathrm{A}}$ is equivalent to $\phi_{-\mathrm{A}}$ (the primary quantum yield) if a product formed in the primary process does not react further with the reactant. However, as much as 10% or more decomposition may be required to obtain sufficient accuracy for $\Phi_{-\mathrm{A}}$. Since a large conversion introduces undesirable complications because of the photolysis of products and subsequent secondary reactions, it is preferable to limit the extent of decomposition to within 1% or less and to measure reaction products rather than the decrease of the amount of reactant molecules. In some cases a primary product is a stable molecule and no other secondary reactions are involved to generate the same molecule. Then the primary quantum yield of that process may be equal to the quantum yield of the stable product. The sum of quantum yields of all primary unimolecular processes, including dissociation, fluorescence isomerization and intersystem crossing must be unity by the Stark-Einstein law. The primary quantum yield of dissociation is unity if direct dissociation occurs, while in the case of predissociation it is less than unity because of the presence of other primary processes such as fluorescence and quenching. If the mode of the primary dissociation process is more than one, the sum of each dissociation process is equal to the primary yield of dissociation.

III–3.2. Calculation of the Primary Quantum Yield

The calculation of the primary quantum yield of dissociation may be illustrated by a classic example of acetone photolysis [Noyes et al. (750),

Calvert and Pitts (4), p. 782]. The primary process of acetone in the near ultraviolet above 125°C at pressures less than 50 torr is

$$CH_3COCH_3 \xrightarrow{h\nu} CO + 2CH_3$$

The quantum yield of CO production is unity. Since we know from various data that CO is produced only from the primary process, we may conclude that

$$\phi = \Phi_{CO} = 1$$

On the other hand, the photolysis of acetone at 3130 Å and at room temperature is mainly

$$CH_3COCH_3 \xrightarrow{h\nu} CH_3 + COCH_3$$

The primary quantum yield of dissociation in this case is the sum of the quantum yields of the products indicated in (III-2a) formed by the following radical–radical reactions

$$2CH_3 = C_2H_6$$
$$2CH_3CO = (CH_3CO)_2$$
$$2CH_3CO = CH_2CO + CH_3CHO$$
$$CH_3 + CH_3CO = CH_3COCH_3$$
$$CH_3 + CH_3CO = CH_4 + CH_2CO$$
$$\phi = \Phi_{C_2H_6} + \Phi_B + \Phi_{CH_3CHO} + \Phi_A + \Phi_{CH_4} \tag{III-2a}$$

where $\Phi_{C_2H_6}$, Φ_B, Φ_{CH_3CHO}, Φ_A, and Φ_{CH_4} are the quantum yields of ethane, biacetyl, acetaldehyde acetone reformed, and methane, respectively. It is necessary to know that products are formed only by radical–radical processes. If, for example, methane is formed by an abstraction process

$$CH_3 + CH_3COCH_3 = CH_4 + CH_2COCH_3$$

the above stoichiometry does not strictly hold. Besides, it is not obvious how to obtain Φ_A formed by recombination of CH_3 with CH_3CO. To circumvent an inherent difficulty in calculating the primary yield from the quantum yields of products, a scavenger technique has been used with some success. For example, the primary yield of acetone photolysis has been obtained by measuring the quantum yield of methyl iodide produced by

$$CH_3 + I_2 = CH_3I + I$$

when I_2 is added to acetone. However, there is some evidence that I_2 deactivates excited acetone molecules. Furthermore, at higher temperatures, I_2 reacts with acetone [see Calvert and Pitts (4), p. 602]. One must be cautious for these reactions in applying a scavenger technique to deduce the primary quantum yield.

By a similar technique Okabe (774) has measured the primary quantum yield of thiophosgene photolysis by adding heptane

$$Cl_2CS \xrightarrow{h\nu} Cl + ClCS$$
$$Cl + C_7H_{16} \rightarrow HCl + C_7H_{15}$$

The quantum yield of HCl formed is equal to the primary quantum yield of dissociation.

$$\phi = \Phi_{HCl}$$

The use of atom and radical trapping agents to probe the photochemical primary process is further discussed in Section III–6.

III–4. ACTINOMETRY

To obtain information on the quantum yield of a photochemical process, one must measure the absolute intensity of the incident monochromatic light and the fraction of light absorbed by a reactant in a reaction cell of a certain length. The absolute intensity of the incident light can, in principle, be measured by a calibrated thermopile–galvanometer system in conjunction with a monochromator of high light gathering power. Although a thermopile is sensitive to light of wavelength from the vacuum ultraviolet to visible region, the system has not been used often by the photochemist since the calibration with the thermopile system is extremely tedious and time consuming, as has been described by Calvert and Pitts (4), p. 771. It is more common practice for the photochemist to use photochemical reactions of certain gases and liquids as a secondary standard for which the quantum yield of a certain product at a given wavelength has been well established. The absolute light intensity can be obtained conveniently by measuring a product formed by photodissociation from a reactant in units of $sec^{-1}\ cm^{-3}$. This method is called chemical actinometry and the photolysis system to be used is called a chemical actinometer. A good chemical actinometer is one for which the quantum yield is well established at a given wavelength, and the photolysis product can be measured easily by a conventional analytical technique. A good chemical actinometer is especially difficult to find in the visible region, since few molecules absorb and dissociate into stable products in this region. Calvert and Pitts [(4), p. 783] suggest nitrosyl chloride photolysis as a gas actinometer and potassium ferrioxalate as a solution actinometer. If a reactant produces n molecules of a stable product M per second at a given wavelength λ, the absolute light intensity I_0^λ at λ can be obtained from

$$I_0^\lambda = \frac{n_M}{\Phi_M(1 - e^{-kpl})} \text{ quanta sec}^{-1} \qquad \text{(III-3)}$$

where Φ_M is the quantum yield of M and $(1 - e^{-kpl})$ is the fraction of light absorbed by the actinometer with an absorption coefficient k at a pressure p and a path length l. The fraction of light absorbed by the chemical actinometer at λ can be either measured directly or can be calculated if k, p, and l are known.

III–4.1. Chemical Actinometers in the Vacuum Ultraviolet Region

Nitrous Oxide. Nitrous oxide is one of the most commonly used chemical actinometers in the vacuum ultraviolet region. The quantum yield of N_2 produced is 1.40 ± 0.1 at 2139 [Simonaitis et al. (884)], at 1849 [Zelikoff and Aschenbrand (1081)], and at 1470 Å [Groth and Schierholz (426)]. At 1236 Å $\Phi_{N_2} = 1.18 \pm 0.03$ [Groth and Schierholz (426)]. N_2O has some drawback since N_2 has to be separately determined in mixtures containing other noncondensable products O_2 and NO. Greiner (423) has concluded that the net number of molecules generated per quantum absorbed in N_2O photolysis is 1.00 ± 0.05 in the 1380 to 2100 Å range and about 0.8 in the range 1200 to 1380 Å. This is based on the stoichiometry that for each quantum absorbed the following overall reaction occurs

$$2N_2O = \tfrac{3}{2}N_2 + \tfrac{1}{2}O_2 + NO$$

Then the accurate measurement of pressure increase by a sensitive pressure indicator should give the absolute intensity of the lamp output.

Ethylene. Borrell et al. (121) have used ethylene as a chemical actinometer at 1849 Å. The quantum yield of hydrogen formation at this wavelength is 0.62 ± 0.01 on the basis of $\Phi_{N_2} = 1.44$ from N_2O photolysis.

On the other hand, Glasgow and Potzinger (399) have obtained $\Phi_{H_2} = 0.42 \pm 0.05$ at three wavelengths, 1850, 1630, and 1470 Å, independent of pressure (0.1 to 100 torr) and temperature (22 to 175°C). Since hydrogen is the only noncondensable product at liquid nitrogen temperature, ethylene requires a simpler analytical procedure than nitrous oxide as an actinometer. At 1849 Å the acetylene to hydrogen ratio is 1.2 in the photolysis of ethylene (87, 121). Thus, $\Phi_{C_2H_2} = 0.50$ (on the basis of $\Phi_{H_2} = 0.42$) may also be used as a standard.

Carbon Dioxide. Carbon dioxide has been frequently used as an actinometer in the region 1000 to 1600 Å. The quantum yield of CO production has recently been redetermined by Inn (513) using the thermopile–galvanometer system. He obtained $\Phi_{CO} = 0.75 \pm 0.05$ at 1470 Å and $\Phi_{CO} = 0.53 \pm 0.08$ in the region 1500 to 1670 Å. Carbon dioxide as an actinometer has some disadvantages since Φ_{CO} is not constant as a function of incident wavelength and CO has to be separately determined in mixtures of the noncondensable products, CO and O_2.

Recently, Slanger et al. (905) have found Φ_{CO} is less than unity in the region 1200 to 1500 Å, while $\Phi_{CO} = 1$ had been commonly assumed in this region. Warneck (1009) has found $\Phi_{CO} = 1.06 \pm 0.1$ at the Ar lines (1048, 1066 Å) in comparison with the NO photoionization yield at these lines. Thus, caution must be exercised in the use of CO_2 as an actinometer.

Hexafluoroacetone. Magenheimer and Timmons (657) have used hexafluoroacetone photolysis as an actinometer at 1470 Å. The quantum yield Φ_{CO} produced from hexafluoroacetone photolysis is 0.97 ± 0.05 on the basis of $\Phi_{N_2} = 1.44$ produced from N_2O photolysis at 1470 Å. Carbon monoxide is the only noncondensable product of the hexafluoroacetone photolysis at liquid N_2 temperature. Hence, it is a simpler actinometer to use than N_2O and CO_2 which produce more than one noncondensable product.

Nitric Oxide. The absorption coefficient and photoionization yield of NO have been carefully measured by Watanabe et al. (1020) in the region 580 to 1350 Å. At 1236 Å the photoionization yield is 77%. As an actinometer one must be cautious to use a proper pressure of NO under properly applied voltage so that all ions and electrons are collected at the electrodes. The ionization potential of NO corresponds to 1340 Å so that nitric oxide can be used at wavelengths below this value.

Alkyl Amines. Salomon and Scala (855) have used various alkyl amines (trimethyl amine, triethyl amine, dipropyl amine, etc.) as physical actinometers at 1470 Å. They have determined the ionization yield on the basis of $\Phi_{CO} = 0.97$ from hexafluoroacetone photolysis at 1470 Å.

III–4.2. Chemical Actinometers in the Ultraviolet Region

Acetone. Acetone has been frequently used as a chemical actinometer in the 2500 to 3200 Å region. $\Phi_{CO} = 1$ has been well established for temperatures above 125°C and pressures below 50 torr. Acetone as an actinometer has some drawback, since the reaction cell has to be heated above 125°C and CO has to be separated from the other noncondensable product CH_4.

Phosgene. Phosgene may be a useful actinometer in the region 2000 to 2800 Å. The only noncondensable product at liquid N_2 temperature is CO with a quantum yield of unity [see Wijnen (1040)]. $\Phi_{CO} = 1.0 \pm 0.1$ at

2537 Å has been confirmed by the author (774) in comparison with $\Phi_{H_2} = 1$ from the photolysis of HBr.

Hydrogen Bromide. Hydrogen bromide has been used for actinometry in the 1800 to 2500 Å region. The quantum yield of H_2 production has been found to be unity in comparison with $\Phi_{N_2} = 1.40$ from N_2O photolysis [Martin and Willard (663)]. The extent of decomposition has to be below 1% to avoid a secondary reaction with the product Br_2.

$$H + Br_2 \rightarrow HBr + Br$$

Hydrogen is the only noncondensable product.

Azomethane. The absorption spectrum of azomethane lies in the region 2700 to 4100 Å with a maximum at about 3400 Å [(4), p. 453] and in the region below 2100 Å (71a). The quantum yield of N_2 is unity in the first absorption region over the range of temperatures 24 to 164°C and is independent of intensity and pressure [Steacie (26)]. The noncondensable products at liquid nitrogen temperature are N_2 and CH_4 from which N_2 must be isolated.

Potassium Ferrioxalate. Parker (796) has developed a liquid chemical actinometer suitable for the region 2537 to 4900 Å. When the potassium ferrioxalate is exposed to light of wavelengths below 4900 Å, weakly absorbing ferrous oxalate is formed, which, when 1,10-phenanthroline is added, is transformed into a strongly absorbing compound. The quantum efficiency increases gradually from about 0.9 near 4800 Å to about 1.2 at 2537 Å. The potassium ferrioxalate actinometer is more sensitive than the uranyl oxalate actinometer and is suitable for measuring low intensity ultraviolet and visible radiation.

III–5. DETERMINATION OF THE ELEMENTARY REACTION RATES

The photochemical process comprises many steps beginning from the initial production of atoms and radicals by absorption of light and ending in the formation of final stable products. To understand the overall photochemistry, the nature of primary products and the subsequent rates involving the primary and secondary reactive products must be known. Each step leading eventually to the final stable products is called the elementary reaction.

In the last decade or so many important radicals in photochemistry have been detected by optical absorption, mass spectrometry, electron spin

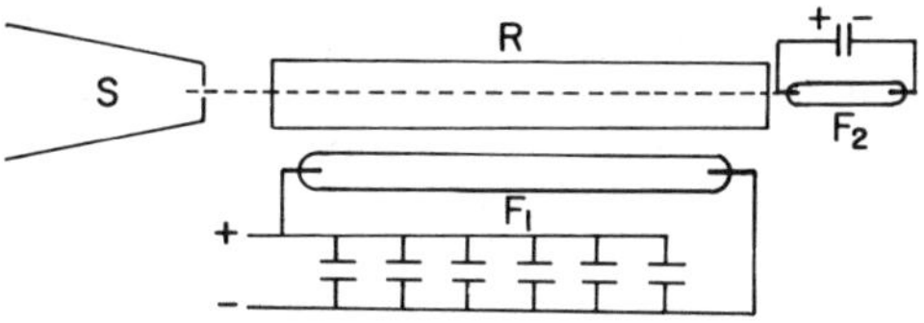

Fig. III–11. Flash photolysis apparatus. A reactant gas is enclosed in the reaction cell *R* and is irradiated by the main intense flash lamp F_1. Immediately after the main flash a second spectro flash F_2 is operated with various intervals from microseconds to milliseconds to give the background continuum for absorption. The absorption spectrum is taken with a spectrograph *S*.

resonance, laser magnetic resonance, and so forth. A particularly important contribution has been made by flash photolysis in combination with optical absorption spectroscopy. Figure III–11 shows a simplified diagram of flash photolysis apparatus. A reactant gas is enclosed in a reaction cell *R* and is irradiated by an intense flash lamp F_1. A second flash tube F_2 is operated immediately after the main flash with variable intervals from microseconds to milliseconds and provides the background continuum. The absorption spectrum is taken with a spectrograph *S*. For the ultraviolet the quartz reaction cell is commonly used, but for the vacuum ultraviolet the cell is equipped with a number of LiF windows. An improved design of flash photolysis apparatus in the vacuum ultraviolet is described by McNesby et al. (685a).

Once the absorption spectra of atoms and radicals are identified, it is possible to obtain their reaction rates by measuring the decay of absorption intensity at a given wavelength as a function of time after the main flash. To obtain relative atom or radical concentration from the measured absorption intensity, care must be taken to examine whether the Lambert-Beer relationship can be applied between radical concentration and its absorption, and deviation from the relationship must be corrected.

Deviation from the law is observed when the instrumental spectral width is broader than those of atomic lines or rotational lines of the band spectra (see Section I–8.2). Instead of the emission continuum with the spectrograph of high resolving power, a light source emitting an atomic resonance line or molecular resonance fluorescence bands corresponding to atoms or radicals of interest can be used to follow their ground state concentrations. The ground state atom or radical concentration is proportional either to the amount of light absorbed or to that of light emitted as a result of resonance absorption. Various atomic line sources that may be used are given in Table III–1. The resonance technique would require a low resolution monochromator or a filter to isolate the resonance line or bands. Bemand and

Clyne (95) discuss the dependence of the types of lamps on the Lambert-Beer law and conclude that the law is obeyed over a wide range of atom concentrations with the lamp having an ideal Doppler shape (see Section I–6.3). Since the law is not always obeyed over a wide range of concentrations with other types of lamps, careful examination is required to determine whether the law is applicable in the range of concentrations to be studied.

In the resonance fluorescence technique, the fluorescence intensity is measured at right angles to the incident beam. The fluorescence method appears more sensitive than the absorption method and thus has an advantage that the low atom concentrations can be used where the Lambert-Beer law can be applicable. If atoms or radicals are formed in electronically excited fluorescing states such as $O(^1D)$, $O(^1S)$, or $O^2(^1\Sigma^+)$, $O_2(^1\Delta)$, the reaction rates of these excited species with the reactant can be followed by measuring the decay time of fluorescence from the excited state, provided that the natural fluorescence decay rate of these species is much slower than that of the reaction.

Many rates of elementary reactions involving atoms and radicals have recently been measured by these and other techniques. The knowledge of primary photolytic products and rates of elementary reactions involving atoms and radicals have greatly advanced our understanding of many photochemical systems for which only the data on product analysis are available.

The decay rate of reactive species follows different equations depending on the type of reaction. Three types of reactions are illustrated below.

III–5.1. Pseudo-First-Order Decay of Reactive Species

If an atom or radical reacts with a reactant molecule rather than with another atom or radical the reaction is first order in the atom or radical when the reactant concentration is much larger than that of the reactive species, which is common in low intensity photolysis systems. The reaction may be represented by

$$R + M \rightarrow \text{products} \qquad \text{(III-4)}$$

where R denotes the atom or the radical and M is the reactant molecule. The rate of decay of the atom is

$$-\frac{d[R]}{dt} = k_5[R][M] \qquad \text{(III-5)}$$

where k_5 is a constant at a given temperature and the brackets signify concentrations. Units of concentration may be given in molec cm^{-3}, mol cm^{-3}, or mol dm^{-3} and the time may be expressed in seconds or minutes. In this book molec cm^{-3} and seconds are used throughout. The conversion factors among various units are given in the Appendix Tables A–4, A–5A, and A–5B. The constant k_5 is called the rate constant and is second order with respect to [R] and [M] with units of cm^3 $molec^{-1}$ sec^{-1}. If [M] is much larger than [R], the rate is first order in [R].

The decay rates of R are usually measured under conditions of $[M] \gg [R]$, that is, [M] remains constant during the decay time t. Hence, (III-5) may be integrated to give

$$\ln [R_0] - \ln [R] = k_5[M]\,t \qquad \text{(III-6)}$$

or

$$[R] = [R_0]e^{-k_5[M]t} \qquad \text{(III-7)}$$

where $[R_0]$ is the concentration of R at $t = 0$.

A plot of ln [R] against t is shown in Fig. III–12. The slope of the curve corresponds to $k_5[M]$ from which k_5 may be obtained. If a reaction proceeds with unit efficiency k_5 is of the order of 10^{-10} cm^3 $molec^{-1}$ sec^{-1}, that is, the gas kinetic collision rate constant. [The typical collision rate constant is 2.5×10^{-10} cm^3 $molec^{-1}$ sec^{-1} from (IV-6), assuming $\sigma = 10^{-8}$ cm, $\mu = 25$ g mol^{-1}]. The reactions of many ground state atoms (Cl, Br, I, O, N, S, H, etc.) and radicals (OH, CH, CH_2, CH_3, NH_2, etc.) have been measured by this method.

III–5.2. Second-Order Decay of Reactive Species

In some photochemical systems the reaction rate of the radical R with the reactant molecule M (III-4) is found to be so slow that the dominant fate of the radical is combination.

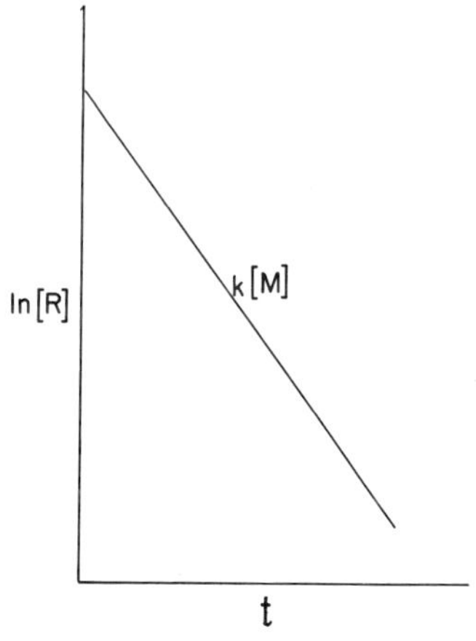

Fig. III–12. Typical pseudo-first-order decay rate of reactive species. The reaction is R + M → products where R is an atom or a radical and M is a reactant molecule whose concentration remains constant during the decay time *t*. The decay curve is represented by $\ln[R_0] - \ln[R] = k[M]t$ and the slope corresponds to $k[M]$, where k is the second-order rate constant (cm^3 $molec^{-1}$ sec^{-1}), [M] is the concentration of the reactant in molec cm^{-3}, and $[R_0]$ is [R] at $t = 0$.

$$R + R \rightarrow \text{products} \tag{III-8}$$

The rate of radical decay is

$$-\frac{d[R]}{dt} = 2k_9[R]^2 \tag{III-9}$$

where k_9 is the rate constant of radical combination.

Integrating (III-9), we obtain

$$\frac{1}{[R]} - \frac{1}{[R]_0} = 2k_9 t \tag{III-10}$$

where $[R]_0$ is the concentration of the radical at $t = 0$. One can estimate the radical concentrations in a typical low intensity photolysis system where the radical disappears by combination, assuming the rate of light absorption $I_a = 10^{11}$ quanta cm^{-3} sec^{-1} and $k_9 = 10^{-11}$ cm^3 molec^{-1} sec^{-1}

$$[R] = \sqrt{I_a/2k_9} = 0.7 \times 10^{11} \text{ molec cm}^{-3}$$

For (III-4) to be competitive with (III-8) we must have a relation $k_5[R][M] = 2k_9[R]^2$. Assuming $[M] = 10^{17}$ molec cm^{-3} and $k_9 = 10^{-11}$ cm^3 molec^{-1} sec^{-1}, $k_5 = 1.4 \times 10^{-17}$ cm^3 molec^{-1} sec^{-1} is obtained.

Examples in which the rate of radical combination is just as important as that of the radical–reactant reaction are the photolyses of acetone and of ketene–nitrogen mixtures. (Nitrogen is used here to collisionally deactivate singlet CH_2 to triplet CH_2.) The reaction rates of methyl with acetone and of triplet methylene with ketene are both slow with $k_5 = 10^{-19}$ and $\leq 10^{-16}$ cm^3 molec^{-1} sec^{-1}, respectively (623a). Braun et al. (143) have measured the rate of triplet methylene association in the flash photolysis of ketene in the presence of He. They have obtained the decay rate from a curve similar to Fig. III–13. In this type of experiment the absolute concentration of radicals must be known. They have determined the triplet methylene concentration from the amounts of ketene decomposed. Since the association products in (III-8) are vibrationally excited, they decompose further into radicals or stable molecules unless stabilized by collisions as exemplified by

$$2CH_3 \rightarrow C_2H_6 \rightarrow 2CH_3$$
$$2CH_2 \rightarrow C_2H_4 \rightarrow C_2H_2 + H_2 \text{ (Ref. 143)}$$
$$CH_2 + CH_3 \rightarrow C_2H_5 \rightarrow C_2H_4 + H \text{ (Ref. 809a)}$$

The decomposition of ethane into methyl radicals occurs below a total pressure of 1 torr (990). Very high total pressures are required to stabilize diatomic products.

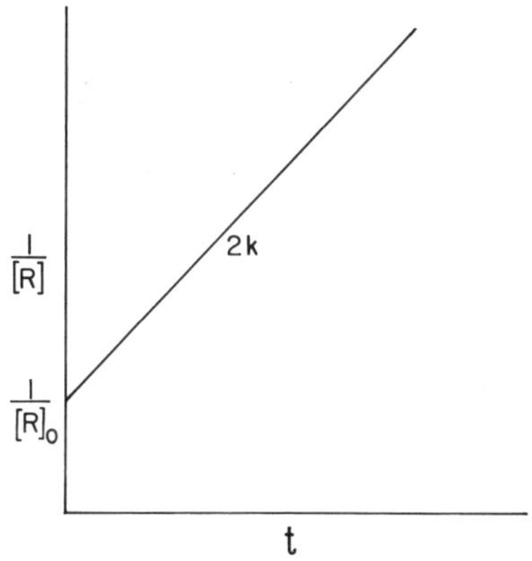

Fig. III–13. Second-order decay of radical by association

$$\mathrm{R + R \rightarrow products}$$

The decay rate follows the equation

$$\frac{1}{[\mathrm{R}]} - \frac{1}{[\mathrm{R}]_0} = 2kt$$

where $[\mathrm{R}]_0$ is the concentration of radicals at time zero. The slope corresponds to $2k$. Braun et al. (143) obtained $k = 5 \times 10^{-11}$ and 7×10^{-11} cm^3 molec^{-1} sec^{-1} for triplet methylene and methyl associations, respectively.

III–5.3. Time Dependent Radical Concentration by Consecutive Reactions

In this scheme the reactive species R_2 is formed by the pseudo-first-order reaction

$$\mathrm{R_1 + M_1 \rightarrow R_2 + P_1} \qquad k_{11} \tag{III-11}$$

and is eliminated by another pseudo-first-order reaction

$$\mathrm{R_2 + M_2 \rightarrow products} \qquad k_{12} \tag{III-12}$$

where R_1 and R_2 signify reactive species, M_1 and M_2 are reactants, and P_1 is a product.

During measurement the concentrations of M_1 and M_2 remain constant. The rate equations for R_1 and R_2 may be given as

$$-\frac{d[\mathrm{R_1}]}{dt} = k_{11}[\mathrm{M_1}][\mathrm{R_1}] \tag{III-13}$$

$$\frac{d[\mathrm{R_2}]}{dt} = k_{11}[\mathrm{M_1}][\mathrm{R_1}] - k_{12}[\mathrm{M_2}][\mathrm{R_2}] \tag{III-14}$$

where the brackets signify concentrations.

Integrating (III-13), we find

$$[\mathrm{R_1}] = [\mathrm{R_1}]_0 e^{-k_{11}[\mathrm{M_1}]t} \tag{III-15}$$

where $[\mathrm{R_1}]_0$ is the concentration of R_1 at $t = 0$. Substituting (III-15) into (III-14) and rearranging, we obtain

$$\frac{d[\mathrm{R_2}]}{dt} + k_{12}[\mathrm{M_2}][\mathrm{R_2}] = k_{11}[\mathrm{M_1}][\mathrm{R_1}]_0 e^{-k_{11}[\mathrm{M_1}]t} \tag{III-16}$$

Multiplying by the integrating factor, $e^{k_{12}[M_2]t}$, we have

$$e^{k_{12}[M_2]t} \cdot \frac{d[R_2]}{dt} + e^{k_{12}[M_2]t} \cdot k_{12}[M_2][R_2] = k_{11}[M_1][R_1]_0 e^{(k_{12}[M_2]-k_{11}[M_1])t}$$

$$d([R_2]e^{k_{12}[M_2]t}) = k_{11}[M_1][R_1]_0 e^{(k_{12}[M_2]-k_{11}[M_1])t}\,dt$$

$$[R_2]e^{k_{12}[M_2]t} = \frac{k_{11}[M_1][R_1]_0}{k_{12}[M_2]-k_{11}[M_1]}\left[e^{(k_{12}[M_2]-k_{11}[M_1])t}-1\right]$$

and for $k_{12}[M_2] > k_{11}[M_1]$

$$[R_2] = \frac{k_{11}[M_1][R_1]_0}{k_{12}[M_2]-k_{11}[M_1]}\left[e^{-k_{11}[M_1]t}-e^{-k_{12}[M_2]t}\right] \qquad \text{(III-17a)}$$

if $k_{11}[M_1] > k_{12}[M_2]$,

$$[R_2] = \frac{k_{11}[M_1][R_1]_0}{k_{11}[M_1]-k_{12}[M_2]}\left[e^{-k_{12}[M_2]t}-e^{-k_{11}[M_1]t}\right] \qquad \text{(III-17b)}$$

Equation (III-17) shows that the concentration–delay time profile is expressed as the difference between the slow exponential decay (the first term in the bracket) and the fast exponential decay (the second term in the bracket). The concentration of R_2 starts at zero and rises to a maximum at short delay times and falls exponentially at long delay times as shown in Fig. III–14, where (III-17a) is reduced to

$$[R_2] \cong \frac{k_{11}[M_1][R_1]_0}{k_{12}[M_2]-k_{11}[M_1]}\,e^{-k_{11}[M_1]t} \qquad \text{(III-18)}$$

The fast exponential decay curve may be obtained by subtraction of $[R_2]$ from the extrapolated portion of the long decay curve as shown in Fig. III–14.

Gilpin et al. (398) have measured the quenching rate constant of $O_2(^1\Sigma_g^+)$ by O_3 using this technique. The $O_2(^1\Sigma^+)$ molecules are produced by the flash photolysis of O_3 in the presence of O_2

$$O_3 \xrightarrow{h\nu} O(^1D) + O_2(^1\Delta)$$

$$O(^1D) + O_2 \rightarrow O_2(^1\Sigma_g^+) + O(^3P) \qquad k_{19} \qquad \text{(III-19)}$$

The $O_2(^1\Sigma_g^+)$ molecules formed are subsequently quenched by O_3

$$O_2(^1\Sigma_g^+) + O_3 \rightarrow 2O_2 + O(^3P) \qquad k_{20} \qquad \text{(III-20)}$$

Hence, time dependence of $[O_2(^1\Sigma_g^+)]$ may be given from (III-17b) where $M_1 = O_2$, $M_2 = O_3$, and $R_1 = O(^1D)$

$$[O_2(^1\Sigma_g^+)] = \frac{k_{19}[O_2][O(^1D)]_0}{k_{19}[O_2]-k_{20}[O_3]}\left[e^{-k_{20}[O_3]t}-e^{-k_{19}[O_2]t}\right]$$

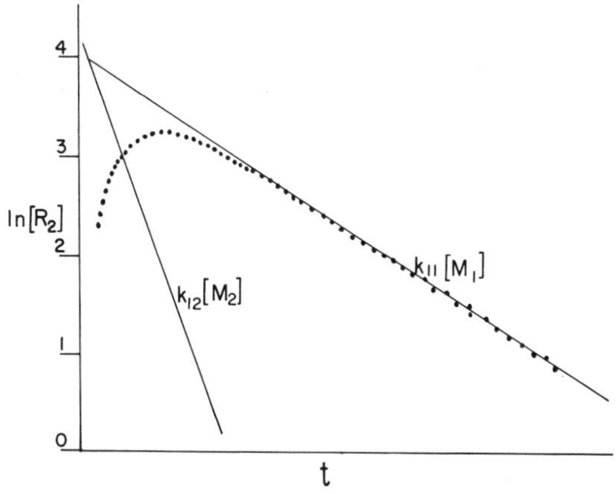

Fig. III–14. Time history of reactive species R_2 by the consecutive reactions

$$R_1 + M_1 \rightarrow R_2 + P_1 \qquad k_{11}$$
$$R_2 + M_2 \rightarrow \text{products} \qquad k_{12}$$

The time history of R_2 is given by

$$[R_2] = \frac{k_{11}[M_1][R_1]_0}{k_{12}[M_2] - k_{11}[M_1]}[e^{-k_{11}[M_1]t} - e^{-k_{12}[M_2]t}]$$

$$k_{12}[M_2] > k_{11}[M_1]$$

where M_1 and M_2 are stable molecules, P_1 is a product, and R_1, R_2 are reactive species. The concentration $[R_2]$ starts from zero, rises to a maximum, and decays exponentially with time. From the rising and falling portions of the curve, k_{12} and k_{11}, respectively, can be obtained. From J. R. McDonald et al. (674b), reprinted by permission.

The concentration of $[O_2(^1\Sigma_g^+)]$ was followed from the emission intensity at 7620 Å (see Section V–6.3). Gilpin et al have obtained $k_{20} = 2.5 \times 10^{-11}$ cm^3 molec^{-1} sec^{-1} from the falling portion of the curve.

III–6. DETERMINATION OF THE PRIMARY PHOTOCHEMICAL PROCESS BY RADICAL TRAPPING AGENTS

The quantitative determination of the primary photoproduct yield has been an extremely difficult problem in photochemistry when the products are atoms or radicals. Even if the atoms or radicals are found by optical absorption in flash photolysis, one must calibrate the observed absorption intensity by generating known amounts of atoms or radicals. Alternatively, the atom concentration may be found from the known oscillator strength by the procedure described in Section I–6.3 if both the atomic line and the light source have Doppler widths, the situation not found in flash photolysis conditions.

A more practical approach is to add small amounts of atom or radical trapping agents with which the reactive species interacts much more rapidly than with the reactant molecules. It is required that the trapping agent does not thermally react with the reactant or product molecules. Then the extent of primary atom or radical formation can be found from the amounts of reaction products or from the change of reaction products. Some trapping agents, such as O_2 and NO, form unstable products with atoms or radicals that may further decompose or react with each other or with other atoms and radicals. For example, in the H_2O–O_2 photolysis at 1470 Å (984) the HO_2 radicals formed by $H_2O \xrightarrow{h\nu} H + OH$, $H + O_2 \xrightarrow{M} HO_2$ would disappear by several reactions, such as $H + HO_2 \xrightarrow{M} H_2O_2$, $OH + HO_2 \rightarrow H_2O + O_2$, $HO_2 + HO_2 = H_2O_2 + O_2$. The products of triplet methylene reacting with O_2 are believed to be $CO_2 + H_2$ (or 2H) and $CO + H_2O$ (or H + OH) [Rowland et al. (844a)].

On the other hand, other trapping agents, such as HI and H_2S, react with H atoms and CH_3 radicals to form stable products, H_2 and CH_4, respectively. Hence, the yield of a particular stable product indicates the extent of the corresponding atom or radical formation process. In applying the radical trapping technique to understand the atom or radical formation process, caution must be exercised that the trapping agent does not quench the electronically excited state. If the lifetime of the excited state is on the order of microseconds, quenching is a distinct possibility. If the primary photochemical process is direct dissociation of the reactant into atoms or radicals, the trapping technique can be safely applied to obtain information on the primary process.

III–6.1. Examples

Several examples may be cited to illustrate how the trapping agents have aided in an understanding of the primary processes.

1. Oxygen has been used extensively to scavenge triplet methylene, because triplet methylene reacts much faster with O_2 than with the reactant (ketene) or with hydrocarbons (see Table VI–8). Thus, Zabransky and Carr (1078) were able to obtain the fraction of singlet methylene in the 3130 Å photolysis of ketene from the ratio of the photolytic products of CH_2CO–paraffin (RH) mixtures with and

without O_2. From the mechanism

$$CH_2(^1A_1) + CH_2CO = C_2H_4 + CO$$
$$CH_2(^1A_1) + RH = RCH_3$$
$$CH_2(^3B_1) + RH = CH_3 + R$$
$$CH_3 + R = RCH_3$$
$$CH_3 + R = CH_4 + \text{olefin}$$
$$CH_3 + RH = CH_4 + R$$
$$2CH_3 = C_2H_6$$

one obtains the fraction of singlet methylene produced in the primary process from the ratio of $(C_2H_4 + RCH_3)$ in the presence of O_2 to $(C_2H_4 + CH_4 + 2C_2H_6 + RCH_3)$ without O_2. The ratio is 0.70. It was assumed that 5% O_2 does not interact with the excited state of ketene.

2. Hydrogen atoms and methyl radicals react efficiently with NO, O_2, HI, and so forth, as shown in Table III–3. Hence, small amounts of these agents can be added to the photochemical system which generates hydrogen atoms or methyl radicals. O_2 and NO are not good trapping agents at low total pressures, particularly for H atoms, and high total pressures are required for efficient trapping while other agents can be used at low total pressures. The photolysis of acetone in the near ultraviolet gives rise to methyl radicals, form methane in the presence of HI or HBr. Hence, the methane quantum yield should be indicative of the primary yield of CH_3. Larson and O'Neal (611a) have found that the CH_4 quantum yield in the acetone–HBr photolysis at 3150 Å decreases rapidly to a constant value with an increase of HBr added and that isopropyl alcohol is concomitantly produced. The results indicate that methyl radicals arise both from direct dissociation and from a long-lived (triplet) acetone and that HBr reacts with the triplet acetone to give isopropyl alcohol

$$HBr + \text{acetone (triplet)} \rightarrow CH_3C(OH)CH_3 + Br$$
$$CH_3C(OH)CH_3 + HBr \rightarrow \text{isopropyl alcohol} + Br$$

The triplet acetone must have a long life (~ 1 μsec) to give the reaction product. Hence, only when the dissociation is instantaneous can we safely obtain the primary yield of methyl production by this technique.

Ausloos et al. (49a) have used HI and H_2S as H atom and alkyl radical scavengers in the vacuum ultraviolet photolysis of propane C_3D_8. They found HI is much more efficient than H_2S as a scavenger as is expected from Table III–3. The yields of HD, CD_3H, and C_2D_5H represent the primary production yields of D, CD_3, and C_2D_5, respectively.

Table III–3. Rate Constants of Reactions of Some Radical Trapping Agents with H and CH_3 at Room Temperature

Reaction	Rate Constant (cm^3 $molec^{-1}$ sec^{-1})
$H + O_2 \xrightarrow{M} HO_2$ (N_2 = 1 atm)	1.5×10^{-12} [a]
$H + NO \xrightarrow{M} HNO$ (Ar = 1 atm)	5.4×10^{-13} [a]
$H + HI \rightarrow H_2 + I$	0.26×10^{-10} [b]
$H + HBr \rightarrow H_2 + Br$	2.7×10^{-12} [b]
$H + I_2 \rightarrow HI + I$	0.7×10^{-9} [b]
$H + H_2S \rightarrow H_2 + HS$	8.4×10^{-13} [a]
$H + C_2H_4 \rightarrow C_2H_5$	2.6×10^{-12} [a]
$H + C_3H_6 \rightarrow C_3H_7$	1.6×10^{-12} [a]
$CH_3 + O_2 \xrightarrow{M} CH_3O_2$ (N_2 = 1 atm)	4×10^{-13} [a]
$CH_3 + NO \xrightarrow{M} CH_3NO$ (He = 1 atm)	7.5×10^{-12} [a]
$CH_3 + HI \rightarrow CH_4 + I$	1×10^{-13} [b]
$CH_3 + HBr \rightarrow CH_4 + Br$	1.5×10^{-14} [b]
$CH_3 + I_2 \rightarrow CH_3I + I$	1.4×10^{-12} [b]
$CH_3 + H_2S \rightarrow CH_4 + HS$	5.4×10^{-15} [b]

[a] From Ref. 1
[b] From Ref. 28a

McNesby et al. (683) have used C_2D_4 as a scavenger of H atoms produced in the photolysis of H_2O and NH_3 in the vacuum ultraviolet. Since H atoms react extremely slowly with H_2O or NH_3 ($<10^{-16}$ cm^3 $molec^{-1}$ sec^{-1}) (ref. 1) they are effectively eliminated by added C_2D_4. Thus, the ratio of H_2 with and without added C_2D_4 shows the extent of molecular hydrogen production in the primary process.

Ung (984) has used O_2 as an H atom scavenger in the 1470 Å photolysis of H_2O. The ratio of H_2 with O_2 present to the overall H_2 yield without O_2 is 23%. From this information Ung estimates that the primary yield of H_2 production is 8%.

chapter IV

Production and Quenching of Electronically Excited Atoms

Electronically excited atoms have been known to play important roles in many photochemical reactions.

In this chapter electronically excited atoms are classified into two groups. The first group of excited atoms are those that are formed by resonance absorption and decay rapidly by fluorescence if not quenched by collisions with foreign gases. Examples are electronically excited Hg, Cd, H, Ar, Kr, and Xe atoms. Of these $Hg(^3P_1)$ atoms and their reactions have been most extensively studied. The mercury sensitized reactions provide a convenient way to generate atoms and radicals in the spectral region where many molecules do not absorb.

The second group of electronically excited atoms consists of metastable atoms such as $O(^1D)$ and $O(^1S)$. These metastable atoms cannot be produced directly from ground state atoms by light absorption. However, they are often formed in photodissociation of molecules. The production of metastable atoms from photodissociation has been known by the different reactivities of the metastables from those of corresponding ground state atoms.

From the spin conservation rules (see Section II–3.1) it is often reasonable to assume the production of metastable atoms in the primary photochemical process, although the direct detection of metastable atoms has succeeded only recently by optical absorption or emission following flash photolysis of molecules. Detection is difficult, since metastable atoms usually react rapidly with the reactant molecules and often the detection of the atoms has to be made in the vacuum ultraviolet. Because of their long radiative lives, the main fate of metastable atoms is physical and chemical quenching by gases present in the system. An excellent review on the reactions of metastable atoms is given by Donovan and Husain (310).

The electronic states and the lifetimes of atoms of photochemical interest are given in the Appendix, Table A–2.

IV–1. FLUORESCENCE QUENCHING IN ATOMS; QUENCHING CROSS SECTIONS

Electronically excited atoms produced by resonance absorption have a short life, on the order of 1 to 100 nsec, after which they return to the ground state. The resulting fluorescence is called resonance fluorescence. In the presence of a foreign gas the intensity of the resonance fluorescence is reduced (quenched). The quenching of the excited atoms, denoted by A*, may result in chemical reactions (chemical quenching) or the quenching may produce metastable or ground state atoms and translationally or internally excited foreign gas molecules (physical quenching).

The sequence of events may be written

$$\mathrm{A} \xrightarrow{h\nu} \mathrm{A}^* \qquad I_a \qquad \text{(IV-1)}$$

$$\mathrm{A}^* \rightarrow \mathrm{A} + h\nu \qquad k_f \qquad \text{(IV-2)}$$

$$\mathrm{A}^* + \mathrm{M} \rightarrow \mathrm{A} + \mathrm{M}' \text{ or } \mathrm{A} + \mathrm{D}_1 + \mathrm{D}_2 \qquad k_\mathrm{M} \qquad \text{(IV-3)}$$

where M signifies a foreign gas, M′ is a foreign gas with excess energy, and D_1 and D_2 are dissociation products.

In steady state conditions we have

$$I_a = k_f(\mathrm{A}^*) + k_\mathrm{M}(\mathrm{A}^*)(\mathrm{M}) \qquad \text{(IV-4)}$$

where I_a is the number of photons absorbed by A per cm^3 sec^{-1}, k_f is the natural decay constant of fluorescence in sec^{-1} and k_M is the quenching rate constant (cm^3 $molec^{-1}$ sec^{-1}) by M. The quantity in parenthesis indicates concentrations. If concentrations of A atoms are large, the primary radiation is absorbed by A forming A*, and A* emits the secondary radiation, which is absorbed again by A atoms and the process is repeated many times. This repeated process of emission and absorption is called radiation imprisonment or radiation trapping. In this case the fluorescence lifetime becomes much longer than the natural lifetime $1/k_f$ and must be replaced by $1/gk_f$ ($g \ll 1$). Blickensderfer et al. (119) have calculated that in $\mathrm{Hg}(^3P_1)$ atoms, imprisonment lifetimes are four times as large as the natural lifetime when the optical thickness is 2 that is, $g = 0.25$. Only when the optical thickness is below 0.1, corresponding to a Hg concentration of 10^{12} cm^{-3} at a path length of 1 cm, we obtain the natural lifetime ($g = 1$). Phillips (807) has recently calculated the ratio of trapping time to natural lifetime for N (1200 Å), H (1216 Å), O (1306 Å). Hg (1849 Å), Cd (2288 Å), Hg (2537 Å), Cd (3261 Å), and Na (5890 Å) resonance lines in the atom concentration range 10^{10} to 10^{17} cm^{-3}. At a concentration of 3×10^{14} atoms cm^{-3} of Hg, the lifetime of $\mathrm{Hg}(^3P_1)$ is about 20 μsec while the natural lifetime is only 0.11 μsec. From $I_f = k_f(\mathrm{A}^*)$, where I_f is the fluorescence intensity (quanta cm^{-3} sec^{-1}), we obtain

$$I_f = \frac{I_a}{1 + (k_\mathrm{M}/k_f)(\mathrm{M})} \qquad \text{(IV-5a)}$$

If I_f^0 and I_f are the fluorescence intensities with and without the foreign gas, respectively, we find

$$\frac{I_f^0}{I_f} = \frac{1}{Q} = 1 + \frac{k_{\mathrm{M}}}{k_f}(\mathrm{M}) \tag{IV-5b}$$

where Q is called quenching. We obtain the linear relationship between $1/Q$ and (M). This relationship is called the Stern-Volmer formula first derived by Stern and Volmer (924) in 1919 for quenching of I_2 vapor. If (A) is large, k_f must be replaced by gk_f ($g \ll 1$)

$$\frac{1}{Q} = 1 + \frac{k_{\mathrm{M}}}{gk_f}(\mathrm{M}) \tag{IV-5c}$$

Hence g must be known (either theoretically or experimentally) to obtain k_{M} from the slope. The number of quenching collisions per excited atom by the foreign gas M per second, Z, can be expressed by the quenching cross section σ^2 (cm^2 molec^{-1}) in analogy with the gas kinetic collision cross section (Note that σ is used before to designate the absorption cross section; see Section I–8.1.)

$$Z = k_{\mathrm{M}}(\mathrm{M}) = \sigma^2 \sqrt{\frac{8\pi RT}{\mu}}(\mathrm{M}) \tag{IV-6}$$

where R is the gas constant equal to 8.3143×10^7 erg K^{-1} mol^{-1}, and μ, the reduced mass, is defined by

$$\frac{1}{\mu} = \frac{1}{M_{\mathrm{A}}} + \frac{1}{M_{\mathrm{Q}}} \tag{IV-7}$$

M_{A} and M_{Q} are masses of the atom A and the quenching gas, respectively. Some workers define the quenching cross section as

$$Z = \sigma_{\mathrm{Q}}^2 \sqrt{\frac{8RT}{\mu\pi}}(\mathrm{M}) \tag{IV-8}$$

where $\sqrt{(8RT/\mu\pi)}$ is the average relative velocity. Hence,

$$\sigma_{\mathrm{Q}}^2 = \pi\sigma^2 \tag{IV-9}$$

At $T = 300°$K, Z, the number of quenching collisions per second for each excited atom by M, is given by

$$Z = \sigma^2 \frac{7.92 \times 10^5}{\sqrt{\mu}}(\mathrm{M}) \tag{IV-10}$$

where (M) is the number of quenching molecules per cm^3.

It has been shown [Mitchell and Zemansky (21), p. 197] that the quenching of the Hg 2537 Å line by foreign gases follows the Stern-Volmer formula only when (1) Lorentz broadening [see Section I–6.1] by foreign gases is absent, and (2) the fluorescence consists of the primary radiation, that is, no absorption and reemission of the primary radiation takes place. The second condition is fulfilled only when the optical thickness $\alpha_0 l$ is near zero. Zemansky performed his experiment under the conditions that reemission of the primary radiation cannot be neglected and Lorentz broadening is absent and he obtained quenching cross sections for various gases by properly correcting the secondary emission. According to Mitchell and Zemansky, the quenching Q is a complicated function of both τZ (τ is the lifetime of the excited atom) and the absorption coefficient α at a given wave number $\bar{\nu}$. Hence, to obtain the quenching cross section from the observed quenching Q, it is necessary to know the absorption coefficient. Since the width of the primary radiation is much broader than the Doppler width of the absorbing gas it is believed that the absorption coefficient changes during the repeated absorption and emission processes, although it is difficult to estimate such changes. Samson [quoted in (21), p. 200], on the other hand, has assumed that at low pressures of the absorbing gas the transmission of light at a given wave number may be expressed as

$$\exp(-\bar{\alpha}l) = \frac{\int_{-\infty}^{\infty} \exp \bar{\omega}^2 \exp[-\alpha_0 l \exp(-\bar{\omega}^2)]\, d\bar{\omega}}{\int_{-\infty}^{\infty} \exp(-\bar{\omega}^2)\, d\bar{\omega}} \qquad \text{(IV-10a)}$$

where $\bar{\omega}$ is $2(\bar{\nu} - \bar{\nu}_0)/\Delta\bar{\nu}_D\sqrt{\ln 2}$ [see (I-60)]. The quantity $\bar{\alpha}$ is the average absorption coefficient and $\bar{\alpha}l$ is called the equivalent opacity. It is now possible to construct theoretical curves of Q as a function of τZ for various $\bar{\alpha}l$ values. Once Q is determined experimentally as a function of pressure p, it is possible to obtain σZ corresponding to a given pressure by comparing the calculated $Q - \sigma Z$ curve with the experimental $Q - p$ curve. From the linear slope of a plot of τZ against various pressures of the quenching gas one can obtain the quenching cross section. The value $\bar{\alpha}l/\alpha_0 l$ is 0.665 for $\alpha_0 l = 1$ and it decreases as $\alpha_0 l$ increases.

Blickensderfer et al. (119) have recently extended the Samson theory to include various vessel geometries under conditions of pure Doppler, pure Lorentz, and Voigt (a combination of Doppler, Lorentz, and natural broadening) line profiles. The quenching cross section can be obtained also by a dynamic method in which the decay of the fluorescence intensity is measured after the exciting light is cut off. If the initial fluorescence intensity is I_f^0 and the intensity after a time t is I_f, we obtain

$$I_f = I_f^0 e^{-[k_f + k_M(M)]t} \qquad \text{(IV-11a)}$$

where k_f is the natural decay rate in $\sec^{-1}$ without the quenching gas and k_M is the rate constant of fluorescence quenching by M. In the case of radiation trapping

$$I_f = I_f^0 \exp\{-[gk_f + k_M(M)]t\} \qquad g \ll 1 \qquad \text{(IV-11b)}$$

From the slope of a plot $\ln(I_f^0/I_f)$ versus decay time t without M we obtain gk_f and with M we obtain $gk_f + k_M(M)$.

Using the decay method Matland (669) has obtained for the quenching cross section of $Hg(^3P_1)$ by N_2 a value of $\sigma^2 = 0.16 \times 10^{-16}$ cm^2, which is in good agreement with $\sigma^2 = 0.19 \times 10^{-16}$ cm^2 obtained by Zemansky (1083) in a static system. Quenching cross sections for $Hg(^3P_1)$ atoms have been measured for a number of gases by Zemansky (21), Michael et al. (402, 702, 703), and Horiguchi and Tsuchiya (481). See reviews by Cvetanović (256) and Calvert and Pitts (4). Some values of the quenching cross section σ^2 are given in Table IV–1. Since it is difficult to achieve sufficiently low concentrations of mercury to avoid radiation imprisonment,

Table IV–1. Quenching Studies of $Hg(^3P_1)$ by Various Simple Molecules

Quenching Molecule	Major Primary Product[a]	Quenching Cross Section $\sigma^2 \times 10^{-16}$ cm^2	Ref.
N_2	$^3P_0(\Phi_{Hg^0} > 0.9)$	0.36	176, 481
		0.274	21
H_2	$\Phi_{HgH} = 0.80$, $\Phi_{Hg_0} < 0.03$	8.6, 9.8	21, 176, 702
		10.8, 11.1	177, 703
CO	$\Phi_{Hg+CO} = 0.2$		
	$\Phi_{Hg^0} = 0.74$, $CO^\dagger(v'' = 1 - 10)$	2.7, 7.4	173, 180, 408, 481, 557, 703
NO	$\Phi_{Hg^0} = 0.20$, $\Phi_{Hg+NO} = 0.61$	28.3, 33	481, 703
HCl	HgCl + H		176
O_2		12,23.9	703
H_2S	Hg + H + SH	33	176, 256
CS_2	$Hg + CS_2^*$	50	176, 256
N_2O	$Hg + N_2 + O$	21.2	176, 703
H_2O	$\Phi_{Hg^0} = 0.38, 0.5$	1.43	21, 173, 180
	$Hg(^3P_0) \cdot H_2O^{*b}$		373
NH_3	$\Phi_{Hg^0} = 0.62, 0.64$	4.2	21, 173, 180
	$Hg(^3P_0) \cdot NH_3^{*b}$		736
	$Hg + NH_2 + H(\Phi \sim 0.3)$		736
CH_4, C_2H_6, C_3H_8	$\Phi_{Hg^0} = 0.1 \sim 0.8$, no HgH		736, 180
	$Hg + R + H$, $Hg(^3P_0) \cdot RH^{*b}$		173, 176, 177
CH_3Cl	$HgCl + CH_3$	34	176, 256
C_2H_2	$\Phi_{Hg^0} < 0.01$, $\Phi_{HgH} = 0.18$	33	176, 177
			256

[a] Φ_{Hg^0} signifies the quantum yield for $Hg(^3P_0)$ formation.
[b] Loose complexes of $Hg(^3P_0)$ with $M(H_2O, NH_3, RH)$ which emit continua.

quenching cross sections have been obtained in most cases under the condition that radiation imprisonment cannot be neglected and corrections are applied accordingly. The values of quenching cross sections depend on the correction factors and are probably accurate to within 50%.

The various processes associated with quenching of $Hg(^3P_1)$ and other excited atoms have been studied extensively and are discussed in the following sections.

IV–2. MERCURY SENSITIZED REACTIONS

Mercury sensitized reactions are predominantly those involving $Hg(^3P_1)$ atoms. Few studies have been made of $Hg(^1P_1)$ atom sensitized reactions. The $Hg(^3P_1)$ state lies 4.886 eV above the ground state. The lifetime is 0.114 μsec. The $Hg(^3P_1)$ atoms are produced from absorption of 2537 Å light by ground state atoms. The reaction of $Hg(^3P_1)$ atoms has been studied extensively since the pioneering work of Cario and Franck (188) in 1922 in which they showed that reaction products of $Hg(^3P_1)$ atoms with H_2 are capable of reacting with metallic oxides, indicating the production of atomic hydrogen.

Apart from its intrinsic interest, $Hg(^3P_1)$ atom sensitized reactions have been used extensively to generate atoms and radicals and their reactions with various molecules have been studied. Detailed quenching processes of $Hg(^3P_1)$ atoms have been of interest to many workers in the past 50 years.

The $Hg(^3P_1)$ sensitized reactions may be represented by the following types of reactions, although the extent of each reaction path is still unknown for many molecules.

1. Direct dissociation:

$$Hg(^3P_1) + RH \rightarrow Hg(^1S_0) + R + H \qquad \text{(IV-12)}$$

2. Formation of mercury hyride:

$$Hg(^3P_1) + RH \rightarrow HgH + R \qquad \text{(IV-13)}$$

3. Spin–orbit relaxation:

$$Hg(^3P_1) + M \rightarrow Hg(^3P_0) + M \qquad \text{(IV-14)}$$

4. Electronic to vibrational energy transfer:

$$Hg(^3P_1) + M \rightarrow Hg(^1S_0) + M^{\dagger} \qquad \text{(IV-15)}$$

   where $M^{\dagger}$ signifies vibrationally excited molecules.

5. Formation of electronically excited molecules, such as

$$Hg(^3P_1) + CS_2 \rightarrow Hg(^1S_0) + CS_2^* \qquad \text{(IV-16)}$$

   where CS_2^* means electronically excited CS_2.

The extent of processes (IV-12) to (IV-16) depends on individual molecules. The quantum yields for these processes have been estimated for some molecules. Some examples of quenching processes are given below.

IV–2.1. $Hg(^3P_1) + H_2$

The production of atomic hydrogen by the $Hg(^3P_1)$ sensitized reaction has been known for many years and the reaction has been used as a clean source of H atoms. Two processes have been postulated without direct evidence of the primary products

$$Hg(^3P_1) + H_2 \rightarrow Hg(^1S_0) + 2H \qquad \text{(IV-17)}$$

$$Hg(^3P_1) + H_2 \rightarrow HgH + H \qquad \text{(IV-18)}$$

Recently the production of HgH has been shown directly by optical absorption in the $Hg(^3P_1)$ sensitized flash photolysis of H_2 by Callear and his coworkers (176, 177, 180). The quantum yield (177) of HgH formation is estimated to be 0.80 ± 0.1. Other minor processes are also possible

$$Hg(^3P_1) + H_2 \rightarrow Hg(^3P_0) + H_2 \qquad \text{(IV-19)}$$

$$Hg(^3P_1) + H_2 \rightarrow Hg(^1S_0) + H_2^{\dagger} \qquad \text{(IV-20)}$$

where $H_2^{\dagger}$ signifies vibrationally excited H_2.

The quantum yield of $(^3P_0)$ production (IV-19) is less than 0.03 (180). Therefore, the most important process must be the formation of HgH. This conclusion is in disagreement with that of Yang et al. (1063), who have concluded from phase space theory that the production of vibrationally excited H_2 is most important.

IV–2.2. $Hg(^3P_1) + N_2$; $Hg(^3P_1) + CO$

The quenching cross sections for N_2, CO, and NO for processes (IV-14) and (IV-15) have been obtained by Horiguchi and Tsuchiya (481) by measuring the steady state concentrations of $Hg(^3P_1)$ and $(^3P_0)$ atoms by optical absorption at 4358 and 4047 Å, respectively in the presence of quenching gases.

The reaction of $Hg(^3P_1)$ with N_2 leads predominantly (481) ($>90\%$) to the formation of $Hg(^3P_0)$ with a quenching cross section of 0.36×10^{-16} cm^2, while collisions with CO (481) produce $Hg(^3P_0)$ and $(^1S_0)$ in a ratio of 3.5:1. Since $Hg(^3P_0)$ atoms are predominantly produced from $(^3P_1)$ in the presence of N_2, the reaction of $Hg(^3P_0)$ with foreign gases M can be studied in Hg–N_2–M mixtures.

The quenching rates for $Hg(^3P_1)$ atoms are in general much faster than those for $(^3P_0)$ atoms (180). Vibrationally excited CO up to $v'' = 9$ has been observed in infrared emission of CO (408, 557).

IV–2.3. $Hg(^3P_1) + H_2O$; $Hg(^3P_1) + NH_3$

More than 50% production of $Hg(^3P_0)$ was observed in the reaction of $Hg(^3P_1)$ atoms with H_2O and NH_3 (180). Charge-transfer complexes between the molecules and $Hg(^3P_0)$ are assumed responsible for continuous emissions near 2900 and 3500 Å for H_2O and NH_3, respectively [Freeman et al. (373, 736)].

IV–2.4. $Hg(^3P_1)$ + Paraffins

The main reactions of $Hg(^3P_1)$ atoms with saturated hydrocarbons are the production of H atoms and alkyl radicals (R) (256):

$$Hg(^3P_1) + RH \rightarrow Hg(^1S_0) + R + H \tag{IV-21}$$

Substantial production of $Hg(^3P_0)$ atoms has been seen by Callear and McGurk (180). Continuous emissions that presumably originate from loose complexes between the $Hg(^3P_0)$ atoms and the hydrocarbons have been observed near 2537 Å (938). No HgH has been detected from the $(^3P_1)$ atom reactions with saturated hydrocarbons (177).

IV–2.5. $Hg(^3P_1)$ + Olefins

The primary products of the $Hg(^3P_1)$ atom sensitized reaction with olefins are electronically excited molecules. For ethylene the process is the production of electronically excited ethylene, $C_2H_4^*$, probably in the triplet state a^3B_{1u} at about 3.56 eV (16).

$$Hg(^3P_1) + C_2H_4 \rightarrow Hg(^1S_0) + C_2H_4^* \tag{IV-22}$$

The excited C_2H_4 further decomposes or is quenched by C_2H_4. Quenching cross sections of $Hg(^3P_1)$ atoms by olefins are usually much larger than those by paraffins (256). Primary products and yields for some quenching reactions are given in Table IV–1.

IV–2.6. $Hg(^1P_1)$ Sensitized Reactions

The $Hg(^1P_1)$ state lies 6.703 eV above the ground state and has a lifetime of 1.31 nsec. The production of $Hg(^1P_1)$ atoms is achieved by absorption of 1849 Å light by ground state atoms. Only a few data are available on the reaction of $Hg(^1P_1)$ atoms and no reliable values have been obtained for quenching cross sections. Relative cross sections for 12 gases were obtained from Stern-Volmer plots by Granzow et al. (418). Quenching cross sections for $Hg(^1P_1)$ atoms are not as widely different from one gas to another as those for $Hg(^3P_1)$ atoms. Emission at 2537 Å was observed when mixtures of Hg and N_2 were irradiated with the 1849 Å line [Gover and

Bryant (414), Granzow et al. (417)]. The formation of $N_2(A^3\Sigma)$ as an intermediate was postulated (417) as

$$Hg(^1P_1) + N_2 \rightarrow Hg(^1S) + N_2(A^3\Sigma)$$
$$N_2(A^3\Sigma) + Hg(^1S_0) \rightarrow N_2 + Hg(^3P_1)$$
$$Hg(^3P_1) \rightarrow Hg(^1S_0) + h\nu \text{ (2537 Å)}$$

IV–3. OTHER ATOM SENSITIZED REACTIONS

IV–3.1. $Cd(^3P_1, {}^1P_1)$ Sensitized Reactions

The $Cd(^3P_1)$ and $(^1P_1)$ states are 3.80 and 5.417 eV above the ground state, respectively. The lifetimes are 2.5 μsec and 1.98 nsec for $(^3P_1)$ and $(^1P_1)$ atoms (21), respectively.

The $Cd(^3P_1)$ and $(^1P_1)$ states are produced from the ground state by absorption of the 3261 and 2288 Å lines, respectively.

The reactions of $Cd(^3P_1)$ and $(^1P_1)$ atoms with H_2, N_2, CH_4, and NH_3 have been studied by Breckenridge and Callear (145) and Morten et al. (719, 720).

Band emission due to a $Cd(^3P_0)$–NH_3 complex has been observed by Morten et al. (719) and similar bands with water, alcohols, and ethers have been observed by Yamamoto et al. (1059).

IV–3.2. $H(^2P)$ Sensitized Reactions

The $H(^2P)$ state is energetic by 10.2 eV with respect to the ground state and it has a lifetime of 1.60 nsec (32). The $H(^2P)$ state can be produced by irradiating flowing ground state H atoms with the 1216 Å resonance line, where the ground state atoms are made in a microwave discharge.

The quenching of $H(^2P)$ atoms by small molecules has been studied extensively by Phillips et al. (1021–1023) and by Tanaka et al. (583, 960). The quenching rate constant ranges from 10^{-8} [Shukla et al. (874)] to 10^{-12} [Wauchop et al. (1023)] cm^3 $molec^{-1}$ sec^{-1} both for N_2 and O_2. The discrepancy is partially due to resonance trapping of the 1216 Å line in the optically thick system. In the optically thin system the value is about 10^{-9} cm^3 $molec^{-1}$ sec^{-1} [Braun et al. (142)]. The reaction of $H(^2P)$ atoms with H_2 appears to produce three ground state H atoms [Van Volkenburgh et al. (992)]. The reactions of $H(^2P)$ atoms with O_2 and N_2 yielded electronically excited OH and NH [Wauchop and Phillips (1022)], respectively.

IV–3.3. $Na(^2P)$ Sensitized Reactions

The $Na(^2P)$ state is 2.10 eV above the ground state (^2S). The lifetime of the (^2P) state is 15.9 nsec. The $Na(^2P_{3/2})$ and $(^2P_{1/2})$ states emit 5890 and 5896 Å resonance lines, respectively. The Na(^2P) states can be generated from ground state atoms by absorption of the resonance lines or by photodissociation ($\lambda = 1900$–2500 Å) of NaI (332, 441). The quenching of Na(^2P) atoms by N_2 yields vibrationally excited molecules with a rate more than the gas kinetic rate (359). The quenching of Na(^2P) atoms by H_2 also gives rise to vibrationally excited H_2 [Lee et al. (623)]. The quenching cross sections are dependent both on the translational energy of Na(^2P) atoms and temperature. It decreases at higher temperatures (632) and at larger translational energies. Quenching cross sections by inert gases are in general much smaller than those by diatomic molecules (359). Quenching of Na(^2P) by CO and O_2 has also been studied (359, 633).

IV–3.4. Ar($^3P_1$, $^1P_1$) Sensitized Reactions

The Ar($^3P_1$, $^1P_1$) levels are 11.623 and 11.827 eV, respectively, above the ground (^1S) level. The lifetimes are 8.4 and 2.0 nsec (33), respectively. The Ar($^3P_1$, $^1P_1$) states are formed by absorption of the Ar resonance lines at 1067 and 1048 Å. In the 1 to 100 mtorr concentration range the lifetime of Ar($^3P_1$, $^1P_1$) atoms is of the order of 10 μsec [Hurst et al. (494)], which is 1000 times as long as that of isolated atoms because of imprisonment of resonance radiation. If the ionization potential of a molecule is below 11.6 eV, it is possible to increase the photoionization yield (sensitize) by adding Ar to the sample. The increase of the ionization yield is caused by collisional energy transfer between Ar($^3P_1$, $^1P_1$) atoms and the molecule before the excited atoms return to the ground state by resonance emission. Yoshida and Tanaka (1065) have found such an increase in the Ar–propane, and Ar–ammonia mixtures when they are excited by an Ar resonance lamp. Boxall et al. (123) have measured quenching rate constants for Ar($^3P_1$) atoms by N_2, O_2, NO, CO, and H_2. They are on the order of the gas kinetic collision rate.

IV–3.5. Kr($^3P_1$, $^1P_1$) Sensitized Reactions

The Kr($^3P_1$) and Kr($^1P_1$) levels lie 10.032 and 10.643 eV, respectively, above the ground level. The lifetimes are 3.7 and 3.2 nsec, respectively (424). The Kr($^3P_1$) and Kr($^1P_1$) states are produced by absorption of the resonance lines at 1236 and 1165 Å, respectively. Sensitized reactions of N_2 and CO by Kr($^3P_1$, $^1P_1$) atoms have been studied by Groth et al. (427).

IV–3.6. Xe($^3P_1$) Sensitized Reactions

The Xe($^3P_1$) state is 8.436 eV above the ground (^1S) state and has a lifetime of 3.7 nsec [Wilkinson (1043)]. The Xe($^3P_1$) state is produced by absorption of the 1470 Å resonance line. The reaction of $Xe(^3P_1)$ with CO yields various triplet states of $CO(d^3\Delta, e^3\Sigma, a'^3\Sigma)$, as well as the $CO(A^1\Pi)$ state, and emission bands from these states have been observed by Slanger and Black (898, 899). Of the total singlet ($^1\Pi$) plus triplet emissions $60 \pm 15\%$ is from the $A^1\Pi$ state. On the other hand, direct excitation of CO by the 1470 Å line produces only the $CO(d^3\Delta, v = 7)$ state (898). Thus, Xe sensitized reaction produces various triplet states indiscriminately. The initially formed collision complex XeCO* has been found to have a lifetime of 25 ± 6 μsec before it dissociates to give CO* [Freeman and Phillips (374)]. The quenching of $Xe(^3P_1)$ atoms by H_2 has been found by Shimokoshi to produce H atoms (873). VonBunau and Schindler (997) have shown that $Xe(^3P_1)$, as well as $Kr(^3P_1)$, atoms sensitize an exchange between H_2 and D_2. The photolysis of methane (881), ethane, and propane (996) sensitized by $Xe(^3P_1)$ and $Kr(^3P_1)$ atoms has been studied. The detailed mechanism of photosensitization is complicated since $Xe(^3P_2)$ or Xe_2^* (electronically excited molecule) may be formed at higher Xe pressures. Direct photolysis of molecules may partially occur by the imprisoned resonance lines.

IV-4. REACTIONS OF METASTABLE O ATOMS

IV-4.1. $O(^1D)$ Atoms

The electronic energy of $O(^1D)$ is 1.967 eV above the ground state. The transition to the ground state $O(^3P)$ is forbidden by electric dipole and spin. The lifetime is 150 sec (32). The $O(^1D)$ production can be detected either by the emission (398) at 6300 Å $[O(^1D) \rightarrow O(^3P)]$ or by the absorption at 1152 Å $[2p^3\ 3s'\ (^1D^0) \leftarrow 2p^4\ (^1D)]$ (456) [Heidner et al. (455), Heidner and Husain (457, 458)].

Photochemical Production of $O(^1D)$ *Atoms.* The production of $O(^1D)$ atoms has been observed in the photolysis of O_2, NO_2, O_3, CO_2, and N_2O

$$O_2 \xrightarrow{h\nu} O(^3P) + O(^1D) \qquad <1750\ \text{Å}$$
$$NO_2 \xrightarrow{h\nu} NO + O(^1D) \qquad <2440\ \text{Å}$$
$$O_3 \xrightarrow{h\nu} O_2(^1\Delta) + O(^1D) \qquad <3100\ \text{Å}$$
$$CO_2 \xrightarrow{h\nu} CO + O(^1D) \qquad <1670\ \text{Å}$$
$$N_2O \xrightarrow{h\nu} N_2 + O(^1D) \qquad <3400\ \text{Å}$$

Physical and Chemical Quenching of $O(^1D)$ *Atoms.* Although the direct detection of $O(^1D)$ has been made only recently, it has been known that O atoms produced from the photolysis of N_2O at 1849 Å react rapidly with N_2O to form N_2 and NO, while $O(^3P)$ atoms do not react with N_2O (794, 864). Furthermore, it has been known that O atoms produced from the photolysis of N_2O at 1849 Å, NO_2 at 2288 Å, and CO_2 at 1470 Å exchange with O atoms in CO_2 (73, 815, 1060), indicating the formation of an intermediate complex CO_3, which dissociates into $CO_2 + O(^3P)$ in 10^{-11} to 10^{-12} sec (227). An exchange of O atoms with those in CO_2 does not take place for the ground state $O(^3P)$ atoms.

The production of $O(^1D)$ atoms is also evidenced by the formation of neopentanol in the photolysis of N_2O–neopentane mixtures at 2138 Å. On the other hand, $O(^3P)$ atoms react with 1-butene to yield 1,2–butene oxide [Paraskevopoulos and Cvetanović (791)].

The reactive O atoms produced from N_2O at 1849 Å can either be $O(^1S)$ or $O(^1D)$. However, O atoms produced from the photolysis of NO_2 at 2288 Å, where the production of only $O(^1D)$ is energetically possible, show the same chemical reactivity as those from N_2O at 1849 Å (815, 1060). The results indicate that reactive O atoms must be $O(^1D)$.

Quenching rates of $O(^1D)$ with Xe, O_2, O_3, CO, CO_2, N_2, NO, N_2O, NO_2, H_2, H_2O, and CH_4 have been measured (10, 262). Rate constants are on the order of 10^{-10} cm^3 molec^{-1} sec^{-1}, that is, the reactions proceed with almost unit collision efficiency. Noxon (745) and Clark and Noxon (216) have measured the absolute emission intensity at 6300 Å in the steady state photolysis of O_2 and CO_2 at 1470 Å. From the rate of production of $O(^1D)$ Noxon was able to obtain a rate constant of 6×10^{-11} cm^3 molec^{-1} sec^{-1} for $O(^1D)$ quenching by O_2. Gilpin et al. (398) were the first to detect $O(^1D)$ atoms directly and measure their decay following the flash photolysis of O_3 in the Hartley band. The 6300 Å emission disappeared within 200 μsec after the flash at an O_3 pressure of 3 mtorr. From the exponential decay curve integrated over 600 shots, the authors have obtained a rate constant of $2.5 \pm 1 \times 10^{-10}$ cm^3 molec^{-1} sec^{-1} for the reaction $O(^1D) + O_3$. Rate constants of $O(^1D)$ atoms with various atmospheric gases have been obtained also by measuring the decay of $O(^1D)$ after the flash photolysis of O_3 in the presence of atmospheric gases (455, 456). The decay of $O(^1D)$ was monitored by optical absorption at 1152 Å.

It has been found (791) that while N_2 [DeMore and Raper (274)], Xe, CO, and CO_2 merely deactivate $O(^1D)$ to $O(^3P)$ atoms, H_2 and CH_4 react with $O(^1D)$ atoms to form respective products.

Overend et al. (787) have found little effect of excess kinetic energy of $O(^1D)$ atoms on reaction rates.

Reactions of $O(^1D)$ *Atoms with* H_2, O_2, N_2O, H_2O, CH_4, *and Chlorofluoromethanes.* $O(^1D) + H_2$ (*276*). No physical quenching was found (791). DeMore (276) has studied the photolysis at 2537 Å of mixtures of O_3–H_2 dissolved in liquid argon at 87°K. The quantum yield of ozone decomposition is near unity when O_2 is added to the mixtures, while without O_2 at least 25O_3 molecules per H atom are dissociated.

DeMore proposed that processes with O_2 added are

$$O(^1D) + H_2 \rightarrow H + OH \quad k_{23} = 2.9 \times 10^{-10}\ \text{cm}^3\ \text{molec}^{-1}\ \text{sec}^{-1} \tag{IV-23}$$

$$H + O_2 \xrightarrow{M} HO_2 \tag{IV-23a}$$

The OH and HO_2 radicals produced do not react further with O_3. Hence, the quantum yield of O_3 decomposition is near unity.

On the other hand, without O_2 the mechanism is (IV-23) followed by chain reactions

$$H + O_3 \rightarrow OH^\dagger + O_2 \tag{IV-23b}$$

$$OH^\dagger + O_3 \rightarrow H + 2O_2 \tag{IV-23c}$$

where $OH^\dagger$ indicates vibrationally excited OH (v'' up to 9).

$O(^1D) + O_2$ (392, 918, 1068). The reaction to form $O_2(^1\Sigma)$ was first suggested by Young and Black (1068)

$$O(^1D) + O_2 \rightarrow O(^3P) + O_2(^1\Sigma) \qquad k_{24} = 7 \times 10^{-11}\ \text{cm}^3\ \text{molec}^{-1}\ \text{sec}^{-1} \tag{IV-24}$$

They observed the $O_2(^1\Sigma)$ emission band at 7620 Å by the steady state photolysis of O_2 at 1470 Å. In the photolysis of O_3–O_2 mixtures at 2537 Å in a flow system, Snelling and Gauthier (918) and Snelling (919) have found that the yield of $O_2(^1\Sigma)$ production is $85 \pm 15\%$.

$O(^1D) + N_2O$ (794, 816, 864). The reactions of $O(^1D)$ atoms with N_2O are

$$O(^1D) + N_2O \rightarrow N_2 + O_2 \qquad k_{25} \tag{IV-25}$$

$$O(^1D) + N_2O \rightarrow 2NO \qquad k_{26} \tag{IV-26}$$

with a 1:1 ratio, while $O(^3P)$ atoms do not react with N_2O. The reaction is fast with a value of 2×10^{-10} cm^3 molec^{-1} sec^{-1} for $k_{25} + k_{26}$.

$O(^1D) + H_2O$. The flash photolysis of O_3–H_2O mixtures in the near ultraviolet yields OH radicals [Engleman (335), Biedenkapp et al. (109)].

$$O_3 \xrightarrow{h\nu} O(^1D) + O_2(^1\Delta)$$

$$O(^1D) + H_2O \rightarrow 2OH \tag{IV-27}$$

Paraskevopoulos and Cvetanović (793) have studied the photolysis of N_2O–neopentane–H_2O mixtures at 2139 Å. They have obtained the reaction rate of $O(^1D)$ with H_2O relative to that with neopentane. The deactivation of $O(^1D)$ to (^3P) by H_2O was examined by adding 1-butene to N_2O–neopentane–H_2O mixtures. No products characteristic of $O(^3P)$

addition to 1-butene were found. The process

$$O(^1D) + H_2O \rightarrow 2OH \qquad k_{27} = 3.5 \times 10^{-10}\ \text{cm}^3\ \text{molec}^{-1}\ \text{sec}^{-1} \qquad \text{(IV-27a)}$$

must be predominant (>90%) (888). It has been observed that hydroxyl radicals formed are vibrationally excited up to $v'' = 2$ (335). The newly formed OH bond acquires most of the available energy while the other OH bond remains vibrationally cold (335). The reaction of $O(^3P)$ atoms with H_2O, however, does not form OH (312).

$O(^1D) + CH_4$ (312, 856). The formation of OH radicals is found in the flash photolysis of N_2O–CH_4 mixtures, indicating the occurrence of the reaction

$$O(^1D) + CH_4 \rightarrow OH + CH_3 \qquad k_{28} = 4 \times 10^{-10}\ \text{cm}^3\ \text{molec}^{-1}\ \text{sec}^{-1} \qquad \text{(IV-28)}$$

No physical quenching of $O(^1D)$ to $O(^3P)$ was found (791). Greenberg and Heicklen (420) have estimated the OH production yield to be 95% of the total. Lin and DeMore (636) have concluded the reaction

$$O(^1D) + CH_4 \rightarrow H_2 + CH_2O \qquad \text{(IV-29)}$$

occurs to an extent of 9%.

$O(^1D)$ + *Chlorofluoromethanes.* The reactions of $O(^1D)$ with chlorofluoromethanes have been studied by Donovan et al. (316), Gillespie and Donovan (396), and Fletcher and Husain (362). The reactions yield predominantly ClO radicals

$$O(^1D) + CF_xCl_y \rightarrow CF_xCl_{y-1} + ClO(X^2\Pi) \qquad \text{(IV-29a)}$$

The rate constants are large ($\sim 5 \times 10^{-10}$ cm^3 molec^{-1} sec^{-1}).

$O(^1D)$ *Atoms in the Upper Atmosphere.* The presence of $O(^1D)$ atoms in the upper atmosphere has been recognized by the observation of the air glow emission at 6300 Å. The formation of $O(^1D)$ atoms must be due to the photolysis of O_2 by light of wavelengths below 1750 Å (>100 km) and to the photolysis of O_3 in the Hartley band (<100 km).

The reaction $O(^1D) + O_2$ may produce $O_2(^1\Sigma)$ in the stratosphere (918, 1085).

Reaction rates and products of $O(^1D)$ with various gases are given in Table IV–2.

IV–4.2. $O(^1S)$ Atoms

The electronic energy of $O(^1S)$ atoms is 4.189 eV above the ground state. The transitions of $O(^1S)$ to the ground and to the $O(^1D)$ states are forbidden.

Table IV–2. Reaction Rates and Products of $O(^1D)$ with Various Gases

Reactant	Products[a]	Rate Constant[b] (10^{-10} cm^3 molec^{-1} sec^{-1})	Ref.
Xe	$Xe + O(^3P)$	1.4 ± 0.3	791
O_2	$O_2(b^1\Sigma_g^+) + O(^3P)$	0.75 ± 0.15	348, 610, 745, 918
	$[O_2(^3\Sigma_g^-) + O(^3P)]$		108, 545, 1068
N_2	$N_2 + O(^3P)$	0.55 ± 0.15	791
	$\Phi_{N_2O} < 10^{-6}$		556, 883
	$N_2^\dagger + O(^3P)$		904
CO	$CO + O(^3P)$	0.75 ± 0.15	313, 791
	$CO^\dagger + O(^3P)$		904
NO	$[NO + O(^3P)]$	2.1 ± 0.4	10
H_2	$H + OH$	2.9 ± 0.5	276, 791
O_3	$O_2 + O_2$ ($a^1\Delta$ or $X^3\Sigma$)	5.0 ± 2.5	392, 458, 1027
	$2O + O_2$		
CO_2	$CO_2 + O(^3P)$	1.8 ± 0.3	313, 791
N_2O	$N_2 + O_2$ } 1:1	2.2 ± 0.4	864
	$2NO$ }		
NO_2	$(NO + O_2)$	2.8 ± 0.5	10
H_2O	$2OH$	3.5 ± 0.6	335, 458, 888
CH_4	$OH + CH_3$	4.0 ± 0.7	312, 636, 791
	$H_2 + CH_2O$		275, 420
neo-C_5H_{12}	Neopentanol	12.3 ± 2.3	791, 792

[a] † indicates a vibrationally excited molecule.
[b] Ref. 1012.

The lifetime of $O(^1S)$ is about 0.71 sec (32). The $O(^1S)$ state can be identified by the emission at 5577 Å $[O(^1S) \rightarrow O(^1D)]$ or by the absorption at 1218 Å $[3s''(^1P^o) \leftarrow 2p^4(^1S)]$ [see McConkey and Kernahan (673)].

Photochemical Production. The formation of $O(^1S)$ atoms has been observed in the photolysis of O_2, N_2O, and CO_2

$$O_2 \xrightarrow{h\nu} O(^3P) + O(^1S) \qquad < 1332 \text{ Å}$$
$$N_2O \xrightarrow{h\nu} N_2 + O(^1S) \qquad < 2115 \text{ Å}$$
$$CO_2 \xrightarrow{h\nu} CO + O(^1S) \qquad < 1286 \text{ Å}$$

Reactions of $O(^1S)$ *Atoms.* In spite of the higher electronic energy of $O(^1S)$ than of $O(^1D)$ the quenching rate constant of the former atoms by many gases is in general much smaller than that of the latter atoms. Table IV–3 shows some results obtained mainly by Welge et al. (48, 349, 646, 949, 1035). A mechanism to explain a large difference in physical

Table IV–3. Comparison of Quenching Rate Constants of $O(^1D)$ and $O(^1S)$ Atoms at Room Temperature (10, 646)

Reactant	k (cm^3 molec^{-1} sec^{-1}) $O(^1D)$	$O(^1S)$
Xe	$1.4 \times 10^{-10\,a}$	6.7×10^{-15}
N_2	$0.55 \times 10^{-10\,a}$	$<5 \times 10^{-17}$
CO	$0.75 \times 10^{-10\,a}$	4.9×10^{-15}
O_2	$0.75 \times 10^{-10\,a}$	2.6×10^{-13}
H_2	$2.9 \times 10^{-10\,b}$	2×10^{-16}
CO_2	$1.8 \times 10^{-10\,a}$	3.7×10^{-13}
O_3	$5 \times 10^{-10\,b}$	5.8×10^{-10}

[a] Physical quenching of $O(^1D)$ to $O(^3P)$.
[b] Chemical quenching.

quenching has been proposed by Donovan and Husain (310) and by Fisher and Bauer (360). The observed quenching properties of $O(^1S)$ and $O(^1D)$ by N_2, for example, may be explained by a correlation diagram show in Fig. IV–1 in which C_s symmetry is assumed for the O–N_2 collision complex. From Table II–4 it can be shown that three reaction surfaces ($^3A' + 2^3A''$) are available for $O(^3P) + N_2$, five surfaces ($3^1A' + 2^1A''$) are available for $O(^1D) + N_2$, and only one surface ($^1A'$) correlates with $O(^1S) + N_2$. Quenching of $O(^1S)$ by N_2 is probably physical, since an $O(^1S) + N_2$ surface does not correlate with $NO + N(^4S)$ surfaces. Physical quenching of $O(^1S)$ would involve a high energy barrier (10^5 collisions are necessary for quenching). On the other hand, $O(^1D)$ quenching would proceed with the initial formation of highly vibrationally excited ground state N_2O and its subsequent dissociation into $O(^3P) + N_2$ by nonadiabatic crossing from $^1A'$ to $^3A' + {}^3A''$ surfaces. The reaction $O(^1D) + N_2 \rightarrow NO + N$ does not occur since it is highly endothermic.

A similar trend has been observed for quenching of $S(^1S)$ and $S(^1D)$ by noble gases. It is not known whether quenching of $O(^1S)$ by O_3 is physical or chemical. It is likely that the products would be $2O_2$ or $O_2 + 2O$, since the rate constant is very large in this case.

Intensity Enhancement of the $O(^1S) \rightarrow O(^1D)$ *Transition.* It has been found by Filseth et al. (349) and by Hampson and Okabe (436) that gases such as H_2, N_2, Ar, Kr, and Xe enhance the $O(^1S) \rightarrow O(^1D)$ emission intensity. This enhancement has been attributed to the shortened lifetime of the

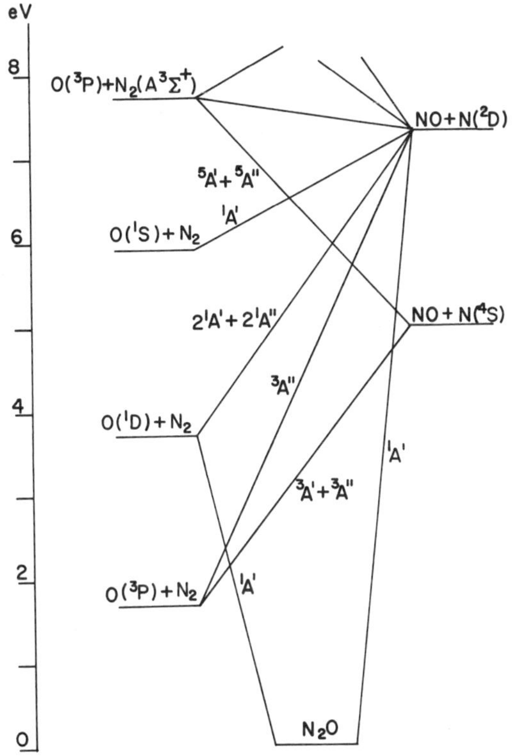

Fig. IV–1. Correlation diagram for the reactions O + N_2 and NO + N. C_s symmetry is assumed for the reaction intermediate. Three reaction paths are available for $O(^3P)$ + N_2, five for $O(^1D)$ + N_2, and one for $O(^1S)$ + N_2. Quenching of $O(^1S)$ by N_2 is probably physical and would require a high energy barrier, while quenching of $O(^1D)$ to $O(^3P)$ is facilitated by nonadiabatic crossing from $^1A'$ to $^3A' + 2^3A''$ surfaces. Reprinted with permission from Donovan and Husain, *Chem. Rev.*, 70, 489 (1970). Copyright by the American Chemical Society.

emission. It has been known that the atomic emission line at 5577 Å is accompanied by diffuse bands towards shorter wavelengths in the presence of Ar (240), N_2 (238), and Xe (254). A proposed mechanism by Hampson and Okabe (436) is that while the atomic transition is forbidden by the selection rules, $\Delta J = 0, \pm 1$, it becomes allowed as a result of the loosely bound molecular formation M—$O(^1D_2)$ (M, an added gas). The transition is now allowed since M—$O(^1D_2)$ has components $\Omega = 2, 1, 0$ along the internuclear axis, and the selection rules are $\Delta\Omega = 0, \pm 1$.

The intensity of the collision induced emission follows the order Xe > Kr > Ar > N_2 > H_2 > He (349, 436). According to Cunningham and Clark (254) this molecular emission is 40% of the total quenching process in the case of Xe, but it is only 1% in the case of Ar. From the Stern-Volmer type plot of the intensity ratio of I_f (with added gas) to I_f^0 (without added gas) against the pressure of an added gas, Black et al. (115) have recently obtained the rate constant for induced emission by He, Ar, N_2, H_2, Kr, and Xe.

$O(^1S)$ *in the Upper Atmosphere.* The presence of $O(^1S)$ in the upper atmosphere is indicated by the emission line at 5577 Å in the airglow and aurora. The mechanism of formation and destruction of $O(^1S)$ atoms has been of great interest in aeronomy. Zipf (1085) gives a detailed account of various processes of $O(^1S)$ in the upper atmosphere.

IV–5. REACTIONS OF METASTABLE S ATOMS

IV–5.1. $S(^1D)$ Atoms

The $S(^1D)$ state lies 1.145 eV above the ground $S(^3P)$ state and is metastable with a lifetime of 28 sec. The $S(^1D)$ state can be detected by absorption in the vacuum ultraviolet at 1667 or 1448 Å (33). However, in spite of much effort $S(^1D)$ atoms have not been detected by optical absorption in the vacuum ultraviolet flash photolysis of OCS because of a rapid reaction of $S(^1D)$ with OCS

$$S(^1D) + OCS \rightarrow S_2(a^1\Delta) + CO$$

The appearance of absorption bands near 1900 Å, due to the transition $S_2(g^1\Delta_u \leftarrow a^1\Delta_g)$ indicates $S(^1D)$ atoms are formed in the primary process since $S(^3P)$ atom reaction with OCS yields $S_2(X^3\Sigma_g^-)$ rather than $S_2(a^1\Delta)$ [Donovan (305)]. The ultraviolet photolysis of OCS is a convenient source of $S(^1D)$ atoms [Gunning and Strausz (430)]. The $S(^1D)$ atoms react with paraffins to produce the corresponding mercaptans, while $S(^3P)$ atoms do not react with paraffins. Thus, the production of mercaptans may be used as diagnosis for $S(^1D)$ atom formation. The reaction of $S(^3P)$ with ethylene yields only ethylene episulfide [$(CH_2)_2S$], while $S(^1D)$ atoms form ethylene episulfide and vinyl mercaptan (430). Donovan et al. (308) have found high quenching efficiencies of $S(^1D)$ atoms by inert gases. Quenching efficiencies for $S(^1D)$ are much higher than those for $S(^1S)$ atoms. The analogous trend has been found for $O(^1D)$ and $O(^1S)$ atom quenching efficiencies. They have explained the higher quenching efficiency of $S(^1D)$ atoms by Xe on the basis of crossings of the potential energy curves for Xe—$S(^1D)$ and Xe—$S(^1S)$ molecules, respectively, with that for a Xe—$S(^3P)$ molecule.

Quenching rates of $S(^1D)$ by atoms and molecules have been measured by Little et al. (642).

IV–5.2. S(^1S) Atoms

The S(^1S) state lies 2.750 eV above the ground state. The S(^1S) state is metastable with a lifetime of 0.47 sec (33). The S(^1S) atoms can be detected by absorption at 1687 or 1782 Å. The emission at 7725 Å ($^1S \rightarrow {}^1D$) or at 4589 Å ($^1S \rightarrow {}^3P$) may also be used to follow the reaction rate of S(^1S) atoms. The photolysis of OCS in the vacuum ultraviolet produces S(^1S) atoms [Dunn et al. (326) Donovan et al. (305, 308)]. Quenching of S(^1S) by various atoms and molecules has been studied by Donovan et al. (308) and by Dunn et al. (326). Quenching rates by NO and NO_2 are extremely fast and may involve chemical reaction (326). Neither the products specific to S(^1S) atom reactions nor the extent of physical quenching is known. Quenching rate constants of S(^1S) are in general much less than those of S(^1D), just like for O atoms.

Collisionally induced emission S(^1S) → S(^1D) has been found to be a major path for deactivation of S(^1S) atoms by rare gases and by N_2 [Black et al. (116)] in analogy with O(^1S) atoms described in Section IV–4.2.

IV–6. REACTIONS OF METASTABLE AND GROUND STATE C ATOMS

IV–6.1. C(^1D) Atoms

The C(^1D) state is 1.263 eV above the ground state C(^3P) and has a lifetime of 53 min (32). The generation of C(^1D) atoms is achieved by the photolysis of carbon suboxide in the vacuum ultraviolet. The concentration of C(^1D) atoms can be monitored by optical absorption at 1931 or 1482 Å [Braun et al. (141), Husain and Kirsch (500)]. Quenching of C(^1D) atoms by noble gases (499), diatomic molecules, and polyatomic molecules has been studied by Braun et al. (141) and Husain and Kirsch. (498–500). Reactions of C(^1D) atoms with molecules are, in general, fast (collision efficiencies of 0.1 to 1).

IV–6.2. C(^1S) Atoms

The C(^1S) state lies 2.683 eV above the ground state C(^3P) with a lifetime of 2 sec (32). The production of C(^1S) atoms is observed in the photolysis of carbon suboxide in the vacuum ultraviolet. The C(^1S) atom production can be detected by absorption at 2479 or 1752 Å. Rate constants of C(^1S) with molecules have been measured by Meaburn and Perner (687), Husain and Kirsch (505), and Braun et al (141). The rate constants are in general much smaller (collision efficiencies 10^{-2} to 10^{-6}) than those for C(^1D), in analogy with the results for O(^1D) and O(^1S) atom quenching rates given in Table IV–3.

The difference in reactivity of C(^1D) from C(^1S) is explained by Donovan and Husain (310) on the basis of symmetry correlations between reactants (for example, carbon atom and hydrogen molecule) and products (methylidyne and atomic hydrogen). Figure IV–2 shows the correlation of C + H_2 with CH + H. The reaction C + H_2 is assumed to form CH_2 of C_s symmetry (or C_{2v} symmetry), which dissociates subsequently into CH + H. Correla-

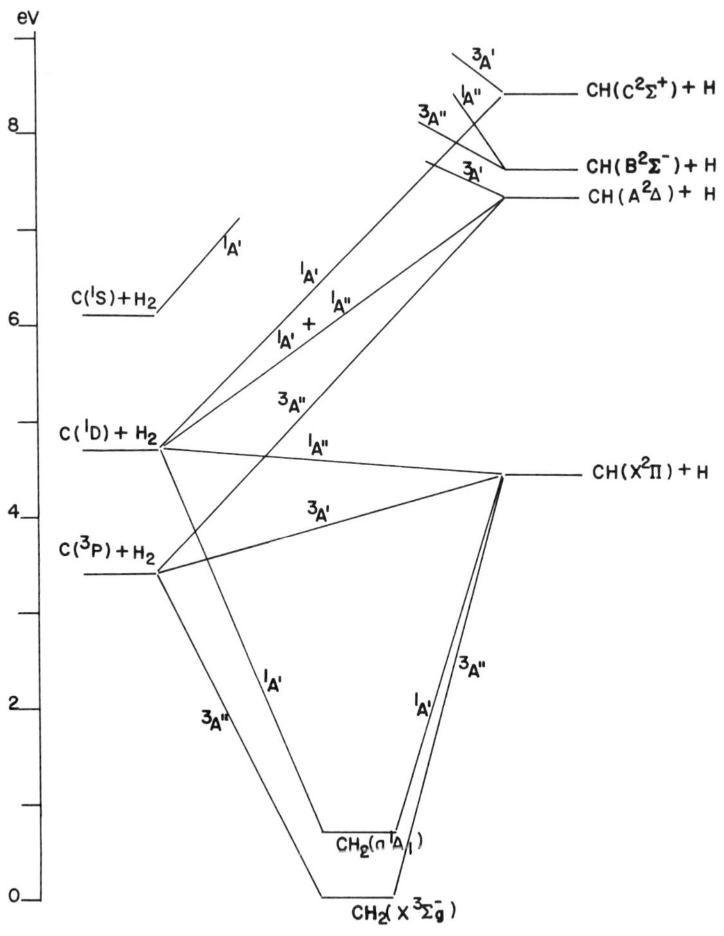

Fig. IV–2. Symmetry correlations between C + H_2 and CH + H. The formation of CH_2 in C_s symmetry (or C_{2v}) is assumed. The reaction C(^1D) + H_2 proceeds through $^1A'$ and $^1A''$ to form CH($X^2\Pi$) + H with high efficiency. On the other hand, C(^1S) + H_2 correlates with highly excited CH + H and the reaction requires a high activation energy. The C(^1S) atoms are probably deactivated to C(^1D). C(^3P) + H_2 forms $CH_2(X^3\Sigma_g^-)$. Reprinted with permission from R. J. Donovan and D. Husain, *Chem. Rev.*, 70, 489 (1970). Copyright by the American Chemical Society.

tions of species $C(^3P, ^1D, ^1S) + H_2$ with those of CH_2 in C_s are given in Table II–4, Correlations of species of CH_2 in C_s with those of $CH(^2\Pi, ^2\Delta, ^2\Sigma^-, ^2\Sigma^+) + H$ are given in Table II–3. The species $C(^3P) + H_2$ becomes $^3A' + 2^3A''$ in C_s. Two of them $(^3A' + ^3A'')$ correlate with $CH(X^2\Pi) + H$ and the third species, $^3A''$, becomes $CH(A^2\Delta) + H$. At low kinetic energy of $C(^3P)$ (<1 eV), CH_2 is the product. For species $C(^1D) + H_2$ five surfaces $(3^1A' + 2^1A'')$ are available, two of which $(^1A' + ^1A'')$ correlate with $CH(X^2\Pi) + H$ and another two $(^1A' + ^1A'')$ are associated with $CH(A^2\Delta) +$ H. The last one $(^1A')$ becomes $CH(C^2\Sigma^+) + H$. The production of $CH(X^2\Pi) + H$ is fast and requires no activation energy. On the other hand, the species $C(^1S) + H_2$ has only one surface $(^1A')$ that correlates with highly excited CH + H. The reaction needs at least an activation energy of 2.5 eV and $C(^1S)$ atoms are more likely to be deactivated to $C(^1D)$ atoms.

IV–6.3. $C(^3P)$ Atoms

The ground state C atoms are produced by the photolysis of carbon suboxide. They can be monitored by optical absorption at 1657 Å. Reactions of $C(^3P)$ atoms with molecules have been studied by Husain and Kirsch (497, 498) and Braun et al. (141).

IV–7. REACTIONS OF OTHER METASTABLE ATOMS

IV–7.1. $N(^2D, ^2P)$ Atoms

The $N(^2D)$ and $N(^2P)$ levels are 2.38 and 3.576 eV, respectively, above the ground level (^4S) and are metastable with lifetimes of about 17 hr and 12 sec, respectively (32). The $N(^2D)$ atoms can be detected by absorption at 1243 and 1493 Å and $N(^2P)$ at 1412 and 1744 Å. In emission the 5199 Å line is due to the transition $^2D \rightarrow ^4S$ and the 3466 and 10,400 Å lines are from the transitions $^2P \rightarrow ^4S$, and $^2P \rightarrow ^2D$, respectively. The $N(^2D, ^2P)$ atoms can be produced from the photolysis of N_2O in the vacuum ultraviolet.

Quenching rates of $N(^2D)$ and (^2P) atoms by H_2, N_2, O_2, CO_2, and N_2O have been measured by Husain et al. (506) by time-resolved attenuation of absorption at 1493 and 1744 Å, respectively. The quenching of $N(^2D)$ atoms by N_2O appears to be chemical reaction to produce $NO(B^2\Pi) + N_2$.

$$N(^2D) + N_2O \rightarrow NO(B^2\Pi) + N_2 \qquad \text{(IV-30)}$$

Quenching of $N(^2D, ^2P)$ by O_2, N_2O, CO_2, NO, N_2, H_2O, Ar, and He has also been studied by Lin and Kaufman (635) and by Slanger and Black (908).

Atmospheric Reactions. The presence of $N(^2D)$ atoms in the upper atmosphere has been known from the 5200 Å emission line observed in the upper atmosphere. The reaction of $N(^2D)$ atoms with O_2 is believed to be a source of NO

$$N(^2D) + O_2 \rightarrow NO + O(^3P) \tag{IV-31}$$

IV–7.2. $Br(^2P_{1/2})$ Atoms

The $Br(^2P_{1/2})$ state lies 0.456 eV above the ground $^2P_{3/2}$ state and is metastable with a lifetime of 1.12 sec. The $Br(^2P_{1/2})$ atoms can be detected by absorption at 1532, 1582, and 1634 Å. The flash photolysis in the vacuum ultraviolet of HBr, Br_2, $CHCl_2Br$, and CF_3Br has produced $Br(^2P_{1/2})$ atoms that have been observed by optical absorption [Donovan and Husain (299)]. The quenching efficiency of $Br(^2P_{1/2})$ produced from the flash photolysis of CF_3Br by various gases has been studied by Donovan and Husain (300).

IV–7.3. $I(^2P_{1/2})$ Atoms

The $I(^2P_{1/2})$ state is 0.942 eV above the ground state $^2P_{3/2}$ and is metastable with a lifetime of about 0.05 sec [Husain and Wiesenfeld (495, 496)]. The $I(^2P_{1/2})$ atoms can be detected either by optical absorption at 1844, 1799, and 2062 Å or by emission at 13152 Å. The $I(^2P_{1/2})$ atom production was observed in the flash photolysis of CF_3I above 2000 Å [Husain and Wiesenfeld (496)]. The flash photolysis of HI above 2000 Å has produced $I(^2P_{1/2})$ and $I(^2P_{3/2})$ with a ratio of 1:5 [Donovan and Husain (297)]. Physical quenching rates of $I(^2P_{1/2})$ atoms by diatomic and polyatomic molecules have been measured by Husain et al. (269, 448, 496) and, Donovan et al. (298, 303). Quenching efficiencies range from 10^{-3} to 10^{-6}. The low quenching efficiency by CF_3I favors the detection of $I(^2P_{1/2})$ atoms. The quenching by HI, H_2O, O_2, NO, and I_2 has been found by Donovan and Husain (298) to be efficient. The quenching by Cl_2, Br_2, ICl, and IBr has been attributed by Donovan et al. (301) to chemical quenching. On the other hand, the quenching of $I(^2P_{1/2})$ by alkyl iodides (RI) is primarily physical, although the reaction to form $I_2 + R$ is exothermic (303). The quenching efficiencies are on the order of 10^{-2} to 10^{-3}. The fraction of $I(^2P_{1/2})$ production from n-C_3H_7I photolysis is twice as much as that from i-C_3H_7I.

IV–7.4. $As(^2D_J, ^2P_J)$ Atoms

The $As(^2D_{3/2})$ and $As(^2D_{5/2})$ states are 1.313 and 1.353 eV, respectively, above the ground (^4S) state. The $As(^2P_{1/2})$ and $As(^2P_{3/2})$ are 2.254 and 2.312 eV, respectively, above the ground state. The $As(^2D_J)$ and $As(^2P_J)$ atoms are generated by the photolysis of $As(CH_3)_3$ and $AsCl_3$, respectively. (105, 172)

IV–7.5. Sn(^1D, ^1S) Atoms

The Sn(^1D) and Sn(^1S) states are 1.068 and 2.128 eV, respectively, above the ground ($^3P_0$) state. The Sn(^1D) and Sn(^1S) atoms are generated by the photolysis of $Sn(CH_3)_4$ and $SnCl_4$, respectively. (151, 152)

IV–7.6. Pb(^1D, ^1S) Atoms

The Pb(^1D) and Pb(^1S) states are 2.660 and 3.653 eV above the ground state, respectively. The Pb(^1D) and (^1S) atoms are generated by the pulsed photolysis of lead tetraethyl. The quenching of the Pb(^1D) and (^1S) atoms by various molecules has been studied by Husain and Littler. (501, 503, 504)

chapter V

Photochemistry of Diatomic Molecules

V–1. HYDROGEN

The ground state of H_2 is $X^1\Sigma_g^+$.

$$D_0(H—H) = 4.4780 \text{ eV (Ref. 468)}$$

Absorption starts at 1108 Å. The banded absorption region 850 to 1108 Å corresponds to the transition $B^1\Sigma_u^+-X^1\Sigma_g^+$ (Lyman bands). The dissociation limit at 844.7 Å corresponds to

$$H_2 \xrightarrow{h\nu} H(1s) + H(2p, 2s) \quad \text{(V-1)}$$

Below 844.7 Å the banded structure ($D^1\Pi_u-X^1\Sigma_g^+$) is superimposed on a continuum [Mentall and Gentieu (695)]. The absorption coefficients in the region 700 to 860 Å have been measured by Mentall and Gentieu (695).

V–2. HYDROGEN HALIDES

V–2.1. Hydrogen Fluoride

The ground state is $X^1\Sigma^+$. D_0(H—F) = 5.86 ± 0.02 eV (286a). Absorption starts at 1613 Å. The absorption coefficients of a weak continuum below 1613 Å are given in Fig. V–1.

V–2.2. Hydrogen Chloride

The ground state is $X^1\Sigma^+$, D_0(H—Cl) = 4.431 ± 0.002 eV (24). The absorption coefficients of the continuous region 1380 to 2000 Å are given in Fig. V–2, and those in the region 1050 to 2100 Å are given in Fig. V–3.

Photolysis at 1849 Å produces H atoms with about 2.27 eV excess kinetic energy. The primary process is most likely the production of H + Cl with a quantum yield of unity since the absorption is continuous. The Cl atoms are in the $^2P_{3/2}$ state [Mulliken (725)]. The photochemical reactions expected are similar to those of HI [Wilson and Armstrong (1051)].

$$HCl \xrightarrow{h\nu} H + Cl(^2P_{3/2}) \quad \text{(V-2)}$$

$$H + HCl \rightarrow H_2 + Cl \quad \text{(V-3)}$$

$$Cl + Cl + M \rightarrow Cl_2 + M \quad \text{(V-4)}$$

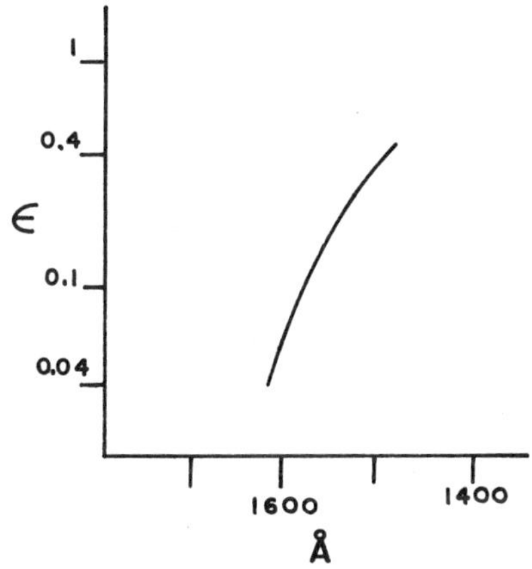

Fig. V–1. Absorption coefficient of HF. ϵ is in units $1\ mol^{-1}\ cm^{-1}$ base 10 (at room temperature). From Safary et al. (849), reprinted by permission. Copyright 1951 by the American Institute of Physics.

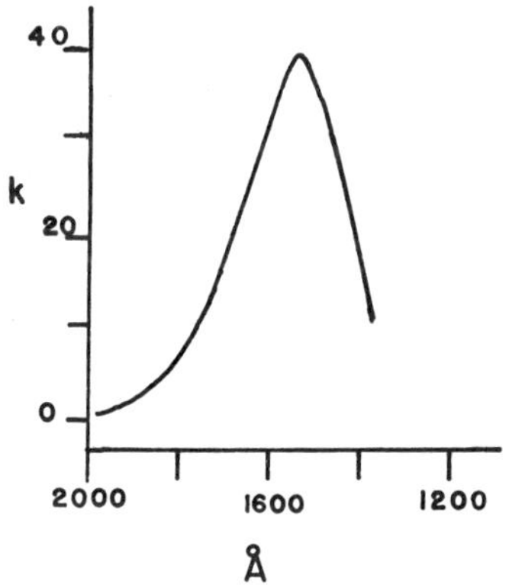

Fig. V–2. Absorption coefficient of HCl. k is in units of $atm^{-1}\ cm^{-1}$, base 10 at room temperature. From J. Romand and B. Vodar (840), reprinted by permission of the Académie des Sciences.

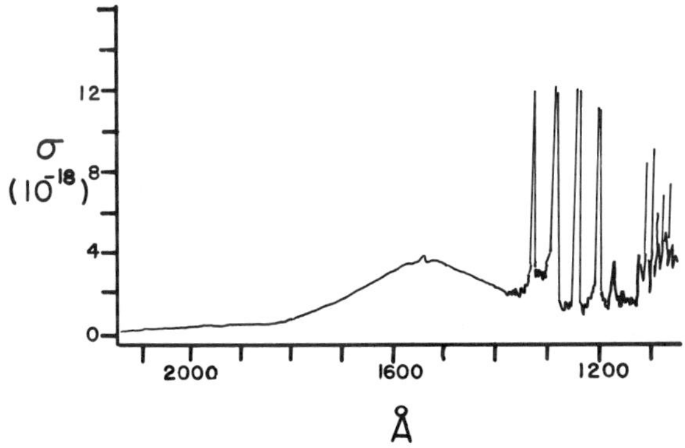

Fig. V–3. Absorption cross sections of HCl in the region 1050 to 2100 Å. σ is in units of $10^{-18}\ cm^2$, base e at room temperature. From Myer and Samson (727), reprinted by permission. Copyright 1970 by the American Institute of Physics.

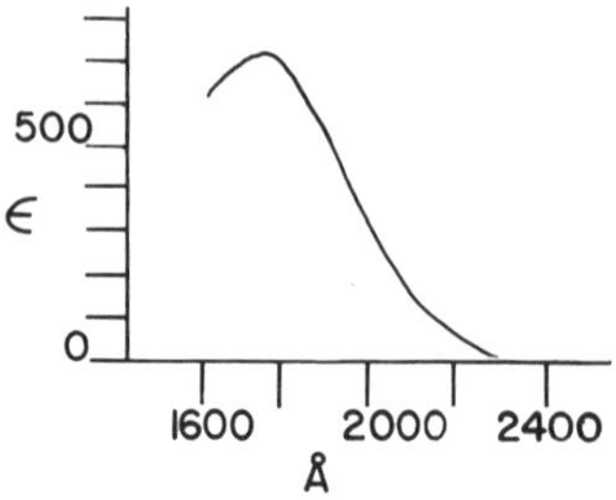

Fig. V–4. Absorption coefficient of HBr. ε is in units of l mol^{-1} cm^{-1}, base 10 at room temperature. Reprinted with permission from B. J. Huebert and R. M. Martin, *J. Phys. Chem.* 72, 3046 (1968). Copyright by the American Chemical Society.

V–2.3. Hydrogen Bromide

The ground state is $X^1\Sigma^+$, D_0(H—Br) = 3.750 ± 0.01 eV (24). Absorption starts at about 2500 Å and is continuous down to about 1600 Å. The absorption coefficients in this region are given in Fig. V–4. The region 1190 to 1580 Å contains various discrete transitions.

The primary photochemical process in the continuous region appears to be the production of the ground state Br atoms $^2P_{3/2}$ [Milliken (725)].

$$\mathrm{HBr} \xrightarrow{h\nu} \mathrm{H} + \mathrm{Br}(^2P_{3/2}) \qquad \text{(V-5)}$$

The H atoms have excess kinetic energy, 2.96 and 1.25 eV at 1849 and 2480 Å, respectively. The secondary processes are [Fass (444)].

$$\mathrm{H} + \mathrm{HBr} \rightarrow \mathrm{H_2} + \mathrm{Br} \qquad \text{(V-6)}$$

$$\mathrm{Br} + \mathrm{Br} + \mathrm{M} \rightarrow \mathrm{Br_2} + \mathrm{M} \qquad \text{(V-7)}$$

The photolysis of HBr has been used for an actinometer in the region 1800 to 2500 Å, as $\Phi_{H_2} = 1$ has been established in comparison with an N_2O actinometer [Martin and Willard (663)].

V–2.4. Hydrogen Iodide

The ground state is $X^1\Sigma^+$, D_0(H—I) = 3.054 ± 0.002 eV. Absorption starts at about 2800 Å and is continuous down to 1800 Å. The absorption coefficients in this region are given in Fig. V–5. The upper state must be repulsive as shown in Fig. V–6. The quantum yields of H_2 and I_2 formation from HI at 1849 Å and at −78°C and 25°C are 1.05 + 0.05 and 1.3 ± 0.3, respectively [Martin and Willard (663)]. The results may be explained by a mechanism

$$\mathrm{HI} \xrightarrow{h\nu} \mathrm{H} + \mathrm{I} \qquad \text{(V-8)}$$

$$\mathrm{H} + \mathrm{HI} \rightarrow \mathrm{H_2} + \mathrm{I} \qquad \text{(V-9)}$$

$$2\mathrm{I} + \mathrm{M} \rightarrow \mathrm{I_2} + \mathrm{M} \qquad \text{(V-10)}$$

The excess energy beyond that required to break the H—I bond is 3.65 eV at 1849 Å. This excess energy appears primarily as the kinetic energy of H

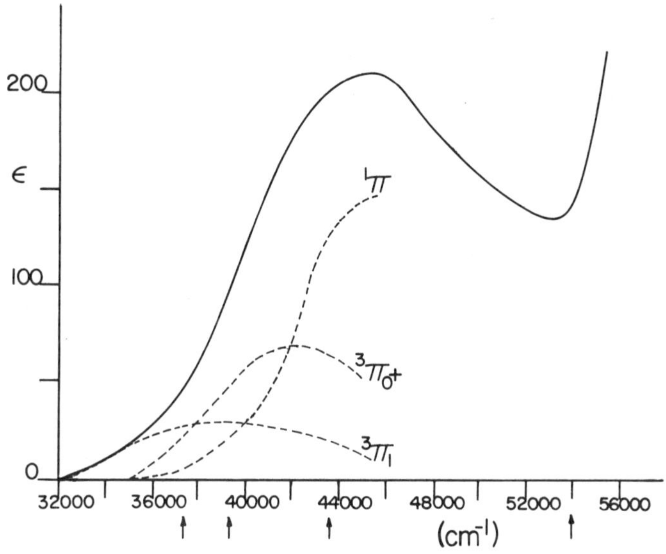

Fig. V–5. Absorption coefficients of HI and contribution of the transitions to the absorption continuum in the ultraviolet region. Solid curve, absorption coefficients ϵ of HI in units of l mol^{-1} cm^{-1} base 10 at room temperature. Reprinted with permission from B. J. Huebert and R. M. Martin, *J. Phys.* Chem. 72, 3046 (1968). Copyright by the American Chemical Society. Dashed curves, absorption coefficients of the transitions $^3\Pi_1$–$^1\Sigma^+$, $^3\Pi_0^+$–$^1\Sigma^+$, and $^1\Pi$–$^1\Sigma^+$. The $^3\Pi$, and $^1\Pi$ states dissociate into H + I($^2P_{3/2}$), while the $^3\Pi_0^+$ state dissociates into H + I($^2P_{1/2}$). The arrows indicate four incident wavelengths (2662, 2537, 2281, and 1850 Å) at which the ratios of I($^2P_{1/2}$) to I($^2P_{3/2}$) are obtained. From Clear et al. (219) reprinted by permission. Copyright 1975 by the American Institute of Physics.

atoms. However, Φ_{H_2} both at 1849 and 2537 Å is unity, showing little or no enhancement of the yield by hot hydrogen atoms. The production of the metastable $^2P_{1/2}$ I atoms is energetically possible below 3100 Å. At 2790 Å Oldershaw et al. (777) estimate the quantum yield of I($^2P_{1/2}$) production to be 0.11 ± 0.14. Estimates by Compton and Martin (231) for the same yield are 0.07 ± 0.1, 0.19 ± 0.1, and 0.0 ± 0.1, respectively, at 2537, 2288, and

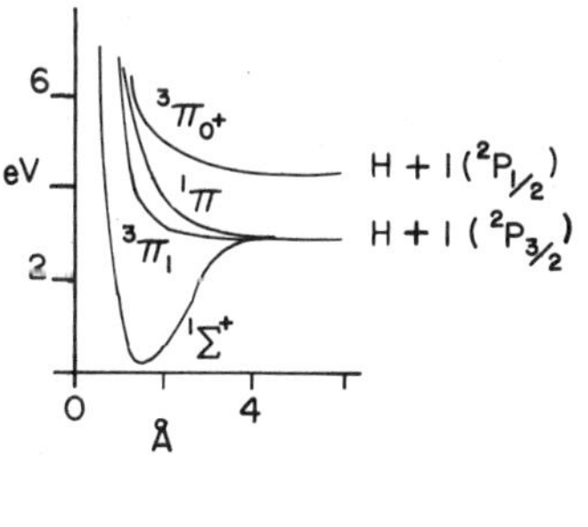

Fig. V–6. Potential energy curves of HI. From Wilson and Armstrong (1051). Originally from Mulliken, *Phy. Rev.* 51, 310 (1937). Reprinted by permission. Copyright 1937 by the American Physical Society.

1850 Å. Thus, in the region 2800 to 1800 Å the ground atom production appears to be predominant. However, Cadman and Polanyi (165) estimate the $I(^2P_{1/2})$ production to be about 0.5 at 2537 Å. Clear et al. (219) have found the production of $I(^2P_{1/2})$ to be 0.36 at 2662 Å by measuring the translational energy of recoil H atoms produced from HI by a pulsed polarized laser.

V–3. CARBON MONOXIDE

The ground state of CO is $X^1\Sigma^+$. D_0(C—O) = 11.09 eV. (24). A weak discrete absorption in the region 1765 to 2155 Å corresponds to the Cameron system $a^3\Pi–X^1\Sigma^+$.

In the vacuum ultraviolet absorption bands in the region 1280 to 1600 Å correspond to the fourth positive system $A^1\Pi–X^1\Sigma^+$. The absorption cross sections of this system are given in Fig. V–7. Since the widths of the CO rotational lines are much smaller than the instrumental resolution ($\sim 10\ cm^{-1}$), it is not possible to obtain the absorption cross section of each rotational line [see Section I-8 for details]. Thus, the cross sections shown in Fig. V–7 are much less than the true cross sections. An estimate of the integrated absorption coefficient of the (0, 0) band is $1.7 \times 10^4\ cm^{-1}\ atm^{-1}$ (899). Various electronic states and transitions are given in Fig. V–8.

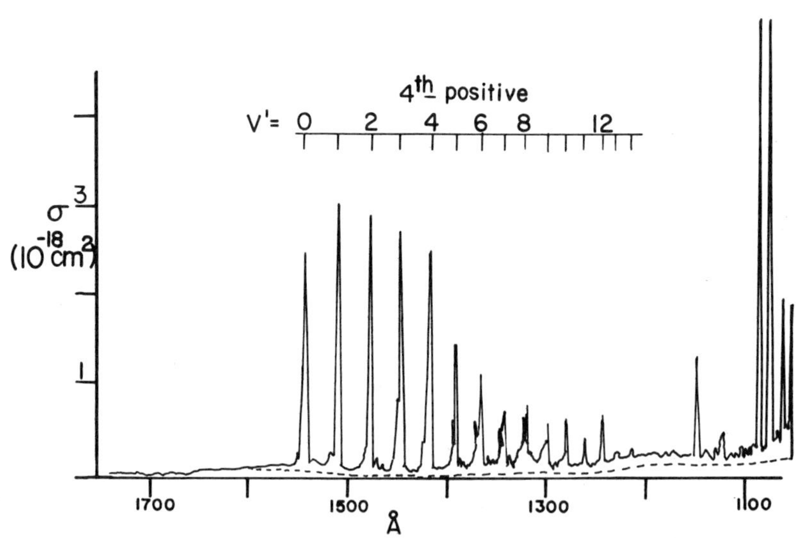

Fig. V–7. Absorption cross sections of CO in the region 1050 to 1750 Å. σ is in units of $10^{-18}\ cm^2$, base *e*, at room temperature. From Myer and Samson (727), resolution, 0.25 Å. Reprinted by permission. Copyright 1970 by the American Institute of Physics.

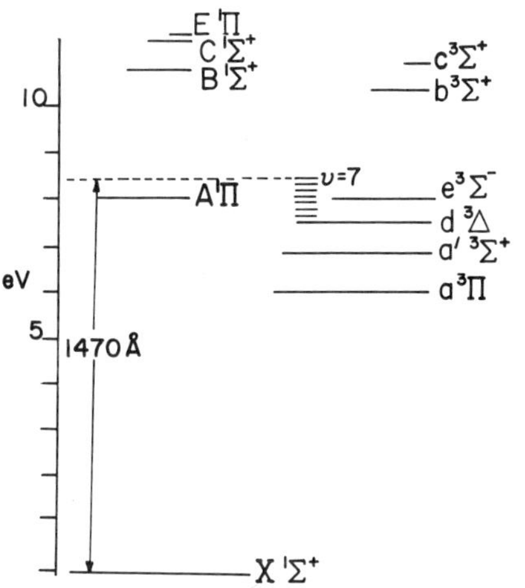

Fig. V-8. Energy level diagram of CO: *A–X*, fourth positive; *B–A*, Angstrom; *C–A*, Herzberg; *B–X*, Hopfield Birge; *a–X*, Cameron; *d–a*, triplet; *e–a*, Herman; *b–a*, third positive; *c–a*, 3A bands; *a′–a*, Asundi. The 1470 Å line is in coincidence with the $d^3\Delta$ ($v' = 7$) (898). From Gaydon (8), p. 210, reprinted by permission of Associated Book Publishers Ltd.

V-3.1. Photochemistry

Since the dissociation energy of CO is 11.09 eV, the photochemical products formed by absorption of light of wavelengths above 1118 Å must be due to the reaction of the electronically excited CO($A^1\Pi$). The photolysis products both at the Xe and Kr lines are CO_2 and C_3O_2 with quantum yields of about 6×10^{-3} and 3×10^{-3}, respectively [Groth et al. (427)]. Since fluorescence quenching takes place with more than unit collision efficiency [Becker and Welge (79)], the excited CO must be deactivated mostly to the ground state. Slanger and Black (906) have concluded that quenching of the $d^3\Delta$ ($v' = 7$) state by CO results in vibrationally excited CO with high efficiency.

A proposed mechanism for product formation is (427)

$$CO \xrightarrow{h\nu} CO^* \tag{V-11}$$

$$CO^* + CO \rightarrow CO_2 + C \tag{V-12}$$

$$C + CO \xrightarrow{M} C_2O \tag{V-13}$$

$$C_2O + CO \xrightarrow{M} C_3O_2 \tag{V-14}$$

where CO* signifies the electronically excited CO.

It has been concluded by Slanger (898) that by absorption of 1470 Å light, CO is excited to $d^3\Delta$ ($v' = 7$) (see Fig. V-8), since only the emission due to the $d^3\Delta$–$a^3\Pi$ (the triplet bands) has been observed. The 1470 Å line is in near coincidence with the transition from $X^1\Sigma^+$ ($v'' = 0, J = 15$) to $d^3\Delta$ ($v' = 7, J = 14$), while the 1470 Å line is off by 367 cm^{-1} from the closest rotational level of the $A^1\Pi$ state in agreement with the observation

Table V–1. Electronic States and Lifetimes of CO

Electronic State (eV)[a]		Lifetime
$A^1\Pi$	8.0275	10 nsec (Ref. 509), 10.5 ± 1 nsec (Ref. 466) 16 ± 1 nsec (Ref. 207) dependent on v'
$B^1\Sigma^+$	10.776	20 nsec
$B^1\Sigma^+ \rightarrow A^1\Pi$		$A = 1.11 \pm 0.12 \times 10^7\ sec^{-1}$ (Ref. 716)[b]
$B^1\Sigma^+ \rightarrow X^1\Sigma^+$		$A = 4.0 \times 10^7\ sec^{-1}$ (Ref. 466)[b]
$a^3\Pi$	6.0099	7.5 ± 1 msec (Ref. 614), 3 ~ 450 msec (J' dependent) 9.51 ± 0.63 msec (Ref. 525), 8.75 msec (average) (Ref. 526)
$d^3\Delta$	7.5192	6 μsec, (Ref. 902) J and Ω dependent
$e^3\Sigma^-$	7.8990	3 μsec (Ref. 903)
$b^3\Sigma^+$	10.394	57.6 ± 1.2 nsec (Ref. 616a), 97 ± 9 nsec (Ref. 716) 53.6 ± 0.3 nsec (Ref. 910) (v' dependent)

[a] From Ref. 24.
[b] Transition probability (sec^{-1}).

that no emission from the $A^1\Pi$ occurs. On the other hand, the Xe sensitized fluorescence of CO (79, 898) consists of emissions from the $d^3\Delta$, $e^3\Sigma^-$, and $a'^3\Sigma^+$, as well as that from the $A^1\Pi$. This indicates that strict resonance of the incident photon energy with a quantum state of the electronically excited CO is required for direct fluorescence while no such restriction is necessary for sensitized fluorescence.

Slanger and Black (907) have also observed emission from the $e^3\Sigma^-$ state by exciting CO with the 1483 Å sulfur line. The transition to this level is facilitated by the perturbation of the nearby $A^1\Pi$ state.

Certain rotational levels of triplet states are perturbed by the $A^1\Sigma$ state and emissions from these levels are seen by excitation of CO with a fourth positive emission lamp [Slanger and Black (900)]. The CO($a^3\Pi$) state is produced by a weak absorption of 2062 Å light [Harteck et al. (443)] or by the Hg($^1P_1$) sensitized reaction [Liuti et al., (644) Simonaitis and Heicklen (885)]. The reaction products of CO($a^3\Pi$) are CO_2 and C_3O_2, which are probably formed by a similar reaction sequence proposed for the photolysis of CO by Xe and Kr lamps. Quenching rates of CO($a^3\Pi$) by various gases have been measured [Taylor and Setser (963)]. The rate constant is on the order of $10^{-10}\ cm^3\ molec^{-1}\ sec^{-1}$. Various electronic states and their lifetimes are given in Table V–1.

V–4. NITROGEN

The ground state is $X^1\Sigma_g^+$ and the bond energy, D_0(N—N), is 9.760 ± 0.005 eV (8), corresponding to the incident wavelength 1270 Å. The Lyman-Birge-

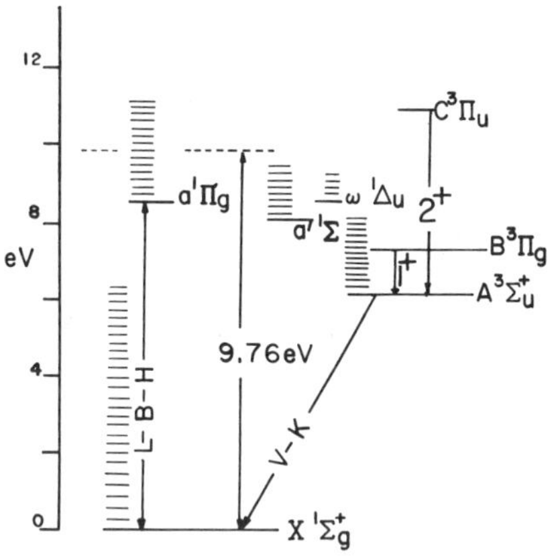

Fig. V–9. Energy level diagram of N_2. D_0(N—N) = 9.76 eV; L-B-H, Lyman-Birge-Hopfield bands; V-K, Vegard-Kaplan bands; 1, first positive bands; 2^+, second positive bands. From Gaydon (8), p. 188, reprinted by permission of Associated Book Publishers Ltd.

Hopfield bands ($a^1\Pi_g$–$X^1\Sigma_g^+$) are the main absorption bands in the region 1000 to 1500 Å. The energy level diagram is shown in Fig. V–9. Although the Lyman-Birge-Hopfield bands are the most intense bands in the region above 1000 Å, the transition by electric dipole is forbidden and the absorption coefficient is less than 0.1 (cm^{-1} atm^{-1}) (31). Table V–2 gives absorption coefficients of some of the bands. The absorption becomes prominent only below 1000 Å. The spectrum shows a strong banded structure between 660

Table V–2. Absorption Coefficients (31) of Some Lyman-Birge-Hopfield Bands of N_2

Band	Å	k (cm^{-1} atm^{-1})
0, 0	1450	0.08
1, 0	1416	0.09
2, 0	1384	0.11
3, 0	1354	0.07
4, 0	1325	0.09
5, 0	1299	0.08
6, 0	1273	0.09
7, 0	1249	0.03
8, 0	1227	0.02
9, 0	1205	0.02
10, 0	1185	0.02
11, 0	1166	0.02

and 1000 Å and a continuum below 660 Å [Cook and Metzger (234)]. Ionization takes place below 795.96 Å of incident wavelength (17).

V–4.1. Photodissociation in the Upper Atmosphere

Photodissociation of N_2 in the region 600 to 1000 Å contributes to the production of the metastable ($^2D^o$, $^2P^o$) and ground state N atoms in the 100 to 300 km region of the earth's upper atmosphere. Cook et al. (235) suggest that the metastable N atoms are produced mainly from the continuous absorption, which amounts up to 25% of the total absorption by N_2 (235). Hudson and Carter (487) have observed that most N_2 absorption bands in the region 800 to 960 Å have widths greater than 2 cm^{-1}. They suggest that the broadened rotational lines are due to predissociation. Then the production of N atoms in the region 100 to 300 km should be much greater than an estimate based on the production rate from the continuous absorption only. Beyer and Welge (106) found the production of N(^4P) (10.3 eV above ground state N atoms) atoms below the incident wavelength, 617 Å, corresponding to

$$N_2 \xrightarrow{h\nu} N(^4P) + N(^4S^o) \qquad \text{(V-15)}$$

The extent of the process is on the order of 1% of the total. The reaction of N(^2D) with O_2 appears to be a source of NO in the earth's upper atmosphere (635)

$$N(^2D) + O_2 \rightarrow NO + O \qquad \text{(V-16)}$$

Table V–3 gives lifetimes of some electronically excited N_2.

Table V–3. Lifetimes of Some Electronically Excited States of N_2

Electronic State	(eV)[a]	Lifetime
$A^3\Sigma_u^+$	6.224	2.0 ± 0.9 sec (Ref. 189) 2.5 sec (F_1, F_3 states) (Ref. 868. 869) 1.36 sec (F_2) (Ref. 868. 869)
$B^3\Pi_g$	7.391	9.1 μsec (Ref. 531)
$a'^1\Sigma_u^-$	8.449	0.7 sec (Ref. 972)
$a^1\Pi_g$	8.589	115 ± 20 μsec (Ref. 122), 140 μsec (Ref. 870) 170 ± 30 μsec (Ref. 631)
$C^3\Pi_u$	11.050	38 nsec (Ref. 531), 39.7 nsec (Ref. 535), 27 nsec (Ref. 352) (v' = 0) 40.5 ± 1.3 nsec (Ref. 181) (v' = 1) 44.4 ± 1.4 nsec (Ref. 181)

[a] Benesch et al. (95a).

V–5. NITRIC OXIDE

The spectroscopy and photochemistry of NO have been extensively studied in recent years. The results are summarized in recent reviews (454, 817). The ground state is $X^2\Pi$. The bond energy D_0(N—O) is 6.496 eV (175), corresponding to the incident wavelength 1908 Å. Absorption by NO begins at about 2300 Å. In the region 1350 to 2300 Å the absorption spectrum is composed of many discrete rotational bands. The region 1960 to 2269 Å

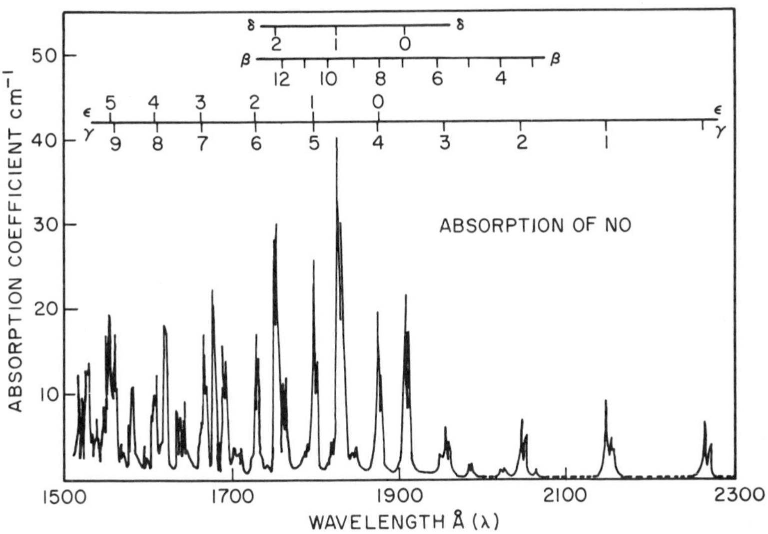

(a)

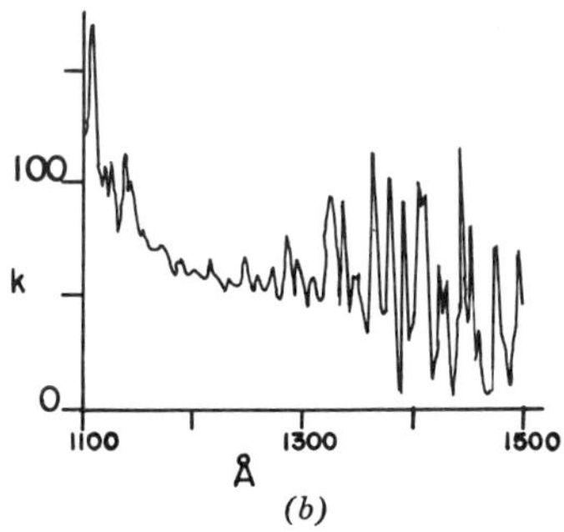

(b)

Fig. V–10. From Marmo (661), reprinted by permission. Copyright 1953 by the American Institute of Physics. (*a*) Absorption coefficients of NO in the region 1500 to 2300 Å, referred to 0°C. Units: atm^{-1} cm^{-1}, base *e*. (*b*) Absorption coefficients of NO in the range 1100 to 1500 Å, referred to 0°C. *k* in units of atm^{-1} cm^{-1}, base *e*.

corresponds to the $A^2\Pi^+-X^2\Sigma$ transition (γ bands). The 1598 to 2063 Å region is associated with the $B^2\Pi-X^2\Sigma$ transition. Absorption bands in the region 1384 to 1915 Å are due to the transitions to the $C^2\Pi$, $D^2\Sigma^+$, and $E^2\Sigma^+$ states.

Figure V–10*a* shows the apparent absorption coefficients of NO in the region 1500 to 2300 Å and Fig. V–10*b* shows those in the region 1100 to 1500 Å. Because of the lack of resolution of the monochromator, the absorption coefficient given in Fig. V–10*a* is a function of the slit width and the pressure of the gas (see Section I–8.1 for details). The true absorption coefficient may be obtained by computation from the oscillar strength and the assumed widths of the rotational lines. According to the computation by Cieslik and Nicolet (214) the true absorption coefficient of individual

Table V–4. Oscillator Strengths for the NO β, γ, δ, and ϵ Band Systems

Transition	Band	Oscillator Strength[a,b] $f(v', v'')$
$B^2\Pi-X^2\Pi$	$\beta(0,0)$	2.46 (−8)[c]
	$\beta(1,0)$	2.25 (−8)[c]
	$\beta(2,0)$	1.55 (−6), 1.19 (−6)[c]
	$\beta(3,0)$	4.61 (−6), 5.3 (−6)[d]
	$\beta(4,0)$	1.38 (−5), 1.0 (−5)[d]
	$\beta(5,0)$	2.64 (−5), 2.4 (−5)[d]
	$\beta(6,0)$	4.62 (−5)
	$\beta(9,0)$	3.58 (−4)
	$\beta(11,0)$	3.62 (−4)
	$\beta(12,0)$	23.1 (−4)
	$\beta(14,0)$	2.006 (−4)
$A^2\Sigma-X^2\Pi$	$\gamma(0,0)$	3.99 (−4), 3.64 (−4)[e]
	$\gamma(1,0)$	7.88 (−4)
	$\gamma(2,0)$	6.73 (−4)
	$\gamma(3,0)$	3.60 (−4)
$C^2\Pi-X^2\Pi$	$\delta(0,0)$	24.9 (−4)
	$\delta(1,0)$	57.8 (−4)
	$\delta(2,0)$	27.4 (−4)
$D^2\Sigma-X^2\Pi$	$\epsilon(0,0)$	25.4 (−4)
	$\epsilon(1,0)$	46.0 (−4)
	$\epsilon(2,0)$	33.2 (−4)

[a] (−8) means $\times 10^{-8}$
[b] From Ref. 103 unless otherwise noted.
[c] From Ref. 445.
[d] From Ref. 342.
[e] From Ref. 806.

rotational lines can be several hundred times greater than the apparent absorption coefficient measured at low resolution. Below about 1350 Å bands become diffuse. The absorption coefficients in the region 580 to 1350 Å are given by Watanabe et al. (1020). The potential energy curves of NO below 9 eV are given in Fig. V–11. Transitions from the $X^2\Pi$ state to the $A^2\Sigma^+$, $B^2\Pi$, $C^2\Pi$, $D^2\Sigma^+$, $B'^2\Delta$, and $E^2\Sigma^+$ states are called the γ, β, δ, ϵ, β', and γ' bands, respectively. The oscillator strengths for the β, γ, δ, and ϵ bands have been measured by many workers, (103, 174, 342, 445, 782, 806) and are given in Table V–4.

V–5.1. Fluorescence

Emission bands from the $A^2\Sigma$, $B^2\Pi$, $C^2\Pi$, and $D^2\Sigma$ states have been observed and decay rates of fluorescence have been measured extensively [Callear et al. (167–171, 174, 175)]. Various spontaneous processes of electronically excited NO are given in Table V–5. These states are quenched to a different degree by various gases. Quenching half pressures, $p_{1/2}$, in torr defined as $p_{1/2.} = (k_q\tau)^{-1}$, where k_q is the quenching rate constant in $\sec^{-1}$ torr^{-1} and τ the mean lifetime in seconds, are given in Table V–6. The $A^2\Sigma$ state is stable for quenching collisions by N_2, although the state is strongly quenched by O_2.

V–5.2. Predissociation

The predissociation of NO above about 6.5 eV is apparent, since the emission bands from $v' \geq 7(B^2\Pi)$, $v' \geq 4(A^2\Sigma)$, and $v' \geq 1(C^2\Pi)$ are missing (see Table V–7) [(8), p. 197]. On the other hand, no corresponding broadening of the lines of the β and δ bands above the dissociation limit has been observed in high resolution absorption studies (462), indicating that the predissociation is weak (see Section II–1.1). From the fluorescence quenching studies of the

Table V–5 Various Spontaneous Processes of Electronically Excited NO (175)

State	Emission to $X^2\Pi$ ($\sec^{-1}$)	Emission to $A^2\Sigma$ ($\sec^{-1}$)	Predissociation ($\sec^{-1}$)
$A^2\Sigma$	0.51×10^7	—	—
$B^2\Pi$	0.316×10^6 (531)	—	—
$C^2\Pi$	5.1×10^7	3.5×10^7	1.65×10^9 [a]
$D^2\Sigma$	4.1×10^7	0.95×10^7	$<0.8 \times 10^7$

[a] Predissociation at the limiting high pressure of Ar.

Table V–6. Quenching Half-Pressures in torr[a] of Various States of NO (175, 185)

State	NO	Ar	CO_2	N_2	CO	O_2[b]	He
$A^2\Sigma$ ($v' = 0$)	0.66	>1880	0.39	1880	7.14	0.8–0.9	9[b]
$C^2\Pi$ ($v' = 0$)	13.1	3760[c]	16	34[d]	28		
$D^2\Sigma$ ($v' = 0$)	0.44	10[e]	1.4	5.5[d]			8.3
	NO	N_2O	CO_2	H_2	H_2O	N_2[f]	CH_4[f]
$B^2\Pi$	0.077[g]	0.22	1.02	0.52	≥0.021	17 ± 3	0.41 ± 0.09

[a] Quenching half-pressure is equal to $(k_q\tau)^{-1}$ where k_q is the rate constant for quenching reaction and τ is the mean lifetime (radiative and predissociative) of excited NO.
[b] Ref. 693.
[c] Ar quenches the $C^2\Pi$ ($v' = 0$).
[d] N_2 induces the transitions $D \to A$, $C \to A$ by the formation of $N_2(A^3\Sigma)$. (169, 171, 175.)
[e] Ar induces the transition $D \to C$ with unit efficiency (174).
[f] Ref. 117.
[g] Ref. 694.

Table V–7. Predissociation of Electronically Excited NO[a]

State	v'	λ (Å)
$A^2\Sigma$	≥4	<1880
$B^2\Pi$	≥7	<1910
$C^2\Pi$[b]	≥1	<1910
$D^2\Sigma$	No predissociation	

[a] No emission appears above indicated levels. [(8), P. 197]
[b] Dissociation takes place through the $a^4\Pi$ state (169, 175).

δ bands, Callear and Pilling (175) suggest that the $C^2\Pi$ ($v = 0$) interacts with the $a^4\Pi$ as shown in Fig. V–11 and that the dissociation energy is 6.496 eV.

V–5.3. Photodissociation

Incident Wavelengths above 1910 Å. The observed photodissociation products must originate from reactions of an electronically excited state since photon energies are not sufficient to break the bond. The reaction products at 2144 and 2265 Å irradiation are N_2, NO_2, N_2O. The quantum yields are $\Phi_{N_2} = 0.19$, $\Phi_{N_2O} = 0.096$, $\Phi_{NO} = 1.05$ (9, 453, 681).

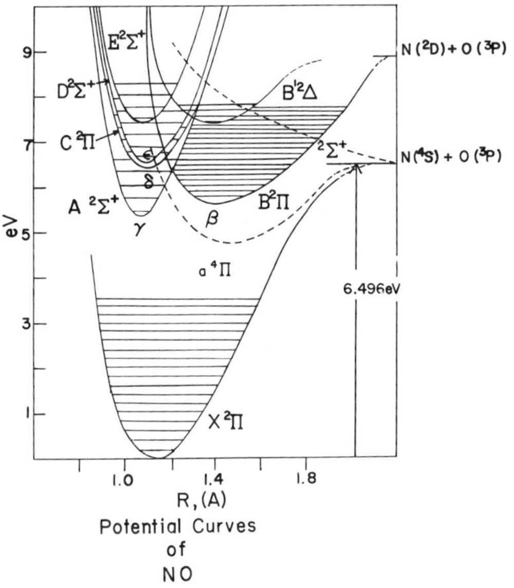

Fig. V–11. Potential energy curves for NO. From Herzberg et al. (462). The $a^4\Pi$ is drawn according to Callear and Pilling (175).

At the 1990, 1930, and 1860 Å lines, the products are N_2, O_2, NO_2, and N_2O (656). The quantum yield of NO disappearance is 1.45. A probable mechanism proposed to explain the photochemistry is

$$NO \xrightarrow{h\nu} NO^* \qquad \text{(V-17)}$$

$$NO^* + NO \rightarrow 2NO \qquad \text{(V-18)}$$

$$NO^* + NO \rightarrow N_2 + O_2 \qquad \text{(V-19)}$$

$$NO^* + NO \rightarrow N_2O + O \qquad \text{(V-20)}$$

$$2NO + O_2 \rightarrow 2NO_2 \qquad \text{(V-21)}$$

$$O + NO + M \rightarrow NO_2 + M \qquad \text{(V-22)}$$

Incident Wavelengths below 1910 Å. Hikida et al. (471, 471a) have observed a weak fluorescence of the $\beta(v' = 9)$ bands when NO is illuminated by the 1849 Å line. Since the incident photon energy is above the dissociation limit of the β system, the major path must be dissociation.

The photolysis products at 1832 Å (366) are N_2 and O_2. Since the molecule is probably raised to the $C^2\Pi$ $(v' = 1)$ level at this line, it would dissociate immediately (<1 nsec) into N + O. A likely mechanism at 1832 Å is

$$NO \xrightarrow{h\nu} N + O \tag{V-23}$$

$$N + NO \rightarrow N_2 + O \tag{V-24}$$

$$O + NO + M \rightarrow NO_2 + M \tag{V-25}$$

$$O + O + M \rightarrow O_2 + M \tag{V-26}$$

The quantum yield of N_2 is expected to be unity, although the yield has not been measured. The photolysis of NO by a hydrogen discharge lamp ($\sim$1600 Å) yields mainly N_2 and NO_2 [Leiga and Taylor (624)] with quantum yields $\Phi_{N_2} = 0.2$–0.4, $\Phi_{NO_2} = 0.3$–0.7, $\Phi_{-NO} = 0.8$–1.8. At 1470 Å (624, 865) the products are N_2, NO_2, N_2O with quantum yields $\Phi_{N_2} = 0.3$–0.5, $\Phi_{NO_2} = 0.6$–1.1, $\Phi_{N_2O} = 0.02$–0.1, and $\Phi_{-NO} = 1.3$–2.7. Quantum yields are dependent on the pressure and the flow rate. Product N_2O is not found in a flow system.

In a static system N_2O is probably formed by (624)

$$N + NO_2 \rightarrow N_2O + O \tag{V-27}$$

The result that $\Phi_{N_2} < 1$ in vacuum ultraviolet photolysis indicates that some electronically excited states produced may not dissociate immediately. This is supported by the results that emission bands from several excited NO states have been observed with 1470 Å excitation [Young et al. (1070)]. On the other hand, Stuhl and Niki (950) have found that O atoms are produced with light of wavelengths above 1600 Å in the pressure region where the reaction of an electronically excited NO can be neglected. It is likely that both the production of excited NO and dissociation take place in the vacuum ultraviolet.

The ionization of NO occurs below 1338 Å (9.266 eV) (1020). The photolysis at the Kr lines (1165, 1236 Å) (624) leads to the formation of ions

$$NO \xrightarrow{h\nu} NO^+ + e \tag{V-28}$$

The NO^+ ions must eventually be neutralized to produce some stable excited states since $\Phi_{N_2} < 1$ is obtained. An increase of Φ_{N_2} with an increase of NO pressure indicates the occurrence of the pressure induced dissociation

$$NO^* + NO \rightarrow N + O + NO \tag{V-29}$$

V–5.4. NO in the Upper Atmosphere (214)

Nitric oxide is a minor constituent in the upper atmosphere (10^8 molec cm^{-3} at an altitude of 105 km) (691). It is probably formed from the reaction

$$O(^1D) + N_2O \rightarrow 2NO \tag{V-30}$$

The ionization of NO by the Lyman-α line is the main source of ions in the D region. The photodissociation of NO in the upper atmosphere occurs from the $A^2\Sigma^+$ ($v' \geq 4$), $B^2\Pi$ ($v' \geq 7$), and $C^2\Pi$ ($v' \geq 0$). The dissociation rate of NO by the solar radiation is proportional to the integrated absorption coefficient of various bands (that is, the oscillator strength). From Table V–4 it can be seen that absorption by the β (12,0) and δ bands is most important in leading to photodissociation.

In the mesophere and stratosphere the effect of the absorption by the Schumann-Runge bands of O_2 on NO dissociation must be considered. Because of the large absorption by O_2 in the region of the β (12,0) band (1760 to 1776 Å), photodissociation of NO above about 50 km is brought about mainly by the absorption of the δ (1,0), (0,0) bands at 1830 and 1900 Å, respectively [Cieslik and Nicolet (214)].

V–6. OXYGEN

V–6.1. $O_2(X^3\Sigma_g{}^-)$

The ground state of oxygen is $X^3\Sigma_g^-$.

$$D_0(\text{O—O}) = 5.115 \pm 0.002 \text{ eV } (24)$$

Very weak absorption bands in the region 2500 to 3000 Å correspond to the forbidden transition $A^3\Sigma_u^+ – X^3\Sigma_g^-$ (see selection rules I-10.2). The band system is called the Herzberg I band. Second absorption bands in the region 1750 to 2000 Å correspond to the $B^3\Sigma_u^- – X^3\Sigma_g^-$ transition and are called the Schumann-Runge system. The region 1300 to 1750 Å is continuous and is called the Schumann-Runge continuum. Below 1300 Å numerous Rydberg transitions have been observed [Yoshino and Tanaka (1066)].

The absorption coefficients in the region 1100 to 2000 Å are given in Fig. V–12 and the absorption cross sections of the O_2 continuum in the 1814 to 2350 Å region are given in Fig. V–13.

The potential energy curves are given in Fig. V–14. Besides the two band systems already described, the two extremely weak systems $b^1\Sigma_g^+ – X^3\Sigma_g^-$ (5380 to 7620 Å) and $a^1\Delta_g – X^3\Sigma_g^-$ (9240 to 15,800 Å) have been observed in atmospheric absorption. The former is called the atmospheric bands and the latter the infrared atmospheric bands.

The convergence limit of the Herzberg I bands is at 2424 Å, corresponding to the production of $O(^3P) + O(^3P)$. Below 2424 Å lies a weak continuum (1920 to 2430 Å) [Hasson and Nicholls (447)].

The Schumann-Runge bands converge to the limit at 1750 Å corresponding to the production of $O(^3P) + O(^1D)$. The integrated absorption coefficients of the Schumann-Runge system from (0,0) to (20,0) have been

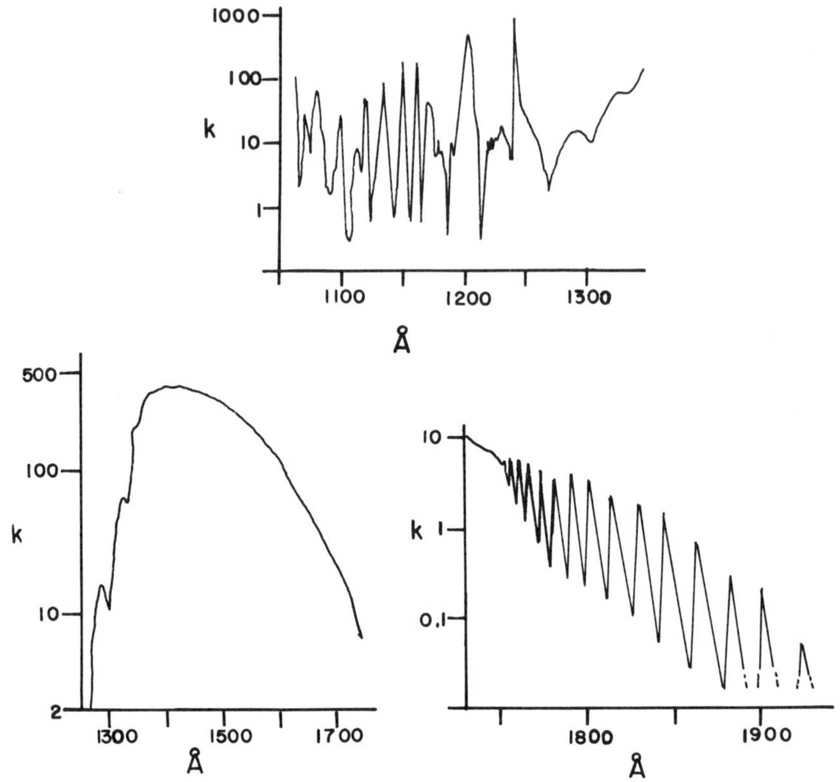

Fig. V–12. Absorption coefficients of O_2 in the region 1100 to 2000 Å. k is in units of $atm^{-1}\ cm^{-1}$, 0°C, base e. 1300 to 1700 Å, Schumann-Runge continuum. 1750 to 1950 Å, Schumann-Runge bands. From Watanabe et al. (1014), reprinted by permission. Copyright 1953 by the American Institute of Physics.

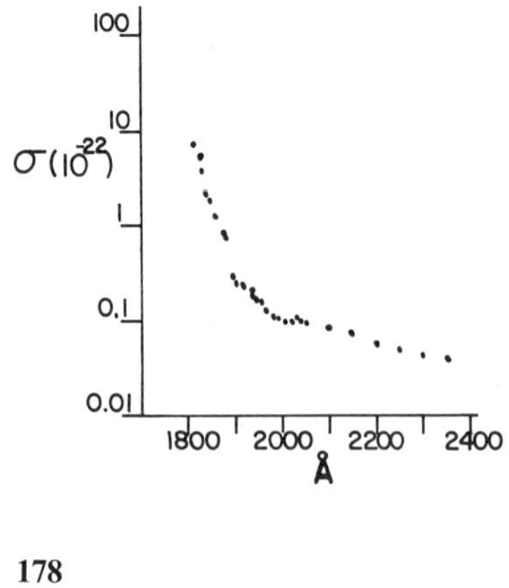

Fig. V–13. The absorption cross sections of the O_2 continuum in the 1814 to 2350 Å region. σ is in units of $10^{-22}\ cm^2\ molec^{-1}$, base e, at room temperature. The absorption cross sections are measured at minima between the well-separated rotational lines of the Schumann-Runge bands for $\lambda < 2025$ Å. σ increases with pressure probably as a result of the formation of O_4. The data are values at the low pressure limit. From Ogawa (755), reprinted by permission. Copyright 1971 by the American Institute of Physics.

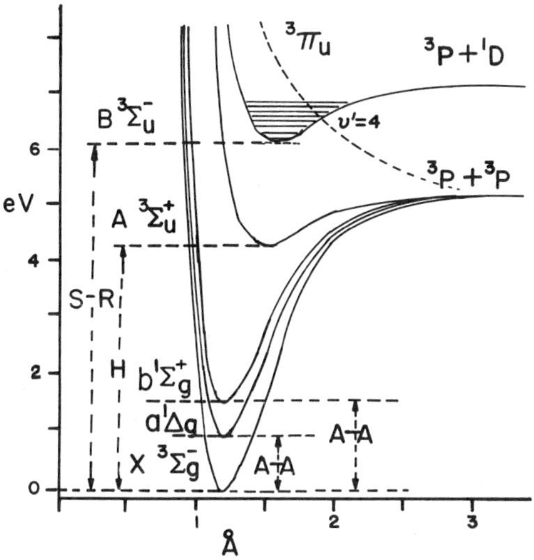

Fig. V–14. Potential energy curves of O_2. S-R, Schumann-Runge bands; H, Herzberg bands; A-A, atmospheric bands. The line-broadening was observed at $v' = 4$ of the $B^3\Sigma_u^-$ state at which point the repulsive $^3\Pi_u$ state crosses the $B^3\Sigma_u^-$ state. See Murrell and Taylor (726). From "Dissociation Energies and the Spectra of Diatomic Molecules" by Gaydon, 3rd Ed. 1968, p. 74, reprinted by permission of Associated Book Publishers Ltd.

determined (104, 341, 444). They are given in Table V–8. Ackerman et al. (38) have measured the absorption coefficients for individual rotational lines in the region 1750 to 2050 Å. The oscillator strengths of the Herzberg I bands from (4, 0) to (11, 0) have been measured by Hasson and Nicholls (446). An absorption coefficient of 0.30 ± 0.03 atm^{-1} cm^{-1} at the 1216 Å line has been determined by several workers (381, 753).

Photochemistry. The threshold wavelengths for the production of $O(^3P)$, $O(^1D)$, and $O(^1S)$ are given below.

	Threshold Wavelength (Å)
$O_2 \xrightarrow{h\nu} O(^3P) + O(^3P)$	2424
$O_2 \xrightarrow{h\nu} O(^3P) + O(^1D)$	1750
$O_2 \xrightarrow{h\nu} O(^3P) + O(^1S)$	1332

Ackerman and Biaume (37) have observed that rotational lines become diffuse at $v' = 4$, 8, and 11 for the Schumann-Runge system. They attribute the diffuseness to predissociation.

A line width of 3 cm^{-1} observed by them implies a lifetime on the order of 10^{-11} sec for the (4, 0) level. It is suggested by Carroll (192) and Murrell and Taylor (726) that predissociation is due to the crossing of the repulsive

Table V–8. Oscillator Strengths of Schumann-Runge Bands of O_2[a]

Band	Oscillator Strength	Band	Oscillator Strength
(0, 0)[b]	3.3 (−10)[c]		
(1, 0)[b]	3.53 (−9)	(11, 0)	2.74 (−5)
(2, 0)	2.69 (−8)	(12, 0)	3.58 (−5)
(3, 0)	1.54 (−7)	(13, 0)	3.66 (−5)
(4, 0)	7.11 (−7)	(14, 0)	3.69 (−5)
(5, 0)	2.80 (−6)	(15, 0)	3.77 (−5)
(6, 0)	4.40 (−6)	(16, 0)	3.31 (−5)
(7, 0)	8.15 (−6)	(17, 0)	3.16 (−5)
(8, 0)	1.22 (−5)	(18, 0)	2.03 (−5)
(9, 0)	1.50 (−5)	(19, 0)	1.74 (−5)
(10, 0)	2.05 (−5)	(20, 0)	1.35 (−5)

[a] From Farmer et al. (341).
[b] From Hasson et al. (444).
[c] (−10) means $\times 10^{-10}$

$^3\Pi_u$ at $v' = 4$ of the $B^3\Sigma_u^-$. This is shown in Fig. V–14. The suggestion is supported by the observation that the Schumann-Runge emission bands are present only up to $v' = 3$. Furthermore, from the effect of added gases on the formation of O_3 produced from the photolysis of O_2 at 1849 Å, Volman (994) concludes that the $O_2(B^3\Sigma_u^-)$ predissociates at $v' \geq 8$. Washida et al. (1011) have found that the quantum yield of O_3 production from the photolysis of O_2 at 1931 Å is only 0.3 at O_2 pressures from 300 to 1300 torr. The low quantum yield of O_3 was explained by collisional deactivation of the $O_2(B^3\Sigma_u^-, v' = 4)$ formed by absorption of the 1931 Å line. However, this conclusion seems contradicted by the finding that the $v' = 4$ level of the $B^3\Sigma_u^-$ has a lifetime of only 10^{-11} sec.

The predissociation probabilities for various vibrational levels of the $B^3\Sigma_u^-$ state have been calculated (726) assuming a repulsive curve that crosses the $B^3\Sigma_u^-$ state near $v' = 4$. Both at 1849 (1011) and at 1470 Å (954) the quantum yield of O_3 formation is 2, indicating a direct dissociation

$$O_2 \xrightarrow{h\nu} O + O \qquad \text{(V-31)}$$

$$O + O_2 + M \rightarrow O_3 + M \qquad \text{(V-32)}$$

Below 1750 Å the production of $O(^1D)$ is energetically possible. The production of $O(^1D)$ in the steady state photolysis of O_2 at 1470 Å was directly demonstrated for the first time by Noxon (745) by detecting a very weak emission at 6300 Å due to the transition $O(^1D) \rightarrow O(^3P)$. The quantum yield for the production of $O(^1D)$ in the primary process at 1470 Å is uncertain.

Noxon calculated the rate constant of $O(^1D)$ quenching by O_2 on the basis of unit quantum yield and of the equilibrium concentration of $O(^1D)$ atoms. His value of 6×10^{-11} cm^3 molec^{-1} sec^{-1} agrees well with 7×10^{-11} cm^3 molec^{-1} sec^{-1} obtained independently (456), indicating that the assumption of unit quantum yield may be justified. Below 1332 Å the production of $O(^1S)$ is energetically possible. Filseth and Welge (348) have observed an emission at 5577 Å due to the transition $O(^1S) \rightarrow O(^1D)$ in the flash photolysis of O_2 below 1340 Å. The intensity is so weak that Xe has to be added to induce the transition. No quantum yield of $O(^1S)$ production has been measured. Recently Stone et al. (937) have measured the flight time of O atoms produced in the flash photolysis of the molecular beam of O_2 in the vacuum ultraviolet. The O atoms are detected by the chemiionization reaction with samarium. The technique is similar to the one described in Section II–4.1.

The released kinetic energies of O atoms by 1470 Å photolysis have a distribution with a maximum at 1.35 eV, indicating the process

$$O_2 \xrightarrow{1470\text{ Å}} O(^3P) + O(^1D)$$

(The photon energy corresponds to the sum of the bond energy, the electronic energy of $O(^1D)$, and kinetic energy of O atoms.) The production of $O(^1S)$ at 1200 and 1240 Å appears to be a minor process.

Photodissociation of O_2 *in the Upper Atmosphere.* The source of O atoms above an altitude of 50 km is mainly from the photolysis of O_2 in the Herzberg I and Schumann-Runge continua (488). The predissociation of O_2 in the Schumann Runge bands ($v' > 3$) [Wray and Fried (1054)] is the additional source of O atoms between 65 and 95 km. Supporting evidence of the predissociation is that no fluorescence of the Schumann-Runge bands above $v' > 1$ has been observed in the upper atmosphere.

A small amount of $O(^1D)$ is produced from the $v'' = 1$ and 2 levels by absorption of the solar radiation below 1850 Å of the Schumann-Runge continuum (40) contributing to $O(^1D)$ atoms in the mesosphere and stratosphere.

V–6.2. $O_2(a^1\Delta_g)$

The electronic energy of $O_2(a^1\Delta_g)$ is 0.977 eV. The ionization potential is 11.086 eV. The $O_2(a^1\Delta_g)$ is metastable with a mean life of 64.6 min [the transition probability $A = 2.58 \times 10^{-4}$ sec^{-1} (54)], since the transition $a^1\Delta_g$–$X^3\Sigma_g^-$ is strongly forbidden by electric dipole. The (0, 0) band at 12,686 Å, known as the infrared atmospheric band, is a prominent emission band in the airglow. The absorption spectrum of $O_2(^1\Delta)$ has been measured in the vacuum ultraviolet (39, 227, 754). A diffuse band at 1442 Å is very strong (absorption coefficient 1548 atm^{-1} cm^{-1}) (754).

Photochemical Production. The direct production of $O_2(^1\Delta)$ from the ground state by light absorption is not possible since the transition is highly forbidden. However, $O_2(^1\Delta)$ is a primary product of the photolysis of O_3 in the Hartley band (see Section VI–11)

$$O_3 \xrightarrow{h\nu} O(^1D) + O_2(^1\Delta) \qquad \text{(V-33)}$$

The $O_2(^1\Delta)$ state can be detected by measuring the ionization current (682) produced by light absorption below 1118 Å

$$O_2(a^1\Delta, v'' = 0) \xrightarrow{h\nu} O_2^+ + e \qquad \text{(V-34)}$$

or by absorption at 1442 Å. The decay of the 1442 Å line has been measured following the flash photolysis of O_3 (227) from which quenching rates of $O_2(^1\Delta)$ by He, Ar, Kr, Xe, N_2, H_2, CO, and O_3 have been reported.

Emission at 12,700 Å also indicates the production of $O_2(^1\Delta)$. The intensity can be measured by a germanium photodiode cooled by liquid nitrogen in conjunction with a proper interference filter (351, 923, 1028).

Reactivities of $O_2(^1\Delta)$. The $O_2(^1\Delta)$ is stable against collisions with most gases (227). The largest quenching rate constant has been obtained for the reaction with O_3 with $k_{35} = 4 \times 10^{-15}$ cm^3 molec^{-1} sec^{-1} (227)

$$O_2(^1\Delta) + O_3 \rightarrow 2O_2 + O \qquad k_{35} \qquad \text{(V-35)}$$

(see Table V-9).

Table V–9. Comparison of the Rate Constant k_q for Quenching $O_2(a^1\Delta)$ and $O_2(b^1\Sigma^+)$ by Various Gases at 300°K

Gas	k_q for $O_2(a^1\Delta)$[a] (cm^3 molec^{-1} sec^{-1})	k_q for $O_2(b^1\Sigma^+)$[b] (cm^3 molec^{-1} sec^{-1})
He	8×10^{-21}	10×10^{-17}
O_2	2×10^{-18} [c]	4.5×10^{-16}
N_2	1.4×10^{-19}	1.8×10^{-15}
H_2	5.3×10^{-18}	1.1×10^{-12}
CO	$<7 \times 10^{-17}$	4.3×10^{-15}
O_3	4.4×10^{-15}	2.5×10^{-11} [d]

[a] From Collins and Husain (227) unless otherwise noted.
[b] From Filseth et al. (350) unless otherwise noted.
[c] From Steer et al. (923).
[d] From Gilpin et al. (398).

$O_2(a^1\Delta)$ *in the Upper Atmosphere.* Emission from the $O_2(^1\Delta)$ has been observed in the upper atmosphere (30 to 80 km) (744). The maximum concentration, located at an altitude of about 50 km, is estimated to be 4×10^{10} molec cm^{-3}. The most likely process for the production in the upper atmosphere is the photolysis of O_3 in the Hartley continuum (2000 to 3000 Å).

V–6.3. $O_2(b^1\Sigma_g{}^+)$

The electronic energy of $O_2(^1\Sigma)$ is 1.626 eV. The ionization potential is 10.437 eV. The mean lifetime of $O_2(^1\Sigma)$ is 6.9 sec ($A = 0.145$ sec^{-1}) (1007). A more recent value is 12 sec ($A = 0.082$ sec^{-1}) (1000). The prominent emission bands at 7619 and 8645 Å in the day glow are the (0, 0) and (0, 1) bands of the transition $^1\Sigma_g^+ \rightarrow {}^3\Sigma_g^-$. The (0, 0) band is called the A band (the atmospheric band). The vacuum ultraviolet absorption by $O_2(^1\Sigma^+)$ has been detected recently by Alberti et al. (39).

Photochemical Production. The production of $O_2(b^1\Sigma^+)$ from the ground state O_2 by light absorption is negligibly small. The $O_2(^1\Sigma^+)$ is produced efficiently from (1068)

$$O(^1D) + O_2 \rightarrow O_2(b^1\Sigma^+) + O(^3P) \qquad \text{(V-36)}$$

with a rate constant of 6×10^{-11} cm^3 $molec^{-1}$ sec^{-1} (745). The $O(^1D)$ atoms can be generated from the photolysis of O_2 or O_3.

Quenching Rates. Quenching rates of $O_2(b^1\Sigma^+)$ by various gases have been measured by following the decay of the 7619 Å band [Filseth et al. (350) and others (80, 282, 398, 515, 745, 752, 949)]. They are shown in Table V–9.

The $O_2(^1\Sigma^+)$ is also produced from the energy pooling reaction (45, 282, 515)

$$O_2(^1\Delta) + O_2(^1\Delta) \rightarrow O_2(^3\Sigma) + O_2(^1\Sigma) \qquad \text{(V-37)}$$

The quenching rates for $O_2(b^1\Sigma^+)$ are in general much faster than those for $O_2(a^1\Delta)$ as shown in Table V–9.

$O_2(b^1\Sigma^+)$ *in the Upper Atmosphere.* The atmospheric band of O_2 observed in the upper atmosphere (40 to 130 km) indicates that the $O_2(^1\Sigma^+)$ is produced by photochemical processes (1000). The most likely process is the photolysis of O_2 in the Schumann-Runge continuum followed by the energy transfer reaction (350)

$$O_2 + h\nu \rightarrow O(^1D) + O(^3P) \qquad \text{(V-38)}$$

$$O(^1D) + O_2 \rightarrow O(^3P) + O_2(^1\Sigma^+) \qquad \text{(V-39)}$$

V–7. SULFUR (S_2)

The ground state of S_2 is $X^3\Sigma_g^-$. The bond energy is $D_0(S—S) = 4.37 \pm 0.01$ eV. Absorption starts at about 3600 Å at 100°C. The absorption spectrum in the region 2420 to 3600 Å corresponds to the transition $B^3\Sigma_u^-$–$X^3\Sigma_g^-$. Second (1650 to 1797 Å) and third (1650 to 1708 Å) absorption spectra correspond to the $C^3\Sigma_u^-$–$X^3\Sigma_g^-$ and $D^3\Pi_u$–$X^3\Sigma_g^-$ transitions, respectively [Rosen (24)].

Ricks and Barrow (830) have obtained the predissociation limit from a rotational analysis of the emission and absorption bands of the $B^3\Sigma_u^-$–$X^3\Sigma_g^-$ system of S_2 vapor. The limit is at 35,999 ± 2.5 cm^{-1}, corresponding to the products $S(^3P_2) + S(^3P_1)$. The predissociating state (similar to R in Fig. II–5) is identified as the 1_u state.

Meyer and Crosley (700) have measured the Franck-Condon factors for the system $B^3\Sigma_u^-$–$X^3\Sigma_g^-$ using resonance excitation to $v' = 3$ and 4 levels. They have also obtained lifetimes of 20.7 ± 1.4 nsec ($v' = 3, N' = 42, J' = 43$) and 18.3 ± 1.4 nsec ($v' = 4, N' = 40, J' = 41$) for the same system (699). Smith (911) has obtained 16.9 ± 3.5 nsec by the phase shift method. The $S_2(X^3\Sigma_g^-)$ is a product in the photolysis of S_2Cl_2 [Donovan et al. (307)].

V–8. HALOGENS

V–8.1. Fluorine

The ground state is $^1\Sigma_g^+$; $D_0(F—F) = 1.56 \pm 0.02$ eV (28). The absorption coefficients in the region 2000 to 4000 Å are given in Fig. V–15 as a function of wavelength. Only a continuum has been seen corresponding to a transition to the repulsive state $A^1\Pi_u$. A series of Rydberg states is observed in the vacuum ultraviolet (24).

V–8.2. Chlorine

The ground state is $^1\Sigma_g^+$; $D_0(Cl—Cl) = 2.479$ eV (626). The very weak banded region 4780 to 6000 Å represents a transition to the $B^3\Pi(0_u^+)$ at

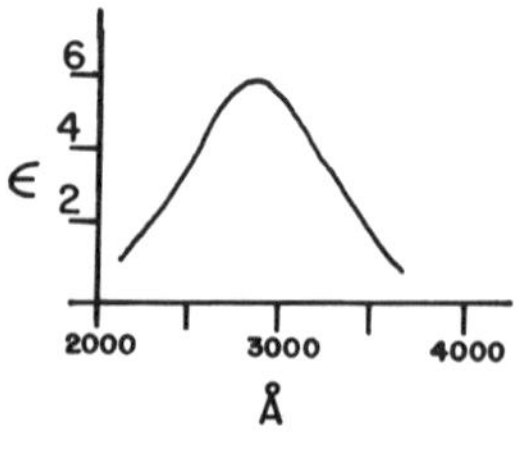

Fig. V–15. Absorption coefficients of F_2 in the region 2000 to 4000 Å. ε is in units of l mol^{-1} cm^{-1}, base 10 at room temperature. Reprinted with permission from R. K. Steunenberg and R. C. Vogel. Copyright by the American Chemical Society.

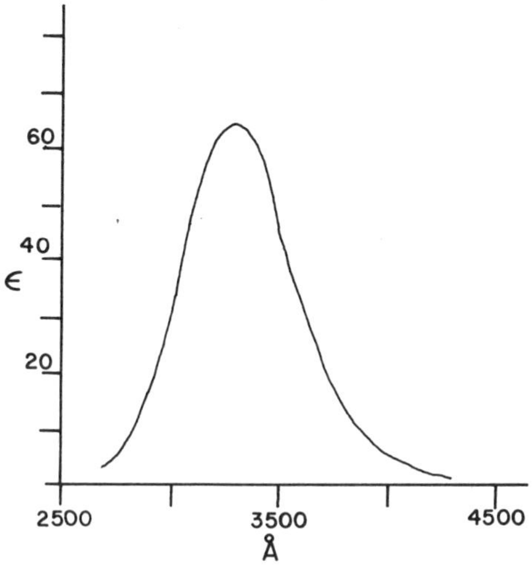

Fig. V–16. Absorption coefficients of Cl_2 in the region 2500 to 4500 Å. ϵ is in units of l mol^{-1} cm^{-1}, base 10 at room temperature. From Gibson and Bayliss (393), reprinted by permission. Copyright 1933 by the American Institute of Physics.

2.188 eV. The banded region is followed by a weak continuum in the region 2500 to 4500 Å. The absorption coefficients in the region 2500 to 4500 Å are given as a function of wavelength in Fig. V–16. The laser photolysis of chlorine molecules at 3471 Å by Busch et al. (159) indicates that the Cl atoms are both produced in the ground state, $^2P_{3/2}$, from a repulsive $^1\Pi(1_u)$ state. The photolysis in the visible banded region is expected to yield $^2P_{3/2} + {}^2P_{1/2}$, corresponding to the $B^3\Pi(0_u^+)$ state.

A second continuum lies in the region 1800 to 1950 Å followed by various Rydberg transitions in the region 1070 to 1870 Å [Lee and Walsh (621)].

V–8.3. Bromine

The ground state is $^1\Sigma_g^+$; D_0(Br—Br) = 1.971 eV (626). The weak banded absoption region 6450 to 8180 Å corresponds to a transition to the $A^3\Pi(1_u)$. The second banded region 5110 to 6400 Å represents a transition to the $B^3\Pi(0_u^+)$. A band progression in this region leads to the convergence limit at 5108 Å followed by a continuum in the region 3000 to 5110 Å. In the region 1560 to 3000 Å another continuum is observed. The absorption coefficients in the region 2000 to 6000 Å are given in Fig. V–17. Potential curves are given in Fig. V–18. The $A^3\Pi(1_u)$ state dissociates into ground state Br atoms, $^2P_{3/2}$, + $^2P_{3/2}$, while the $B^3\Pi(0_u^+)$ state yields one metastable $^2P_{1/2}$ and one ground state $^2P_{3/2}$ atom. Kistiakowsky and Sternberg (571)

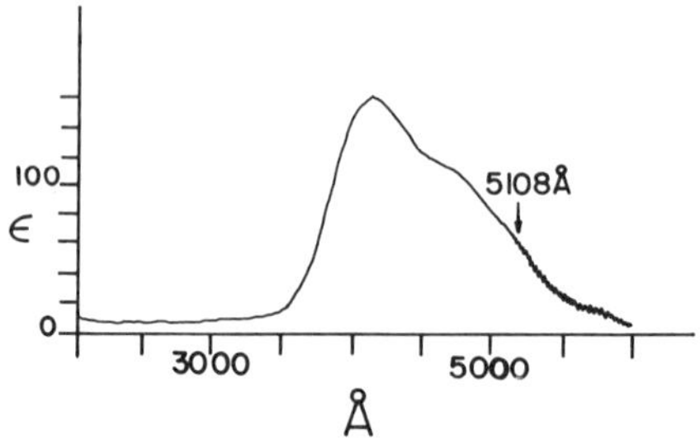

Fig. V–17. Absorption coefficients of Br_2. ϵ is in units of l mol^{-1} cm^{-1}, base 10 at room temperature. The arrow shows the convergence limit in the $^3\Pi(0_u^+)$–$^1\Sigma_g^+$ transition leading to the production $Br(^2P_{1/2})$ + $Br(^2P_{3/2})$. From Calvert and Pitts (4), p. 184, reprinted by permission of John Wiley & Sons.

have shown that the quantum yield of Br atom production is nearly independent of wavelengths between 4800 and 6800 Å. Above 6300 Å the incident photon energy is not sufficient to break the bond of bromine molecules in the lowest vibrational and rotational levels of the ground state. Photodissociation above 6300 Å can be understood if the light absorbing molecules are originally in high vibrational and rotational levels of the ground state. The sum of the photon and internal energies is equal to or exceeds the bond energy in analogy with NO_2 photolysis described in Section I–4.3. At 7150 Å no bromine atoms are produced. With 6940 Å laser light, Tiffany (971) has found that bound excited molecules are formed in the $^3\Pi(1_u)$ state, 1% of which undergo dissociation by subsequent collisions. Oldman et al. (779) have recently studied the photolysis of Br_2 by a polarized pulsed laser in the visible and ultraviolet regions.

At 5324 Å bromine molecules dissociate mainly from the $A^3\Pi(1_u)$ into ground state Br atoms. Apparently the $B^3\Pi(0_u^+)$ state is not formed at this wavelength. At 4662 Å the main dissociation is from $B^3\Pi(0_u^+)$ into $Br(^2P_{3/2})$ and $Br(^2P_{1/2})$. To a smaller extent dissociation from $A^3\Pi(1_u)$ and $^1\Pi(1_u)$ into

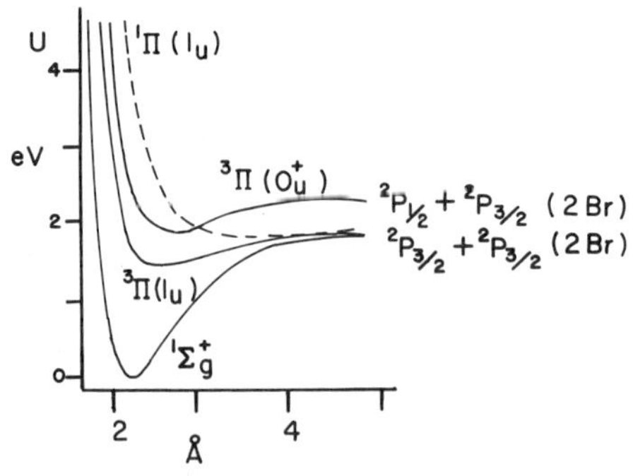

Fig. V–18. Potential energy curves of Br_2. After Kistiakowsky and Sternberg (571), reprinted by permission. Copyright 1953 by the American Institute of Physics. Absorption of light at 5435, 5940, 6140, and 6800 Å leads to the continuum of the $^3\Pi(1_u)$ state which dissociates into two normal Br atoms, $^2P_{3/2}$. Absorption at 4820 Å produces the $^3\Pi(0_u^+)$ state, which may dissociate into two normal atoms by the interaction with the repulsive $^1\Pi_u$ state or produce one normal and one excited atom.

ground state atoms has been found. (The threshold energy for the production $Br(^2P_{3/2}) + Br(^2P_{1/2})$ is 2.43 eV corresponding to 5106 Å.)

Photolysis at 3471 Å yields two ground state atoms from the state $^1\Pi(1_u)$.

Capelle et al. (186) have measured fluorescence lifetimes and quenching cross sections of vibrational levels from $v' = 1$ to 31 of the $B^3\Pi(0_u^+)$ state of Br_2 molecules. The lifetimes vary from 1.3 ($v' = 27$) to 0.14 μsec ($v' = 17$). Lifetimes must be much shorter than the radiative life since in this absorption region (5130 to 6260 Å) photodissociation is predominant (571).

The lifetimes and quenching cross sections of rotational levels in the $B^3\Pi(0_u^+)$ state have been measured near the dissociation limit by McAfee and Hozack (671). The observed lifetime is on the order of 3 μsec.

Bemand and Clyne (94) have found excited Br atoms ($5s$, $^4P_{5/2}$, $^4P_{3/2}$) in the vacuum ultraviolet photolysis of Br_2.

V–8.4. Iodine

The ground state is $^1\Sigma_g^+$; $D_0(I—I) = 1.542$ eV (626). The banded absorption region 8300 to 9300 Å corresponds to a transition to the $A^3\Pi(1_u)$ at 1.463 eV and the banded region 4990 to 8400 Å corresponds to a transition to the $B^3\Pi(0_u^+)$ at 1.949 eV. The convergence limit of the vibrational progression of the $B^3\Pi(0_u^+)$ state is at 4995 Å. The banded region is followed by a continuum in the 4000 to 4990 Å region. The 1800 to 2000 Å region is also a continuum (Cordes bands). The region 1600 to 1800 Å shows various discrete bands. Figures V–19*a* and V–19*b* give the absorption coefficients in the region 1000 to 6000 Å. Figure V–20 shows vibrational structure of the absorption bands in the region 5000 to 6500 Å. The convergence limit at 4995 Å is shown by the vertical arrow. The $A^3\Pi(1_u)$ state dissociates into two normal I atoms ($^2P_{3/2} + {}^2P_{3/2}$) while the $B^3\Pi(0_u^+)$ state yields one normal and one excited atom ($^2P_{3/2} + {}^2P_{1/2}$). This is shown in Fig. V–21.

The underlying continuum between 5000 and 6500 Å may be associated with a transition to the repulsive $^1\Pi(1_u)$ state, which dissociates into two ground state I atoms. A transition to the $B^3\Pi$ below the dissociation limit yields electronically excited I_2, which either predissociates into ground state atoms by way of the repulsive $^1\Pi$ state or returns to the ground state by fluorescence. Brewer and Tellinghuisen (147) have measured the relative concentrations of I atoms by the resonance fluorescence technique in the steady state photolysis in this region. The quantum yield of I atom production varies with the wavelength of incident light, as shown in Fig. V–22, where the quantum yield at 4920 Å is taken as unity. The quantum yields are dependent on the vibrational levels v'. When v' is low the quantum yield is near unity but it decreases at higher v', reaching a minimum near $v' = 15$. This is explained by predissociation by way of the $^1\Pi$, which crosses the

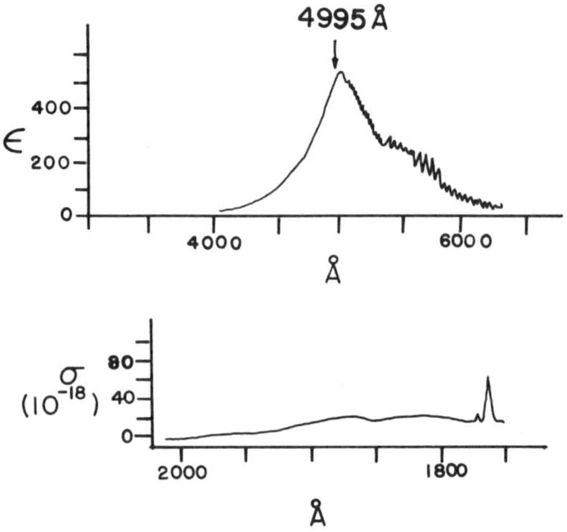

(*a*)

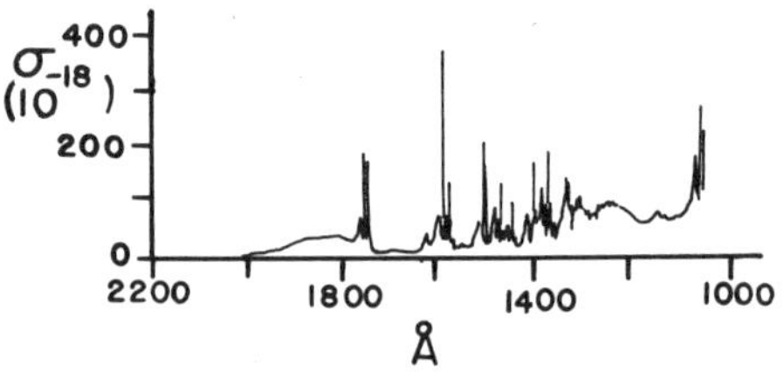

(*b*)

Fig. V–19. (*a*) Absorption coefficients of I_2 as a function of wavelength in the region 4000 to 6000 Å. Units, l mol^{-1} cm^{-1}; 70 to 80°C; base 10. The arrow shows the convergence limit in the $B^3\Pi(0_u^+)$–$X^1\Sigma$ transition leading to the production of I($^2P_{1/2}$) + I($^2P_{3/2}$). From Calvert and Pitts (4), p. 184.

Region 1800 to 2000 Å (Cordes bands): units, (10^{-18}) cm^2; base *e*. From Myer and Samson (728), 10^{-18} (cm^2) corresponds to 27 (atm^{-1} cm^{-1}), 0°C. Reprinted by permission of Wiley and the American Institute of Physics. Copyright 1970 by the American Institute of Physics. (*b*) Absorption cross sections σ of I_2 in the region 1000 to 2200 Å. Units, 10^{-18} cm^2; base *e*, room temperature. From Myer and Samson (728), reprinted by permission. Copyright 1970 by the American Institute of Physics.

$B^3\Pi$ near the bottom of the potential curve as indicated in Fig. V–21. The $^1\Pi$ state must cross the $B^3\Pi$ state again near $v' = 25$ ($\lambda = 5500$ Å) where the dissociation quantum yield shows a subsidiary maximum. The measured lifetime of fluorescence changes accordingly. The fluorescence lifetime is shortest at $v' = 4$ (0.53 μsec) and increases smoothly up to $v' = 13$, then decreases (852). This dependence of lifetime on the vibrational level may be

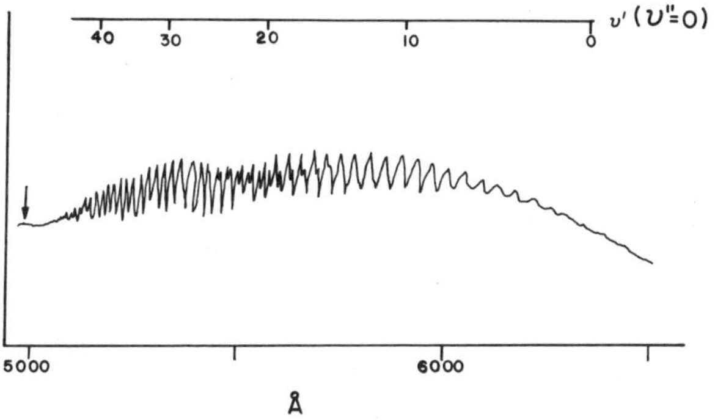

Fig. V–20. Absorption spectrum of I_2 showing the v' progression ($v'' = 0$) leading to the convergence limit at 4995 Å (indicated by the arrow). The transition in this region is $B^3\Pi(0_u^+)$–$X^1\Sigma_g^+$. Above 6000 Å contributions from $v'' = 1$ and 2 become significant. From Capelle and Broida (187), reprinted by permission. Copyright 1973 by the American Institute of Physics.

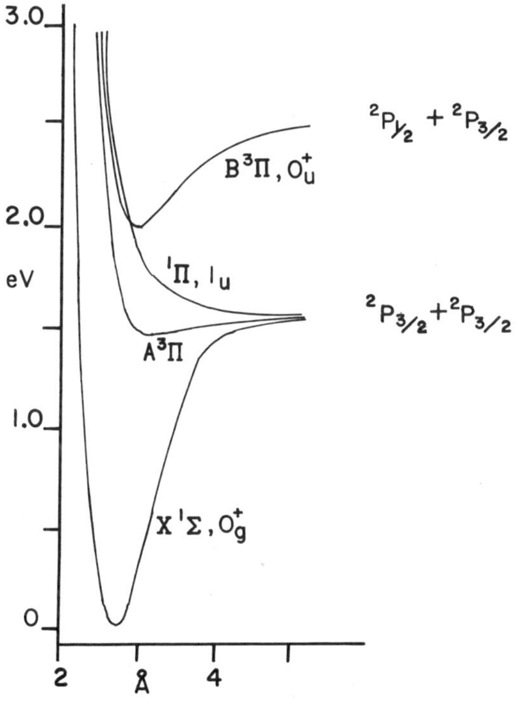

Fig. V–21. Potential energy curves of I_2. The $A^3\Pi(1_u)$ state dissociates into ${}^2P_{3/2}$ + ${}^2P_{3/2}$ atoms while the $B^3\Pi(0_u^+)$ yields ${}^2P_{1/2} + {}^2P_{3/2}$ atoms. The ${}^1\Pi(1_u)$ repulsive state crosses the $B^3\Pi$ state near the lowest vibrational level. From Brewer and Tellinghuisen (147), reprinted by permission. Copyright 1972 by the American Institute of Physics.

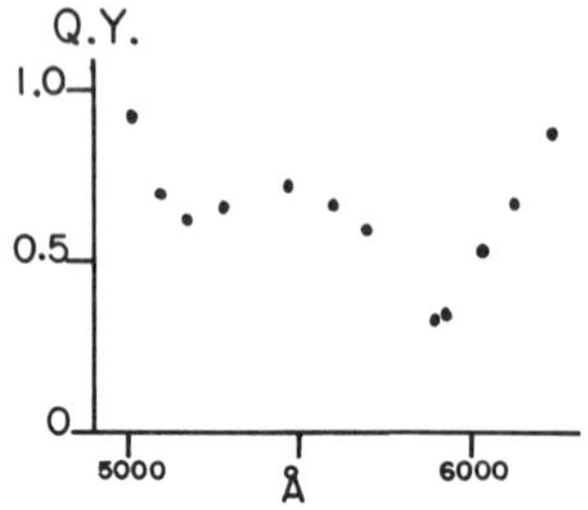

Fig. V–22. Quantum yield for photodissociation of I_2 (the production of I atoms) as a function of wavelength. The quantum yield approaches unity at both ends. The results are explained on the basis of direct dissociation from the $^1\Pi(1_u)$ repulsive state and of predissociation from the $B^3\Pi(0_u^+)$ state, which is v' dependent (a minimum near $v' = 15$, a submaximum near $v' = 25$). From Brewer and Tellinghuisen (147), reprinted by permission. Copyright 1972 by the American Institute of Physics.

explained by the simultaneous occurrence of fluorescence and predissociation as follows.

The measured lifetime τ can be expressed by the pure radiative lifetime and the rate of predissociation k_p

$$\frac{1}{\tau} = \frac{1}{\tau_0} + k_p \qquad \text{(V-40)}$$

Since the radiative lifetime is nearly independent of v' (852), it can be seen that the measured decay rate $1/\tau$ is proportional to k_p, which in turn is proportional to the quantum yield of I atom production. Therefore, the wavelength dependence of decay rate follows approximately the quantum yield curve shown in Fig. V–22, that is, the decay rate is faster when the quantum yield of atom production is larger. However, the exact correspondence may not be expected, since both the $B^3\Pi$ and $^1\Pi$ states contribute to the I atom production, while only the $B^3\Pi$ state gives rise to fluorescence. Then the percent absorption due to a transition to the $B^3\Pi$ state must be known at each wavelength.

Lifetimes and quenching cross sections of rotational levels in the $B^3\Pi$ (O_u^+) state have been measured by Broyer et al. (153) and by Ornstein and Derr (781). The production of iodine atoms $^2P_{1/2}$, $^2P_{3/2}$ was observed by absorption in the vacuum ultraviolet following the flash photolysis of I_2 above 2000 Å (296). While $^2P_{1/2}$ I atoms are produced it is not certain whether the ratio of the metastable to the ground state population is 1:1.

Using pulsed, polarized, monochromatic light of wavelength 4649 Å, Oldman et al. (778) have shown that not only the $B^3\Pi$ continuum corresponding to the production $^2P_{1/2} + {^2P_{3/2}}$, but also the $^1\Pi$ state yielding $^2P_{3/2} + {^2P_{3/2}}$, is formed by light absorption at this wavelength. Accordingly, the production of metastable $^2P_{1/2}$ atoms is much less than that of ground state atoms. Photolysis at 2662 Å by a polarized pulsed laser yields $I(^2P_{3/2}) + I(^2P_{1/2})$ [Clear and Wilson (218)]. The dissociation is probably from the $C^3\Sigma^+(1_u)$ as assigned by Mulliken.

V–9. INTERHALOGENS

V–9.1. Bromine Monochloride

The ground state is $X^1\Sigma^+$; D_0(Br—Cl) = 2.23 ± 0.01 eV. The banded absorption region 5500 to 6771 Å is ascribed to a transition to the $B^3\Pi(0^+)$. A continuum follows in the region 3520 to 3600 Å. Discrete bands (302) are observed in the regions 1613 to 1656, 1539 to 1573, 1483 to 1512, and 1383 to 1415 Å. The primary process by absorption above 2000 Å is (302)

$$BrCl + h\nu \rightarrow Br(^2P_{3/2}) + Cl(^2P_{3/2,1/2}) \qquad \text{(V-41)}$$

The fluorescence lifetime and quenching cross section of the $B^3\Pi(0^+)$ state of BrCl near the lowest vibrational level have recently been measured by Wright et al. (1055). The observed lifetime is 18.5 ± 3 μsec and the quenching cross section by BrCl is 0.1×10^{-16} cm^2.

V–9.2. Iodine Monochloride

The ground state is $X^1\Sigma^+$; D_0(I—Cl) = 2.151 ± 0.001 eV (24). Closely spaced red-degraded absorption bands appear in the region 5500 to 8800 Å. These are attributed to transitions to the $A^3\Pi(1)$ (5730 to 8770 Å), $B^3\Pi(0^+)$ (5500 to 5730 Å), and $B'(0^+)$ (5600 to 5700 Å). A continuum appears in the region 2200 to 2650 Å. Various discrete transitions appear in the vacuum ultraviolet, namely, 1795 to 1910 and 1670 to 1740 Å regions.

Flash photolysis in the region above 2000 Å leads to the production of I($^2P_{3/2}$) and probably Cl($^2P_{1/2}$) (302). Absorption above 2000 Å consists of a continuum and the discrete $B'(0^+)$-$X\ ^1\Sigma^+$ transition. The production of Cl($^2P_{1/2}$) was not observed probably because of the rapid reaction with ICl. Fluorescence is observed with light of wavelengths 5820 to 6100 Å, which lie below the dissociation limit at 5763 Å. The lifetime is on the order of 100 μsec (479) with some variation with excitation wavelengths.

V–9.3. Iodine Monobromide

The ground state is $X^1\Sigma^+$, D_0(I—Br) = 1.817 ± 0.001 eV. Closely spaced absorption bands appear in the region 5450 to 8060 Å. These are attributed to transitions to the $A^3\Pi(1)$ (5740 to 8060 Å), $B^3\Pi(0^+)$ (6200 to 6765 Å), and $B'(0^+)$ (5450 to 6190 Å). A weak continuum lies in the region near 2700 Å followed by discrete transitions in the regions 1867 to 1975 and 1728 to 1800 Å. Numerous bands are observed in the 1300 to 1600 Å region (314). Flash photolysis above 3000 Å produces I($^2P_{3/2}$) and Br($^2P_{1/2}$), which are detected by absorption in the vacuum ultraviolet (302). The $B^3\Pi$ state appears to be responsible for the production of the ground state I and the metastable Br atoms.

The photolysis at 5310 Å by a pulsed polarized laser produces mainly ground state atoms and to a smaller extent $I(^2P_{3/2}) + Br(^2P_{1/2})$ atoms [Busch et al. (160)]. Both processes are associated with the 0^+ state.

V–10. ALKALI IODIDES

V–10.1. Sodium Iodide

The ground state is $X^1\Sigma^+$, $D_0(\text{Na—I}) = 2.97$ eV (24). Three continuous regions of absorption (24) have been found in the ultraviolet with the absorption maxima at 3240, 2580, and 2120 Å. The dissociation products corresponding to the three continua are, respectively, $Na(^2S) + I(^2P_{3/2})$, $Na(^2S) + I(^2P_{1/2})$, and $Na(^2P) + I(^2P_{3/2})$.

The threshold wavelength of incident photons to produce the electronically excited $Na(^2P)$ is 2440 Å. The quenching cross sections by H_2, HCl, CO_2, and H_2O for the sodium D lines have been measured as a function of exciting wavelength above 500°C. The quenching cross sections by HCl and CO_2 decrease with an increase of relative velocities between the excited Na atoms and quenching molecules [Hanson (441), Earl et al. (332, 333)].

V–10.2. Potassium Iodide

The ground state is $X^1\Sigma^+$; $D_0(\text{K—I}) = 3.35$ eV (24). Three continuous regions of absorption have been found with maxima at 3260, 2610, and 2340 Å. The corresponding dissociation products are $K(^2S) + I(^2P_{3/2})$, $K(^2S) + I(^2P_{1/2})$, and $K(^2P) + I(^2P_{3/2})$, respectively.

Ormerod et al. (780) have studied the photolysis of KI with pulsed polarized light of wavelength 3472 Å. From the angular distribution of the product I atoms dissociated from a molecular beam of KI, they have concluded that the recoil I atom direction is nearly perpendicular to the electric vector of the polarized light. The results suggest a transition from the ionic ground state to a neutral excited state.

V–11. ELECTRONIC TRANSITIONS AND LIFETIMES OF SOME DIATOMIC RADICALS

Diatomic radicals are often produced in their ground and electronically excited states as primary photolytic products or as reaction intermediates.

Radicals such as OH and ClO are now believed to play key roles in air pollution of the troposphere and stratosphere, which is discussed in Section VIII–2.

Reaction rates of diatomic radicals have frequently been studied by following time dependent optical absorption immediately after the radicals are formed. Alternatively, radicals formed in the ground state can be brought to fluorescing excited states by suitable light sources whose wavelengths are coincident with major absorption bands of radicals. Since the fluorescence intensity is approximately linearly proportional to the ground state radical concentration, reaction rates can be measured by the time dependent fluorescence intensity (see Section III–5).

The steady state OH concentration in the atmosphere has been measured by the fluorescence technique using a dye laser tuned near 2820 Å [Wang and Davis (1006), Davis et al. (267)] or a microwave excited OH resonance lamp [Anderson (42)].

Rate constants of diatomic radicals such as OH, SH, ClO, and SO with atmospheric constituents are tabulated in recent publications by Hampson and Garvin (10) and by Anderson (1). In the following section we present the main transitions, lifetimes, and a few reaction rates of some diatomic radicals of photochemical interest. The results are summarized in Tables V–10A through V–14.

V–11.1. Diatomic Radicals Containing Hydrogen

CH (*Methylidyne*). The ground state is $X^2\Pi$; $D_0(\mathrm{C—H}) = 3.469 \pm 0.01$ eV. Both absorption and emission have been observed in regions 3143, 3900, and 4315 Å corresponding to transitions to the $C^2\Sigma^+$, $B^2\Sigma^-$, and $A^2\Delta$ states, respectively [Wallace (30)].

Radiative lifetimes of the $A^2\Delta$, $B^2\Sigma^-$, and $C^2\Sigma^+$ states have been measured by Anderson et al. (43), Fink and Welge (353), Hesser and Lutz (470), and Hinze et al. (473). These are given in Table V–10A.

Various transitions in the vacuum ultraviolet have been observed by Herzberg and Johns (467). The CH molecule dissociates by absorption of light below 3200 Å. The CH absorption has been observed in the flash photolysis of diazomethane in the near ultraviolet by Herzberg and Johns (467).

CH($A^2\Delta$) emission has been seen by the photolysis of diazomethane and diazirine in the vacuum ultraviolet [Laufer and Okabe (605, 607)].

Reaction rates of CH($X^2\Pi$) with various molecules have been measured by Bosnali and Perner (122a) and are given in Table V–10B. The reactions are generally fast with efficiencies ranging from 1 to 0.01.

Barnes et al. (61a) have detected ground state CH radicals in flame at atmospheric pressure by measuring the CH($A^2\Delta \rightarrow X^2\Pi$) fluorescence intensities excited by a tunable dye laser at 4315 Å.

NH (*Imidogen*). The ground state is $X^3\Sigma^-$; $D_0(\mathrm{N—H}) = 3.54 \pm 0.1$ eV. Three systems of NH, $A^3\Pi$–$X^3\Sigma^-$, $c^1\Pi$–$a^1\Delta$, and $c^1\Pi$–$b^1\Sigma^+$ have been observed both in absorption and in emission at 3360, 3240, and 4502 Å,

Table V–10A. Electronic Transitions and Lifetimes of CH and NH

Diatomic Radical	State	E_0 (eV)	System (Region of Absorption)	Lifetime	Ref.
CH	$X^2\Pi$	0			
	$A^2\Delta$	2.873	A–X (4315 Å)	500 ± 50 nsec	353, 470, 473
	$B^2\Sigma^-$	3.188	B–X (3600–3900 Å)	400 ± 60 nsec	43, 353, 470, 473
	$C^2\Sigma^+$	3.942	C–X (3144–3160 Å)	100 nsec	473
NH	$X^3\Sigma^-$	0			
	$a^1\Delta$	1.561			394, 762
	$b^1\Sigma^+$	2.633	b–X	≥5 msec	394, 666
			emission		1084
			(4710 Å)	17.8 msec	389
	$A^3\Pi$	3.716	A–X	0.46 μsec	352, 912
			(3200–3400 Å)	0.40–0.45 μsec	916
	$c^1\Pi$	5.374	c–a	0.43 μsec	352
			(3240 Å)	0.41–0.30 μsec	916
			c–b (4502 Å)	0.48 μsec	912
	$d^1\Sigma^+$	10.272	$d \rightarrow c$	18 ± 3 nsec	912
			emission (2530–4700 Å)	46 ± 5 nsec ($v' = 0$, Q branch)	486b

Table V–10B. Rate Constants of CH($X^2\Pi$) Reactions with Various Molecules (122*a*)

Reactant	Products	k (cm^3 $molec^{-1}$ sec^{-1})
NO	CO + NH[a]	?
N_2		1.0×10^{-12}, 7.3×10^{-14}[b]
H_2	$[CH_3]$[c]	1.7×10^{-11}, 1.1×10^{-12}[b]
O_2	CO + OH[d]	4×10^{-11}
CO		4.8×10^{-12}
H_2O		4.5×10^{-11}
NH_3		9.8×10^{-11}
CH_4	$C_2H_4 + H$	3.3×10^{-11}, 2.6×10^{-12}[a]
C_3H_8		1.4×10^{-10}
C_2H_2		7.5×10^{-11}
C_2H_4		1.1×10^{-10}

[a] From Ref. 637a.
[b] From Ref. 139.
[c] Reaction to form $CH_2 + H$ is endothermic by 3 kcal mol^{-1}.
[d] From Ref. 637b.

Table V–10C. Comparison of Reaction Rates between NH($a^1\Delta$) and NH($b^1\Sigma^+$) (cm^3 $molec^{-1}$ sec^{-1})

	NH	
Reactant	$a^1\Delta^a$	$b^1\Sigma^{+b}$
He	—	4.2×10^{-17}
Ar	—	1.8×10^{-16}
N_2	—	6.0×10^{-16}
O_2	—	2.4×10^{-15}
H_2	—	8.6×10^{-13}
H_2O	—	4.9×10^{-13}
HCl	7.9×10^{-11}	—
HN_3	9.3×10^{-11}	—
NH_3	—	4.1×10^{-13}
CH_4	1.2×10^{-11}	1.8×10^{-13}
C_2H_2	—	5.5×10^{-14}
C_2H_4	3.8×10^{-11}	1.4×10^{-13}
C_3H_6	3.6×10^{-11}	4.7×10^{-13}
C_6H_{12}	6.7×10^{-11}	—

[a] From Ref. 674b.
[b] From Ref. 1084a.

respectively [Wallace (30)]. The NH ($X^3\Sigma^-$, $a^1\Delta$, and $c^1\Pi$) states have been observed in the vacuum and near ultraviolet photolysis of ammonia and hydrazoic acid [see Sections VII–1 and VII–9 and Hansen et al. (440)]. The NH($b^1\Sigma^+$) has recently been detected in the vacuum ultraviolet photolysis of ammonia by Masanet et al. (666) and its reaction rate with ammonia has been measured by Zetzsch and Stuhl (1084).

Lifetimes of $A^3\Pi$, $b^1\Sigma^+$, and $c^1\Pi$ have been measured by various workers and are given in Table V–10A. The dependence of lifetime on rotational and vibrational levels of $A^3\Pi$ and $c^1\Pi$ states has been observed by Smith et al. (916).

Quenching of $A^3\Pi$ and $c^1\Pi$ by various gases has been studied by Kawasaki et al. (559).

Rates of reaction of NH($b^1\Sigma^+$) with CH_4, C_2H_4, and C_3H_6 are two orders of magnitude slower than the corresponding rates of NH($a^1\Delta$) as shown in Table V–10C.

The trend that more energetic NH($b^1\Sigma^+$) react less rapidly than NH($a^1\Delta$) is strikingly similar to the behavior of O(^1S), which is less reactive than O(^1D) as shown in Table IV–3.

OH(*Hydroxyl*), SH (*Sulfur Monohydride*), PH (*Phosphorus Monohydride*). The ground state of OH is $X^2\Pi$; D_0(O—H) = 4.394 ± 0.01 eV. The first transition $A^2\Sigma^+$–$X^2\Pi$ has been extensively studied both in absorption and in emission.

The OH ($X^2\Pi$) can be generated from the photolysis of water, hydrogen peroxide, and nitric and nitrous acid. Reactions of OH($X^2\Pi$) with various hydrocarbons are important in understanding photochemical smog formation (see Section VIII–2).

The OH($A^2\Sigma^+$) has been seen in the vacuum ultraviolet photolysis of water, hydrogen peroxide, and nitric acid. The OH($A^2\Sigma^+$) emission produced from OH($X^2\Pi$) by light absorption has been extensively used to measure OH($X^2\Pi$) reaction rates [for example, Stuhl and Niki (951)].

The lifetime and predissociation of OH($A^2\Sigma^+$) have been extensively studied by German (390), Smith (913), Sutherland and Anderson (955); see Table V–11A.

Quenching of the OH $A^2\Sigma^+$ state has been extensively studied by Welge et al. (1034), Hogan and Davis (477), Becker et al. (84), and Kley and Welge (576). The ground state of SH is $X^2\Pi$. D_0(S—H) = 3.60 ± 0.2 eV. The near ultraviolet absorption at 3237, 3241, and 3279 Å corresponds to the $A^2\Sigma^+$–

Table V–11A. Electronic Transitions and Lifetimes of OH, SH, and PH

Diatomic Radical	State	E_0 (eV)	System (Region of Absorption)	Lifetime	Ref.
OH	$X^2\Pi$	0		$f(0.0) = 8 \times 10^{-4}$	913
	$A^2\Sigma^+$	4.017	A–X	0.69 μsec ($N' = 0$)	390, 913
			(3064–3472 Å)	N' dependent	286
				0.82 μsec ($N' = 2$)	84
				Predissociation $> N' = 23$	955
				($N' = 34$ for OD)	1044
	$B^2\Sigma^+$	8.477	$B \rightarrow A$ (4216 Å)		
	$C^2\Sigma^+$	11.087	C → A (2160 Å)	6 nsec	915
			[C–X]	2 nsec	915
SH	$X^2\Pi$	0			
	$A^2\Sigma^+$	3.802	A–X	0.55 μsec	82
			(3240 Å)	(0.28 μsec for SD)	85
PH	$X^3\Sigma^-$	0			
	$A^3\Pi$	3.656[a]	A–X (3400 Å)	0.44 μsec	352

[a] Rostas et al. (842).

Table V–11B. Electronic Transitions of HgH[a]

Diatomic Radical	State	E_0 (eV)	System (Region of Absorption)
HgH	$X^2\Sigma$	0	
	$A^2\Pi_{1/2}$	3.047	A–X (4017 Å)
	$A^2\Pi_{3/2}$	3.503	A–X (3500 Å)
	$B^2\Sigma$	4.200	B–X (2950 Å)
	$C^2\Sigma$	4.414	C–X (2807 Å)

[a] From Refs. 24, 177, and 179.

$X^2\Pi$ transition. The corresponding emission is seen only from $v' = 0$, indicating predissociation of the SH($A^2\Sigma^+$) state for $v' > 0$ [Pathak and Palmer (798a)]. The lifetime of SH($A^2\Sigma^+$) is 0.55 μsec [Becker and Haaks (82)]. Several transitions have been found in the vacuum ultraviolet region (24). Ground state PH is $X^3\Sigma$. D_0(P—H) = 3.0 ± 0.3 eV. Near ultraviolet absorption at 3420 Å is ascribed to the $A^3\Pi$–$X^3\Sigma^-$ transition (24). Three band systems, namely, ${}^1\Phi \leftarrow a^1\Delta$, ${}^1\Pi \leftarrow a^1\Delta$, and ${}^3\Pi \leftarrow X^3\Sigma^-$, have recently been found at 1625, 1595, and 1435 Å, respectively, by Balfour and Douglas (57a). The lifetime of PH($A^3\Pi$) is 0.44 μsec [Fink and Welge (352)].

HgH (*Mercury Hydride*). The ground state is $X^2\Sigma$; D_0(Hg—H) = 0.37 eV (8). Four main transitions have been found near 4017, 3500, 2950, and 2807 Å corresponding to the $A^2\Pi_{1/2}$–$X^2\Sigma$, $A^2\Pi_{3/2}$–$X^2\Sigma$, $B^2\Sigma$–$X^2\Sigma$, and $C^2\Sigma$–$X^2\Sigma$ transitions, respectively, both in emission and in absorption [absorption by Callear et al. (177, 179)]. These transitions are shown in Table V–11B.

New HgH absorption bands have been found by Callear and Wood (179) in the vacuum ultraviolet. The HgH($X^2\Sigma$) has been seen as the main reaction product of Hg($^3P_1$) + H_2

$$\mathrm{Hg}(^3P_1) + \mathrm{H_2} \rightarrow \mathrm{HgH}(X^2\Sigma) + \mathrm{H}$$

The lifetime has apparently not been measured.

V–11.2. Diatomic Radicals Containing Carbon

CN (*Cyano*). The ground state is $X^2\Sigma^+$; D_0(C—N) = 7.85 ± 0.05 eV. Two main transitions, $A^2\Pi$–$X^2\Sigma^+$, $B^2\Sigma^+$–$X^2\Sigma^+$, have been found both in absorption and in emission in the regions 4300 to 15,100 and 3500 to 4800 Å, respectively.

The $X^2\Sigma^+$ state has been observed in the photolysis of various cyanogen compounds in the near and vacuum ultraviolet. The $B^2\Sigma^+$ and $A^2\Pi$ states have been seen in the photolysis of cyanogen compounds in the vacuum ultraviolet [Mele and Okabe (692)]. Lifetimes of the $A^2\Pi$ and $B^2\Sigma^+$ states have been measured by Jeunehomme (532), Cook and Levy (236), Luk and Bersohn (650), Liszt and Hesser (641), and Jackson (518). These values are given in Table V–12. Quenching of the $B^2\Sigma$ state has been measured by Jackson (518) and Luk and Bersohn (650).

C_2 (*Diatomic Carbon*). The ground state is a singlet $X^1\Sigma_g^+$; D_0(C—C) = 6.113 ± 0.05 eV. Seven triplet and six singlet states have been found for C_2. The strongest and most easily excited system is the Swan bands, $d^3\Pi_g \rightarrow a^3\Pi_u$ in the 4300 to 6700 Å region. The emission lifetime has been measured by Fink and Welge (353) and is 0.2 ± 0.05 μsec. Electronic transition moments of various bands of the C_2 molecule have been measured by Cooper and Nicholls (239).

Table V–12. Electronic Transitions and Lifetimes of CN and C_2

Diatomic Radical	State	E_0 (eV)	System (Region of Absorption)	Lifetime	Ref.
CN	$X^2\Sigma^+$	0			
	$A^2\Pi$	1.131	A–X Red	140 nsec ($v' = 10$)	236
			(4374–15,100 Å)	7.0 ± 0.5 μsec ($v' \leq 9$)[a] Slightly v' dependent	532
	$B^2\Sigma^+$	3.199	B–X Violet	39 nsec ($v = 0$)	236
			(3590–4216 Å)	63 ± 3 nsec K' dependent	518, 650 641
			$B \rightarrow A$ Emission (4000–5000 Å)		618
C_2	$X^1\Sigma_g^+$	0			
	$a^3\Pi_u$	0.089[b]			
	$d^3\Pi_g$	2.483[b]	d–a Swan (4383–5165 Å)	0.2 ± 0.05 μsec	353

[a] Collisionally induced intersystem crossing from CN($A^2\Pi \rightarrow B^2\Sigma^+$) observed at low pressures of BrCN appears to support a value of 7 μsec. See Ref. 47a.

[b] From Ref. 24.

The ground state $C_2(X^1\Sigma_g^+)$ is a primary product of acetylene photolysis. The $d^3\Pi_g$ state is formed from the photolysis of bromoacetylene in the vacuum ultraviolet. It is also formed in flame and discharges through carbon containing compounds. The Swan system is a major feature of emission spectrum from the heads of comets.

V–11.3. Diatomic Radicals Containing a Halogen; FO, ClO, BrO, and IO

The ground state of XO (X = F, Cl, Br, I) is $X^2\Pi$; D_0(F—O) = 2.40 ± 0.2 eV (220, 628), D_0(Cl—O) = 2.7504 ± 0.0004 eV (250), D_0(Br—O) = 2.39 ± 0.03 eV (327), D_0(I—O) = 1.8 ± 0.2 eV (327).

The $A^2\Pi$–$X^2\Pi$ transition has been observed in absorption in regions 2600 to 3100 [Coxon and Ramsay (250)] 2890 to 3550, and 4200 to 4600 Å, respectively, for ClO, BrO, and IO. Various electronic transitions of ClO in the vacuum ultraviolet have been found recently by Basco and Morse (68).

Fluorescence from ClO($A^2\Pi$) formed by light absorption of the ground state has not been detected probably because of strong predissociation of the excited state [Clyne et al. (223)].

The ClO($X^2\Pi$) has been detected by optical absorption following the flash photolysis of Cl_2O and ClO_2 or by the reaction of Cl with O_3. The latter reaction is an important source of ClO in the stratosphere [see Section VIII–2.2]. The XO($A^2\Pi$–$X^2\Pi$) transition is given in Table V–13 for ClO, BrO, and IO.

V–11.4. Diatomic Radicals Containing Sulfur

SO (*Sulfur Monoxide*). The ground state is $X^3\Sigma^-$; D_0(S—O) = 5.34 ± 0.02 eV (225, 768). The transitions $B^3\Sigma^-$–$X^3\Sigma^-$, $A^3\Pi$–$X^3\Sigma^-$ have been ob-

Table V–13. Electronic Transitions of ClO, BrO, and IO

Radical	State	E_0 (eV)	System (Region of Absorption)	Ref.
ClO	$X^2\Pi$	0		
	$A^2\Pi$	3.842	$A \leftarrow X$ (2600–3100 Å)	24, 327
BrO	$X^2\Pi$	0		
	$A^2\Pi$	3.462	$A \leftarrow X$ (2890–3550 Å)	24, 327
IO	$X^2\Pi$	0		
	$A^2\Pi$	2.673	$A \leftarrow X$ (4200–4600 Å)	24

served in the 1900 to 2600 and 2400 to 2600 Å regions, respectively. The lifetime of the $B^3\Sigma^-$ state has been determined by Smith (911) (see Table V–14).

The $SO(X^3\Sigma^-)$ is formed as a primary product of the photolysis of SO_2 below 2190 Å. The $SO(A^3\Pi, B^3\Sigma)$ states have been found in the vacuum ultraviolet photolysis of $OSCl_2$ [Okabe (768)].

CS (*Carbon Monosulfide*). The ground state is $X^1\Sigma^+$; $D_0(\text{C—S}) = 7.39 \pm 0.03$ eV. The main transition is $A^1\Pi$–$X^1\Sigma^+$ in the region 2400 to 2800 Å. The lifetime of the $A^1\Pi$ state has been measured by Smith (911) and Silvers and Chiu (882) (see Table V–14).

The $CS(X^1\Sigma^+)$ has been seen in the flash photolysis of CS_2 in the near ultraviolet. The $CS(A^1\Pi)$ has been observed in the vacuum ultraviolet photolysis of CS_2 (769) and $SCCl_2$ (774). Fluorescence from the $CS(a^3\Pi)$ state has been observed in the photolysis of CS_2 in the 1250 to 1400 Å region of absorption. The lifetime and quenching rates of $CS(a^3\Pi)$ by various gases have been determined by Black et al. (118).

Table V–14. Electronic Transitions and Lifetimes of SO and CS

Radical	State	E_0 (eV)	System (Region of Absorption)	Lifetime	Ref.
SO					24, 226
	$X^3\Sigma^-$	0			
	$a^1\Delta$	[0.79]			
	$b^1\Sigma^+$	1.303	$b \rightarrow X$ Emission (9500–10,900 Å)		
	$A^3\Pi_0$	4.748	A–X (2400–2600 Å)		
	$B^3\Sigma^-$	5.161	B–X (1900–2600 Å)	17 ± 3 nsec	911
CS					24
	$X^1\Sigma^+$	0			
	$a^3\Pi$	3.423	$a \rightarrow X$ Emission (3400–3860 Å)	16 ± 3 msec	118
	$A^1\Pi$	4.810	A–X (2400–2800 Å)	255 ± 25 nsec 176 ± 14 nsec	911 882

chapter VI

Photochemistry of Triatomic Molecules

The photochemical processes of triatomic molecules have been extensively studied in recent years, particularly those of water, carbon dioxide, nitrous oxide, nitrogen dioxide, ozone, and sulfur dioxide, as they are important minor constituents of the earth's atmosphere. (Probably more than 200 papers on ozone photolysis alone have been published in the last decade.) Carbon dioxide is the major component of the Mars and Venus atmospheres. The primary photofragments produced and their subsequent reactions are well understood for the above-mentioned six triatomic molecules as the photodissociation involves only two bonds to be ruptured and two fragments formed in various electronic states. The photochemical processes of these six molecules are discussed in detail in the following sections. They illustrate how the knowledge of primary products and their subsequent reactions have aided in interpreting the results obtained by the traditional end product analysis and quantum yield measurements.

VI–1. WATER (H_2O)

The ground state of H_2O is $\tilde{X}\,^1A_1$ with an H—O—H angle of 105.2° (16); the bond energy, D_0(H—OH) = 5.118 ± 0.01 eV (118.02 ± 0.2 kcal mol^{-1}) (28).

The absorption spectrum of water in the vacuum ultraviolet has been studied by Johns (533) and by Bell (92). Sharp rotational structure has been observed only below 1240 Å (533). The 1240 Å bands have been assigned to the $^1B_1$—$^1A_1$ transition and is the first member of the Rydberg series. The absorption coefficients of water in the vacuum ultraviolet have been measured by Watanabe et al. (1016, 1018) and are shown in Fig. VI–1. The absorption coefficients of D_2O have been measured by Laufer and McNesby in the region 1300 to 1800 Å (601).

VI–1.1. Photodissociation

The vacuum ultraviolet photolysis of water has been reviewed by McNesby and Okabe (684) and more recently by Dixon (289).

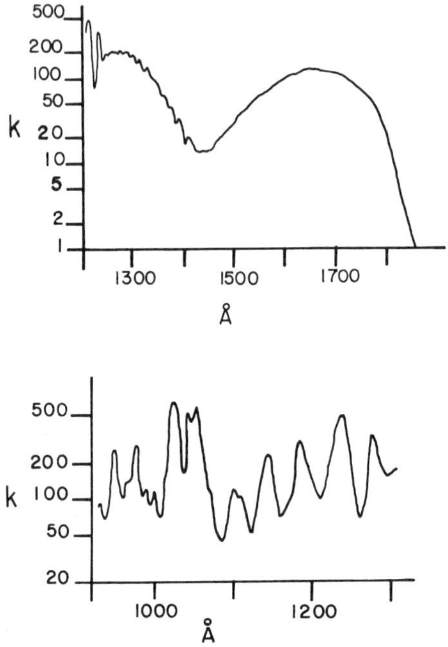

Fig. VI–1. Absorption coefficients of water in the vacuum ultraviolet region. *k* is given in units of atm^{-1} cm^{-1}, base *e*, 0°C. From Watanabe and Zelikoff (1016), reprinted by permission. Copyright 1953 by the American Institute of Physics.

1400 to 1900 Å *Region.* A major primary process in this region is the production of H and OH($X^2\Pi$)

$$H_2O \xrightarrow{h\nu} H + OH(X^2\Pi) < 2420 \text{ Å} \tag{VI-1}$$

The flash photolysis of water in this region has produced OH($X^2\Pi$), which according to Welge and Stuhl (1033), is rotationally excited only up to $N'' = 5$ and no vibrational excitation is found. The rotational distribution of OH is practically equal to that at room temperature, suggesting that the excess energy, the difference between $h\nu$ and D_0(H—OH), is distributed between translational energies of H and OH [also see Masanet et al. (241, 665)]. The excited state of water responsible for dissociation in this region is considered to be the unstable $\tilde{A}(^1B_1)$ state [Horsley and Fink (485), Miller et al. (704)].

The electron configuration of ground state water is

$$(1a_1)^2(2a_1)^2(1b_2)^2(3a_1)^2(1b_1)^2$$

and that of the first excited state $\tilde{A}(^1B_1)$ is

$$(1a_1)^2(2a_1)^2(1b_2)^2(3a_1)^2(1b_1)(4a_1)$$

The $\tilde{A}(^1B_1)$ state is derived from the promotion of a nonbonding electron in the $1b_1$ orbital to the $4a_1$ orbital. Since the bond angle and the geometry change very little by this promotion, the $OH(X^2\Pi)$ acquires very little angular momentum as the H atom flies apart from the molecule as has been described in Section II–4.2. The second primary process that is energetically feasible and spin-allowed is

$$H_2O \xrightarrow{h\nu} H_2 + O(^1D) \qquad <1770\ \text{Å} \tag{VI-2}$$

The production of molecular hydrogen to an extent of 6% of the primary process at 1470 Å has been suggested by Stief (929) from the photolysis of mixtures of water and ethylene. A similar conclusion is reached by Ung (984). At 1470 Å process (VI–2) is less than 0.3% of (VI–1) [Chou et al. (210)]. A more recent estimate for the ratio of (VI-1) to (VI-2) is 0.99:0.01 for $\lambda >$ 1450 Å and 0.89:0.11 for the 1050 to 1450 Å region [Stief et al (935)]. Stuhl and Welge (948) have obtained, from flash photolysis of mixtures of H_2O and a large excess of H_2, higher concentrations of OH than those from pure H_2O. They attribute the production of excess OH to process (VI-2) followed by

$$O(^1D) + H_2 \rightarrow OH + H \tag{VI-3}$$

1200 to 1400 Å Region. The processes (VI-1) and (VI-2) represent main primary processes in this region, although (VI-2) appears to gain more importance in the second continuum [Stief et al. (935)]. Below 1350 Å the following process occurs to an extent of up to 5%

$$H_2O \xrightarrow{h\nu} H + OH(A^2\Sigma^+) \qquad <1357\ \text{Å} \tag{VI-4}$$

The energy above the minimum required for (VI-4) is transformed predominantly into rotational excitation of $OH(A^2\Sigma^+)$ [Carrington (191)]. The observed rotational excitation of OH may be qualitatively explained from the electron configuration of an excited state of water responsible for dissociation. The excited state of water from which $OH(A^2\Sigma^+)$ dissociates is considered to be the $\tilde{B}(^1A_1)$ state with the configuration [Horsley and Fink (485), Miller et al. (704)].

$$(1a_1)^2(2a_1)^2(1b_2)^2(3a_1)(1b_1)^2(3sa_1)$$

That is, a bonding electron in the $3a_1$ orbital is excited to the $3sa_1$ orbital. The promotion would result in an increase of an H—O—H angle. The transition to the $\tilde{B}(^1A_1)$ from the ground state would therefore produce a highly excited bending vibration. A combination of antisymmetric stretching and bending vibration would yield the necessary torque to strongly rotate the $OH(A^2\Sigma)$ as the H atom flies apart. See Section II–6.5 p. 96.

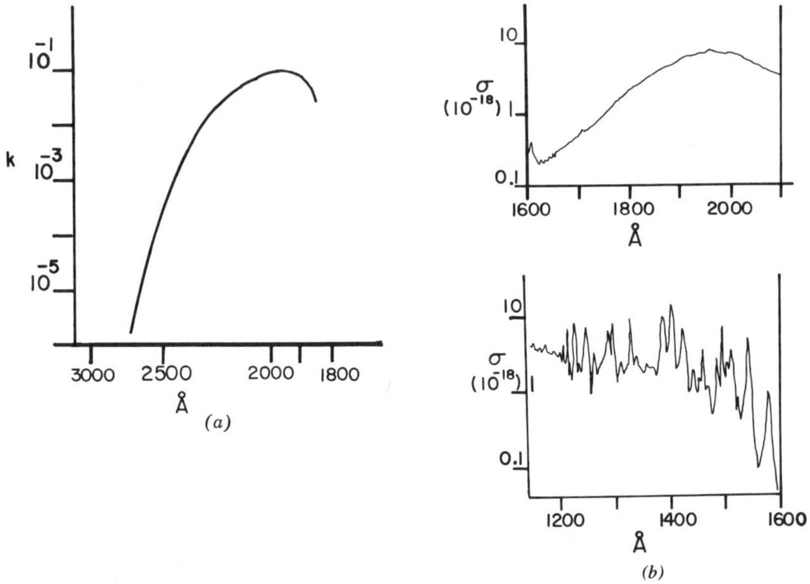

Fig. VI–2. (*a*) **Absorption coefficients of H_2S in the region 1800 to 3000 Å. k in units of $mm^{-1}\ cm^{-1}$, base 10, room temperature. 10^{-3} ($mm^{-1}\ cm^{-1}$), base 10 corresponds to 1.91 ($atm^{-1}\ cm^{-1}$), base *e*. From Goodeve and Stein (410), reprinted by permission of The Chemical Society. (*b*) Absorption cross sections of H_2S in the region 1200 to 2000 Å. σ is given in units of $10^{-18}\ cm^2$, base *e*, room temperature. From Watanabe and Jursa (1018), reprinted by permission. Copyright 1964 by the American Institute of Physics.**

VI–2. HYDROGEN SULFIDE (H_2S)

The ground state is $\tilde{X}^1A_1$ with an H—S—H angle of 92.2° (16). The bond energy D_0(H—SH) is 3.91 ± 0.2 eV (28). Absorption starts at about 2500 Å with a maximum at about 1870 Å. The absorption spectrum in this region is nearly continuous. The absorption coefficients in the near ultraviolet have been measured by Goodeve and Stein (410) and in the vacuum ultraviolet by Watanabe and Jursa (1018). They are shown in Figs. VI–2*a* and VI–2*b*. Gallo and Innes (382) have recently confirmed the 1391 Å band as being due to the $^1B_1$–$^1A_1$ transition previously assigned by Price.

VI–2.1. Photodissociation

The primary process by light absorption in the near ultraviolet appears predominantly to be the production of H atoms and SH radicals

$$H_2S \xrightarrow{h\nu} H + SH \qquad \lambda < 3170\ \text{Å} \tag{VI-5}$$

Evidence of this process is provided by the observation of SH in the flash photolysis of H_2S by Porter (813). In addition to SH, the bands due to S_2 also have been observed [Fowles et al. (371), Langford and Oldershaw (599)]. Infrared absorption of SH has been observed by Barnes et al. (61) in the ultraviolet photolysis of H_2S in low temperature matrices. Below 2000 Å S atoms have been observed in flash photolysis [Kurylo et al. (592)], indicating the occurrence of

$$H_2S \xrightarrow{h\nu} H + H + S \qquad \lambda < 1650 \text{ Å} \qquad \text{(VI-6)}$$

or

$$H_2S \xrightarrow{h\nu} H_2 + S \qquad \lambda < 4060 \text{ Å} \qquad \text{(VI-7)}$$

The quantum yield of H_2 production is about 1.2 at 2288 Å [Darwent and Roberts (259)]. The secondary reactions to explain the results are

$$H + H_2S \rightarrow H_2 + HS \qquad k_8 \qquad \text{(VI-8)}$$

$$2HS \rightarrow S_2 + H_2 \qquad \text{(VI-9)}$$

$$\rightarrow H_2S + S \qquad \text{(VI-10)}$$

$$\rightarrow HS_2 + H \qquad \text{(VI-11)}$$

$$\rightarrow H_2S_2 \qquad \text{(VI-12)}$$

The radical, HS_2, has been seen in the flash photolysis of H_2S (371, 813). Darwent et al. (260) did not detect H_2 from reactions of HS radicals. Thus, they concluded that (VI-10) is more important than (VI-9). A rate constant, k_8, of $(1.29 \pm 0.15) \times 10^{-11} \exp[(-1709 \pm 60)/1.987T]$ cm^3 molec^{-1} sec^{-1} has recently been measured by Kurylo et al. (592).

VI–2.2. Energy Partitioning in Photodissociation of H_2S

Gann and Dubrin (383) have photolyzed mixtures of $H_2S + C_4D_{10}$ at 2138 Å. The initial average kinetic energy of H atoms is found to be 1.8 eV using a technique given in Section II–4.1. This value nearly corresponds to the difference between the incident photon energy, 5.80 eV, and the bond energy, 3.9 eV; that is, in H_2S photolysis at 2138 Å nearly all excess energy appears as the translational energy of H atoms. Sturm and White (952), using a similar technique, have demonstrated that HS may be internally excited at 1849 Å, while Compton et al. (230) have found that 75% of the excess energy appears as the kinetic energy of H atoms. Oldershaw et al. (777) have found that nearly all the excess energy appears in translation of hydrogen for H_2S photolysis at 2480 Å. Compton and Martin (232) have found D atoms from D_2S photolysis have kinetic energies of 2.6 and 1.4 eV, respectively, at 1850 and 2288 Å. See p. 82.

VI–3. HYDROGEN CYANIDE (HCN)

The ground state is $X^1\Sigma^+$; D_0(H—CN) = 5.20 ± 0.05 eV (264). Hydrogen cyanide has no absorption in the visible and near ultraviolet regions. It starts to absorb weakly at about 1900 Å. Herzberg and Innes (463) have found three band systems in the region 1350 to 1900 Å, corresponding to the γ, β, and α systems. The upper states are all bent.

VI–3.1. Photochemistry

Mizutani et al. (710) have photolyzed HCN at 1849 Å. They have found cyanogen and hydrogen as major products and methane, ammonia, ethane, hydrazine, and methylamine as minor products. Mele and Okabe (692) have found CN($A^2\Pi$) and CN($B^2\Sigma$) radicals when HCN was irradiated in the vacuum ultraviolet. The vibrational and rotational energy distributions of CN($B^2\Sigma$) have been measured.

VI–4. CYANOGEN HALIDES

The ground states of cyanogen halides are $X^1\Sigma^+$; D_0(F—CN) = 4.80 ± 0.04 eV, D_0(Cl—CN) = 4.20 ± 0.05 eV, D_0(Br—CN) = 3.60 ± 0.05 eV, D_0(I—CN) = 3.16 ± 0.05 eV (264). The absorption spectra of some cyanogen halides are shown in Figs. VI–3*a* to VI–3*c* [see King and Richardson (568), Myer and Samson (727)]. They are characterized by (1) weak continuous absorption in the 1800 to 2600 Å region (the A system) resulting from the $A^1\Pi$–$X^1\Sigma^+$ transition, (2) a second weak continuous absorption at shorter wavelengths (the α system) resulting from a transition to either the second $^1\Pi$ state or a bent state $^1A'$ or $^1A''$ symmetry, (3) the intense discrete absorption in the 1300 to 1700 Å region (the *B* and *C* systems) (569), (4) Rydberg bands.

VI–4.1. Photochemistry

Donovan and Konstantatos (315) have made flash photolysis studies of ICN in the region above 2000 Å. They have found that I($^2P_{1/2}$) atoms are less than 5% of the total I atoms produced and have concluded CN radicals carry over 80% of the excess energy (about 2 eV) as translational energy. Ling and Wilson (638) have measured translational energies of the fragments, CN and I, produced from the laser photolysis of ICN at 2662 Å.

Contrary to a conclusion (315) that CN carries most of the excess energy as translational energy, Ling and Wilson (638) have found that CN radicals are produced in two different internally excited states, one probably in the $A^2\Pi$ state (60%) and the other in the vibrationally and rotationally excited

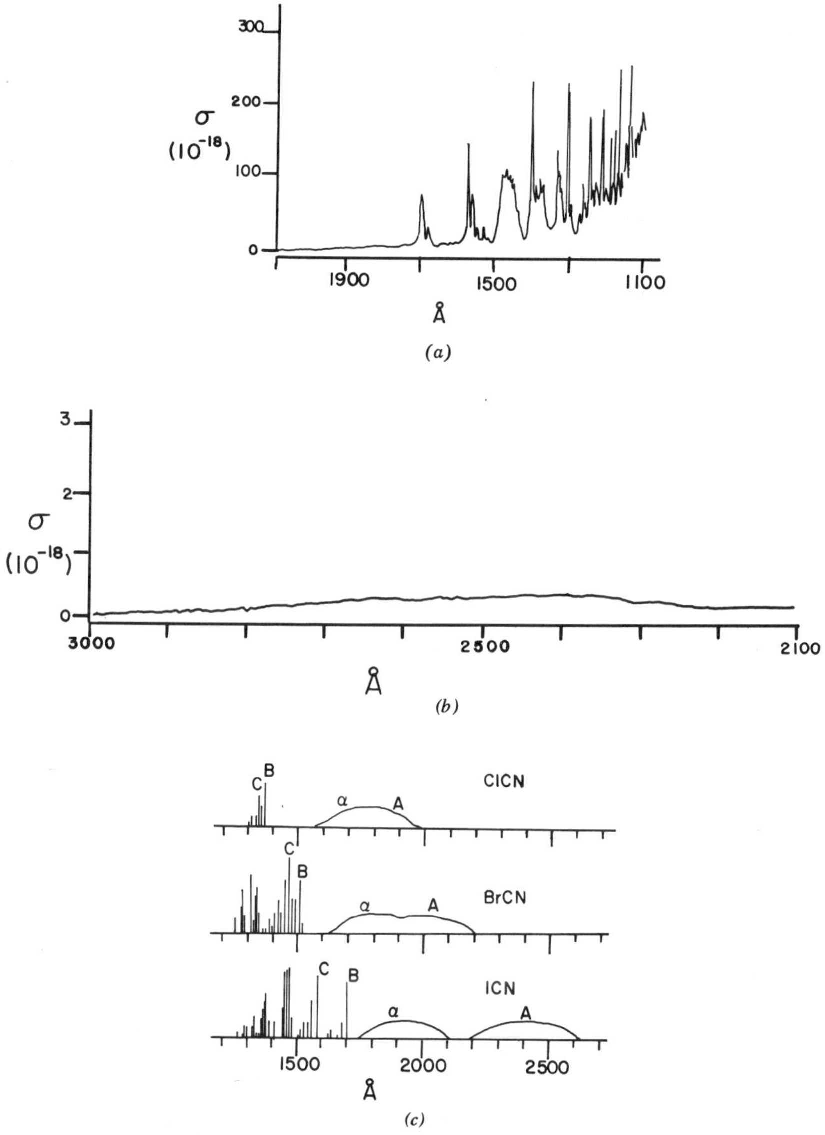

Fig. VI–3. (*a*) Absorption cross sections of ICN in the region 1100 to 2100 Å. $\sigma(10^{-18}$ $cm^2)$, base *e*, room temperature. From Myer and Samson (727), reprinted by permission. Copyright 1970 by the American Institute of Physics. (*b*) Absorption cross sections of ICN in the region 2100 to 3000 Å. σ is in units of 10^{-18} cm^2, base *e*, room temperature. From Myer and Samson (727), reprinted by permission. (*c*) Electronic absorption spectra of cyanogen halides (schematic). From King and Richardson (568), reprinted by permission of Academic Press, Inc.

ground state (40%). The angular distribution of CN radicals by polarized light indicates that dissociation takes place parallel to the transition moment. Hence, the excited state A of ICN at 2662 Å cannot be $A^1\Pi$ or $^1\Sigma^-$ previously assigned from absorption spectroscopy.

Engleman (336) has found vibrationally excited CN radicals in the flash photolysis of BrCN in the near ultraviolet. The vibrational excitation is considered to arise from the following reaction sequence

$$\mathrm{BrCN} \xrightarrow{h\nu} \mathrm{Br} + \mathrm{CN}(X^2\Sigma, v'' = 0) \qquad \text{(VI-13)}$$

$$\mathrm{CN}(X^2\Sigma) \xrightarrow{h\nu} \mathrm{CN}(B^2\Sigma) \qquad \text{(VI-14)}$$

$$\mathrm{CN}(B^2\Sigma) \rightarrow \mathrm{CN}(X^2\Sigma, v'' = 0, 1, 2\ldots) \qquad \text{(VI-15)}$$

In the vacuum ultraviolet, halogen halides partially dissociate into $CN(B^2\Sigma)$ and halogen atoms (264, 692).

Ashfold and Simons (47a) have recently shown that both $CN(A^2\Pi)$ and $(B^2\Sigma^+)$ states are formed in the vacuum ultraviolet photolysis of BrCN. At the low pressure limit $CN(B^3\Sigma^+)$ shows a vibrational population inversion at the 1236 Å photolysis (a maximum at $v' = 2$), while at higher pressures the population shows a monotonic decrease with an increase of v' observed before by Mele and Okabe (692). They attribute the pressure effect to the collisionally induced intersystem crossing between the $A^2\Pi$ ($v' \geq 10$) and neighboring $B^2\Sigma^+$ ($v' \geq 0$) levels. Because of the long radiative life of $A^2\Pi$ state ($\sim 7\ \mu$sec) (532), it is susceptible to collisions even at a pressure of 10 mtorr.

VI–5. CARBON DIOXIDE (CO_2)

The ground state of CO_2 is $X^1\Sigma_g^+$ (linear); the bond energy $D_0(\mathrm{OC{-}O}) = 5.453 \pm 0.002$ eV (28). Absorption begins at about 1700 Å. The absorption coefficients in the region 1050 to 1750 Å have been measured by Inn et al. (511), and more recently by Nakata et al. (730). The absorption coefficients in the region 1050 to 1750 Å are given in Fig. VI–4*a* and in the 1720 to 2160 Å region in Fig. VI–4*b*.

The three peaks observed at 1474, 1332, and 1119 Å are assigned by Winter et al. (1052) to the $^1\Delta_u$, $^1\Pi_g$, and $^1\Sigma_u^+$ states, respectively, on a theoretical basis. Recently the measurement has been extended beyond 1700 Å [Ogawa (755), Heimerl (461), Shemansky (871)] as the importance of the photochemistry of CO_2 in the lower atmosphere of Mars and Venus has been recognized.

The temperature dependence of the absorption coefficients in the region 1700 to 2000 Å has been measured by DeMore and Patapoff (279). The results suggest that the CO_2 absorption coefficients in the range 1700 to

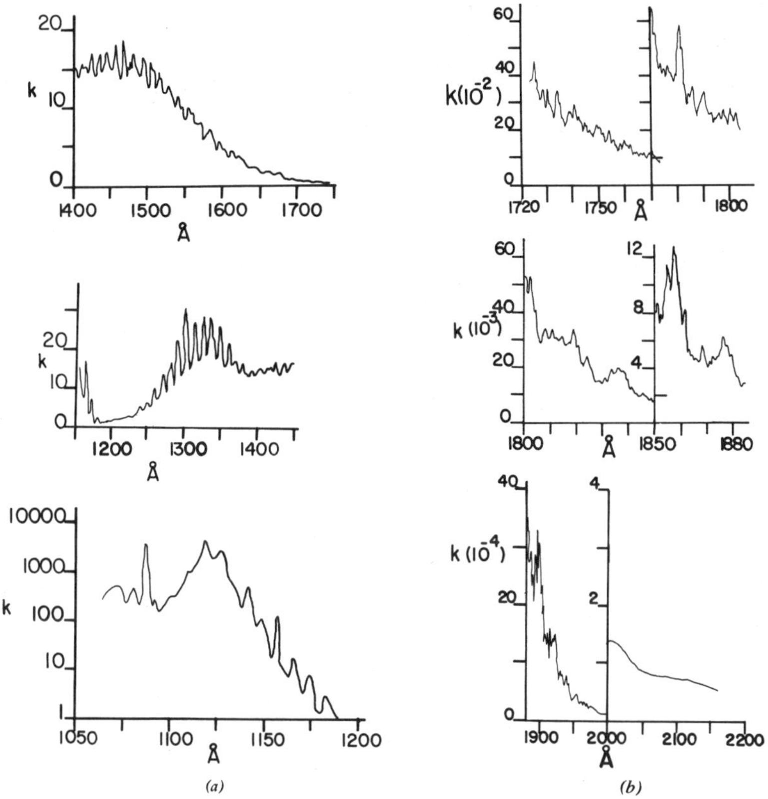

Fig. VI–4. (*a*) Absorption coefficients of CO_2 in the region 1050 to 1750 Å. *k* is in units of atm^{-1} cm^{-1}, base *e*, 0°C. From Inn et al (511), reprinted by permission. Copyright 1953 by the American Institute of Physics. (*b*) Absorption coefficients of CO_2 in the region 1720 to 2160 Å. *k* is given in units of atm^{-1} cm^{-1}, base *e*, 0°C. From Ogawa (755), reprinted by permission. Copyright 1971 by the American Institute of Physics.

2000 Å in the Mars atmosphere (200°K) are approximately one half those at room temperature. Table VI–1 shows the threshold wavelengths below which indicated reactions are energetically possible.

VI–5.1. Photochemical Reactions

The products of the photolysis of CO_2 are CO, O_2, and small amounts of O_3 at all incident wavelengths. However, the quantum yield of CO is not always unity. Furthermore, the ratio O_2/CO is usually less than half that expected from material balance. It appears that wall conditions are an

Table VI–1. Photodissociation Thresholds of CO_2 in Angstrom Units for the Production of $CO(X^1\Sigma^+, a^3\Pi, A^1\Pi)$ and $O(^3P, ^1D, ^1S)$

CO/O	^3P	^1D	^1S
$X^1\Sigma^+$	2275[a]	1672	1286
$a^3\Pi$	1082	923	792
$A^1\Pi$	920	803	702

[a] The threshold wavelength below which the dissociation of CO_2 into $CO(X^1\Sigma^+) + O(^3P)$ is energetically possible.

important factor to determine the yield of CO and O_2. Table VI–2 summarizes the results of the CO_2 photolysis. The photochemistry of CO_2 may be conveniently discussed for three regions of absorption: (*a*) above 1672 Å where only the production of $O(^3P)$ atoms is energetically allowed, (*b*) the 1200 to 1672 Å region where the $O(^1D)$ atom production is predominant, (*c*) below 1200 Å where the production of both $O(^1D)$ and $O(^1S)$ is energetically possible.

Above 1672 Å. The absorption by CO_2 in this region is extremely small (<0.4 cm^{-1} atm^{-1}) (755). The only energetically possible primary process is the production of $O(^3P)$

$$CO_2 \xrightarrow{h\nu} CO + O(^3P) \tag{VI-16}$$

The primary photodissociation yield is unity at 1849 Å (measured by the yield of $O(^3P)$ atom production), although the quantum yields of CO and O_2 are much less than 1 and 0.5 respectively [DeMore and Mosesman (278)]. The quantum yield of CO is 0.2 to 1, depending on wall conditions (278). The O_2 to CO ratio is 0 to 0.4, which is less than the 0.5 expected from (VI-16) followed by the combination of O atoms

$$O + O + CO_2 \rightarrow O_2 + CO_2 \tag{VI-17}$$

Inn and Heimerl (512) and Krezenski et al. (585), on the other hand, obtained Φ_{CO} of near unity in the 1750 to 2100 Å region and at 1849 Å, respectively. The CO yield at 2139 Å is 0.16 (585), which may indicate the production of a nondissociating excited state, although the results are much less conclusive than those at 1849 Å because the absorption at 2139 Å is only 1% of that at 1849 Å.

Table VI–2. The Quantum Yields of CO in the Photolysis of CO_2 at Various Wavelengths

Exciting Wavelength (Å)	Quantum Yield	O_2/CO	Actinometry	Ref.
2139	$\Phi_{CO} = 0.16 \pm 0.05$		$\Phi_{N_2} = 1.4$ from N_2O photolysis	585
1849	$\Phi_{CO} = 1.08 \pm 0.12$	0.44 ± 0.05	$\Phi_{N_2} = 1.4$ from N_2O	585
	$\Phi_0 = 1.0$[a]		Φ_{O_3} from $O + O_2 + M = O_3 + M$	278
	$\Phi_{CO} = 0.3–0.9$	0–0.4		
1750–2100	$\Phi_{CO} = 1.0 \pm 0.2$		Calibrated thermopile	512
1633		0.3–0.4 ($O_3/CO = 0.04–0.1$)		649
1500–1670	$\Phi_{CO} = 0.5–0.6$[b]		Calibrated thermopile	513
1200–1500	$\Phi_{CO} = 0.2–0.85$		$\Phi_{CO} = 0.75$ at 1470 Å	905
1470	$\Phi_{CO} = 1.0$	0.35 ± 0.02	$\Phi_{N_2} = 1.4$ from N_2O	1062
	$\Phi_{CO} = 1.0$	0.1–0.3	$\Phi_{N_2} = 1.4$ from N_2O	932
	$\Phi_{CO} = 0.27$	0–0.7	$\Phi_{N_2} = 1.4$ from N_2O	983
	($\Phi_{CO} = 1$)	Pressure independent (0.002–20 torr)		346
	$\Phi_{CO} = 0.6$	0.4–0.6	$\Phi_{O_3} = 2$ from O_2	846
	$\Phi_0 = 1.0$[a]		$\Phi_{O_3} = 2$ from O_2	901
	$\Phi_{CO} = 0.7–0.8$[b]		Calibrated thermopile	513
		0.05–0.2		866
1236		0–0.5		897
	$\Phi_{CO} = 0.4–0.5$	0.5–0.6	$\Phi_{O_3} = 2$ from O_2	846
		0.04–0.2		866
1048, 1066	$\Phi_{CO} = 1.06 \pm 0.1$		Photoionization yield of NO	1009
		0.02–0.2		808

[a] Quantum yield of O atoms.
[b] Probably the most reliable value, ±20% error limit.

1200 to 1672 Å Region. The production of $O(^1D)$ atoms is energetically possible in this region

$$CO_2 \xrightarrow{h\nu} CO + O(^1D) \qquad \text{(VI-18)}$$

The direct detection of $O(^1D)$ by the emission at 6300 Å in the steady state photolysis of CO_2 near 1470 Å has failed [Young and Ung (1067), Clark and Noxon (216)], although $O(^1D)$ atoms from O_2 photolysis near 1470 Å have been detected. The failure to detect the emission must be due to the weak absorption of CO_2, rapid quenching of $O(^1D)$ by CO_2, and the long radiative life of $O(^1D)$ atoms. However, the production of $O(^1D)$ in the CO_2 photolysis is strongly indicated by the following observations:

1. O atoms produced from O_3 photolysis at 2537 Å are capable of exchanging with O atoms in CO_2, while O atoms produced from O_3 by absorption of visible light do not exchange (558).
2. O atoms from N_2O photolysis at 1849 Å exchange with those in CO_2 (1060).
3. When $C^{16}O^{16}O$–$C^{18}O^{18}O$ mixtures are irradiated by 1470 Å light, $C^{16}O^{18}O$ is produced (73). These observations suggest that O atoms produced from CO_2 at 1470 Å must be in the same state as those produced from O_3 at 2537 Å and from N_2O at 1849 Å, namely, $O(^1D)$.
4. The product, neopentanol, of the photolysis of CO_2–neopentane mixtures at 1633 Å indicates that $O(^1D)$ atoms are produced. Furthermore, the ratio neopentanol/CO = 0.65 obtained in a large excess of neopentane suggests that the quantum yield of $O(^1D)$ production is close to unity [Quick and Cvetanović (821)], since the same ratio of 0.65 is obtained for N_2O photolysis at 2139 Å. To explain the rapid exchange of $O(^1D)$ with CO_2, Katakis and Taube (558) postulated the intermediate formation of CO_3. In fact, new infrared absorption bands found in the 2537 Å photolysis of O_3 in a CO_2 matrix at 50 to 60°K have been assigned to CO_3 absorption (713, 1028).

Jacox and Milligan (520a) favor the three-membered ring structure with an O—C—O angle of 65° from the analysis of infrared spectra of isotopic species of CO_3 in low temperature matrices. A new broad and weak absorption band at 4060 Å with an absorption coefficient of 1.1 ± 0.3 atm^{-1} cm^{-1} is also found in O_3—CO_2 matrix (551). However, no corresponding infrared absorption bands have been found in the gas phase photolysis [DeMore and Dede (277)]. The photochemistry of CO_2 in this region may be summarized as follows.

$$CO_2 \xrightarrow{h\nu} CO + O(^1D)$$

$$O(^1D) + CO_2 \rightarrow CO_3^*$$

$$CO_3^* \rightarrow CO_2 + O(^3P)$$

$$2O(^3P) + CO_2 \rightarrow O_2 + CO_2$$

The lifetime of CO_3^* with respect to dissociation into $CO_2 + O(^3P)$ is about 10^{-11} to 10^{-12} sec [DeMore and Dede (277)], which corresponds to 10 to 100 vibrations during its lifetime. Arvis (46), on the other hand, photolyzed 1 torr of CO_2 at 1470 Å and found, by infrared absorption, a product, CO_3, captured on a cooled LiF window. He estimates the lifetime of CO_3^* to be 0.04 sec, which is much longer than the estimated value of 10^{-11} to 10^{-12} sec. It is probable that CO_3 may be formed *in situ* on the cooled window rather than in the gas phase. Slanger (897) believes O_2 is formed by the combination of CO_3

$$2CO_3 \rightarrow 2CO_2 + O_2 \qquad \text{(VI-19)}$$

As in the photolysis above 1672 Å the ratio O_2/CO is generally much less than 0.5, a value expected from material balance. Low values obtained at low CO_2 pressures indicate the loss of $O(^3P)$ atoms on the walls (649, 897). At high CO_2 pressures O_3 is formed from $O(^3P) + O_2 \xrightarrow{M} O_3$, which partially explains the O_2 deficiency [Loucks and Cvetanović (649)].

Slanger et al. (905) have measured the relative yield of CO production at several wavelengths in the range 1200 to 1500 Å. The yields are 0.57 ± 0.11, at 1216 Å, 0.21 ± 0.07 at 1302 to 1306 Å, 0.46 ± 0.05 near 1390 Å, and 0.58 ± 0.06 at 1492 to 1495 Å, using $\Phi_{CO} = 0.75$ at 1470 Å obtained by Inn (513). The authors have concluded that only the direct dissociation from an excited CO_2 produced by light absorption in the continuous portion of the absorption spectrum may contribute to the CO production. A good material balance ($O_2/CO = 0.5$) was obtained in the Xe sensitized photolysis of CO_2 at 1470 Å (932, 933).

Photolysis of CO_2 *below 1200* Å. The production of $O(^1S)$ is energetically possible below 1286 Å

$$CO_2 \xrightarrow{h\nu} CO + O(^1S) \qquad \text{(VI-20)}$$

Lawrence (616) and Koyano et al. (584) have detected the production of $O(^1S)$ by the emission at 5577 Å ($^1S{-}^1D$ transition) in the entire region of absorption, 800 to 1220 Å. The $O(^1S)$ yield increases to a maximum at about 1150 Å and starts to decrease below 1080 Å where the production of $CO(a^3\Pi)$ begins. At 1048 Å the $O(^1S)$ quantum yield is $75 \pm 25\%$ (616).

The $CO_2(^1\Sigma_u^+)$ state must be responsible for the production of $O(^1S)$. Below 1082 Å the production of $CO(^3\Pi)$ is possible

$$CO_2 \xrightarrow{h\nu} CO(a^3\Pi) + O(^3P) \qquad \text{(VI-21)}$$

Lawrence (615) has confirmed the production of $CO(a^3\Pi)$ from the emission of the Cameron bands ($a^3\Pi{-}X^1\Sigma$). The absolute yield of $CO(a^3\Pi)$ has been measured in the absorption region 850 to 1100 Å. The yield increases smoothly from threshold to a maximum of about 60% near 900 Å. The production of metastable O atoms [$O(^1S)$ and $O(^1D)$] was detected by electron emission from metal surfaces when CO_2 was irradiated in the region 1050 to 1700 Å [Welge and Gilpin (1036)]. By measuring the time-of-flight of metastable O atoms to reach a detector after flash photolysis, they have concluded that more than 50% of dissociation leads to internally excited $CO(X^1\Sigma)$. Vibrationally excited $COa'^3\Sigma^+$, $d^3\Delta$, and $e^3\Sigma^-$ are produced by excition of CO_2 with light of wavelengths below 923 Å (552, 622). The vibrational population distributions are found to follow a Poisson formula. The quantum yield of CO, $\Phi_{CO} = 0.20 \pm 0.05$, was obtained for the photolysis of CO_2 at 584 Å (965). Emission spectra from the $CO_2^+(B^2\Sigma_u^+, A^2\Pi_g)$ states have been observed by Wanchop and Broida (1024) in the illumination of CO_2 at 584 Å.

VI–5.2. Stability of CO_2 in the Mars and Venus Atmosphere

It is known that the main constituent of the atmospheres of Mars and Venus is CO_2. The results of the photochemical studies of CO_2 in the laboratory indicate that CO_2 should be converted into CO and O_2 with solar radiation below 2275 Å. The atmospheres of Mars and Venus should thus contain substantial amounts of CO and O_2. Yet it has been observed that the mixing ratio of CO and O_2 relative to CO_2 is only on the order of 10^{-3} on Mars (677) and 10^{-5} to 10^{-6} on Venus (678). This unusual stability of CO_2 toward photolysis has been a mystery. McElroy and Donahue (677) and Parkinson and Hunten (798) have proposed an $OH–HO_2$ cycle to catalytically recombine CO + O to form CO_2

$$OH + CO \rightarrow CO_2 + H \qquad \text{(VI-22)}$$

$$H + O_2 + CO_2 \rightarrow HO_2 + CO_2 \qquad \text{(VI-23)}$$

$$O + HO_2 \rightarrow OH + O_2 \qquad \text{(VI-24)}$$

$$\text{net } CO + O \rightarrow CO_2$$

Hydroxyl radicals are produced by the photolysis of H_2O, which is present to an extent of 0.2% in the Mars and Venus atmosphere. Besides water, HCl, a minor constituent (6×10^{-7} mixing ratio) in the Venus atmosphere, may provide additional H atoms [McElroy et al. (678)].

$$HCl \xrightarrow{h\nu} H + Cl$$

$$Cl + H_2 \rightarrow HCl + H$$

$$\text{net } H_2 \rightarrow 2H$$

More details are given in Section VIII–3.

VI–6. CARBONYL SULFIDE (OCS)

The ground state is $X^1\Sigma^+$ (linear). The bond dissociation energies are D_0(OC—S) = 3.12 ± 0.03 eV and D_0(O—CS) = 6.81 ± 0.13 eV (28, 769). The absorption coefficients of OCS in the near and vacuum ultraviolet have been measured by Sidhu et al. (880), Ferro and Reuben (347), and Matsunaga and Watanabe (670). These are shown in Figs. VI–5*a* and 5*b*. The temperature dependence of absorption in the near ultraviolet has been obtained by Ferro and Reuben (347). The near ultraviolet absorption spectrum starts at about 2550 Å and is continuous.

VI–6.1. Photodissociation in the Near Ultraviolet (1900 to 2550 Å)

The quantum yield of CO formation has been found by Sidhu et al. (880) to be 1.81 both at 2537 and 2288 Å. The yield of CO has been reduced to one half (to a value of 0.9) by the addition of sufficient amounts of olefins. The primary process must be

$$\mathrm{OCS} \xrightarrow{h\nu} \mathrm{CO} + \mathrm{S} \qquad \text{(VI-25)}$$

with a quantum yield of 0.9. It has been shown by Gunning and Strausz (430) that the reactions of $S(^1D)$ with paraffins produce corresponding alkyl mercaptans. It has been found (880) that 74% of the S atoms produced in (VI-25) form mercaptans with alkanes at 2288 Å. Therefore, at this wavelength

$$\mathrm{OCS} \xrightarrow{h\nu} \mathrm{CO} + \mathrm{S}(^1D) \qquad \phi \simeq 0.67 \qquad \text{(VI-25a)}$$

$$\mathrm{OCS} \xrightarrow{h\nu} \mathrm{CO} + \mathrm{S}(^3P) \qquad \phi \simeq 0.24 \qquad \text{(VI-25b)}$$

The $S(^3P, ^1D)$ atoms react with OCS to produce CO + S_2 [Langford and Oldershaw (600)]

$$\mathrm{S}(^3P) + \mathrm{OCS} \rightarrow \mathrm{CO} + \mathrm{S}_2(X^3\Sigma_g^-) \qquad \text{(VI-26)}$$

$$\mathrm{S}(^1D) + \mathrm{OCS} \rightarrow \mathrm{CO} + \mathrm{S}_2(^1\Delta, ^1\Sigma) \qquad k_{27} \qquad \text{(VI-27)}$$

The deactivation process

$$\mathrm{S}(^1D) + \mathrm{OCS} \rightarrow \mathrm{S}(^3P) + \mathrm{OCS} \qquad \text{(VI-28)}$$

has been found to be minor. From the flash photolysis of OCS above 2200 Å, Fowles et al. (371) have found indirect evidence that S_2 in (VI-27) is either in the $^1\Delta$ or $^1\Sigma$ metastable state. The metastable S_2, although not detected directly, may be collisionally deactivated to the ground state $X^3\Sigma_g$, which was detected by optical absorption

$$\mathrm{S}_2(^1\Delta, ^1\Sigma) + \mathrm{M} \rightarrow \mathrm{S}_2(X^3\Sigma_g) + \mathrm{M} \qquad \text{(VI-29)}$$

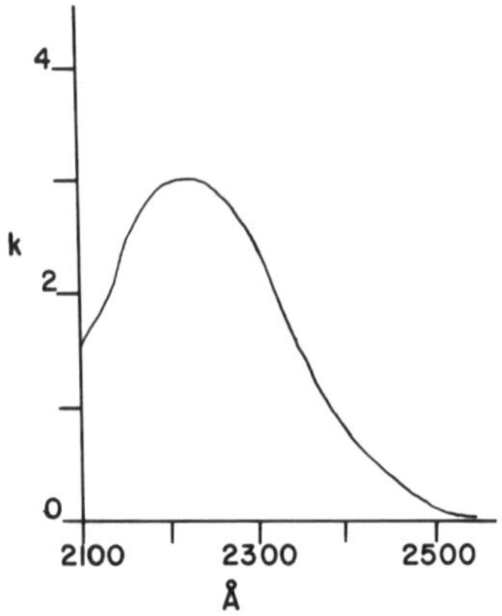

(a)

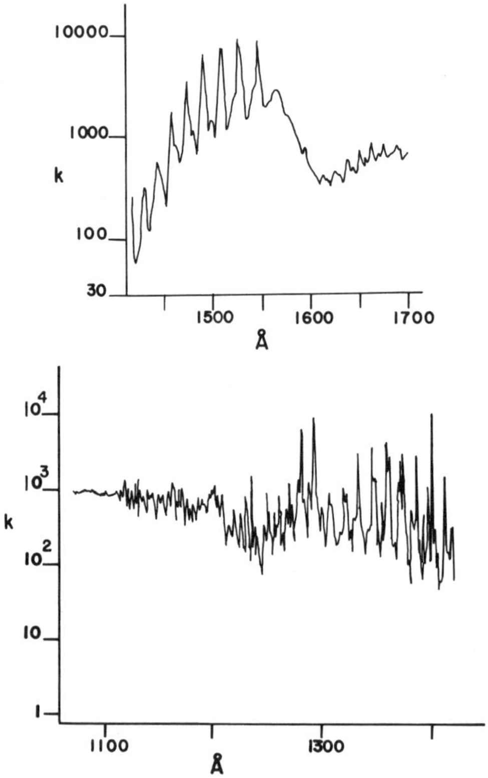

(b)

Fig. VI–5. **(*a*) Absorption coefficients of OCS in the region 2100 to 2500 Å. *k* is given in units of atm^{-1} cm^{-1}, 0°C, base *e*. From Ferro and Reuben (347), reprinted by permission of The Chemical Society. (*b*) Absorption coefficients of OCS in the vacuum ultraviolet. *k* is given in units of atm^{-1} cm^{-1}, 0°C, base *e*. From Matsunaga and Watanabe (670), reprinted by permission. Copyright 1967 by the American Institute of Physics.**

Breckenridge and Taube (144) have studied the photolysis of $OCS + CS_2$ and $OCS + N_2O$ mixtures at 2288 and 2537 Å. They have demonstrated that the primary yield of production of $S(^1D)$ [process (VI-25a)] is 0.74 ± 0.04 and 0.25 for $S(^3P)$ production [process (VI-25b)] in agreement with the results of Gunning and Strausz (430). The deactivation process [process (VI-28)] must be about one third that of the total reaction of $S(^1D)$ with OCS in order to be consistent with their finding that 50% of the S atoms formed in the primary process react as $S(^3P)$ (144).

VI–6.2. Photodissociation in the Vacuum Ultraviolet

Both $S(^3P)$ and $S(^1S)$ atomic absorption lines have been observed in the vacuum ultraviolet photolysis of OCS by Donovan et al. (304, 305), indicating the occurrence of (VI-25b) and

$$OCS \xrightarrow{h\nu} CO + S(^1S) \qquad \lambda < 2110 \text{ Å} \qquad \text{(VI-30)}$$

In spite of strong chemical evidence for the production of $S(^1D)$, absorption lines of $S(^1D)$ have not been found (304), presumably because of the rapid reaction with OCS molecules. Black et al. (114) have found that $S(^1S)$ atoms are produced with a quantum yield of almost unity in the incident wavelength region 1420 to 1600 Å, the $COS(^1\Sigma^+)$ state presumably dissociating into $CO + S(^1S)$. In the vacuum ultraviolet photolysis, the processes

$$OCS \xrightarrow{h\nu} CS + O(^3P) \qquad \lambda < 1820 \text{ Å} \qquad \text{(VI-31a)}$$

$$OCS \xrightarrow{h\nu} CS + O(^1D) \qquad \lambda < 1410 \text{ Å} \qquad \text{(VI-31b)}$$

are energetically possible below the indicated wavelengths. However, Donovan (305) has found them to be of minor importance. Donovan et al. (304) also have found the $S_2(a^1\Delta)$ absorption bands in the vacuum ultraviolet flash photolysis of OCS. From the rate of increase of $S_2(a^1\Delta)$ they have concluded that the rate constant k_{27} is larger than 0.7×10^{-10} cm^3 molec^{-1} sec^{-1}. As $S_2(a^1\Delta)$ decays with time $S_2(X^3\Sigma_g)$ starts to increase, indicating that $S_2(X^3\Sigma_g)$ is formed by the collisional deactivation of $S_2(a^1\Delta)$ [process (VI-29)]. Klemm et al. (574) have observed that the S atom production in the primary process is more than 50 times as large as O atom production in the vacuum ultraviolet flash photolysis.

VI–7. CARBON DISULFIDE (CS_2)

The ground state is $X^1\Sigma_g^+$ (linear). The bond energy, D_0(SC—S), is 4.463 ± 0.014 eV, corresponding to the incident wavelength 2778 ± 10 Å [Okabe (769)]. The absorption spectrum of CS_2 in the near ultraviolet consists of two distinct regions of absorption, one extending from 2900 to 3800 Å and

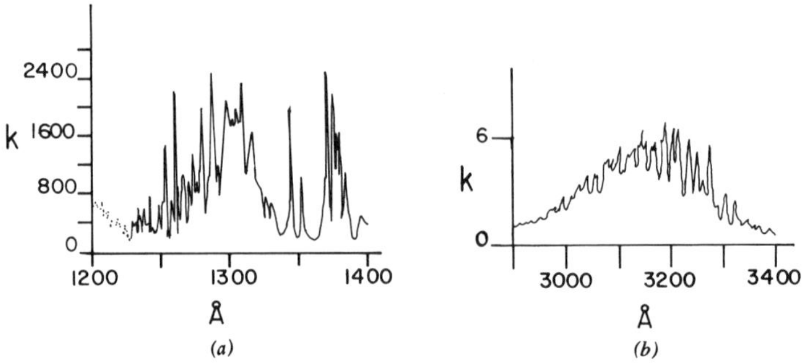

Fig. VI–6. (*a*) The absorption coefficients of CS_2 in the region 1200 to 1400 Å. *k* is in units of atm^{-1} cm^{-1}, base *e*, 23°C. From Okabe (769), reprinted by permission. Copyright 1972 by the American Institute of Physics. (*b*) The absorption spectrum of CS_2 in the region 2900 to 3400 Å. Approximate absorption coefficients are given in units of atm^{-1} cm^{-1}, base *e*. See Jungen et al. (554), and Treiber et al. (975).

the other much stronger absorption extending from 1850 to 2300 Å. The 3300 to 4300 Å absorption bands have been assigned to the $^3A_2$–$^1\Sigma_g^+$ transition by Douglas and Milton (320). The 2900 to 3200 Å bands are very complex. Jungen et al. (554) have made a rotational and vibrational analysis from which they have concluded that the bands belong to the $^1B_2$–$^1\Sigma_g^+$ transition. The 1850 to 2300 Å bands have been partially analyzed by Douglas and Zanon (321). The upper state is $^1B_2$. Only a vibrational analysis has been made for the 1650 to 1750 Å system (818). Two Rydberg series have been found by Price and Simpson (818) below 1400 Å. The absorption coefficients in the region 1200 to 1400 Å have been measured by Okabe (769) and are shown in Fig. VI–6*a*. The absorption spectrum in the region 2900 to 3400 Å is shown in Fig. VI–6*b*. The photochemistry of CS_2 may be discussed above the incident wavelength 2778 Å where the electronically excited state is important and below 2778 Å where photodissociation may be important.

VI–7.1. Photochemistry above 2778 Å

Heicklen (451) was the first to observe fluorescence in the region 4200 to 6500 Å when CS_2 was excited by incident light of wavelengths 2800 to 3600 Å. Douglas (324) has measured a lifetime of 15 μsec, which is somewhat longer than the 3 μsec calculated from the integrated absorption coefficient. Brus (155) has measured the lifetime of the fluorescence excited by the 3371 Å laser line. Two collision-free lifetimes, 2.9 ± 0.3 and 17 ± 2 μsec, have been found. Jungen et al. (553) have studied the absorption spectrum near 3371 Å and have assigned $^1A_2$ and a triplet state as the two fluorescing states.

Lambert and Kimbell (596) have investigated quenching effects of various gases on fluorescence. No S atoms have been found in the photolysis in the region above 2300 Å [deSorgo et al. (283)]. Instead, CS and S_2 have been found in the flash photolysis of CS_2 and N_2 mixtures. The proposed reactions are

$$CS \xrightarrow{h\nu} CS_2^* \qquad \text{(VI-32)}$$

$$CS_2^* + CS_2 \rightarrow 2CS + S_2 \qquad \text{(VI-33)}$$

where CS_2^* signifies an electronically excited state.

VI–7.2. Photochemistry below 2778 Å

The photolysis of $CS_2 + C_2H_4$ mixtures in the region 1950 to 2250 Å have produced ethylene episulfide, an indication of $S(^3P)$ production [deSorgo et al. (283)].

$$CS \xrightarrow{h\nu} CS + S(^3P)$$

$$S(^3P) + C_2H_4 \rightarrow \underset{\diagdown S \diagup}{H_2C\text{—}CH_2} \qquad \text{(VI-34)}$$

In the flash photolysis of CS_2 in the region 1900 to 2100 Å, Callear (166) has observed the production of vibrationally excited $CS(X^1\Sigma)$ and $S(^3P)$ but not $S(^1D)$. Apparently CS and $S(^3P)$ are predissociated from the $^1B_2$ state in violation of the spin conservation rules because of the presence of heavy S atoms. In the vacuum ultraviolet photolysis a major primary process is

$$CS_2 \xrightarrow{h\nu} CS(A^1\Pi) + S(^3P_2) \qquad \text{(VI-35a)}$$

in apparent violation of the spin conservation rules [Okabe (769)]. The dissociation below 1337 Å apparently takes place from Rydberg states.

Recently, Black et al. (118) have found the production of $CS(a^3\Pi)$ in the 1250 to 1400 Å region with high efficiencies

$$CS_2 \xrightarrow{h\nu} CS(a^3\Pi) + S(^3P) \qquad \text{(VI-35b)}$$

The lifetime of $CS(a^3\Pi)$ is 16 ± 3 msec.

VI–8. NITROUS OXIDE (N_2O)

The ground state is $X^1\Sigma^+$ (linear). Absorption starts at about 2400 Å. The absorption coefficients in the region 1080 to 2400 Å have been measured by Zelikoff et al. (1079), by Thompson et al. (967), and recently by Johnston and Selwyn (544). They are given in Figs. VI–7*a* through VI–7*e*. Winter (1053) has assigned the 1809, 1455, and 1291 Å absorption bands to the $^1\Delta$, $^1\Pi$, and $^1\Sigma^+$ states respectively, on a theoretical basis. The bond dissociation energies are $D_0(N_2\text{—}O) = 1.672 \pm 0.005$ and $D_0(N\text{—}NO) = 4.992 \pm 0.005$ eV. Table

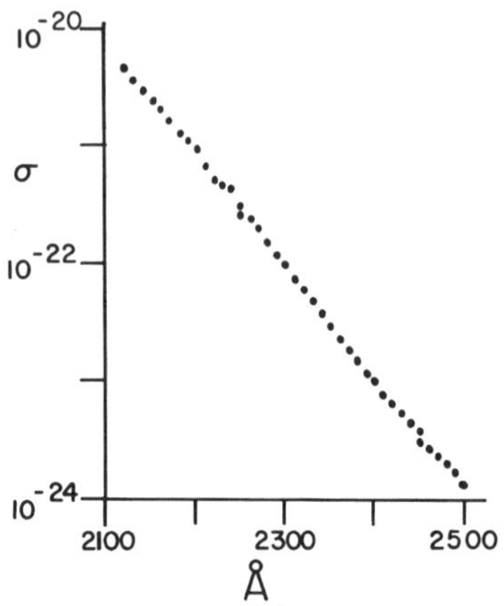

(a)

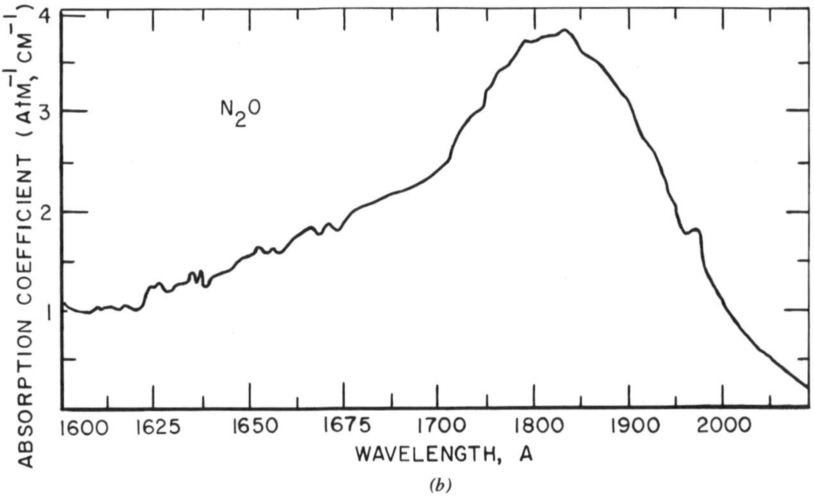

(b)

Fig. VI–7. (*a*) The absorption cross sections of N_2O in the region 2100 to 2500 Å. σ is in units of cm^2 $molec^{-1}$, base *e*, room temperature. From H. S. Johnston and G. S. Selwyn Geophys. Res. Lett. 2, 549 (1975). Reprinted with permission. Copyright by American Geophysical Union. (*b*)–(*e*) Absorption coefficients of N_2O in the region 1080 to 2100 Å. σ is in units of atm^{-1} cm^{-1}, base *e*, 0°C. From Zelikoff et al. (1079), reprinted with permission. Copyright 1953 by the American Institute of Physics.

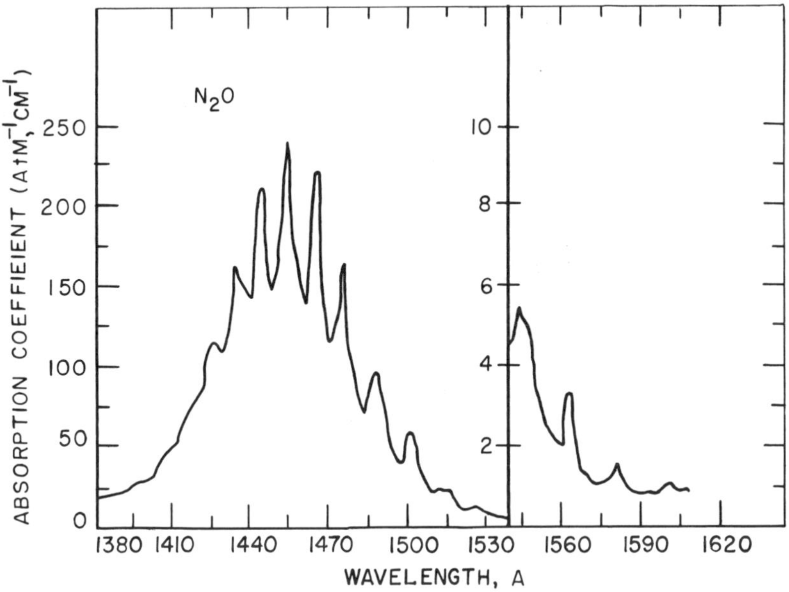

(c)

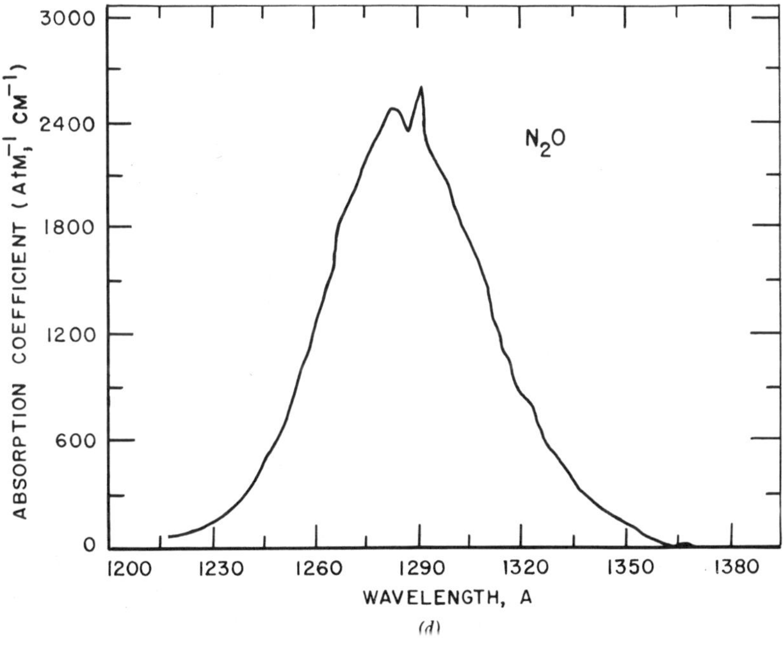

(d)

Fig. VI–7. (*cont.*)

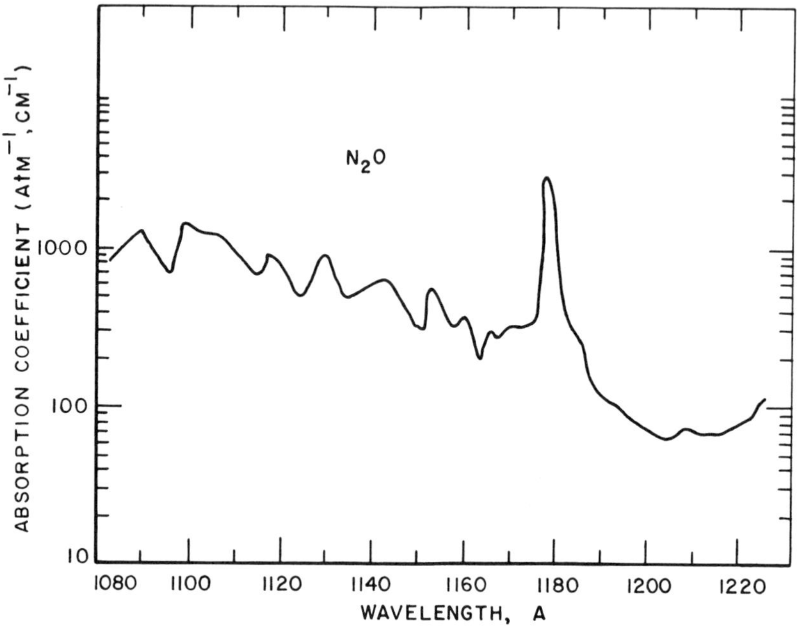

(e)

Fig. VI–7. (*cont.*)

Table VI–3a. Threshold Wavelengths (Å) below Which Indicated Reactions are Energetically Possible in the Photolysis of N_2O

N_2/O	^3P	^1D	^1S
$X^1\Sigma$	7415[a]	3407	2115
$A^3\Sigma_u$	1581	1264	1031
$B^3\Pi_g$	1374	1128	938
$B'^3\Sigma_u^-$	1260	1050	884

[a] The threshold wavelength below which the dissociation of N_2O into $N_2(X^1\Sigma)$ + $O(^3P)$ is energetically possible.

Table VI–3b. Threshold Wavelengths (Å) below Which Indicated Reactions are Energetically Possible in the Photolysis of N_2O.

NO/N	^4S	^2D	^2P
$X^2\Pi$	2519	1698	1459
$A^2\Sigma^+$	1192	970	887
$B^2\Pi$	1174	958	877

VI–3*a* and VI–3*b* gives threshold wavelengths below which indicated reactions are energetically possible.

VI–8.1. Photochemical Reactions

The photodecomposition products are known to be N_2, O_2, NO, and NO_2. The formation of NO_2 from NO and O_2 is slow in the gas phase but the reaction $2NO + O_2 \rightarrow 2NO_2$ appears to be accelerated when mixtures of $NO + O_2$ are repeatedly cooled to $-196°C$ and warmed again to room temperature (864).

Primary Processes. Two primary processes, both spin-forbidden, are energetically possible below 2500 Å

$$N_2O \xrightarrow{h\nu} N_2 + O(^3P) \tag{VI-36}$$
$$N_2O \xrightarrow{h\nu} NO + N(^4S) \tag{VI-37}$$

It has been found by Cvetanović et al. (257) that O atoms produced from the photolysis of N_2O at 1849 and 2138 Å are metastable O* [$O(^1D)$ or $O(^1S)$] atoms, since the production of N_2 by the reaction $O^* + N_2O$ is suppressed by the addition of CO_2 or Xe, which quenches O* to nonreactive $O(^3P)$ atoms. Subsequent studies suggest [see Paraskevopoulos and Cvetanović (791)] that, at least in the region 1850 to 2300 Å, the $O(^1D)$ atoms are formed from the photolysis of N_2O

$$N_2O \rightarrow N_2 + O(^1D) \tag{VI-38}$$

The possible occurrence of (VI-37) has been studied by Preston and Barr (816). If the primary process is in part

$$N_2O \xrightarrow{h\nu} NO + N$$

the N atoms formed would react with ^{15}NO added initially to N_2O to form $^{29}N_2$.

$$N + {}^{15}NO \rightarrow {}^{29}N_2 + O \qquad \text{(VI-39)}$$

The results of the photolysis of mixtures of N_2O + 1% ^{15}NO at 2288, 2139, and 1849 Å show the production of less than 1.7% $^{29}N_2$, indicating that (VI-37) is less than 2% of the primary process.

Secondary Processes. Two secondary processes may be proposed to explain the products N_2, NO, O_2, and NO_2

$$O(^1D) + N_2O \rightarrow N_2 + O_2 \qquad k_{40} \qquad \text{(VI-40)}$$
$$O(^1D) + N_2O \rightarrow 2NO \qquad k_{41} \qquad \text{(VI-41)}$$

The ratio k_{40}/k_{41} can be computed from Φ_{O_2} and Φ_{NO} provided there is no reaction between O_2 and NO.

$$\frac{k_{40}}{k_{41}} = \frac{2\Phi_{O_2}}{\Phi_{NO}} \qquad \text{(VI-42)}$$

However, because some O_2 and NO tend to react with each other to form NO_2 (or N_2O_3) during the analysis, the ratio k_{40}/k_{41} cannot be obtained reliably from the measured ratio of Φ_{O_2} to Φ_{NO}.

Scott et al. (864) have measured instead the ratio of N_2 to NO_2 produced from the photolysis of O_3–N_2O mixtures in the region where only O_3 absorbs. From the reaction sequence

$$O_3 \xrightarrow{h\nu} O_2 + O(^1D) \qquad \text{(VI-43)}$$
$$O(^1D) + N_2O \rightarrow N_2 + O_2 \qquad k_{44} \qquad \text{(VI-44)}$$
$$O(^1D) + N_2O \rightarrow 2NO \qquad k_{45} \qquad \text{(VI-45)}$$
$$NO + O_3 \rightarrow NO_2 + O_2 \qquad \text{(VI-46)}$$

they have obtained the ratio, as NO molecules are all converted to NO_2 molecules.

$$\frac{k_{44}}{k_{45}} = \frac{2[N_2]}{[NO_2]} = 0.99 \pm 0.06$$

The ratios obtained by various methods range from 0.6 [Greenberg and Heicklen (419), Ghormley et al. (391), Simonaitis et al. (884)] to 1 [Greiner (423)]. It is most likely that the ratio is near unity since three independent methods agree with each other (864).

The photolysis of N_2O above 1850 Å may be summarized as follows

$$N_2O \xrightarrow{h\nu} N_2 + O(^1D) \qquad \text{(VI-47)}$$

$$O(^1D) + N_2O \rightarrow N_2 + O_2, \qquad k_{48} \qquad \text{(VI-48)}$$

$$O(^1D) + N_2O \rightarrow 2NO \qquad k_{49} \qquad \text{(VI-49)}$$

$$k_{48} = k_{49}$$

In spite of various energetically possible reactions given in Tables VI–3a and 3b, the quantum yields of the products are almost independent over the wavelength region studied except at 1236 Å, suggesting that the above processes are predominant above 1470 Å. The quantum yields of various products of N_2O photolysis are given in Table VI–4.

VI–8.2. Production of Metastable Species by the Photolysis of N_2O

Tables VI–3*a* and 3*b* shows that the photolysis above 1200 Å can produce electronically excited atoms $O(^1D, ^1S)$ $N(^2D, ^2P)$ and molecules $N_2(A^3\Sigma, B^3\Pi)$. Of these $O(^1S)$ and $N_2(B^2\Pi)$ are directly observed by the emission at

Table VI–4. Quantum Yields of Products in the N_2O Photolysis

Wavelength (A)	Product	Quantum Yield
1236	N_2	1.18 ± 0.03 (Ref. 426), 1.34 ± 0.04 (Ref. 293)
1470		1.40 ± 0.06 (Refs. 426 and 1061), 1.48 (Ref. 292), 1.44 (Ref. 1080)
1849		(1.44) (Ref. 1081)
2139		1.51 ± 0.11 (Ref. 419)
1236	O_2	0.2 ± 0.01 (Ref. 426), 0.19 ± 0.01 (Ref. 1061)
1470		(0.58 ± 0.03) (Ref. 1061), 0.15 ± 0.01 (Ref. 426)
		0.5 (Ref. 1080)
1849		(~0.4) (Ref. 423)
2139		0.059 (Ref. 419)
1470	NO	(0.78 ± 0.03) (Ref. 1061)
1849		0.81 ± 0.08 (Ref. 423)
1236	NO_2	0.52 ± 0.02 (Ref. 426)
1470		0.78 ± 0.03 (Ref. 1061), 0.74 ± 0.05 (Ref. 426)
	$-N_2O$	
1236		1.45 ± 0.04 (Ref. 426)
1470		1.76 ± 0.08 (Ref. 426), (Ref. 1080)
1849		(1.71 ± 0.13) (Ref. 423)

Note: 1849 Å, $\Phi_{\Delta n} = \Phi_{N_2} + \Phi_{NO} + \Phi_{O_2} - \Phi_{N_2O} = 1.00 \pm 0.05$ (Ref. 423) ($-2N_2O = \frac{3}{2}N_2 + \frac{1}{2}O_2 + NO$ for each photon).

5577 Å and by the first positive bands, respectively [Hampson and Okabe (436) Young et al. (1070, 1073)]. The production of $O(^1D)$ atoms is suggested by the reactivity of O* atoms with N_2O as described before. The $N(^2D)$ atoms are responsible for the production of $NO(B^2\Pi)$ [Welge (1029) Young et al. (112, 1071), Okabe (761)] by

$$N(^2D) + N_2O \rightarrow NO(B^2\Pi) + N_2 \qquad \text{(VI-50)}$$

The formation of $N_2(A^3\Sigma)$ is proposed from the observation of NO γ bands by

$$N_2(A^3\Sigma) + NO \rightarrow NO(A^2\Sigma) + N_2 \qquad \text{(VI-51)}$$

[Welge (1029), Okabe (761), Young et al. (1071, 1072)] and by electron emission due to collisions of $N_2(A^3\Sigma)$ on a metal surface (397). Recently, $N(^2D, ^2P)$ atoms produced by the photolysis of N_2O above 1050 Å have been directly observed by absorption at 1493 and 1745 Å, respectively [Husain et al. (502)]. The quantum yields of metastable species at 1470 Å estimated by Young et al. (1071) and Black et al. (113) are shown below.

Species	Quantum Yield
$O(^3P)$	0.08
$O(^1D)$	0.55
$O(^1S)$	0.5, 0.1 (Ref. 680)
$N(^2D)$	0.1
$N_2(A^3\Sigma)$	0.08

Recently quantum yields of $O(^1S)$ from N_2O have been determined as a function of incident wavelength. The yield is near unity at 1290 Å [McEwan et al. (680), Black et al. (113)]. Because the photolysis of N_2O is a convenient source for the production of $O(^1S)$, $N(^2D)$, $N_2(A^3\Sigma)$, and $N_2(B^3\Pi)$, their quenching rates by many gases have been measured by monitoring emissions produced by the photolysis of N_2O and quenching gas mixtures. Chamberlain and Simons (203) believe that in the region 1400 to 1550 Å NO is produced mostly from two reactions

$$O(^1D) + N_2O \rightarrow 2NO \qquad \text{(VI-52)}$$

$$N_2(A^3\Sigma) + N_2O \rightarrow N_2 + N + NO \qquad \text{(VI-53)}$$

VI–8.3. N_2O in the Upper Atmosphere

The concentration of N_2O in the lower stratosphere is about 0.2 ppm (405). The reaction of $O(^1D)$ with N_2O to produce NO is considered by Nicolet and Peetermans (739a) to be an important source of NO in the stratosphere.

$$O(^1D) + N_2O \rightarrow 2NO \qquad \text{(VI-54)}$$

See Section VIII–2.2 for further discussion.

VI–9. NITROGEN DIOXIDE (NO_2)

The ground state is $\tilde{X}^2A_1$ with an O—N—O angle of 134.1° (16). D_0(ON—O) = 3.118 ± 0.01 eV (28). The absorption spectrum of NO_2 in the near ultraviolet and visible is extremely complex and for the most part has no

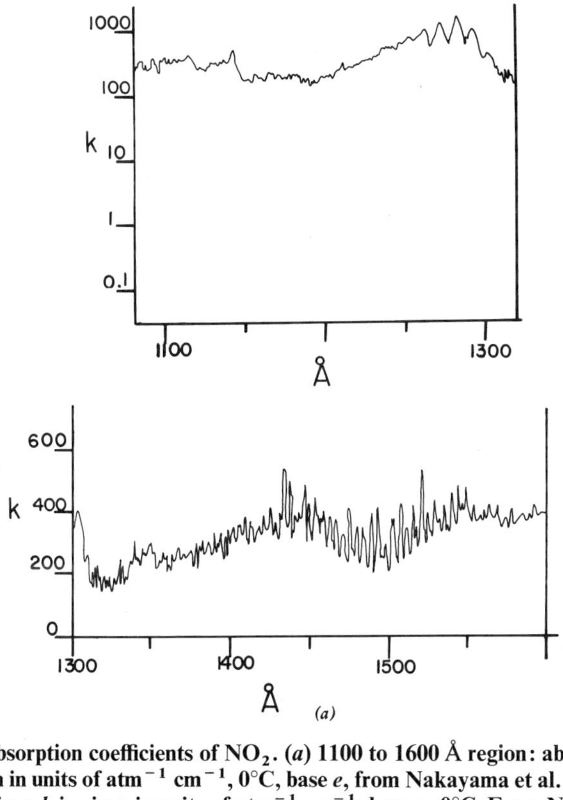

Fig. VI–8. Absorption coefficients of NO_2. (*a*) 1100 to 1600 Å region: absorption coefficient *k* is given in units of atm^{-1} cm^{-1}, 0°C, base *e*, from Nakayama et al. (731). (*b*) 1600 to 2700 Å region: *k* is given in units of atm^{-1} cm^{-1}, base *e*, 0°C. From Nakayama et al. (731). (*c*) 2500 to 5000 Å region: absorption coefficient *k* is given in units of mm^{-1} cm^{-1}, 25°C, base 10, from Hall and Blacet (431). 10^{-3} (mm^{-1} cm^{-1}) base 10 corresponds to 1.91 (cm^{-1} atm^{-1}) base *e*. Reprinted with permission. Copyright 1952 and 1959 by the American Institute of Physics.

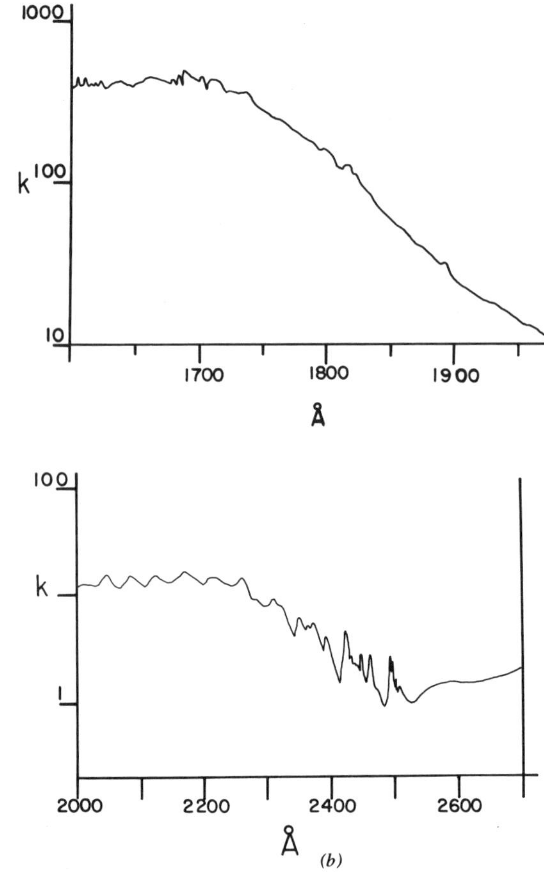

(b)

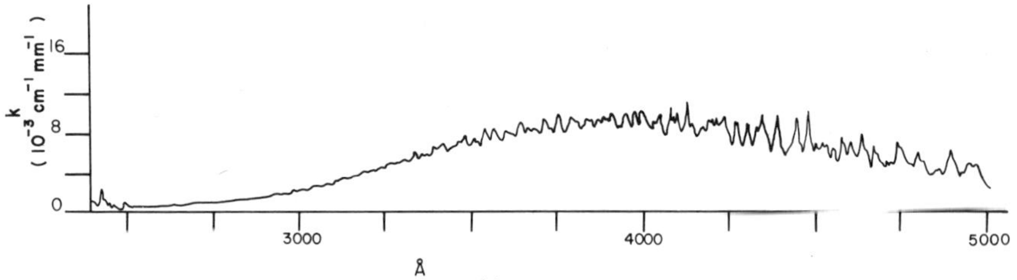

(c)

Fig. VI–8. (*cont.*)

apparent regularity in rotational and vibrational structure. Electronic states identified are $E^2\Sigma_u^+$, in the region 1350 to 1650 Å by Ritchie and Walsh (832), $\tilde{B}^2B_2$ at 2491 Å by Ritchie et al. (831), and $\tilde{A}'^2B_1$ or $^2\Pi_u$ in the region 3700 to 4600 Å by Douglas and Huber (322). Recently, Hardwick and Brand (442) assigned the origin of the transition $\tilde{A}'^2B_1$–$\tilde{X}^2A_1$ at 14,743.5 cm^{-1} (1.828 eV). Brand et al. (132) have determined the 0–0 transition of the $\tilde{A}^2B_2$–$\tilde{X}^2A_1$ system at 11,956 cm^{-1} (1.482 eV).

The rotational analysis of the 8000 to 9000 Å band system, $\tilde{A}^2B_2$–$\tilde{X}^2A_1$, has been made by Brand et al. (133). The $\tilde{A}^2B_2$ state is severely perturbed by the $\tilde{X}^2A_1$ state. The $^2B_2$ state has an O–N–O bond angle of 102° [Gillispie et al. (395)]. Smalley et al. (909) have measured the fluorescence excitation spectrum of the cooled NO_2 in the region 5708 to 6708 Å. The $^2B_2$ state is the only upper state in this region.

The absorption coefficients in the vacuum ultraviolet, near ultraviolet, and visible regions have been measured by Nakayama et al. (731) and by Hall and Blacet (431) and are given in Figs. VI–8*a* through VI–8*c*. The absorption coefficients in the region 1850 to 4100 Å at higher resolution have recently been measured by Bass et al. (72). The electronic energies, potential energy surfaces, and oscillator strengths of various upper states have been calculated by Fink (354, 355), Gangi and Burnelle (384, 385), and Gillispie et al. (395).

Nitrogen dioxide is one of a few simple molecules in which the primary quantum yield near the dissociation limit (3980 Å) has been measured nearly continuously as a function of incident wavelength. The energetics of photodissociation is given in Table VI–5. The thermochemical threshold at 0°K for the reaction, $NO_2 \rightarrow NO + O(^3P)$, corresponds to the incident wavelength 3978 Å, which nearly coincides with the wavelength 3979 ± 1 Å below which

Table VI–5. Threshould Wavelengths (Å) Below Which Indicated Reactions are Energetically Possible in NO_2 Photolysis

NO/O	^3P	^1D	^1S
$X^2\Pi$	3978[a]	2439	1697
$A^2\Sigma^+$	1442	1174	970

Note: $NO_2 \rightarrow N + O_2$; $\Delta H = 103.9$ kcal mol^{-1} corresponding to 2751 Å.

[a] The threshold wavelength below which the dissociation of NO_2 into $NO(X^2\Pi)$ and $O(^3P)$ is energetically possible.

rotational structure of the NO_2 absorption spectrum becomes diffuse [Douglas and Huber (322)]. Even above 3980 Å internal energy is known to contribute to dissociation [Pitts et al. (810), Jones and Bayes (550)]. Above 3980 Å it has been known that NO_2 fluoresces strongly in the visible region. The lifetime, the spectral analysis, and quenching properties of the fluorescing state have been studied by many workers. The photodissociation of NO_2 may be conveniently discussed above and below 3980 Å.

VI–9.1. Photodissociation above 3980 Å

The primary quantum yield of NO_2 dissociation obtained from Φ_{O_2} falls off rapidly above 3980 Å and is 0.005 at 4358 Å [Pitts et al. (810)]. Since the extent of isotopic scrambling of O_2 by illuminating mixtures of NO_2 and $^{18,18}O_2$ closely follows the quantum yield of NO production, there is little doubt that O atoms are formed by direct dissociation of NO_2 at four wavelengths tested, 3660, 4020, 4060, and 4120 Å [Jones and Bayes (550)].

The falloff curve of the quantum yield is explained by the contribution of the internal (mostly rotational) energy to supplement the incident photon energy (550, 810), (see Section I–4.3 for details). Above 4358 Å a small ($\sim$0.01) but significant yield of NO was observed, which is attributed to reactions of electronically excited NO_2 (NO_2^*) by Jones and Bayes (550).

$$NO_2^* + NO_2 \rightarrow 2NO + O_2 \qquad \text{(VI-55)}$$

$$NO_2^* + NO_2 \rightarrow NO_3 + NO \qquad \text{(VI-56)}$$

Creel and Ross (251) have studied the NO_2 photolysis in the region 4580 to 6300 Å. They also conclude the occurrence of (VI-55) by measuring the production of O_2. Hakala et al. (433) have found the production of O_2 from NO_2 by irradiation with the 6943 Å laser that has insufficient energy to induce dissociation. Since the O_2 production was dependent on the square of the laser intensity, they postulated the consecutive absorption of two photons.

VI–9.2. Photodissociation below 3980 Å

The electronically excited NO_2 formed by absorption of light wavelengths below 3980 Å must have a lifetime on the order of 10^{-12} sec, since the absorption spectrum shows diffuse rotational structure (but not the vibrational structure) [Douglas and Huber (322)]. Experimentally, a lifetime on the order of 5×10^{-13} sec is obtained from the photolysis yield of NO_2 at various added N_2 pressures in the region 3100 to 4100 Å [Gaedtke et al. (378, 379)].

Busch and Wilson (162, 163) irradiated a molecular beam of NO_2 with a pulsed laser of wavelength 3471 Å and measured the flight times of the photodissociation fragments to a mass spectrometer–detector. They obtained information on the energy partitioning between translational and internal degrees of freedom of the recoiling O and NO fragments. When a polarized light source was used, the angular distribution of recoiled O atoms peaks along the direction of the electric vector of the polarized light. This indicates the predominant state produced by absorption at 3471 Å is $^2B_2$. However, this assignment is not in accord with that of $^2B_1$ by a rotational analysis (322) [see also Section II–5]. The main photochemical reactions of NO_2 in the region 2439 to 3978 Å are the production of $O(^3P)$ atoms and the rapid reaction of O atoms with NO_2:

$$NO_2 \xrightarrow{h\nu} NO + O(^3P) \qquad \text{(VI-57)}$$

$$O(^3P) + NO_2 \rightarrow NO + O_2 \qquad k_{58} \qquad \text{(VI-58)}$$

$$k_{58} = 9.1 \times 10^{-12}\ \text{cm}^3\ \text{molec}^{-1}\ \text{sec}^{-1}\ \text{(Ref 9)}$$

From the mechanism the quantum yield of the primary process is equal to the quantum yield of O_2 production (Φ_{O_2}) or one half of NO production [$\frac{1}{2}\Phi_{NO}$]. It has been shown by several workers [Pitts et al. (810), Ford and Jaffe (369), Jones and Bayes (550)] that Φ_{O_2} or $\frac{1}{2}\Phi_{NO}$ is near unity at 3130 Å and gradually decreases as the incident wavelength increases. There is some indication that another electronically excited state may be formed by light absorption in the region 3300 to 5900 Å to an extent of several percent (372, 549). This state, being sufficiently long-lived, is capable of transferring its electronic energy to O_2 to produce $O_2(^1\Delta)$ [Jones and Bayes (548, 549), Frankiewicz and Berry (372)].

At 2288 Å three dissociation processes are energetically possible

$$NO_2 \xrightarrow{h\nu} NO + O(^3P) \qquad \text{(VI-59)}$$

$$NO_2 \xrightarrow{h\nu} NO + O(^1D) \qquad \text{(VI-60)}$$

$$NO_2 \xrightarrow{h\nu} N + O_2 \qquad \text{(VI-61)}$$

Process (VI-61) contributes little, if any, since N_2O and N_2, the expected products from the reaction of N with NO_2, are minor products in NO_2 photolysis at 2288 Å [Preston and Cvetanović (815)].

Since isotopic scrambling of O_2 (production of $^{32}O_2$ and $^{34}O_2$ mixtures) has been seen in the photolysis of mixtures of $NO_2 + C^{18}O_2$ at 2288 Å, but not at 2537 or 3660 Å, Preston and Cvetanović (815) have concluded the production of $O(^1D)$ from (VI-60) followed by the reactions

$$O(^1D) + C^{18}O_2 \rightarrow CO_2 + {}^{18}O \qquad \text{(VI-62)}$$

$$^{18}O + NO_2 \rightarrow NO + {}^{34}O_2 \qquad \text{(VI-63)}$$

They estimated about 40% $O(^1D)$ is produced in the primary processes at 2288 Å.

Table VI–5 shows that the dissociation process, $NO_2 \rightarrow NO + O(^1D)$ takes place energetically below 2439 Å. Uselman and Lee (985) have measured the production of $O(^1D)$ as a function of incident wavelength near 2439 Å. They have found that the contribution of rotational energy to dissociation is insignificant near the second threshold in contrast to the case near the first threshold at 3980 Å where the contribution of rotational energy is substantial. They attribute the lack of rotational contribution to the presence of large rotational barriers at high J values in the excited state (987). The quantum yield of $O(^1D)$ production increases to a plateau of about 0.5 ± 0.1 towards shorter wavelengths, indicating that at least two processes, (VI-59) and (VI-60), occur concurrently below the second threshold wavelength.

VI–9.3. Photodissociation in the Vacuum Ultraviolet

Welge (1030) has observed that electronically excited NO in the $A^2\Sigma^+$ and $B^2\Pi$ states was produced from NO_2 by radiation with the Kr (1165, 1236 Å) and Xe (1236 Å) lines. Lenzi and Okabe (625) have measured a fluorescence yield of about 2% at 1216 Å. The production of excited states other than $A^2\Sigma^+$ and $B^2\Pi$ has been suggested.

VI–9.4. Fluorescence

When NO_2 is irradiated with light of wavelengths above 3980 Å, fluorescence from NO_2 has been observed. The fluorescence spectrum lies in the region from the exciting wavelength to above 7500 Å. Many rotational and vibrational lines are superimposed on a continuum. The collision-free lifetime, measured by Neuberger and Duncan (733), is 44 μsec when excited at 4358 Å, while the mean life calculated from the integrated absorption coefficient is about 0.26 μsec. This discrepancy between the measured lifetime and the lifetime calculated from the integrated absorption coefficient has been explained by Douglas (324) (see Section II–2.1) on the basis of the interaction of the excited and the ground state. Keyser et al. (562) have found no significant variation of lifetime of fluorescence in the region of absorption 4360 to 6000 Å, indicating that there is probably only one excited state. On the other hand, Abe (35) and Abe et al. (34, 36) have concluded that both the $^2B_1$ and $^2B_2$ states are responsible for fluorescence in the visible region of the absorption spectrum. The excited state formed by the 4420 Å line is $^2B_1$ with a lifetime of 36 μsec [Schwartz and Senum (863)]. Stevens et al. (926) have found evidence of the two excited states, $^2B_2$, $^2B_1$ in the 5934 to 5940 Å region of absorption from the rotational analysis of the fluorescence spectra and the lifetime of fluorescence. They have concluded that the $^2B_2$ has a lifetime of 30 ± 5 μsec and the $^2B_1$ has a lifetime of 115 ± 10 μsec. Some rotational levels of the NO_2 electronic state in the visible region appear to be unperturbed by other states, since the lifetime of the fluorescence from these levels is indeed very short (0.5–3.7 μsec) (847), approaching that predicted from the integrated absorption coefficient. These levels have been found in the region 4544 to 4550 Å by Sackett and Yardley (847) and by Solarz and Levy (920) at 4880 Å.

A more recent work by Paech et al. (789) on the collision-free lifetimes of NO_2 excited by a tunable laser near 4880 and 5145 Å states that although only a single level is excited, three different lifetimes of fluorescence, 3, 28, and 75 μsec, have been observed. The results lead them to conclude that the initially formed $^2B_1$ state crosses over rapidly to another state, $^2B_2$, with higher level density. The $^2B_2$ state can have two different lifetimes (28 and 75 μsec), depending on the extent of interaction with the ground state. The short life observed, 3 μsec, is determined primarily by the rate of internal conversion from the $^2B_1$ to $^2B_2$ state. The results of some reported collision-free fluorescence lifetimes are given in Table VI–6*a*.

The NO_2 fluorescence is quenched by almost all gases. However, the usual Stern-Volmer plot, assuming one excited state that may either radiate or be collisionally deactivated to the ground state, shows a linear relationship for self-quenching only at a given exciting and fluorescence wavelength. That is, the self-quenching constant a_A defined by (II-1) is a function of both the exciting and fluorescence wavelength. Furthermore, quenching by foreign gases, He, N_2, and O_2 does not follow the linear relationship given by (II-1) [Myers et al. (729), Braslavsky and Heicklen (136)].

The fluorescence spectrum shows a red shift at higher pressures (729), indicating that quenching is a multistep process with consecutive vibrational

Table VI–6a. Radiative Lifetime of the Electronically Excited NO_2

Lifetime (μ sec)	Exciting Light Wavelength (A)	Method	Ref.
44	4358	Pulsed Light	733
55 ± 5	4360–6000	Phase Shift	562
55–90	3980–6000	Phase Shift	862
42 ± 6	4420, 4510, 4840	Pulsed Light	850
0.5–3[a]	4545–4550	Pulsed Light	847
62–75	4515–4605	Pulsed Light	848

[a] Observed only at several excitation wavelengths.

deexcitation of NO_2^* and radiation from each of several vibrational levels. When the fluorescence wavelength is farther apart from the exciting wavelength, the quenching constant decreases rapidly. Keyser et al. (562) have concluded that the dependence of the quenching constant a_M on $\Delta\bar{\nu}$ (the difference of the wavenumber of the exciting and fluorescing light) is best explained by assuming a single electronically excited state of a lifetime of about 50 μsec, an efficient vibrational relaxation (almost unit efficiency), and slow electronic quenching (1 in 100 collisions). The vibrational quantum transferred per quenching collision is 1000 ± 500 cm^{-1}. Some quenching constants at the 4358 Å excitation and the fluorescence wavlengths above 4600 Å are given in Table VI–6*b* [Myers et al. (729)].

The fluorescence yield as a function of incident wavelength has been measured by Lee and Uselman (619). The yield starts to increase from 0 at 3979 Å to nearly 100% above 4150 Å. The decrease of the yield below 4150 Å is attributed to an increase of predissociation supplemented by the rotational energy of the molecule, since incident light of wavelengths above 3979 Å does not have sufficient energy to dissociate the molecule at 0°K (550, 619).

Uselman and Lee (986) have observed that the fluorescence lifetime is constant (70 μsec) in the absorption region 3980 to 4200 Å. From the results they suggest two kinds of excited NO_2, one fluoresces with a constant lifetime and the other predissociates within less than 100 nsec.

The fluorescence from NO_2 excited by the 4416 and 4880 Å lines is used for measuring NO_2 concentrations in air in the parts per billion range (388).

Table VI–6b. Quenching Constants,[a] a_M of NO_2 at the 4358 Å Excitation and the Fluorescence Wavelength Near 5500 Å [by Myers et al. (729)]

Quenching Gas	Quenching Constant a_M (torr^{-1})	Quenching Gas	Quenching Constant a_M (torr^{-1})
He	29	CH_4	82
Ar	30	N_2O	91
N_2	44	NO_2	100
O_2	48	CO_2	105
H_2	62	SF_6	155
NO	82	CF_4	160
		H_2O	280

[a] The quenching constant is k_M/k_f defined in Section II–1.2, where k_M is the quenching rate constant in torr^{-1} sec^{-1} and k_f is the fluorescence decay rate in sec^{-1}.

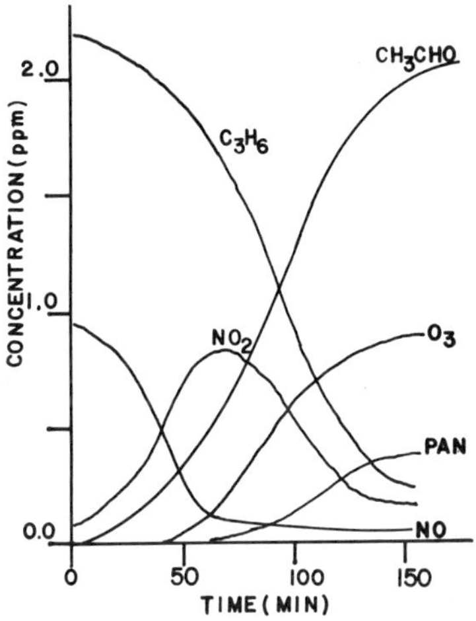

Fig. VI–9. Concentration–time history of reactants and some products in the photooxidation of C_3H_6. Mixtures of C_3H_6 + NO + NO_2 in air were irradiated by simulated sunlight in smog chamber. Reprinted with permission from H. Niki, E. E. Daby, and B. Weinstock, in *Photochemical Smog and Ozone Reactions*, R. F. Gould, Ed., American Chemical Society, Washington, D. C. Copyright by the American Chemical Society, 1972.

VI–9.5. Nitrogen Dioxide in the Atmosphere

It has been recognized that NO_2 plays a central role in the formation of photochemical air pollution [see Niki et al. (743), p. 16]. Absorption of sunlight by NO_2 in the spectral region 3000 to 3890 Å leads to the production of O atoms. By the combination of O atoms with O_2, O_3 is formed. The reactions of O atoms, OH radicals, and O_3 with hydrocarbons (mainly olefins) initiate photochemical air pollution. Figure VI–9 shows a typical time history of the concentrations of reactants and products when mixtures of C_3H_6, NO, and NO_2 in air were irradiated by simulated sunlight in a smog chamber (743). The concentrations of C_3H_6 and NO decrease with irradiation time while NO_2 and an oxidation product of C_3H_6, acetaldehyde, start to increase. At a later time, O_3 and PAN (peroxyacetylnitrate) start to appear when the NO_2 concentration reaches a maximum.

According to Bufalini (158) $O_2(^1\Delta)$ produced partially by energy transfer from the electronically excited NO_2 to O_2 molecules, postulated by Frankiewicz and Berry (372a), does not contribute to air pollution.

Further discussion of photochemical air pollution is given in Section VIII-2.

VI–10. NITROSYL HALIDES

VI–10.1. Nitrosyl Chloride (NOCl)

The ground state NOCl is bent with an O—N—Cl angle of 116°. D_0(ON—Cl) = 1.61 ± 0.01 eV (28). Absorption starts at about 6500 Å. The absorption

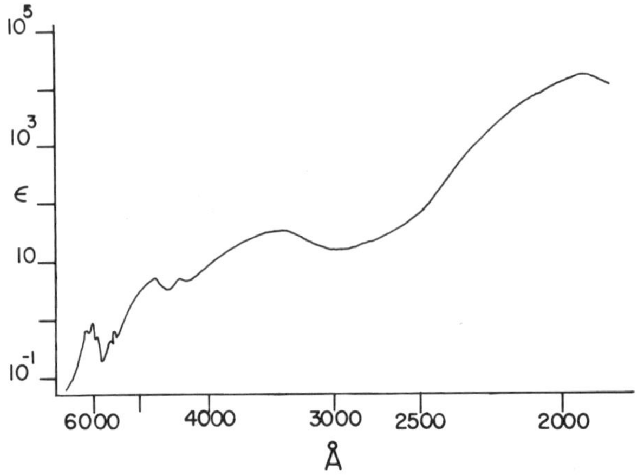

Fig. VI–10. Absorption coefficients of NOCl in the visible and ultraviolet region. ϵ is given in units of 1 mol^{-1} cm^{-1}, base 10, room temperature. [Goodeve and Katz (411), revised by Ballash and Armstrong (58)], reprinted by permission of Pergamon Press.

coefficients in the region 1900 to 4000 Å have recently been measured by Illies and Takacs (508). The absorption coefficients in the region 1800 to 6500 Å have been measured by Goodeve and Katz (411) and more recently by Ballash and Armstrong (58). They are shown in Fig. VI–10. The absorption coefficients in the region 1100 to 2000 Å have been measured by Lenzi and Okabe (625) and are shown in Fig. VI–11.

Photochemistry. The quantum yield of NOCl decomposition is 2 over the incident wavelength region from 3650 to 6300 Å [Kistiakowsky (570)] and at 2537 Å [Wayne (1025)]. Since the absorption spectrum in this region is continuous, the photochemical process must be

$$\mathrm{NOCl} \xrightarrow{h\nu} \mathrm{NO} + \mathrm{Cl} \qquad \text{(VI-64)}$$

$$\mathrm{Cl} + \mathrm{NOCl} \rightarrow \mathrm{NO} + \mathrm{Cl_2} \qquad k_{65} \qquad \text{(VI-65)}$$

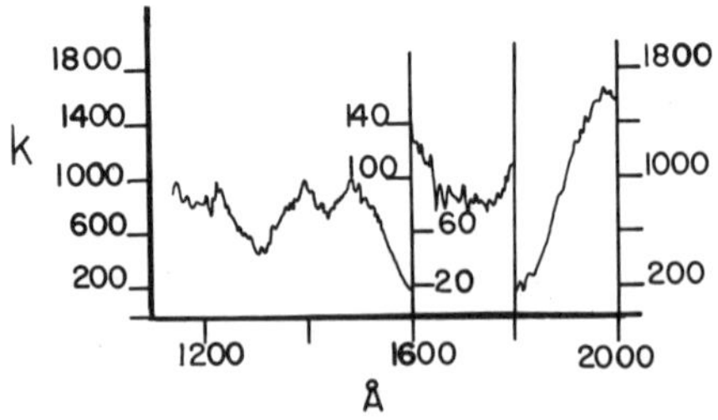

Fig. VI–11. The absorption coefficients of NOCl in the region 1100 to 2000 Å. *k* is given in units of atm^{-1} cm^{-1}, base *e* at room temperature (23°C). [Lenzi and Okabe (625)], reprinted with permission of Verlag Chemie.

The rate constant, $k_{65} = 3 \pm 0.5 \times 10^{-11}$ cm^3 molec^{-1} sec^{-1}, has been measured by Clyne and Cruse (221). The photodissociation of NOCl at 3471 Å has been studied by Busch and Wilson (164) using a polarized monochromatic pulsed laser. They concluded that the photodissociation occurs from a state with A' symmetry since dissociation into NO + Cl is induced by light polarized in the molecular plane. The excess energy beyond that required to break the N—Cl bond goes predominantly (70%) into the translational energy of the recoiling fragments. Basco and Norrish (63) have observed vibrationally excited NO up to $v'' = 11$ in the flash photolysis of NOCl in the near ultraviolet. The mechanism of production is either directly

$$\mathrm{NOCl} \xrightarrow{h\nu} \mathrm{NO}(X^2\Pi, v'' \leq 11) + \mathrm{Cl} \qquad \text{(VI-66)}$$

or through NO($^4\Pi$)

$$\mathrm{NOCl} \xrightarrow{h\nu} \mathrm{NO}(^4\Pi) + \mathrm{Cl} \qquad \text{(VI-67)}$$

$$\mathrm{NO}(^4\Pi) + \mathrm{M} \rightarrow \mathrm{NO}(X^2\Pi, v > 0) + \mathrm{M} \qquad \text{(VI-68)}$$

Equation (VI-67) involves a crossover from the initially formed singlet to a repulsive triplet state of NOCl. The photolysis of NOCl in the vacuum ultraviolet produces NO($A^2\Sigma^+, v' = 0, 1, 2$) [Welge (1030)] and possibly other excited states of NO [Lenzi and Okabe (625)]. The fluorescence yield near 1500 Å is about 4% (625).

VI–10.2. Nitrosyl Fluoride (NOF)

The ground state NOF is $\tilde{X}^1A'$ with an O—N—F angle of 110° (16). D_0(ON—F) = 2.38 ± 0.03 eV (28).

The absorption spectrum in the near ultraviolet lies in the region 2600 to 3350 Å (16) with some vibrational structure.

Photolysis with an unfiltered medium pressure Hg lamp has been made in the presence of ethylene and other hydrocarbons. The photolysis rate decreased when inert gases were added, indicating the formation of electronically excited NOF (364), which may decompose or be deactivated by an added gas.

VI–11. OZONE (O_3)

Ozone is a bent molecule with an angle of 116.8°. The bond energy D_0(O—O_2) is 1.05 ± 0.02 eV (28). Absorption of light starts at about 9000 Å. The absorption in the ultraviolet and visible regions consists of the Hartley bands (2000 to 3200 Å), the Huggins bands (3000 to 3600 Å), and the Chappuis bands (4400 to 8500 Å). The absorption coefficients of these bands are given in Figs. VI–12*a* and VI–12*b*. Figure VI–12*c* shows the absorption coefficients in the vacuum ultraviolet region.

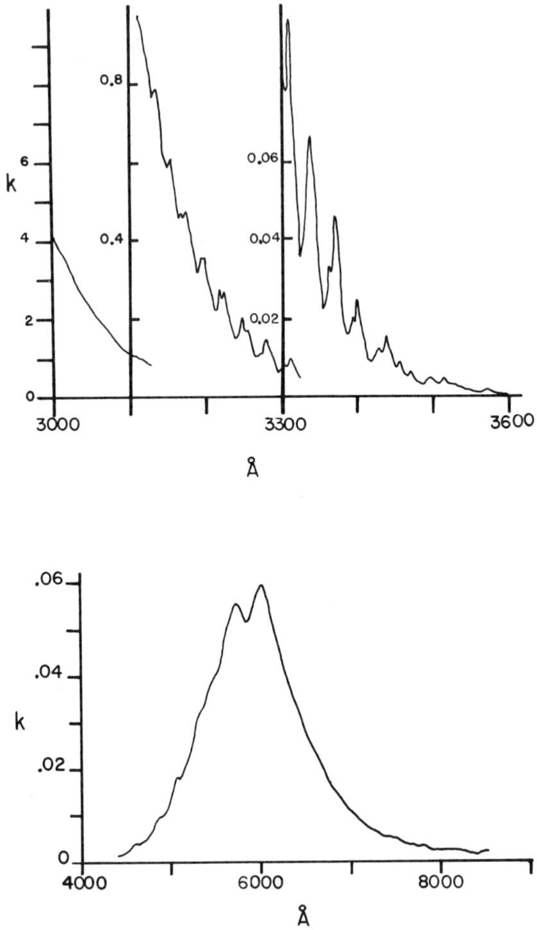

(a)

Fig. VI–12. (*a*) Absorption coefficients of O_3 Huggins bands (3000 to 3600 Å) and Chappuis bands (4400 to 8500 Å); k (atm^{-1} cm^{-1}), 0°C, base 10. From Griggs (425), reprinted with permission. Copyright 1968 by the American Institute of Physics. (*b*) Absorption coefficients of O_3 Hartley bands (2000 to 3000 Å). k is in units of atm^{-1} cm^{-1}, 0°C, base 10. From Griggs (425), reprinted by permission. Copyright 1968 by the American Institute of Physics. (*c*) Absorption coefficients of O_3 in the region 1000 to 2200 Å. k is in units of (atm^{-1} cm^{-1}), 0°C, base e. From Tanaka et al. (961), reprinted by permission. Copyright 1953 by the American Institute of Physics.

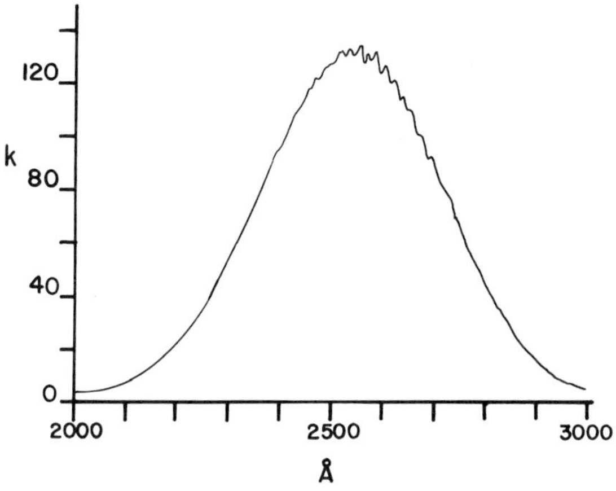

(b)

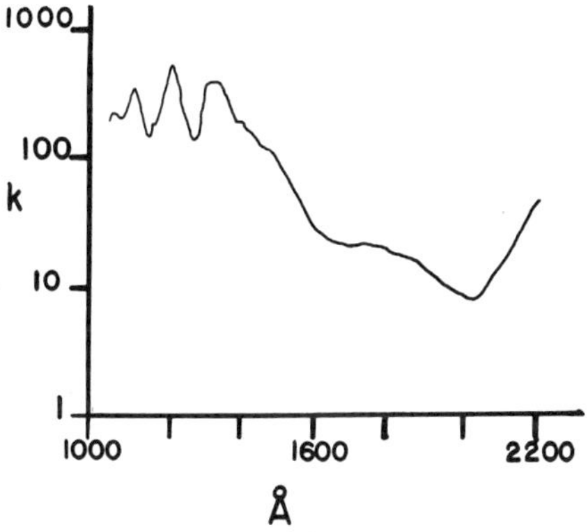

(c)

Fig. VI–12. (*cont.*)

Since the absorption spectra are diffuse over the entire spectral region, no assignments of transitions have been made from the analyses of the spectra. Hay and Goddard (449) have recently assigned the Hartley bands to a transition ${}^1B_2$–${}^1A_1$ (origin at 3515 Å) and the Chappuis bands to ${}^1B_1$–${}^1A_1$.

An excellent review on O_3 photochemistry up to 1971 has been given recently by Schiff (857). Ozone dissociates by absorption of light of wavelengths below 9000 Å. Table VI–7 gives the threshold wavelengths below which designated photochemical processes are energetically possible. We discuss photodissociation processes separately for the three absorption regions in the visible and near ultraviolet.

VI–11.1. Photodissociation in the Chappuis Bands (4400 to 8500 Å)

The quantum yield of O_3 disappearance by absorption of red light ($\sim$ 6000 Å) is 2 (193). The result is best explained by the primary process

$$O_3 \xrightarrow{h\nu} O(^3P) + O_2(X\,^3\Sigma) \qquad \text{(VI-69)}$$

followed by

$$O(^3P) + O_3 \rightarrow 2O_2 \qquad k_{70} \qquad \text{(VI-70)}$$

$k_{70} = 2 \times 10^{-11} \exp\left[-(4522 \text{ cal mol}^{-1})/RT\right] \text{ cm}^3 \text{ molec}^{-1} \text{ sec}^{-1}$ (265). Although the production of $O_2(^1\Delta)$ is energetically possible below 6110 Å (see Table VI–7) by a spin-forbidden process, there is no evidence that $O_2(^1\Delta)$ is produced, since the overall quantum yield of ozone decomposition remains the same throughout the entire region. [If $O_2(^1\Delta)$ is produced for $\lambda < 6110$ Å, Φ_{-O_3} would increase from 2 to 4. See the following section.]

VI–11.2. Photodissociation in the Huggins Bands (3000 to 3600 Å)

The quantum yield of O_3 decomposition at 3340 Å is 4 (196, 546), indicating that one of the products must be an excited species capable of decomposing O_3 further. Castellano and Schumacher (196) have found no effect on the quantum yield even when 500 torr of N_2 was added to 50 torr of O_3. If the primary process is

$$O_3 \xrightarrow{h\nu} O(^1D) + O_2(X\,^3\Sigma) \qquad \text{(VI-71a)}$$

it is known, as is described later, that the $O(^1D)$ atom reaction with O_3 is

$$O(^1D) + O_3 \rightarrow O_2 + 2O(^3P) \qquad \text{(VI-71b)}$$

Table VI–7. Wavelength Thresholds (Å) below Which Indicated Reactions are Energetically Possible in O_3 Photolysis

O/O_2	$^3\Sigma_g^-$	$^1\Delta_g$	$^1\Sigma_g^+$	$^3\Sigma_u^+$	$^3\Sigma_u^-$
^3P	11,800[a]	6110	4630	2300	1730
^1D	4,110	3100	2660	1680	1360
^1S	2,370	2000	1800	1290	1090

[a] The threshold wavelength below which the dissociation of O_3 into $O_2(^3\Sigma_g^-) + O(^3P)$ is energetically possible.

The addition of N_2 should quench $O(^1D)$ to $O(^3P)$ quite efficiently (Section IV–4.1). Therefore, it is expected that the quantum yield is reduced from 4 to 2 when N_2 is added. Since no change of quantum yield is found, the excited species must be either $O_2(^1\Delta)$ or $O_2(^1\Sigma)$. It is not apparent from the effect of N_2 alone which species is formed.

The primary process of the O_3 photolysis at 3340 Å must be a spin forbidden process

$$O_3 \xrightarrow{h\nu} O(^3P) + O_2(^1\Sigma \text{ or } ^1\Delta) \qquad \text{(VI-72)}$$

followed by

$$O(^3P) + O_3 \rightarrow O_2 + O_2 \qquad \text{(VI-73)}$$

and

$$O_2(^1\Sigma \text{ or } ^1\Delta) + O_3 \rightarrow O_2 + O_2 + O(^3P) \qquad \text{(VI-74)}$$

From the mechanism the overall quantum yield of O_3 decomposition is 4.

Another process, $O_3 \xrightarrow{h\nu} O(^1D) + O(^1\Delta)$, becomes important with light of wavelengths below 3200 Å. The process is described below.

VI–11.3. Photodissociation in the Hartley Bands (2000 to 3200 Å)

The Primary Process. The primary process of O_3 photolysis in the Hartley bands appears to have been well established. The primary molecular product has recently been directly identified as $O_2(^1\Delta)$ by detecting the infrared atmospheric band at 12,700 Å during the 2537 Å photolysis (547). Jones and Wayne (547) obtained a quantum yield of nearly unity (0.9 ± 0.2) for the $O_2(^1\Delta)$ production. The $O_2(^1\Delta)$ produced by O_3 photolysis was also detected by its absorption in the vacuum ultraviolet (311) (see Section V–6.2), although an earlier study (516) failed to detect the atmospheric band emission from the $O_2(^1\Delta)$. The electronic state corresponding to the Hartley bands is probably $^1B_2$, which can dissociate by a spin-allowed process into $O_2(^1\Delta) + O(^1D)$ below 3100 Å

$$O_3(^1B_2) \xrightarrow{h\nu} O_2(^1\Delta) + O(^1D) \qquad \text{(VI-75)}$$

The direct detection of $O(^1D)$ may be made either by absorption at 1152 Å or emission at 6300 Å immediately after flash photolysis of O_3 (see Table A–2).

The detection of $O(^1D)$ by the emission at 6300 Å would be extremely difficult because the emission life is about 150 sec (32). Even at a pressure of 1 mtorr of O_3 each $O(^1D)$ atom would undergo about 10^6 collisions with O_3 molecules during its lifetime. Consequently, the emission intensity would be reduced by a million times by collision quenching. In spite of the difficulties, Gilpin et al. (398) have succeeded in following the decay of the extremely weak emission at 6300 Å by accumulating signals from 600 flashes. The decay rate of $O(^1D)$ is governed by the reaction rate of $O(^1D)$ with O_3 from which Gilpin et al. obtained a rate constant of $2.5 \pm 1 \times 10^{-10}$ cm^3 molec^{-1} sec^{-1}, which corresponds to almost unit collision efficiency. They have shown also that the production of $O_2(^1\Sigma)$ in the primary process is not more than 5%, although the process is energetically allowed below 2660 Å (see Table VI–7).

Secondary Processes. Snelling et al. (917) followed the consumption of O_3 as a function of time after flash photolysis of O_3. The results show that the ozone consumption occurs in two distinct stages, a very fast process (less than 50 μsec) followed by a slow process lasting many milliseconds. The fast process is attributed to the reaction

$$O(^1D) + O_3 \rightarrow O_2 + 2O \text{ or } 2O_2 \qquad \text{(VI-76)}$$

Giachardi and Wayne (392), using a flow system, have found that the yield of O_3 decomposition at the end of the fast process in a He–O_3 system is twice as large as that in a N_2–O_3 system. The results agree with the observation by Snelling et al. (917). The decrease of the O_3 consumption in the N_2–O_3 system is attributed to an efficient quenching of $O(^1D)$ by N_2 to $O(^3P)$. The slow consumption process would involve the reactions

$$O(^3P) + O_3 \rightarrow 2O_2 \qquad \text{(VI-77)}$$

$$O_2(^1\Delta) + O_3 \rightarrow 2O_2 + O \qquad \text{(VI-78)}$$

The reaction of $O(^1D)$, produced in the primary process, with O_3 would be (392, 1027) either

$$O(^1D) + O_3 \rightarrow 2O(^3P) + O_2 \qquad \text{(VI-79)}$$

or

$$O(^1D) + O_3 \rightarrow 2O_2 \qquad \text{(VI-80)}$$

The reaction could also produce vibrationally excited O_2 (55, 361, 1027) which does not appear to induce further decomposition of O_3 [Fitzsimmons and Bair (361)]. Giachardi and Wayne (392) have estimated the occurrence of (VI-79) to be one third the total reactive collisions of $O(^1D)$ with O_3. Another estimate of over 90% for (VI-79) is given by Webster and Bair (1027) and Bair et al. (55).

The primary molecular product, $O_2(^1\Delta)$, reacts further with O_3 by the process

$$O_2(^1\Delta) + O_3 \rightarrow O(^3P) + 2O_2 \qquad \text{(VI-81)}$$

followed by

$$O(^3P) + O_3 \rightarrow 2O_2 \qquad \text{(VI-82)}$$

The maximum quantum yield of ozone decomposition, Φ_{-O_3}, has been measured to be 4 (546), 5.5 (640), and 6 (998). Norrish and Wayne (748) found that Φ_{-O_3} increases at higher O_3 pressures, although no pressure effect on Φ_{-O_3} has been found by others (640, 998). This discrepancy is probably due to experimental conditions such as the presence of impurities and the walls, which would deactivate excited species.

The quantum yield, $\Phi_{-O_3} = 6$, can be explained on the basis of (VI-75), (VI-79), (VI-81), and (VI-82), while $\Phi_{-O_3} = 4$ can be explaned on the basis of (VI-75) and (VI-80) through (VI-82). The intermediate value $\Phi_{-O_3} = 5.5$ indicates (VI-75) and (VI-79) through (VI-82). Thus, one obtains the various values of the quantum yield of O_3 decomposition ranging from 4 to 6 by changing the ratio of (VI-79) to (VI-80).

Instead of the reactions (VI-79) and (VI-80) for the $O(^1D)$ with O_3 reaction, some workers (998, 1026) propose the reaction

$$O(^1D) + O_3 \rightarrow O_2^* + O_2 \qquad \text{(VI-83)}$$

followed by

$$O_2^* + O_3 \rightarrow O(^3P) + 2O_2 \qquad \text{(VI-84)}$$

where O_2^* is an unspecified excited state of O_2. The sequence (VI-75), (VI-81), (VI-82), (VI-83), and (VI-84) gives $\Phi_{-O_3} = 6$.

The photolysis at 3130 Å gives conflicting results. Castellano and Schumacher (194, 195) obtained $\Phi_{-O_3} = 6$ and concluded that the same primary process [process (VI-75)] proposed for the 2537 Å photolysis occurs at 3130 Å with unit quantum efficiency.

Lin and DeMore (637) have irradiated mixtures of O_3 and isobutane with monochromatic light of wavelengths from 2750 to 3340 Å at −40°C. The bandwidth was 16 Å. The relative quantum yields of $O(^1D)$ production were obtained from the yield of isobutyl alcohol, a product of the reaction $O(^1D)$ + isobutane. The results are shown in Fig. VI–13. The quantum yields are constant below 3000 Å and show a sharp cutoff at 3080 Å, the thermochemical threshold wavelength for the production of $O(^1D) + O_2(^1\Delta)$.

According to Moortgat and Warneck (717) the process

$$O_3 \xrightarrow{h\nu} O_2(^1\Delta) + O(^1D) \qquad \text{(VI-85)}$$

becomes important below the incident wavelength 3200 Å at room temperature. The thermochemical threshold at 0°K for (VI-85) is 3100 Å. The quantum yield of $O(^1D)$ production at 3130 Å at room temperature is about 0.3 [Moortgat and Warneck (717), Kuis et al. (591)] and it decreases at lower temperatures (591) [Lin and DeMore (637) Kajimoto and Cvetanović (555)]. The quantum yield of $O(^1D)$ production estimated by other workers is 0.1 (546) and 0.5 (887) at 3130 Å.

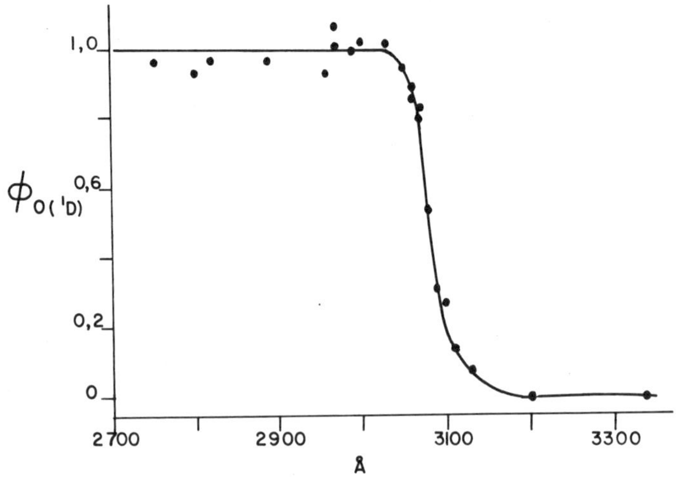

Fig. VI–13. The quantum yield of $O(^1D)$ from O_3 photolysis at $-40°C$ as a function of incident wavelength. Bandwidth 16 Å. From Lin and DeMore (637), reprinted by permission of Elsevier Sequoia, S.A.

The $\Phi_{-O_3} = 6$ obtained by Castellano and Schumacher (194, 195) at 3130 Å indicates $\phi_{O(^1D)}$ to be unity. To be consistent with other results, the effective wavelengths for O_3 photolysis in their experiment must be below 3100 Å. At room temperature internal energy of O_3 contributes to dissociation at 3130 Å. See p. 22.

VI–11.4. Photolysis of O_3 in the Presence of Other Gases

$O_3 + H_2O$ *system.* Quantum yields of O_3 photodecomposition at 2537 Å in the presence of H_2O are much larger than those of pure O_3. They increase linearly with $\sqrt{P_{H_2O}}$ (640, 749). The results are interpreted on the basis of the following chain mechanism

$$O_3 \xrightarrow{h\nu} O(^1D) + O_2(^1\Delta) \qquad \text{(VI-86)}$$

$$O(^1D) + H_2O \rightarrow 2OH \qquad \text{(VI-87)}$$

$$OH + O_3 \rightarrow HO_2 + O_2 \qquad \text{(VI-88)}$$

$$HO_2 + O_3 \rightarrow OH + 2O_2 \qquad \text{(VI-89)}$$

However, there is no direct evidence for the chain reactions (VI-88) and (VI-89). DeMore (276) presents evidence that (VI-89) does not occur at 87°K.

He also proposed the alternative chain mechanism

$$O(^1D) + H_2O \rightarrow OH(v \leq 2) + OH \quad \text{(VI-90)}$$

$$OH(v \geq 1) + O_3 \rightarrow H + 2O_2 \quad \text{(VI-91)}$$

$$H + O_3 \rightarrow OH(v \leq 9) + O_2 \quad \text{(VI-92)}$$

where v indicates the vibrational quantum number of OH radicals. Since the reaction $OH + O_3 \rightarrow H + 2O_2$ is endothermic by 8 kcal mol^{-1}, OH must be vibrationally excited ($v \geq 1$) for the reaction to be energetically feasible. The reaction $H + O_3$ is known to produce OH in $v \leq 9$ (206).

Reaction (VI-90) has been observed to yield OH in $v = 0, 1$, and 2 (335). However, (VI-91) has not been verified. The chain reaction has not been observed in the flash photolysis of O_3–H_2O mixtures (109, 370, 609). On the other hand Simonaitis and Heicklen (889) have found that the photolysis of O_3–H_2O mixtures at 2537 Å leads to the chain decomposition of O_3. They have found that the addition of O_2 does not inhibit the chain appreciably. Since O_2 reacts with H to form HO_2 they have concluded that the sequence (VI-91), (VI-92), which involves the H—OH chain, is not operative in the O_3–H_2O photolysis. The sequence (VI-88), (VI-89), which is the OH—HO_2 chain, is favored [see also DeMore and Tschuikow-Roux (280)].

O_3–RH *System.* Norrish and Wayne (749) have studied the O_3–RH system where RH = H_2O, H_2, CH_4, and HCl. In each case the quantum yield of O_3 decomposition is greater than that for pure O_3 photolysis. The results are explained by the following mechanism

$$O_3 \xrightarrow{2537\ \text{Å}} O_2 + O(^1D) \quad \text{(VI-93)}$$

$$O(^1D) + RH \rightarrow OH + R \quad \text{(VI-94)}$$

$$OH + O_3 \rightarrow HO_2 + O_2 \quad \text{(VI-95)}$$

$$HO_2 + O_3 \rightarrow OH + 2O_2 \quad \text{(VI-96)}$$

$$R + O_3 \rightarrow RO + O_2 \quad \text{(VI-97)}$$

VI–11.5. Ozone in the Atmosphere

Because of absorption by O_3 in the stratosphere, solar ultraviolet radiation reaching the surface of the earth is limited to wavelengths above about 3000 Å.

Since the possible partial destruction of the ozone layer by the injection of pollutants such as NO_x(NO and NO_2) and chlorofluoromethanes would induce global temperature changes and have an adverse health effect, such as an increase of skin cancer, the subject has been studied by many workers and is further discussed in Section VIII–2. An extensive review of O_3 and O atom reactions with hydrogen, nitrogen, and chlorine compounds in the stratosphere is given by Nicolet (740) and by Dütsch (328).

Briefly, the four following reactions control the ozone profile in the stratosphere.

$$O_2 \xrightarrow{h\nu} O + O \qquad \lambda < 2400\ \text{Å} \tag{VI-98}$$

$$O + O_2 + M \rightarrow O_3 + M \qquad M = N_2, O_2 \tag{VI-99}$$

$$O + O_3 \rightarrow 2O_2 \qquad k_{100} \tag{VI-100}$$

$$O_3 \xrightarrow{h\nu} O + O_2 \qquad \lambda < 11{,}800\ \text{Å} \tag{VI-101}$$

The equilibrium concentration of ozone is established with a maximum at an altitude of about 25 km, depending on the intensity of the solar flux, the rates of formation [processes (VI-98) and (VI-99)] and those of destruction of ozone [processes (VI-100) and (VI-101)]. The concentration profile of ozone as a function of altitude is given in Fig. VI-14. It has been recognized, however, that the four reactions (VI-98) to (VI-101) are not sufficient to account for the global ozone balance. About 80% of the ozone produced by sunlight must be destroyed by reactions other than those proposed by Chapman.

Johnston (543) and others have proposed that the most important catalytic cycle responsible for ozone destruction is a NO–NO_2 cycle

$$NO + O_3 \rightarrow NO_2 + O_2 \tag{VI-102}$$

$$O + NO_2 \rightarrow NO + O_2 \qquad k_{103} \tag{VI-103}$$

$$\text{net } O + O_3 \rightarrow 2O_2$$

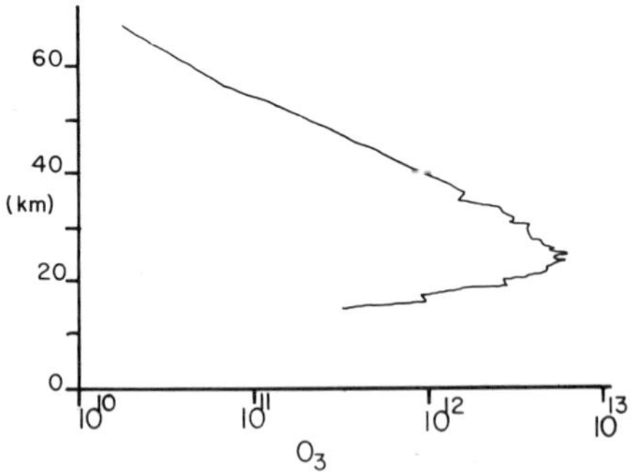

Fig. VI–14. Ozone concentration (molec cm^{-3}) as a function of altitude. From Randhawa (822), reprinted by permission of Birkhäuser Verlag.

Solar ultraviolet radiation also partially destroys NO_2 by

$$NO_2 \xrightarrow{h\nu} NO + O \qquad \lambda < 4000 \text{ Å} \tag{VI-104}$$

The ratio of the rate of ozone destruction with and without $NO_x(NO + NO_2)$, the catalytic ratio ρ, is (208, 537, 538)

$$\rho = 1 + \frac{k_{103}(NO_2)}{k_{100}(O_3)} \tag{VI-105}$$

Since $k_{103} \approx 10{,}000 \times k_{100}$ at 230°K, an approximate temperature in the stratosphere, even a very small fraction of NO_2, 0.01% of O_3, present in the stratosphere is as effective as destroying ozone by (VI-100). Besides the NO–NO_2 cycle, another effective catalytic cycle is a Cl–ClO chain [Molina and Rowland, (711) Crutzen (252)].

The Cl atoms are produced from the photolysis of chlorofluoromethanes used as refrigerants and as aerosol propellants.

$$CF_2Cl_2 \xrightarrow{h\nu} CF_2Cl + Cl \qquad \lambda < 2300 \text{ Å} \tag{VI-106}$$

The Cl atoms produced react with ozone

$$Cl + O_3 \rightarrow ClO + O_2 \tag{VI-107}$$

followed by

$$\begin{array}{l} ClO + O \rightarrow Cl + O_2 \\ \hline \text{net } O + O_3 \rightarrow 2O_2 \end{array} \tag{VI-108}$$

Further discussion of the Cl–ClO chain is given in Section VIII–2.

VI–12. SULFUR DIOXIDE (SO_2)

The ground state SO_2 is $\tilde{X}\,^1A_1$ with an O—S—O angle of 119.5°. The bond energy D_0(OS—O) is 5.65 ± 0.01 eV (767). Sulfur dioxide exhibits complex absorption spectra in the near ultraviolet as well as in the vacuum ultraviolet regions. There are three main regions of absorption in the ultraviolet, namely, an extremely weak absorption in the 3400 to 3900 Å, a weak absorption in the 2600 to 3400 Å, and a strong absorption in the 1800 to 2350 Å region. Figures VI–15*a* through VI–15*c* show the absorption coefficients of SO_2 in the near ultraviolet as well as in the vacuum ultraviolet regions. The dissociation of SO_2 to SO + O starts below about 2190 Å. Above this wavelength SO_2 exhibits strong fluorescence and phosphorescence.

In spite of many studies the complete analysis of the ultraviolet absorption spectrum has not been successful [Herzberg (16), p. 511]. The spectroscopy and photochemistry of SO_2 may be conveniently discussed for four regions of absorption:

3400 to 3900 Å region
2600 to 3400 Å region
1800 to 2400 Å region
1050 to 1800 Å region

VI–12.1. Spectroscopy and Photochemistry of SO_2 in the 3400 to 3900 Å Region

Absorption is extremely weak in this region (see Fig. VI–15*a*). A rotational analysis indicates that the upper state is $\tilde{a}^3B_1$ [Brand et al. (128)] with an electronic origin at 3.194 eV. An additional state $^3A_2$ near 3700 Å may be present in this region (135). The triplet emission lifetime of about 1 msec has been measured by Collier et al. (228) and by Sidebottom et al. (876) when SO_2 was excited by the 3828.8 Å laser line. This lifetime is much shorter than 7 msec measured by Caton and Duncan (197). A collision-free lifetime of 2.7 msec has been obtained by Briggs et al. (149). The lifetime of a few milliseconds is much shorter than that expected from the integrated absorption coefficient (1.7×10^{-2} sec), indicating that the radiationless transition

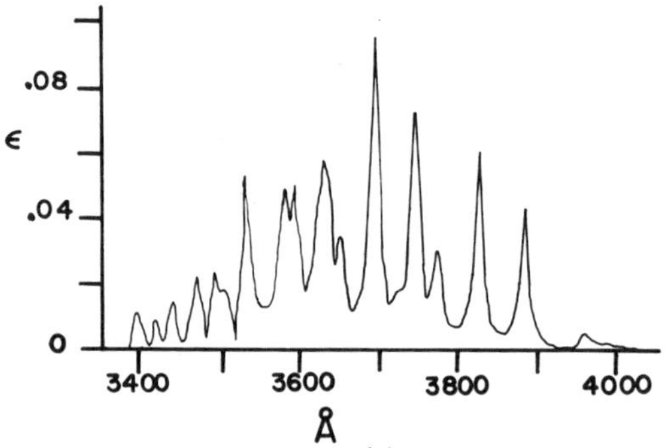

(*a*)

Fig. VI–15. **(*a*) Absorption coefficients of SO_2. 3400 to 4000 Å region; ε in units of liter mol^{-1} cm^{-1}, base 10, room temperature. Reprinted with permission from H. W. Sidebottom, C. C. Badcock, G. E. Jackson, J. G. Calvert, G. W. Reinhardt, and E. K. Damon, *Environ. Sci. Technol.* 6, 72 (1972). Copyright by the American Chemical Society. (*b*) Absorption coefficients of SO_2. 1900 to 2150 Å region. From Golomb et al. (407). 2100 to 3100 Å region. From Warneck et al. (1008a); *k* is given in units of atm^{-1} cm^{-1}, base e, 0°C. Reprinted by permission. Copyright 1964 by the American Institute of Physics. (*c*) Absorption coefficients of SO_2 in the region 1050 to 1900 Å. *k* is given in units of atm^{-1} cm^{-1}, base *e*, 0°C. From Golomb et al. (407), reprinted by permission. Copyright 1962 by the American Institute of Physics.**

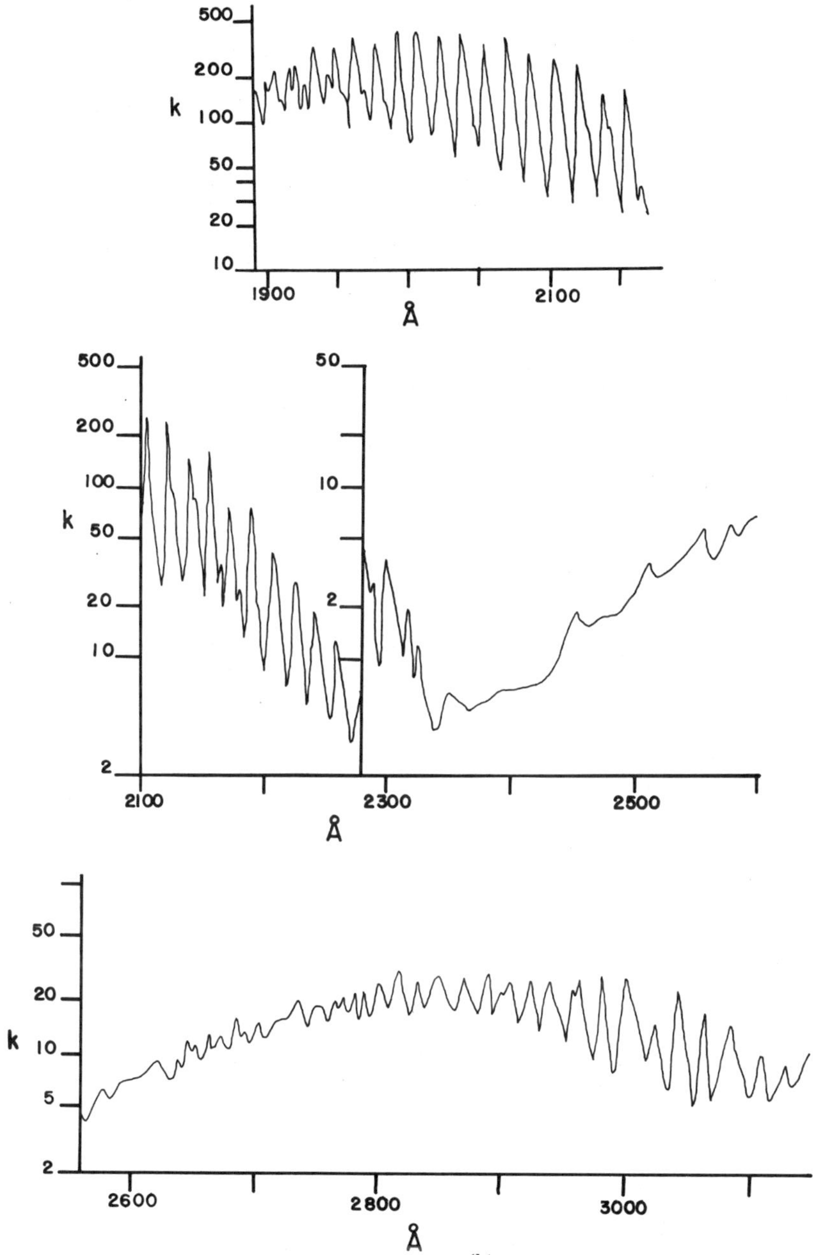

(b)

Fig. VI–15. (*cont.*)

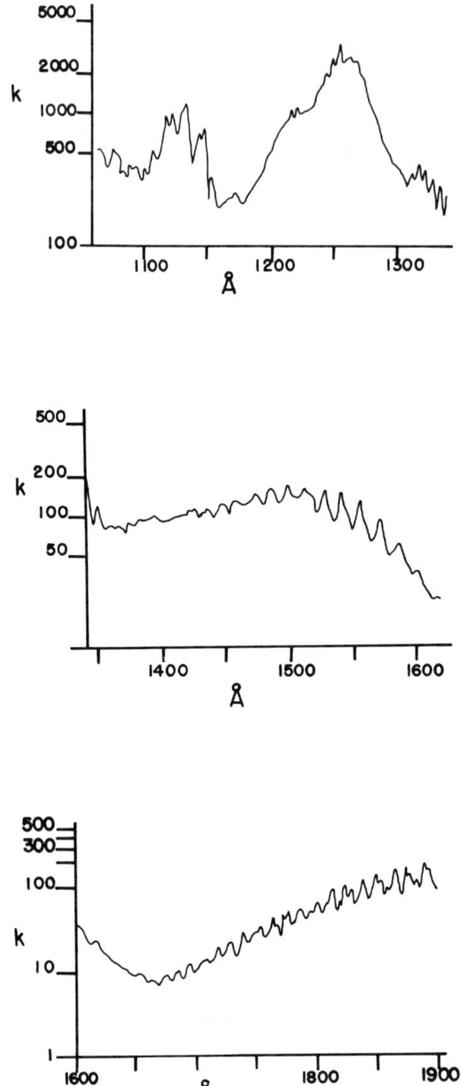

(c)

Fig. VI–15. (*cont.*)

to the ground state is a predominant process. The quantum yield of phosphorescence is estimated to be 0.12 ± 0.09 in comparison with the phosphorescence quantum yield of biacetyl–cyclohexane mixtures [Sidebottom et al. (876)]. On the other hand, Strickler et al. (941) have obtained a quantum yield of 0.07 from the measured phosphorescence decay rate at zero pressure, $1.12 \pm 0.20 \times 10^3$ sec^{-1}, and the calculated radiative rate constant (79 ± 5 sec^{-1}). Nelson and Borkman (732) have measured the lifetime and quantum yield of solid SO_2 at low temperatures. They have found $S^{18}O_2(^3B_1)$ has a somewhat longer lifetime than $S^{16}O_2(^3B_1)$.

Very recently Su et al. (953a) obtained a collision-free lifetime of $SO_2(^3B_1)$ of 8.1 ± 2.5 msec, which is considerably longer than that reported previously. The new lifetime is in reasonable agreement with that calculated from the integrated absorption coefficient using (I-91c), indicating that the phosphorescence quantum yield is near unity. Apparently, intersystem crossing to the ground state is not important as was predicated by Bixon and Jortner on a theoretical basis (110).

The self-quenching rate constant (876) is 6.64×10^{-13} cm^3 molec^{-1} sec^{-1}. The quenching products may be SO_2 or $SO + SO_3$

$$SO_2(^3B_1) + SO_2 \rightarrow 2SO_2 \qquad \text{(VI-109)}$$

$$SO_2(^3B_1) + SO_2 \rightarrow SO + SO_3 \qquad \text{(VI-110)}$$

The quenching rates of the triplet by CO, N_2, O_2, H_2O, Ar, He, Xe, CO_2, O_3, NO, and hydrocarbons have been measured by Calvert et al. (53, 517, 824, 877, 879, 1005). The quenching rate by NO is almost gas kinetic, while that by O_2 is unexpectedly small (10^{-13} cm^3 molec^{-1} sec^{-1}) (824, 879). The quenching by CO produces substantial amounts of CO_2 at higher temperatures, while at room temperature only 8% of the bimolecular collision results in CO_2 (517). The rate constants for N_2, CO, CO_2, CH_4, and the rare gases are all very similar (10^{-13} cm^3 molec^{-1} sec^{-1}).

VI–12.2. Spectroscopy and Photochemistry in the 2600 to 3400 Å Region

The rotational structure in this region is very complex and no conclusive analysis has been made. The $^1B_1$ state is tentatively assigned by Herzberg [(16), p. 605]. Another state, $^1A_2$, has been suggested (472) near the $^1B_1$ state. Brand and Nanes (130) have recently assigned $^1B_1$ for the 3000 to 3400 Å region. At shorter wavelengths may lie the $^1A_2$–$^1A_1$ transition (130). On the other hand, Hamada and Merer, (434, 435) and Dixon and Halle (288) have concluded that the upper state is $^1A_2$ and the forbidden transition $^1A_2$–$^1A_1$ becomes allowed by the excitation of the ν_3 antisymmetric stretching vibration of b_2 symmetry. The origin of the $\tilde{B}(^1B_1)$–$\tilde{X}(^1A_1)$ transition lies in the region 3100 to 3160 Å (435).

According to a recent study by Brand et al. (135), a quasicontinuous absorption underlying the structured $\tilde{A}^1A_2$–$\tilde{X}^1A_1$ bands is the $\tilde{B}^1B_1$–$\tilde{X}^1A_1$ transition and the $^1B_1$ state is strongly coupled with the $\tilde{X}^1A_1$ state. The $^1A_2$ state is vibrationally coupled with $^1B_1$ to have the necessary oscillator strength in the $^1A_2$–$^1A_1$ transition.

Greenough and Duncan (421) have obtained a fluorescence lifetime of 42 μsec when SO_2 is excited by light of wavelengths near 3000 Å. Sidebottom et al. (878) have obtained a zero pressure lifetime of 36 μsec at the 2662 Å excitation, but it appears to decrease with less excess vibrational energy.

On the other hand, Brus and McDonald indicate (156, 157) that at least two states appear to be involved in fluorescence by light absorption in the 2600 to 3200 Å region, one with a short collision-free lifetime ($\tau_S = 50$ μsec) and the other with a long collision-free lifetime ($\tau_L = 80$–530 μsec). Furthermore, τ_L is longer at the longer exciting wavelength, while τ_S appears to be independent of the exciting wavelength. The state with the short lifetime is quenched very rapidly ($>10 \times 10^{-10}$ cm^3 molec^{-1} sec^{-1}), while the state with the long lifetime is quenched with unit collision efficiency. The measured lifetime τ_L is much longer than that calculated from the integrated absorption coefficient (0.2 μsec) (421). This anomalous lifetime, according to Douglas (324) (see Section II–2.1), can be explained on the basis of the interaction of the singlet state with a low-lying metastable state or the ground state. As a result of this interaction the singlet state is "diluted" by the state that does not combine with the ground state. Therefore, the lifetime is much longer than that calculated from the integrated absorption coefficient. The lengthening of the lifetime, as well as the efficient quenching of the singlet state, is explained by the interaction of the vibrational levels of the singlet state with those of the ground state (386). The fluorescence quantum yield appears to be near unity (156, 183). Brand et al. (135) suggest that the two lifetimes τ_s and τ_L are, respectively, from the $\tilde{A}^1A_2$ and $\tilde{B}^1B_1$ states.

At low pressures of SO_2 a broad structureless emission with a maximum at about 3600 Å was excited by light of wavelengths near 3000 Å [Strickler and Howell (940), Mettee (698)]. This emission is due to fluorescence from the lowest singlet state. At higher pressures some new relatively sharp vibrational structure starts to appear in the region 3900 to 4900 Å. The bands can be ascribed to the transition from a vibrationally equilibrated triplet state to the ground in comparison with the absorption spectrum (698, 940). Phosphorescence is seen by the 3130 Å excitation even at the lowest possible pressure [Caton and Gangadharan (198)]. The following reaction mechanism may be presented for photochemical reactions in this region

$$SO_2 \xrightarrow{h\nu} {}^1SO_2 \qquad \text{(VI-111)}$$

$$^1SO_2 + SO_2 \rightarrow 2SO_2 \qquad \text{(VI-112)}$$

$$^1SO_2 + SO_2 \rightarrow {}^3SO_2 + SO_2 \qquad \text{(VI-113)}$$

$$^1SO_2 \rightarrow SO_2 + h\nu \qquad \text{(VI-114)}$$

$$^1SO_2 \rightarrow SO_2 \qquad k_{115} \qquad \text{(VI-115)}$$

$$^1SO_2 \rightarrow {}^3SO_2 \qquad k_{116} \qquad \text{(VI-116)}$$

$$^3SO_2 \rightarrow SO_2 + h\nu \qquad k_{117} \qquad \text{(VI-117)}$$

$$^3SO_2 \rightarrow SO_2 \qquad k_{118} \qquad \text{(VI-118)}$$

$$^3SO_2 + SO_2 \rightarrow 2SO_2 \qquad k_{119} \qquad \text{(VI-119)}$$

$$^3SO_2 + SO_2 \rightarrow SO_3 + SO \qquad k_{120} \qquad \text{(VI-120)}$$

The $^1SO_2$ and $^3SO_2$ represent the first singlet and triplet states, respectively. The reaction $^1SO_2 + SO_2$ produces $^3SO_2$ with about 8% efficiency [Rao et al. (823)].

The photolysis of pure SO_2 in the region 2500 to 3200 Å has produced SO_3 ($\Phi_{SO_3} = 0.08$) (775, 785) and SO (69, 523). Otsuka and Calvert (785) conclude that SO_3 is produced from (VI-120) and $k_{120} > k_{119}$, while James et al. (523) believe that $^3SO_2$ is not $\tilde{a}\,^3B_1$ but a nonphosphorescent $^3A_2$. The triplet SO_2 decays only 10% by emission, that is, $10k_{117} \simeq k_{118}$, as described earlier in Section VI–12.1. (According to very recent results by Su et al. (953a), $k_{117} = 1.2 \times 10^2\ \text{sec}^{-1}$, $k_{118} \approx 0$.) The main unimolecular reaction of $^1SO_2$ appears to be fluorescence (156, 183, 482) that is, k_{115}, $k_{116} \simeq 0$.

The quenching of $^1SO_2$ and $^3SO_2$ by foreign gases has been extensively studied by Calvert and Heicklen and their coworkers. Quenching gases include biacetyl (345, 482, 823), CO_2 (824), N_2 (483), CO (200, 201, 1003), C_2F_4 (200), hydrocarbons (53, 651), and other simple gases (785, 936, 1004). To explain the pressure dependence on the yield of the sensitized biacetyl emission (345, 484) and on CO oxidation (200, 201, 1003), it appears necessary to assume that singlet and triplet states other than $^1B_1$ and $^3B_1$ are involved. The singlet excited state of SO_2 seems important only as a source of $^3SO_2$ molecules through collision-induced intersystem crossing. The $^3SO_2$ formed subsequently participates in chemical reactions with CO, SO_2, C_2F_4, and olefins (272). The reaction of $^3SO_2$ with C_2F_4 produces CF_2O as a product (0.5 of the total quenching) (200, 877). The reaction of $^3SO_2$ with olefins produces sulfinic acids with collision efficiencies ranging from 0.14 to 0.5 (877). The reaction products may be formed from a common intermediate triplet diradical of a SO_2–RH complex. The reactions of $^3SO_2$ with 2-butenes (*cis*- and *trans*-) form isomers in the ratio *trans*-2-butene/*cis*-2-butene = 1.8 [Demerjian et al. (272)]. The quenching of $^3SO_2$ by ethane and higher paraffinic hydrocarbons is predominantly chemical, probably H atom abstraction reactions [Wampler et al. (1005)].

Both the $^1B_1$ and $^3B_1$ states induce isomerization of *cis*-2-butene to *trans*-2-butene (273). *cis*-2-Butene is very efficient in collisionally inducing a $^1B_1$–$^3B_1$ transition.

Hellner and Keller (461a) and Bottenheim and Calvert (122b) have observed a transient continuous absorption in the 2600 to 3500 Å region lasting 1 sec when SO_2 (~ 1 torr) and SO_2–Ar and SO_2–He mixtures are subjected to flash photolysis. The continuous absorption is due to a SO_2 dimer or an isomer of SO_2.

VI–12.3. Spectroscopy and Photochemistry in the 1800 to 2350 Å Region

Two band systems, α_1 and α_2, have tentatively been assigned in this region [Herzberg (16), pp. 512, 605]. However, Brand et al. (129, 131, 134) believe that only one excited state, $\tilde{C}^1B_2$, 5.279 eV above the ground, is involved in transition over the entire region. The bond distance increases from $r_0(SO) = 1.432$ Å of the ground to $r_0(SO) = 1.560$ Å and the bond angle decreases from 119.5 to 104.3° (129) in the $^1B_2$–$^1A_1$ transition. In accordance with a large shift in bond length the fluorescence spectrum extends to 4300 Å with a broad maximum at about 3200 Å, when excited near the 2100 Å line [Lotmar (648), Okabe et al. (770)].

Fluorescence and predissociation apparently compete in this region. Figure VI–16 shows the fluorescence efficiency as a function of the incident wavelength. A sudden decrease of the fluorescence yield near 2190 Å is taken as evidence of predissociation [Okabe (767)]. The fluorescence lifetime is dependent on the exciting wavelength [Hui and Rice (490)]. Below about 2200 Å the lifetime is shortened from about 40 (2200 Å) to 8 nsec (2150 Å), indicating the occurrence of predissociation. The fluorescence quantum yield appears to be near unity (490) above 2200 Å. Few photochemical studies have been done in this region. The photolysis of pure SO_2 at 1849 Å (325) produced SO_3 with a quantum yield of 0.50 ± 0.07. The addition of O_2 up to 70% increased Φ_{SO_3} to a plateau value of 1.04 ± 0.13, but with a further increase of O_2 the quantum yield Φ_{SO_3} decreased again to 0.5.

VI–12.4. Photochemistry in the 1100 to 1800 Å Region

Lalo and Vermeil (594, 595) have photolyzed SO_2 in the vacuum ultraviolet with and without H_2. H_2 is added to avoid sulfur deposition on the lamp

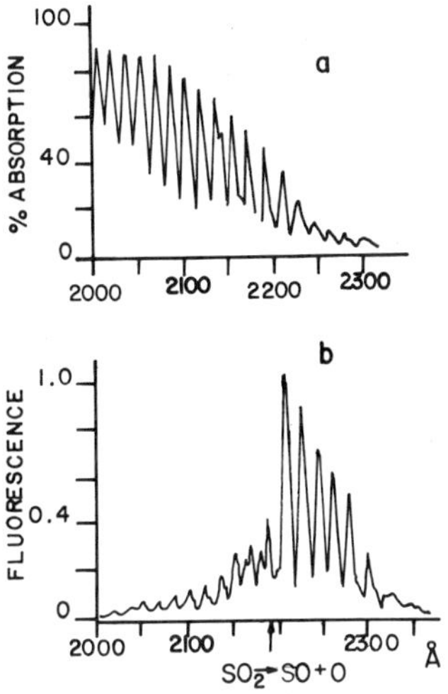

Fig. VI–16. (*a*) **The absorption spectrum of 0.3 torr of SO_2. Path length, 6.95 cm; resolution, 3 Å. (*b*) Fluorescence intensity (undispersed) of 0.3 torr SO_2 as a function of incident wavelength. Resolution, 3 Å; the arrow shows the thermochemical threshold of dissociation corresponding to 2192 Å. Reprinted with permission from H. Okabe J. Am. Chem. Soc. 93, 7095 (1971). Copyright by the American Chemical Society.**

window. Lalo and Vermeil have concluded that a primary process

$$SO_2 \xrightarrow{h\nu} S + O_2 \qquad \text{(VI-121)}$$

is important at 1236 Å and less so at 1470 Å. The S atoms are most likely in the ^1D state. At 1165 Å the $SO(A^3\Pi)$ is produced from the primary process. The phosphorescence from $SO_2(\tilde{a}^3B_1)$ is also observed at the Kr (1165, 1236 Å) lines.

VI–12.5. Photooxidation of SO_2 in the Atmosphere

It has been known that SO_2 in the atmosphere is gradually photooxidized to sulfuric acid and in the presence of NH_3 it is oxidized to ammonium sulfate with a typical rate of 0.1 to 0.7% per hour [Cox and Penkett (243)].

The quantum yield of SO_3 production in pure SO_2 with light of wavelengths 2500 to 4000 Å ranges from 8×10^{-2} [Okuda et al. (775)] to 3×10^{-3} [Cox (244), Skotnicki et al. (896)]. The photooxidation process in pure SO_2 is attributed to

$$^3SO_2 + SO_2 \rightarrow SO_3 + SO \qquad \text{(VI-122)}$$

Recently, Chung et al. (212) found that the quantum yields of SO_3 production increase with an increase of the flow rate, indicating reactions

$$SO + SO_3 \rightarrow 2SO_2 \qquad \text{(VI-123)}$$

$$2SO \rightarrow SO_2 + S\ [\text{or } (SO)_2] \qquad \text{(VI-124)}$$

are important. The limiting quantum yield is about 0.1.

In the atmosphere, however, the photooxidation process is likely to be [Sidebottom et al. (879)]

$$^3SO_2 + O_2 \rightarrow SO_3 + O(^3P) \qquad \text{(VI-125)}$$

An expected product, O_3, following reaction (VI-125), however, has not yet been found. Another possible reaction to form sulfuric acid in NO_x–RH contaminated atmospheres is

$$RO_2 + SO_2 \rightarrow RO + SO_3 \qquad \text{(VI-126)}$$

Sidebottom et al. (879) conclude that in polluted atmospheres $^3SO_2$ is mainly formed by collisions of N_2 and O_2 with initially produced $^1SO_2$ by absorption of sunlight in the 2900 to 3400 Å region and is partially (up to 20%) generated by the direct absorption of sunlight in the wavelength region 3400 to 3900 Å. It is also possible that SO_2 is oxidized by the heterogeneous reaction or by atomic oxygen in polluted air near smoke stacks. The nature of chemical reactions such as $^3SO_2 + H_2O$, $^3SO_2 + NO$, important in the atmosphere, has not been established.

Calvert and McQuigg (184) speculate that the initial free radical products of the addition reactions of HO_2, RO_2, HO, and RO to SO_2 would ultimately lead to sulfuric acid, peroxysulfuric acid, alkyl sulfates, and various other precursors to sulfuric acid, nitric acid, and salts of these acids, probably in aerosol particles.

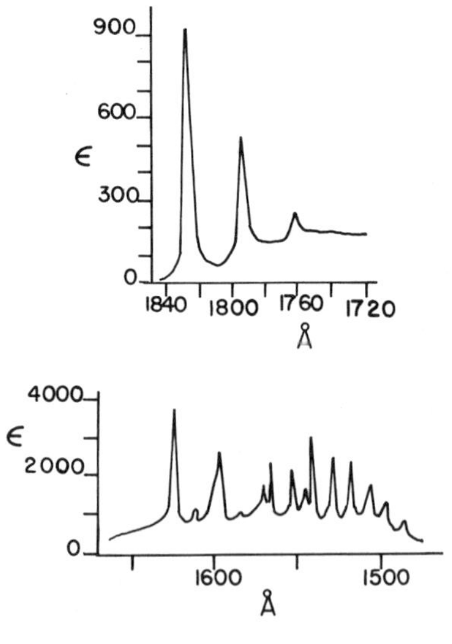

Fig. VI–17. The absorption coefficients of ClO_2. ε is given in units of $l\ mol^{-1}\ cm^{-1}$, base 10, room temperature. 1720 to 1850 Å region: $\tilde{C}$–$\tilde{X}$ system; 1500–1650 Å region: $\tilde{D}$–$\tilde{X}$, $\tilde{E}$–$\tilde{X}$ systems. From Basco and Morse (70), reprinted by permission of the Royal Society.

VI–13. CHLORINE OXIDES

VI–13.1. Chlorine Dioxide (ClO_2)

The ground state is $\tilde{X}^2B_1$ with an O—Cl—O angle of 117.6° [Herzberg (16)]; $D_0(\text{OCl—O}) = 2.50 \pm 0.07$ eV (28).

Absorption starts at about 5100 Å. The spectrum in the region 2700 to 5100 Å contains many vibrational bands, which have been analyzed by Coon and Ortiz (237). The spectrum corresponds to the $\tilde{A}^2A_2$–$\tilde{X}^2B_1$ transition (16).

Predissociation is apparent below 3750 Å. The absorption coefficients in the vacuum ultraviolet region 1300 to 1840 Å have recently been measured by Basco and Morse (70). This is shown in Fig. VI–17.

Photochemistry. Basco and Dogra (65) have studied the flash photolysis of ClO_2 above 3000 Å. They have proposed the following mechanism:

$$ClO_2 \xrightarrow{h\nu} ClO + O \qquad \text{(VI-127)}$$

$$O + ClO_2 \rightarrow ClO + O_2^{\dagger} \qquad k_{128} \qquad \text{(VI-128)}$$

$$2ClO \rightarrow Cl_2 + O_2 \qquad k_{129} \qquad \text{(VI-129)}$$

$$O + ClO \rightarrow Cl + O_2^{\dagger} \qquad k_{130} \qquad \text{(VI-130)}$$

where $O_2^{\dagger}$ denotes vibrationally excited O_2 molecules. They have obtained the rate constants $k_{128} = 5 \times 10^{-11}$, $k_{129} = 4.5 \times 10^{-14}$, and $k_{130} = 1.16 \times 10^{-11}$ all in cm^3 molec^{-1} sec^{-1}.

Fluorescence has been observed by laser excitation in the region 4579 to 4880 Å [Sakurai et al. (851)], corresponding to the transition $\tilde{A}^2A_2$–$\tilde{X}^2B_1$. Vibrational constants of the ground state have been calculated from the fluorescence bands.

Curl et al. (255) have analyzed the fluorescence spectrum excited by the 4765 Å Ar^+ laser line. The $\tilde{A}^2A_2$ state produced must predissociate appreciably at this wavelength since fluorescence is very weak and some decomposition products are found. Since the photon energy used (2.601 eV) is higher than $D_0(\text{OCl—O}) = 2.50$ eV, dissociation is expected.

VI–13.2. Chlorine Monoxide (Cl_2O)

The ground state of Cl_2O is bent with a Cl—O—Cl angle of 111°. $D_0(\text{ClO—Cl}) = 1.36 \pm 0.03$ eV (28) corresponds to the incident wavelength 9116 Å.

Absorption starts at about 7000 Å. The absorption coefficients in the 2200 to 7000 Å region have been measured by Goodeve and Wallace (409) and are shown in Fig. VI–18. The spectrum is continuous with three regions of absorption.

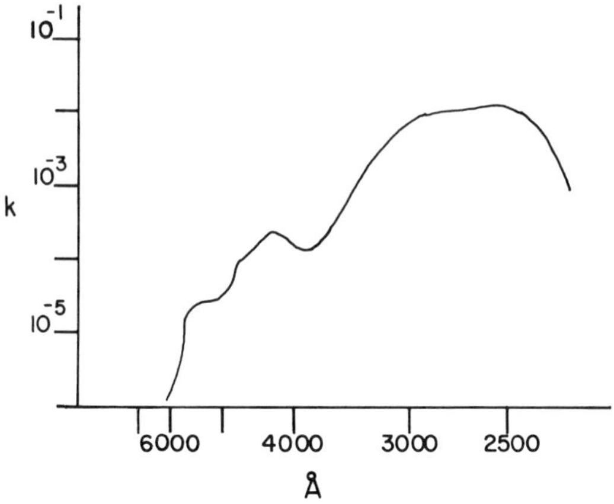

Fig. VI–18. The absorption coefficients of Cl_2O in the visible and near ultraviolet regions. k is given in units of mm^{-1} cm^{-1}, base 10, room temperature. From Goodeve and Wallace (409), reprinted by permission of the Chemical Society.

Photochemistry. The quantum yield of Cl_2O disappearance has been measured at 3130, 3650, and 4358 Å at 10°C. The yield is about 3.4. The yield in the region 2350 to 2750 Å is 4.5 [Finkelnburg et al. (356) and Schumacher and Townend (858)].

The flash photolysis of Cl_2O in the near ultraviolet has been studied by Edgecombe et al. (334) and Basco and Dogra (66). A mechanism proposed is

$$Cl_2O \xrightarrow{h\nu} ClO + Cl \qquad \text{(VI-131)}$$

$$Cl + Cl_2O \rightarrow Cl_2 + ClO \qquad k_{132} \qquad \text{(VI-132)}$$

$$2ClO \rightarrow Cl_2 + O_2 \qquad k_{133} \qquad \text{(VI-133)}$$

$$ClO + Cl_2O \rightarrow ClO_2 + Cl_2 \qquad k_{134} \qquad \text{(VI-134)}$$

$$ClO + Cl_2O \rightarrow Cl_2 + O_2 + Cl \qquad k_{135} \qquad \text{(VI-135)}$$

The rate constants obtained are $k_{132} = 6.8 \times 10^{-13}$, $k_{133} = 4.6 \times 10^{-14}$, $k_{134} = 4.3 \times 10^{-16}$, and $k_{135} = 1.1 \times 10^{-15}$ all in cm^3 $molec^{-1}$ sec^{-1}.

VI–14. TRIATOMIC RADICALS; PHOTOCHEMICAL PRODUCTION, DETECTION, AND REACTIVITIES

VI–14.1. Methylene (CH_2)

The ground state is the triplet $\tilde{X}^3B_1$ with an H—C—H angle of about 140° [Herzberg and Johns (469)]. Recent electron spin resonance work on CH_2 in solid matrices by Wasserman et al. (1013a) indicates that the H—C—H angle is 136°. $D_0(HC—H) = 4.36 \pm 0.05$ eV.

The $\tilde{X}^3B_1$ state can be identified by strong $\tilde{B}^3A_2$–$\tilde{X}^3B_1$, $\tilde{C}$–$\tilde{X}$, and $\tilde{D}$–$\tilde{X}$ absorption bands respectively at 1415, 1410, and 1397 Å [Herzberg (16)].

The oscillator strengths of the $\tilde{B}$–$\tilde{X}$, $\tilde{C}$–$\tilde{X}$, and $\tilde{D}$–$\tilde{X}$ transitions are 2.1×10^{-3}, 3.1×10^{-4}, and 2.2×10^{-4}, respectively [Pilling et al. (809)].

The first singlet state is $\tilde{a}^1A_1$ with an H—C—H angle of about 102°. The $\tilde{a}^1A_1$ state can be identified by the transition $\tilde{b}^1B_1 \leftarrow \tilde{a}^1A_1$ in the region 5000 to 9000 Å [Herzberg and Johns (465)]. Recently, emission bands due to the transition $\tilde{b}^1B_1 \rightarrow \tilde{a}^1A_1$ have been reported by Masanet and Vermeil (667) in the photolysis of CH_4 at 1048 and 1236 Å.

The difference in energy between $\tilde{a}^1A_1$ and $\tilde{X}^3B_1$ has been estimated to be in the range from about 0.1 to about 1 eV. See a recent review by Chu and Dahler (211). Very recently a value of 0.27 eV was obtained as an upper limit of this energy difference from the occurrence of the process $CH_2CO \xrightarrow{h\nu} CH_2(\tilde{a}^1A_1) + CO$ at 3370 Å (258a). It has been known for many years that chemical reactivities of $\tilde{X}^3B_1$ and $\tilde{a}^1A_1$ states are very different. The $^1A_1$ state reacts 3 orders of magnitude faster than the $^3B_1$ with H_2 and CH_4 [Braun et al. (143)]. Both states are generated by the photolysis of ketene or diazomethane. Methylene is also a primary product of hydrocarbon photolysis in the vacuum ultraviolet [Ausloos and Lias (49)].

The intersystem crossing from $CH_2(^1A_1)$ to $(^3B_1)$ is induced by inert gases (143). A theory dealing with the collision-induced singlet to triplet transition of methylene is developed by Chu and Dahler (211).

Reactions of CH_2 with inorganic and organic molecules have been extensively studied for the last two decades. They were summarized by Kirmse (17a) in 1971 to which review the reader is referred for detailed information. The low intensity photolysis of ketene or diazomethane in the presence of hydrocarbons and various quenching gases has been studied by many workers. The results of the end product analysis indicate that the two states of methylene, singlet and triplet, react quite differently with hydrocarbons. The rates of singlet and triplet methylene reactions and those of the collision induced transition from singlet to triplet by inert gases were first measured by Braun et al. (143) from the decay rates of triplet methylene at 1415 Å produced in flash photolysis. Subsequently, the reaction rates of CH_2 with various gases have been measured by Laufer and Bass (608).

The pertinent results are summarized below and in Table VI–8.

1. Singlet methylene, $^1CH_2$, inserts into the C—H bonds of paraffins with rates, tertiary C—H > secondary C—H > primary C—H

$$^1CH_2 + RH \rightarrow R—CH_3$$

Table VI–8. Comparison of the Rate Constants for $CH_2(^1A_1)$ and $CH_2(^3B_1)$

Reactant	Products	Rate Constant (cm^3 $molec^{-1}$ sec^{-1})	Ref.
	A. Singlet Methylene $(^1A_1)$		
He	$CH_2(^3B_1) + He$	3×10^{-13}	143
Ar	$CH_2(^3B_1) + Ar$	6.7×10^{-13}	143
Xe	$CH_2(^3B_1) + N_2$	1.8×10^{-12}	91a
N_2	$CH_2(^3B_1) + N_2$	9×10^{-13}	143
H_2	$CH_3 + H$	7×10^{-12}	143
	$CH_2(^3B_1) + H_2$	$<1.5 \times 10^{-12}$	143
NO		$<4 \times 10^{-11}$	608
O_2	$[CH_2(^3B_1) + H_2]$	$<3 \times 10^{-11}$	608
		4.0×10^{-12}	91a
CO		$<9 \times 10^{-12}$	608
CH_4	$CH_3 + CH_3$	1.9×10^{-12}	143
	$CH_2(^3B_1) + CH_4$	1.6×10^{-12}	143
C_3H_8	$n\text{-}C_4H_{10}$	4.4×10^{-12}	91a
	$i\text{-}C_4H_{10}$	1.9×10^{-12}	91a
	$CH_2(^3B_1) + C_3H_8$	2.4×10^{-12}	91a
CH_2N_2	$C_2H_4 + N_2$	3.1×10^{-11}	91a
CH_2CO	$C_2H_4 + CO$	3.2×10^{-11}	608
	B. Triplet Methylene $(^3B_1)$		
H_2	$CH_3 + H$	$<5 \times 10^{-14}$	143
NO		1.6×10^{-11}	608
O_2	CO, CO_2, H_2	1.5×10^{-12}	608
CO		$\leq 1.0 \times 10^{-15}$	608
CO_2	$HCHO + CO$	3.9×10^{-14}	608a
CH_2	$C_2H_2 + H_2$ (or 2H)	5.3×10^{-11}	143
CH_3	$C_2H_4 + H$	5×10^{-11}	809a
CH_4	$CH_3 + CH_3$	$<5 \times 10^{-14}$	143
C_2H_2	C_3H_4	7.5×10^{-12}	608
C_2H_4		$<10^{-15}$	623a
CH_2N_2	$CH_3 + CHN_2$	1.0×10^{-12}	608a
CH_2CO	$[CH_3 + CHCO]$	$<10^{-17}$	623a

while triplet methylene abstracts H atoms

$$^3CH_2 + RH \rightarrow CH_3 + R$$

2. The reactions of $^1CH_2$ with paraffins proceed with efficiencies of 0.02 to 0.05, while those of $^3CH_2$ occur with efficiencies of 10^{-5} to 10^{-6}.
3. Singlet CH_2 is partially deactivated to triplet CH_2 by collisions with paraffins.

4. Both $^1CH_2$ and $^3CH_2$ react with O_2 at a comparable efficiency of about 0.02. Hence, the addition of O_2 preferentially suppresses the products formed by reactions of $^3CH_2$ with paraffins.
5. Singlet CH_2 primarily adds to the carbon–carbon double bond of olefin and, to a small extent, inserts into the C—H bonds.
6. Singlet CH_2 adds to the double bond of *cis*-2-butene to form predominantly *cis*-1,2,-dimethylcyclopropane, while the triplet CH_2 addition to *cis*-2-butene results in both *cis*- and *trans*-1,2-dimethylcyclopropane. The former type of reaction, that is, the reaction product retains the same geometrical configuration as the reactant, is called stereospecific. Thus, the singlet methylene addition to *trans*-2-butene is also stereospecific, that is, *trans*-1,2-dimethylcyclopropane is mainly observed. Since the initially formed 1,2-dimethylcyclopropane is vibrationally excited, it rearranges to its structural isomers (pentenes) and the geometrical isomer at low total pressures and is stable only at high total pressures.

The singlet to triplet methylene ratio initially produced in the photolysis of CH_2CO has been estimated by various workers as a function of incident wavelength on the basis of stereospecific addition of methylene to butenes. Although the results are in general agreement that less triplet methylene is formed at shorter wavelengths, the estimated fraction of triplet methylene ranges from 0.15 to 0.37 at 3130 Å [Eder and Carr (333a)].

The discrepancy may arise partly from the yet uncertain *cis*- to *trans*-1,2-dimethylcyclopropane ratio resulting from the triplet methylene addition to 2-butene and from the collision induced transition of methylene from singlet to triplet that may occur to a different degree depending on substrate molecules.

Eder and Carr (333a) have obtained values of 0.29 and 0.87 at 3130 and 3660 Å, respectively, for the fraction of triplet methylene by comparing the total product ratio with and without O_2, which preferentially scavenges triplet methylene. However, they assumed that the initially formed $^1CH_2$ are not quenched to $^3CH_2$ by reactant molecules. Further work is required to remove these ambiguities.

VI–14.2. Amidogen (NH_2)

The ground state is $\tilde{X}^2B_1$ with an H—N—H angle of 103° (16); D_0(HN—H) $= 4.1 \pm 0.2$ eV.

The $NH_2(\tilde{X}^2B_1)$ state can be identified by the absorption $\tilde{A}^2A_1 \leftarrow \tilde{X}^2B_1$ in the region 4300 to 9000 Å or by the induced fluorescence $\tilde{A} \rightarrow \tilde{X}$ with a proper light source [Kroll (587) Hancock et al. (437)].

A collision-free fluorescence lifetime of $NH_2(\tilde{A}^2A_1)$ of 10 μsec has been measured by Halpern et al. (432). The quenching of $^2A_1$ by various gases has also been studied. Quenching rates are almost equal to or above the gas kinetic value. The ground state NH_2 is the major product of the ammonia photolysis. The $NH_2(\tilde{A}^2A_1)$ is produced in the vacuum ultraviolet photolysis of ammonia. The reaction rate of NH_2 $(\tilde{X}^2B_1)$ with NO has been measured by Hancock et al. (437) by laser induced fluorescence. The rate constant is $2.1 \pm 0.2 \times 10^{-11}$ cm^3 molec^{-1} sec^{-1}. The reactions of $NH_2(\tilde{X}^2B_1)$ with NO and O_2 have been studied by Jayanty et al. (530). Possible reactions are

$$NH_2 + NO \rightarrow N_2 + H_2O \qquad \text{(VI-136)}$$

$$NH_2 + O_2 \rightarrow NH_2O_2 \qquad \text{(VI-137)}$$

VI–14.3. Phosphorus Hydride (PH_2)

The ground state of PH_2 is $\tilde{X}^2B_1$ with an H—P—H angle of 92° (16); D_0(H—PH) $= 3.4 \pm 0.4$ eV. The absorption spectrum corresponding to the transition $\tilde{A}^2A_1 \leftarrow \tilde{X}^2B_1$ is in the 3600 to 5500 Å region and the corresponding emission is in the 4540 to 8520 Å region. The PH_2 radical is the major primary product of the photolysis of PH_3 (620, 747).

VI–14.4. Ethynyl (C_2H)

Graham et al. (416) have concluded that the ground state is linear $^2\Sigma$ on the basis of electron spin resonance of C_2H in solid matrices. The absorption spectra in the gas phase have apparently not been observed. Graham et al. (416) have seen two weak absorption bands at about 3300 and 10,000 Å in solid Ar and have assigned the former to a $\tilde{B}^2A' \leftarrow \tilde{X}^2\Sigma$ transition and the latter to an $A^2\Pi \leftarrow X^2\Sigma$ transition. However, Gilra (398a) believes that the 10,000 Å transition is part of the C_2 Phillips system. $D_0(C_2\text{—}H) = 5.33 \pm 0.05$ eV.

The ground state C_2H is a major primary product of the acetylene and haloacetylene photolysis. Okabe (773) has observed the production of an electronically excited C_2H that fluoresces in the region 4000 to above 5500 Å in the vacuum ultraviolet photolysis of acetylene and bromoacetylene. The lifetime of this fluorescence is about 6 μsec and the fluorescence is quenched readily by C_2H_2, H_2, N_2, and Ar [Becker et al (81)]. On a theoretical basis, Shih et al. (872a) speculate that the fluorescence arises from a transition $^4\Sigma^+ \rightarrow X^2\Sigma$.

Ethynyl radicals have recently been detected in interstellar medium by microwave spectroscopy [Tucker et al. (980a)].

The reactions of the ground state C_2H with hydrocarbons have been extensively studied by Cullis et al. (253) and by Tarr et al. (962). The reaction with alkanes is the hydrogen abstraction to form acetylene

$$C_2H + RH \rightarrow C_2H_2 + R \qquad \text{(VI-138)}$$

VI-14.5. Formyl (HCO)

The ground state is $\tilde{X}^2A'$ (C_s symmetry) with an H—C—O angle of 120°. D_0(OC—H) = 0.9 ± 0.3 eV.

Absorption bands are in the region 4600 to 8600 Å, corresponding to the $\tilde{A}^2A'' \leftarrow \tilde{X}^2A'$ transition (16). The corresponding emission $^2A'' \rightarrow {}^2A'$ has not been observed presumably because of the strong predissociation observed for the $^2A''$ state [Johns et al. (534)].

The formyl radical is a major primary product of the photolysis of formaldehyde in the near ultraviolet. The formyl radicals produced in the atmosphere by sunlight may react with O_2 to form CO and HO_2

$$HCO + O_2 \rightarrow CO + HO_2 \qquad k_{139} \tag{VI-139}$$

Thus, the reaction may contribute to the formation of photochemical smog which is further discussed in Section VIII-2.

The rate constant k_{139} has recently been measured by Washida et al. (1013), who report a value of $5.7 \pm 1.2 \times 10^{-12}$ cm^3 molec^{-1} sec^{-1}.

Osif and Heicklen (784) suggest two other reaction paths

$$HCO + O_2 \rightarrow HCO_3 \tag{VI-140}$$

$$HCO + O_2 \rightarrow CO_2 + OH \tag{VI-141}$$

VI-14.6. Nitroxyl Hydride (HNO)

The ground state is $\tilde{X}^1A'$ (C_s symmetry) with an H—N—O angle of 105°. D_0(H—NO) ≤ 2.11 eV (16).

The transition $\tilde{A}^1A''$–$\tilde{X}^1A'$ has been observed in absorption in the region 7300 to 7750 Å and in emission in the 6000 to 10,000 Å region. Recently, Callear and Wood (178) have found absorption bands in the 1980 to 2080 Å region.

The HNO radical is not the primary product of photolysis. It is formed by a third body combination of H atoms with NO.

Lewis et al. (629) have used the chemiluminescence reaction

$$H + NO + M \rightarrow HNO(^1A'') + M \tag{VI-142}$$

as a probe for measuring the concentrations of H atoms produced from the photolysis of formaldehyde. Ishiwata et al. (514) have seen HNO($^1A''$) in reactions of O(^3P)–O_2 with NO–hydrocarbon mixtures.

VI-14.7. Hydroperoxyl (HO_2) and HSO Radical

The ground state of the HO_2 radical is $\tilde{X}^2A''$ (C_s symmetry) with an H—O—O angle of 99° [Ogilvie (756)]. D_0(H—O_2) = 2.0 ± 0.1 eV. The ultraviolet absorption spectrum of HO_2 in the region 1800 to 2700 Å is continuous with a maximum at about 2100 Å [Hochanadel et al. (474) Kijewski and Troe (563), Paukert and Johnston (799)] (see Fig. VI–19).

The HO_2 radicals are generated by the flash photolysis of mixtures of water, helium, and oxygen. The observed continuous spectrum is indicative of a repulsive upper state.

In addition, near infrared absorption bands at 1.255 and 1.425 μm have recently been found by Hunziker and Wendt (493), who have attributed the bands to a transition $^2A' \leftarrow {}^2A''$. The band at 1.504 μm corresponds to the $^2A''(200) \leftarrow {}^2A''(000)$ transition. The corresponding emission bands of HO_2 have been detected recently by Becker et al. (83, 86). The HO_2 radical is an important reaction intermediate in combustion, in polluted atmospheres, and in the photolysis of H_2O_2. The reaction of HO_2 with NO is considered as a key reaction in photochemical smog formation, which is discussed in Section VIII–2.

The reaction

$$HO_2 + NO \rightarrow NO_2 + OH \qquad k_{143} \qquad \text{(VI-143)}$$

has been studied by Cox and Derwent (247), who obtained indirectly a value of $k_{143} = 1.2 \times 10^{-12}$ cm^3 molec^{-1} sec^{-1}. The reaction has also been studied by Simonaitis and Heicklen (886, 890) and by Payne et al. (801). The rate constant k_{143} has recently been measured directly using laser magnetic resonance by Howard and Evenson (486a), who obtained a value of 8×10^{-12} cm^3 molec^{-1} sec^{-1}.

The ground state of the HSO radical is $\tilde{X}^2A''$ with an H—S—O angle of 102° (859a). D_0(H—SO) $\approx$ 1.6 eV, D_0(HS—O) $\approx$ 3.4 eV. Chemiluminescence found in the region 5200 to 9600 Å in flowing O—H_2S—O_3 mixtures is ascribed by Schurath et al. (859a) to the $^2A'$–$^2A''$ transition of HSO. The $^2A'$ state is presumably formed by

$$SH + O_3 \rightarrow HSO(^2A') + O_2$$

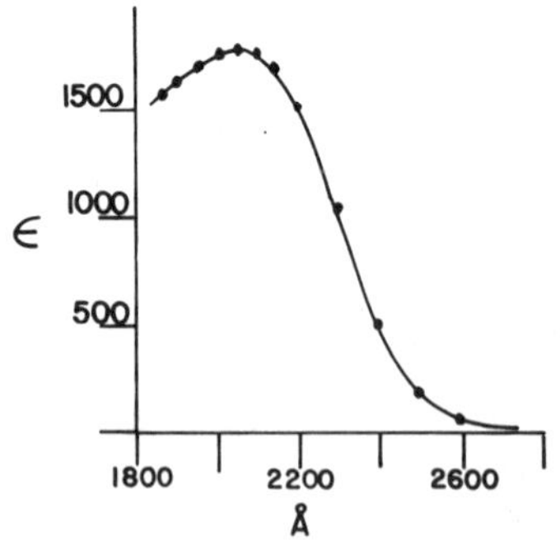

Fig. VI–19. Absorption coefficients of HO_2. ϵ is given in units of 1 mol^{-1} cm^{-1}, base 10, room temperature. The absorption spectrum is obtained by the flash photolysis of mixtures H_2O, He, and O_2. The continuous nature of the spectrum indicates a repulsive upper state. From Hochanadel et al. (474), reprinted by permission. Copyright 1972 by the American Institute of Physics.

From the highest v' level (= 7) observed in emission, they deduced an upper limit of 14.9 kcal mol^{-1} for $\Delta Hf°$(HSO).

VI–14.8. Triatomic Carbon (C_3); CCO Radical

The ground state of C_3 is $X^1\Sigma_g^+$ (linear). Both the absorption and the emission bands of C_3 have been detected in the region 3400 to 4100 Å with prominent bands at 4050 Å. The transition is ascribed to $A^1\Pi_u \leftarrow X^1\Sigma_g^+$. $D_0(C_2{-}C) = 7.31 \pm 0.02$ eV. Stief (934) suggests that C_3 radicals observed in comets originate from the photolysis of propyne

$$C_3H_4 \xrightarrow{h\nu} C_3 + 2H_2 \qquad \text{(VI-144)}$$

The ground state of C_2O radical is $X^3\Sigma^-$ (285). $D_0(C{-}CO) = 2.2 \pm 0.3$ eV.

The absorption spectrum in the region 5000 to 9000 Å has been analyzed by Devillers and Ramsay (285). They have assigned the bands to the $A^3\Pi$–$X^3\Sigma^-$ transition with the 0–0 band near 8580 Å.

The most common source of C_2O is the photolysis of carbon suboxide above 2000 Å. Bayes (76) and Williamson and Bayes (1046) believe that both triplet and singlet C_2O are produced from the photolysis of C_3O_2. Triplet C_2O is formed by absorption of light of wavelengths above 2900 Å and singlet at about 2500 Å. Bayes (76) suggests that triplet C_2O is the ground $X^3\Sigma^-$ state and singlet C_2O is probably the $a^1\Delta$ state 0.5 eV above the ground state.

The reactivities of C_2O radicals towards various molecules appear to depend strongly on their electronic state. Bayes and coworkers have found that the relative reaction rates of ground (triplet) C_2O with various olefins increase in a similar manner from ethylene to more complex olefins such as those of O(^3P) and S(^3P), while singlet C_2O shows similar reaction rates with various olefins as shown in Table VI–9. Reactions of triplet C_2O with O_2 and NO are much faster than with ethylene and hence the major reaction product, allene,

$$C_2O + C_2H_4 \rightarrow C_3H_4 + CO$$

is strongly reduced by the addition of O_2 or NO.

On the other hand, the reaction rate of singlet C_2O with O_2 is not much different from that of ethylene or other olefins and the allene is not much reduced by the addition of O_2 in the 2500 Å photolysis of $C_3O_2 + C_2H_4$ mixtures. The reactivities of singlet and triplet C_2O are similar in trend to those of singlet and triplet CH_2 shown in Table VI–8. The reactions of singlet C_2O with olefins must be much faster than those of triplet C_2O. It is not known, however, whether triplet C_2O is formed at least partially from singlet C_2O by collisional deactivation in analogy with methylene.

Table VI–9. Comparison of Relative Reaction Rates of $C_2O(X^3\Sigma)$ and $(a^1\Delta)$ at Room Temperature

Reactant	Products (Ref. 1047)	$X^3\Sigma$	$a^1\Delta^c$
Oxygen	CO_2 + CO (Ref. 1045a) O + 2CO	135	≤0.5
Nitric oxide	N_2, N_2O, CO, CO_2 (Ref. 1045a)	20,000	
Hydrogen	CH_2 + CO (Ref. 368)		0.009 (Ref. 368)
Carbon suboxide	CO, polymer (Ref. 368)		1.4 (Ref. 74)
Ethylene	C_3H_4 (allene) + CO	1.0	1.0
Propylene	1,2-Butadiene + CO	6.1	1.2
cis-2-Butene	2,3-pentadiene + CO	10.1	1.9
2,3-Dimethyl-2-butene	2,4-Dimethyl-2,3-pentadiene + CO	250	2.1
1,3-Butadiene		210	2.4

[a] Rates are relative to ethylene (1046).
[b] The $a^1\Delta$ state has not been observed spectroscopically.
[c] $a^1\Delta$ is 0.5 eV above the ground $^3\Sigma$ (76).

Since no absorption spectrum due to the singlet C_2O is known, the direct comparison of reactivities of the two C_2O electronic states is at present not possible.

The photolysis of ketene (604) in the vacuum ultraviolet yields C_2O to an extent of several percent by

$$CH_2CO \xrightarrow{h\nu} H_2 + C_2O$$

VI–14.9. Azide (N_3); NCN Radical; NCO Radical

The ground state of N_3 is $X^2\Pi_g$ (linear); $D_0(N_2—N) = 0.56 \pm 0.2$ eV. The absorption spectrum has been observed in the 2600 to 2725 Å region corresponding to the transition $B^2\Sigma_u^+ - X^2\Pi_g$ (16). The N_3 radicals are produced in the photolysis of HN_3 [Douglas and Jones (323)] and NCN_3 (590).

$$HN_3 \xrightarrow{h\nu} H + N_3 \qquad \text{(VI-145)}$$

$$NCN_3 \xrightarrow{h\nu} CN + N_3 \qquad \text{(VI-146)}$$

The chemiluminescent reactions of N_3 with Cl, Br, O, and N atoms have been studied by Clark and Clyne (215).

$$X + N_3 \rightarrow NX^* + N_2 \qquad \text{(VI-147)}$$

where NX^* signifies $NCl(b^1\Sigma^+)$, $NBr(A^1\Sigma^+)$, $NO(A^2\Sigma^+, B^2\Pi)$, and $N_2(B^3\Pi_g)$.

The ground state of the NCN radical is $X^3\Sigma_g^-$ (linear); D_0(N—CN) = 4.3 ± 0.2 eV (764). The absorption spectrum of the NCN radical has been found in the region near 3290 Å, which is associated with the transition $A^3\Pi_u - X^3\Sigma_g^-$. The second absorption bands near 3327 Å have been found by Kroto et al. (590), who assigned the bands to the $^1\Delta_u \leftarrow a^1\Delta_g$ transition. Emission bands corresponding to the transition $A^3\Pi \rightarrow X^3\Sigma^-$ have been observed in the vacuum ultraviolet photolysis of cyanogen azide [Okabe and Mele (764)]. Apparently they are produced by the sequence

$$NCN_3 \xrightarrow{h\nu} NCN(X^3\Sigma) + N_2(A^3\Sigma) \quad \text{(VI-148)}$$

followed by

$$N_2(A^3\Sigma) + NCN_3 \rightarrow NCN(A^3\Pi) + 2N_2 \quad \text{(VI-149)}$$

Kroto (588) suggests that the primary process in the near ultraviolet photolysis of cyanogen azide is

$$NCN_3 \xrightarrow{h\nu} NCN(^1\Delta) + N_2 \quad \text{(VI-150)}$$

The NCN($^1\Delta$) produced is deactivated to the ground triplet state by collisions with parent molecules

$$NCN(^1\Delta) \xrightarrow{NCN_3} NCN(X^3\Sigma^-) \quad \text{(VI-151)}$$

The energy separation between $^1\Delta$ and $^3\Sigma^-$ is unknown.

The ground state of the NCO radical is $X^2\Pi$ (linear); D_0(N—CO) = 2.10 ± 0.15 eV. The absorption bands of NCO have been observed in the regions near 4400 Å and 2650 to 3200 Å (16). They are assigned to the transitions $A^2\Sigma^+ \leftarrow X^2\Pi$ and $B^2\Pi \leftarrow X^2\Pi$, respectively.

The NCO radicals are produced in the flash photolysis of isocyanic acid

$$HNCO \xrightarrow{h\nu} H + NCO(X^2\Pi) \quad \text{(VI-152)}$$

The emission bands due to $A^2\Sigma^+ \rightarrow X^2\Pi$ (strong) and $B^2\Pi \rightarrow X^2\Pi$ (weak) have been observed in the vacuum ultraviolet photolysis of isocyanic acid by Okabe (765).

$$HNCO \xrightarrow{h\nu} H + NCO(A^2\Sigma^+) \quad \text{(VI-153)}$$

$$HNCO \xrightarrow{h\nu} H + NCO(B^2\Pi) \quad \text{(VI-154)}$$

The NCO($X^2\Pi$) radicals probably recombine to give N_2 and CO (125, 1057)

$$2NCO \rightarrow N_2 + 2CO \quad \text{(VI-155)}$$

The addition of O_2 up to 5 torr has no effect on the CO and N_2 production in the near ultraviolet photolysis of HNCO, suggesting NCO does not react with O_2 [Back and Ketcheson (51)].

VI–14.10. Carbon Difluoride (CF_2)

The ground state of CF_2 is $\tilde{X}^2A_1$ with an F—C—F angle of 105°; D_0(F—CF) = 5.3 ± 0.1 eV. The absorption spectrum has been found in the region near 2500 Å as a result of the transition $\tilde{A}^1B_1$–$\tilde{X}^1A_1$ [Mathews (668)]. The corresponding emission bands have been seen in the region 2450 to 3220 Å.

The CF_2 radicals are produced in the vacuum ultraviolet photolysis of CF_2Cl_2 [Rebbert and Ausloos (828)] and in the near ultraviolet photolysis of CF_2Br_2 and CF_2HBr [Simons and Yarwood (891)].

$$CF_2Cl_2 \xrightarrow{h\nu} CF_2 + Cl_2 \text{ (or 2Cl)} \qquad \text{(VI-156)}$$

$$CF_2Br_2 \xrightarrow{h\nu} CF_2 + Br_2 \qquad \text{(VI-157)}$$

$$CF_2HBr \xrightarrow{h\nu} CF_2 + HBr \qquad \text{(VI-158)}$$

The CF_2 radical appears to be unreactive with itself and with O_2. It is thermochemically stable ($\Delta Hf_0^\circ = -43.6$ kcal mol^{-1}).

VI–14.11. Disulfur Monoxide (S_2O)

The ground state is $\tilde{X}^1A'$ (C_s symmetry) with an S—S—O angle of 118° (16); D_0(S—SO) = 3.45 ± 0.01 eV. The S_2O is not a primary photolytic product. It is produced by an electric discharge through a mixture of sulfur and sulfur dioxide.

The absorption spectrum is observed in the region 2500 to 3400 Å. The reactions of S_2O with O, H, N, Ar(^3P), Cl, and O_3 have been studied by Stedman et al. (922a).

The reaction with O atoms provides a clean source of SO radicals

$$O + S_2O \rightarrow SO + SO \qquad \text{(VI-159)}$$

chapter VII

Photochemistry of Polyatomic Molecules

FOUR-ATOM MOLECULES

VII–1. AMMONIA (NH_3)

The ground state of ammonia is pyramidal $\tilde{X}^1A_1$ of C_{3v} symmetry. The bond energy $D_0(H—NH_2)$ is approximately 4.40 eV (102 kcal mol^{-1}). The absorption spectrum of NH_3 has been reviewed by Herzberg [(16), p. 515]. An absorption in the region 1700 to 2170 Å corresponds to the transition $\tilde{A}^1A_2''$–$\tilde{X}^1A_1$. (Ammonia is planar in the $\tilde{A}$ state belonging to D_{3h} symmetry.) A second absorption in the region 1400 to 1690 Å is due to the transition $\tilde{B}^1E''$–$\tilde{X}^1A_1$. In the region 1150 to 1500 Å several discrete bands appear corresponding to transitions to $\tilde{C}^1A_1'$, $\tilde{D}^1A_2''$, and $\tilde{E}^1A_2''$. The absorption coefficients in the region 1100 to 2200 Å have been measured by Watanabe (1017) and more recently by Watanabe and Sood (1019). They are shown in Fig. VII–1.

The minimum energy required for the process

$$NH_3 \rightarrow NH(X^3\Sigma) + H_2 \tag{VII-1}$$

is most likely to be 3.98 eV (91.7 kcal mol^{-1}) using the threshold photon energy of 9.35 eV for $NH_3 \rightarrow NH(c^1\Pi) + H_2$(762) and the energy difference of 1.56 eV between $NH(a^1\Delta)$ and $NH(X^3\Sigma)$ states (394, 666).

VII–1.1. Primary Processes

Photochemistry of ammonia has been reviewed recently by McNesby and Okabe (684). The photodissociation process of ammonia appears to follow the spin conservation rules [Okabe and Lenzi (762)] in that the spin-allowed process

$$NH_3 \xrightarrow{h\nu} NH(c^1\Pi) + H_2 \tag{VII-2}$$

has been observed, while the spin-forbidden process

$$NH_3 \xrightarrow{h\nu} NH(A^3\Pi) + H_2$$

has not been seen [Becker and Welge (77)].

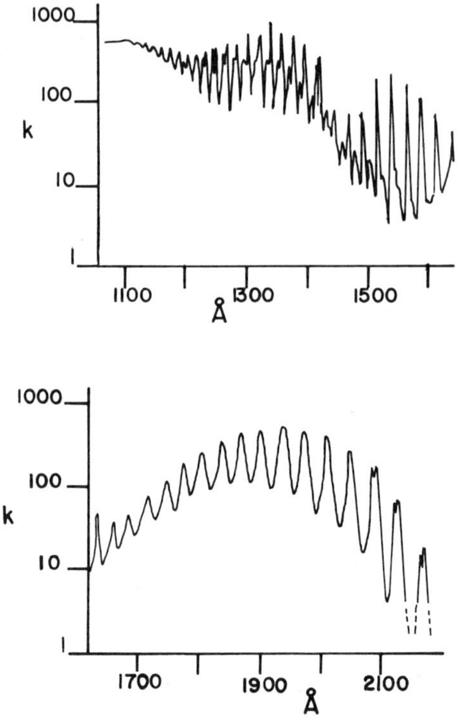

Fig. VII–1. Absorption coefficients of ammonia in the region 1100 to 2200 Å. *k* is given in units of atm^{-1} cm^{-1}, 0°C, base *e*. From Watanabe (1017), reprinted by permission. Copyright 1954 by the American Institute of Physics.

Three main primary processes are found to occur in the near and vacuum ultraviolet photolysis.

$$NH_3 \xrightarrow{h\nu} NH_2(\tilde{X}^2B_1) + H \qquad <2800 \text{ Å}$$
$$NH_3 \xrightarrow{h\nu} NH(a^1\Delta) + H_2 \qquad <2240 \text{ Å}$$
$$NH_3 \xrightarrow{h\nu} NH(X^3\Sigma) + H + H \qquad <1470 \text{ Å}$$

The 1700 to 2200 Å Region ($\tilde{A}$–$\tilde{X}$ *System*). The absorption spectrum of NH_3 consists of a long progression arising from the excitation of the ν_2' out-of-plane vibration and is too diffuse to show rotational structure [Douglas (319)]. On the other hand, the ND_3 spectrum shows diffuse rotational structure for the $v_2' = 0$ and 1 bands. Weak fluorescence has been observed for ND_3 with excitation at the 2139 and 2144 Å lines [Koda et al (580)].

The upper electronic state of ammonia is $^1A_2''$ in D_{3h} symmetry. The dissociation from this state may be represented by

$$NH_3 \xrightarrow{h\nu} NH_2(\tilde{X}^2B_1) + H \qquad \text{(VII-3)}$$

in accordance with the observation of the ground state NH_2 absorption in the flash photolysis of NH_3.

The quantum yield of (VII-3) is near unity at 2062 Å [Groth et al. (429), Schurath et al. (859)] and also at 1849 Å [McNesby et al. (683)]. Although the process

$$NH_3 \xrightarrow{h\nu} NH(a^1\Delta) + H_2 \qquad \text{(VII-4)}$$

is energetically and spin allowed, it occurs with a quantum yield less than 0.005 (429, 859) at 2062 Å and less than 0.04 (683) at 1849 Å.

The 1400 to 1700 Å Region ($\tilde{B}$–$\tilde{X}$ *System*). The spectrum in this region consists of a progression of out-of-plane vibrational bands. Rotational structure is diffuse but it can be resolved. The upper electronic state is planar $^1E''$ in D_{3h}(318).

Dissociation of ammonia in this region appears to yield $NH_2 + H$ as the major process and $NH + H_2$ as a minor process [Groth et al (428)]. The quantum yield of the $ND + D_2$ formation has been found to be 0.032 at 1470 Å [Lilly et al (634)].

Electronically excited $NH_2(\tilde{A}^2A_1)$ is observed below the incident wavelength 1640 Å with an efficiency less than 0.1% [Okabe and Lenzi (762)].

Below 1400 Å Region. Major primary processes in this region are (VII-3) and

$$NH_3 \xrightarrow{h\nu} NH(X^3\Sigma) + H + H \qquad \text{(VII-5)}$$

while (VII-4) is about 14% at 1236 Å [McNesby et al. (683)]. Lilly et al. (634) have found that (VII-4) increases with decreasing incident wavelength, that is, 3.2% at 1470 Å, 24% at 1236 Å, 31% at 1048-1067 Å. The $NH(X^3\Sigma)$ has been observed in the flash photolysis of NH_3 by Bayes et al. (75) Although $NH(c^1\Pi)$ is produced below the incident wavelength 1325 Å (762), $NH(a^1\Delta)$ has not been observed in the flash photolysis of NH_3 by Stuhl and Welge (947). It is likely that the absorption by $NH(a^1\Delta)$ is small and that it is below the detection limit. The $NH(b^1\Sigma)$ is also produced by irradiation with the Kr and Ar resonance lamps [Masanet et al. (666)]. Below 1220 Å NH_3 ionizes

$$NH_3 \xrightarrow{h\nu} NH_3^+ + e \qquad \text{(VII-6)}$$

VII–1.2. Secondary Reactions

The reactions of NH formed in the flash photolysis of NH_3 have been studied by several workers. Mantei and Bair (660) have obtained the rate constant for the reaction

$$NH + NH_3 \xrightarrow{M} N_2H_4 \qquad k_7 \qquad \text{(VII-7)}$$

At the high pressure limit,

$$k_7 = 1.7 \times 10^{-11} \text{ cm}^3 \text{ molec}^{-1} \text{ sec}^{-1}$$

is found.

Meaburn and Gordon (688) have calculated an upper limit for the reaction rate of two NH radicals

$$NH + NH \rightarrow N_2 + 2H \qquad k_8 \qquad \text{(VII-8)}$$
$$k_8 \leq 1 \times 10^{-9} \text{ cm}^3 \text{ molec}^{-1} \text{ sec}^{-1}$$

The rate constant for the reaction

$$NH(b^1\Sigma^+) + NH_3 \rightarrow \text{products} \qquad \text{(VII-9)}$$

has been found to be 4.1×10^{-13} cm^3 molec^{-1} sec^{-1} by Zetzsch and Stuhl (1084).

The reaction of NH_2 with NH_3 is slow and hence NH_2 radicals disappear by combination,

$$NH_2 + NH_2 \rightarrow N_2H_4 \qquad \text{(VII-10)}$$

with a rate constant of 3.9×10^{-12} cm^3 molec^{-1} sec^{-1}, independent of the total pressure in the range 0.4 to 0.85 torr [Hanes and Bair (439)]. The H atoms produced disappear by combination reactions

$$H + NH_2 \xrightarrow{M} NH_3 \qquad \text{(VII-11)}$$

and

$$H + H \xrightarrow{M} H_2 \qquad \text{(VII-12)}$$

The reaction of H with NH_3,

$$H + NH_3 \rightarrow NH_2 + H_2 \qquad k_{13} \qquad \text{(VII-13)}$$

is slow ($k_{13} < 10^{-16}$ cm^3 molec^{-1} sec^{-1}) (10).

VII–2. PHOSPHINE (PH_3)

The ground state is pyramidal $\tilde{X}^1A_1$ of C_{3v} symmetry. $D_0(H—PH_2) = 3.4 \pm 0.1$ eV. A first absorption region is 1600 to 2300 Å with a maximum at 1800 Å, and is continuous. The absorption coefficients in the region 2000 to 2300 Å are given in Fig. VII–2 [Kley and Welge (575)].

VII–2.1. Photolysis

The primary processes in the near ultraviolet must be

$$PH_3 \xrightarrow{h\nu} PH(X^3\Sigma) + H_2 \qquad \text{(VII-14)}$$
$$PH_3 \xrightarrow{h\nu} PH_2(\tilde{X}^2B_1) + H \qquad \text{(VII-15)}$$

since $PH(X^3\Sigma)$ and $PH_2(\tilde{X}^2B_1)$ radicals, initially formed in the vibrationally excited levels, have been found in the flash photolysis of PH_3 [Berthou et al. (102), Kley and Welge (575)]. Process (VII-14) is in violation of the spin conservation rules since the electronically excited state of PH_3 is most likely a singlet. (Absorption in the region 1600 to 2300 Å is large with $k_{max} > 100$

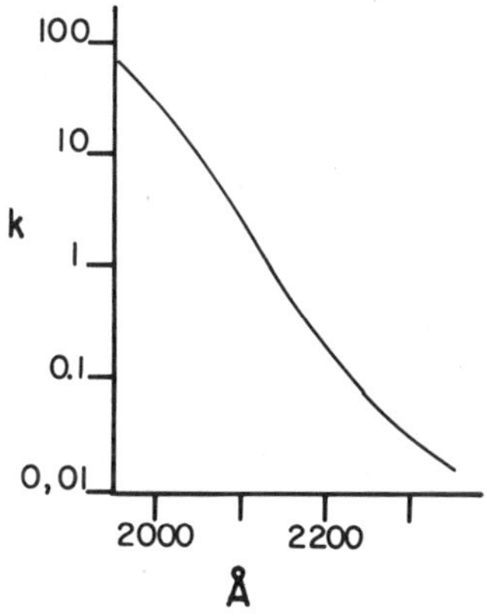

Fig. VII-2. The absorption coefficient of phosphine in the region 2000 to 2300 Å. *k* is given in atm^{-1} cm^{-1}, base *e*, room temperature. From Kley and Welge (575) reprinted by permission of Zeitschrift für Naturforschung.

atm^{-1} cm^{-1}). The process

$$PH_3 \xrightarrow{h\nu} PH(X^3\Sigma) + 2H \qquad \text{(VII-16)}$$

is energetically possible only below 1800 Å. Norrish and Oldershaw (747) and Lee et al. (620) have proposed the following secondary reactions from the results of the flash photolysis of PH_3.

$$H + PH_3 \rightarrow PH_2 + H_2 \qquad k_{17} \qquad \text{(VII-17)}$$

$$2PH_2 \rightarrow PH + PH_3 \qquad \text{(VII-18)}$$

$$2\,PH \rightarrow P_2 + H_2 \qquad \text{(VII-19)}$$

$$H + H \xrightarrow{M} H_2 \qquad \text{(VII-20)}$$

A rate constant, k_{17}, of $4.52 \times 10^{-11} \exp(-740/T)$ cm^3 molec^{-1} sec^{-1} has recently been measured by Lee et al. (620).

In the vacuum ultraviolet photolysis of PH_3, Becker and Welge (78) have found the production of $PH(A^3\Pi)$ both at the Kr and Xe resonance lines, indicating the process

$$PH_3 \xrightarrow{h\nu} PH(A^3\Pi) + H_2 \qquad \text{(VII-21)}$$

in apparent violation of the spin conservation rules.

VII–3. ACETYLENE AND HALOACETYLENES

VII–3.1. Acetylene (C_2H_2)

The ground state of C_2H_2 is ${}^1\Sigma_g^+$ (linear). The bond energy is $D_0(H—C_2H) = 5.38 \pm 0.05$ eV (771).

Absorption starts at about 2370 Å. The absorption coefficients in the region 1100 to 2000 Å have been measured by Nakayama and Watanabe

(731a) and are shown in Fig. VII–3. The bands in the region 2100 to 2370 Å show rotational structure and are assigned to the transition ^1A_u–$^1\Sigma_g^+$ (16). The 1500 to 2000 Å bands are diffuse and have been assigned to the ^1B_u–$^1\Sigma_g^+$ transition by Foo and Innes (367). The bands 1403 to 1519 Å probably belong to a $^1\Pi_u$ state. Below 1403 Å several transitions have been found with vibrational structure. Most bands belong to the Rydberg series. Demoulin and Jungen (281) have made theoretical assignments of the acetylene spectrum.

Photochemistry. The Hg($^3P_1$) sensitized photolysis of C_2H_2 has produced benzene, hydrogen, and polymer [Shida et al. (872)]. Since the Hg($^3P_1$) state does not have sufficient energy required to dissociate the H—C_2H bond, the products must be produced by reactions of an electronically excited C_2H_2. The photolysis of acetylene at 1849 Å has produced hydrogen, ethylene, vinylacetylene, diacetylene, benzene, and solid polymers [Tsukada and Shida (979), Zelikoff and Aschenbrand (1082)].

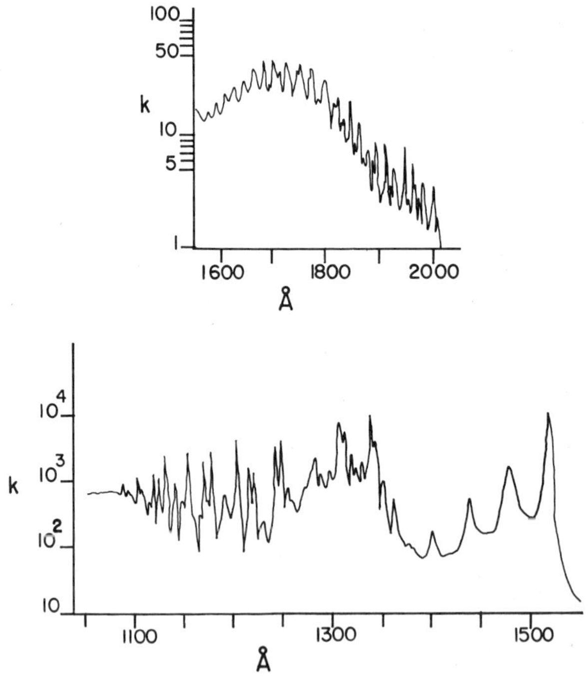

Fig. VII–3. Absorption coefficients of acetylene in the region 1100 to 2000 Å. *k* is given in units of atm^{-1} cm^{-1}, base *e*, 0°C. From Nakayama and Watanabe (731a), reprinted with permission. Copyright 1964 by the American Institute of Physics.

The primary process of C_2H_2 photolysis at 1236 Å in the low pressure region (<1 torr) appears to be [Stief et al. (928), Payne and Stief (802)]

$$C_2H_2 \xrightarrow{h\nu} C_2H_2^* \rightarrow C_2 + H_2 \qquad \text{(VII-22)}$$
$$C_2H_2 \xrightarrow{h\nu} C_2H_2^* \rightarrow C_2H + H \qquad \text{(VII-23)}$$

where $C_2H_2^*$ indicates an electronically excited C_2H_2.

The $C_2H_2^*$ must have a sufficiently long life (~ 1 μsec) to be deactivated by CO_2 and N_2, since Φ_{H_2} decreases with an increase of CO_2 or N_2 pressure:

$$C_2H_2^* + M \rightarrow C_2H_2 + M \qquad \text{(VII-24)}$$

where M signifies CO_2 or N_2. In the high pressure region of C_2H_2 (>1 torr), the following reactions involving the electronically excited C_2H_2 appear important [Takita et al. (957, 958)].

$$C_2H_2 \xrightarrow{h\nu} C_2H_2^* \qquad \text{(VII-25)}$$
$$C_2H_2^* + C_2H_2 \rightarrow C_4H_2 + 2H \qquad \text{(VII-26)}$$
$$H + C_2H_2 \rightarrow C_2H_3 \quad k_{27} \qquad \text{(VII-27)}$$
$$C_2H_3 + C_2H_2 \rightarrow C_4H_5 \quad \text{etc.} \qquad \text{(VII-28)}$$

A rate constant of $k_{27} = 5 \times 10^{-14}$ cm^3 molec^{-1} sec^{-1} (high pressure limit) has been determined by Payne and Stief (802).

Flourescence has been observed in the photolysis of C_2H_2 at the 1236 Å line [Stief et al. (928), Becker et al. (81)]. It has been concluded that the emitter is an electronically excited C_2H [Okabe (773)],

$$C_2H_2 \xrightarrow{h\nu} C_2H^* + H \qquad \text{(VII-29)}$$
$$C_2H^* \rightarrow C_2H + h\nu' \qquad \text{(VII-30)}$$

Okabe (773) has derived the electronic energy $E_0(C_2H) \leq 4.11 \pm 0.05$ eV from the threshold incident wavelength for the production of C_2H^*, which is predissociated from the electronically excited C_2H_2. The C_2H^* has a lifetime of 6 μsec (81).

VII–3.2. Chloroacetylene (ClC_2H)

The ground state is linear. The bond energy is estimated to be $D_0(Cl–C_2H) = 4.51$ eV [Okabe (773)].

Absorption starts at 2550 Å. Above 2000 Å there are two transitions leading to nonlinear excited states. Below 2000 Å a number of Rydberg transitions have been found in which upper states are linear [Thomson and Warsop (969), Evans et al. (338)].

Photochemistry. The photolysis of ClC_2H at 2300 Å is given by the sequence [Tarr et al. (962)]

$$ClC_2H \xrightarrow{h\nu} Cl + C_2H \qquad (VII\text{-}31)$$

$$Cl + ClC_2H \rightarrow ClC_2HCl \qquad (VII\text{-}32)$$

$$C_2H + ClC_2H \rightarrow C_4H_2 + Cl \qquad (VII\text{-}33)$$

$$2C_2H \rightarrow C_4H_2 \qquad (VII\text{-}34)$$

Evans and Rice (337) and Evans et al. (338) have found that ClC_2H fluoresces with exciting wavelengths 2330 to 2475 Å with a quantum yield of about 0.2. They propose the following scheme for fluorescence and non-radiative processes (dissociation):

$$ClC_2H \xrightarrow{h\nu} ClC_2H^* \qquad (VII\text{-}35)$$

$$ClC_2H^* \rightarrow ClC_2H + h\nu' \qquad (VII\text{-}36)$$

$$ClC_2H^* \rightarrow ClC_2H^{**} \qquad (VII\text{-}37)$$

$$ClC_2H^{**} \rightarrow Cl + C_2H \qquad (VII\text{-}38)$$

where ClC_2H^* signifies an upper singlet state and ClC_2H^{**} signifies the vibrationally excited ground state.

Although the photon energy is much larger than the bond energy $D_0(Cl—C_2H) = 4.51$ eV, corresponding to 2750 Å, the fluorescence, with observed lifetimes in the range 8 to 50 nsec, still competes with dissociation. Evans and Rice attribute the slow decomposition to the poor acceptance of internal energy by ν_3 (the C—Cl stretching vibration), the effective vibrational mode for dissociation.

VII–3.3. Bromoacetylene (BrC_2H)

Bromoacetylene is linear in the ground state; $D_0(Br—C_2H) = 3.95 \pm 0.05$ eV (773). Its absorption starts at 2800 Å. Above 1730 Å there are two continuous absorption bands with maxima at 2115 and 1780 Å, while below 1730 Å a number of Rydberg transitions are observed with complicated vibrational structure [Thomson and Warsop (970), Evans et al. (338)].

Photochemistry. The primary photochemical process at 2537 Å is the production of C_2H and Br [Tarr et al. (962)].

$$BrC_2H \xrightarrow{h\nu} C_2H + Br \qquad (VII\text{-}39)$$

followed by

$$C_2H + BrC_2H \rightarrow C_4H_2 + Br \qquad (VII\text{-}40)$$

$$C_2H + BrC_2H \rightarrow C_2H_2 + C_2Br \qquad (VII\text{-}41)$$

$$Br + BrC_2H \rightarrow BrC_2HBr \qquad (VII\text{-}42)$$

$$2C_2H \rightarrow C_4H_2 \qquad (VII\text{-}43)$$

Evans and Rice (337) and Evans et al. (338) have observed fluorescence with a lifetime of about 13 nsec with incident wavelengths of 2440 to 2700 Å. Quantum yields are 0.024 to 0.079. A similar mechanism proposed for ClC_2H is also operative for the photochemistry of BrC_2H in this region.

In the vacuum ultraviolet both $C_2(d^3\Pi_g)$ and C_2H^* (electronically excited C_2H) are produced [Okabe (773)]:

$$BrC_2H \xrightarrow{h\nu} C_2H^* + Br \quad \text{(VII-44)}$$

$$BrC_2H \xrightarrow{h\nu} C_2(d^3\Pi_g) + HBr \quad \text{(VII-45)}$$

The C_2H^* gives rise to quasi-continuous emission in the region 4000 to above 5500 Å. Processes (VII-44) and (VII-45) are predissociative.

VII–3.4. Iodoacetylene (IC_2H)

The ground state is linear; $D_0(I—C_2H) = 3.3$ eV [estimated by Okabe (773)]. Absorption starts at about 3000 Å [Salahub at Boschi (854)]. The absorption coefficients have been measured in the region 1050 to 3000 Å (854). The bands are assigned to $\sigma^* \leftarrow n$, $\pi^* \leftarrow n$ and $\Pi^* \leftarrow \Pi$ transitions, as well as to members of seven Rydberg series.

VII–4. FORMALDEHYDE (HCHO)

The ground state of formaldehyde is planar (C_{2v}) with an H—C—H angle of 121°. The ultraviolet absorption spectrum consists of many sharp bands in the region 2400 to 3600 Å. The excited state responsible for the absorption is the near planar $^1A_2$ at 3.495 eV above the ground. The transition is forbidden by the electric dipole selection rules [see a review by Moule and Walsh (723)]. The absorption coefficients in the near ultraviolet have been measured by McQuigg and Calvert (686) and in the vacuum ultraviolet by Mentall and Gentieu (696). They are shown in Fig. VII–4. In addition, a very weak absorption in the region 3600 to 3967 Å, due to a transition to the $\tilde{a}\,^3A_2$ state, is present. The dissociation energy, D_0(H—CHO), has not been definitely established. If the recent values for the heat of formation, $\Delta Hf^\circ_0(H_2CO) = -25.05 + 0.11$ kcal mol^{-1} (363) and $\Delta Hf^\circ_0(HCO) = 9.0 \pm 2$ kcal mol^{-1} (1001), are used, $D_0(H—CHO) = 86.0 \pm 2$ kcal mol^{-1} or 3.7 ± 0.1 eV is obtained. On the other hand, Brand and Reed (126) have concluded that the breaking-off of the fluorescence bands above 28736 cm^{-1} or 3.56 eV excited by discharges must be due to a dissociation into H + CHO. In this case, $D_0(H—CHO) \leq 3.56$ eV, corresponding to the incident wavelength 3483 Å.

VII–4.1. Photochemistry in the Near Ultraviolet

Fluorescence (*707, 853, 1074, 1076*). Weak fluorescence has been observed in the region 3400 to 5000 Å by absorption of light of wavelengths

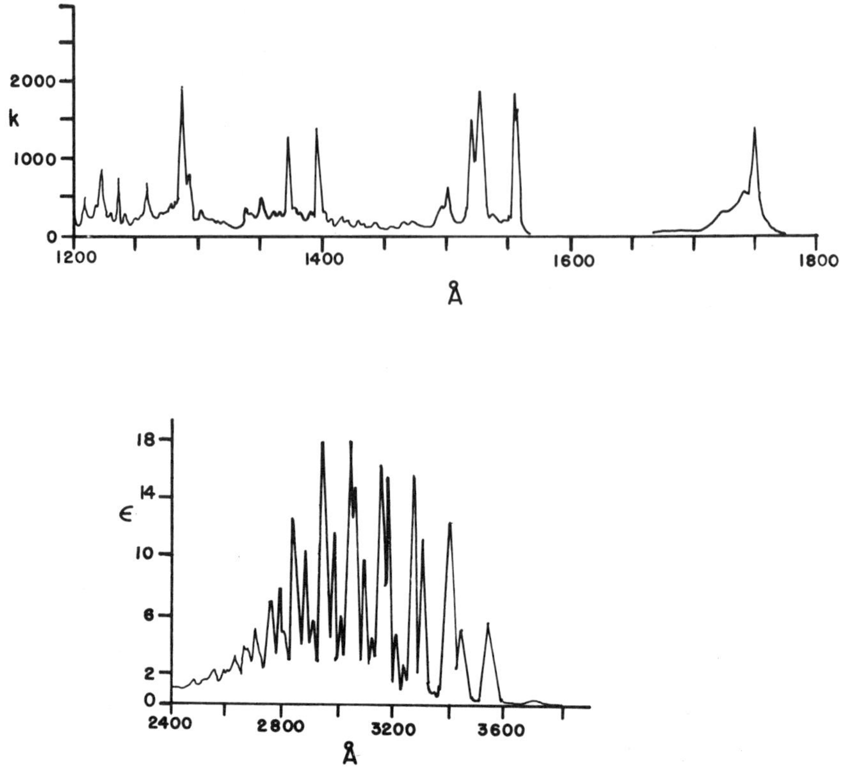

Fig. VII–4. Absorption coefficients of HCHO in the regions 1200 to 1800 (k) and 2400 to 3700 Å (ϵ). 1200 to 1800 Å region: units, atm^{-1} cm^{-1}; 0°C; base *e*. From Mentall et al. (696), reprinted by permission. Copyright 1971 by the American Institute of Physics. 2400 to 3700 Å region: units, mol^{-1} cm^{-1}; base 10, room temperature. Reprinted with permission from R. D. McQuigg and J. G. Calvert, *J. Am. Chem. Soc.* 91, 1590 (1969). Copyright by the American Chemical Society.

3000 to 3600 Å [Miller and Lee (707), Yeung and Moore (1076)]. The quantum yield of fluorescence of H_2CO is 0.03 at 3532 Å (707) and the yield decreases as the incident wavelength decreases. The observed fluorescence lifetime of H_2CO near the 0–0 transition is about 100 nsec, while near 3100 Å excitation the lifetime is about 5 nsec. The radiative lifetime estimated from the integrated absorption coefficient is about 5 μsec. The lifetime of D_2CO fluorescence ranges from 4.3 μsec (3535 Å) to 53 nsec (3082 Å) (1076).

Miller and Lee (707) have found that the nonradiative (that is, dissociative) rate increases with excess vibrational quanta of the upper $^1A_2$ state and that the extent of increase is much larger for the ν_4' out-of-planar bending mode than for the ν_5' asymmetric C—H stretching mode.

The lifetime of the $^1A_2$ state produced by absorption of light in the 2685–2851 Å region has been determined to be 4 to 14 psec [Baronavski et al. (62)].

Photodissociation. The main photochemical primary processes are

$$H_2CO \xrightarrow{h\nu} H + HCO \qquad \phi_{46} \tag{VII-46}$$

$$H_2CO \xrightarrow{h\nu} H_2 + CO \qquad \phi_{47} \tag{VII-47}$$

The threshold wavelength for (VII-46) is about 3500 Å. McQuigg and Calvert (686) have measured H_2, HD, and D_2 products produced from the high intensity photolysis of H_2CO–D_2CO mixtures. They have concluded that the primary quantum yield, $\phi_{46} + \phi_{47}$, is near unity over the entire absorption region in the near ultraviolet. They have also found that CO and H_2 are formed in nearly equal amounts. The results may be explained on the basis of the following secondary reactions occurring after (VII-46) and (VII-47)

$$H + HCO \rightarrow H_2 + CO \tag{VII-48}$$

$$2H + M \rightarrow H_2 + M \tag{VII-49}$$

$$2HCO \rightarrow 2CO + H_2 \tag{VII-50}$$

On the other hand, DeGraff and Calvert (271) have found that the CO yield is significantly greater than the H_2 yield at the low intensity photolysis. They have attributed a CO excess to reactions such as

$$H + H_2CO \rightarrow H_2COH \tag{VII-51}$$

$$H_2COH + H_2CO \rightarrow CH_3OH + HCO \tag{VII-52}$$

$$H_2CO^* + H_2CO \rightarrow H_2COH + HCO \qquad k_{53} \tag{VII-53}$$

where H_2CO^* indicates an electronically excited H_2CO. At low intensity and higher temperatures the reaction

$$H + H_2CO \rightarrow H_2 + HCO \qquad k_{54} \tag{VII-54}$$

may become important. The rate constant k_{54} is $5.4 \pm 0.5 \times 10^{-14}$ cm^3 molec^{-1} sec^{-1} at 297°K. The activation energy is about 2 kcal mol^{-1} (146, 833). The quantum yield of (VII-47), the molecular production process, appears to be predominant at 3660 Å and the yield of process (VII-46), the radical production process, increases at shorter wavelengths. At 2800 Å (VII-46) is five times as important as (VII-47) [see McQuigg and Calvert (686), Sperling and Toby (922)].

However, other results suggest that the molecular process is more important at shorter wavelengths (271, 966).

The quenching of the electronically excited state, (VII-53), is very efficient [$k_{53} = 6 \times 10^{-10}$ cm^3 molec^{-1} sec^{-1} (1076)] and may be important in the formation of products.

Recently, Houston and Moore (486) have measured the CO production rate following the pulsed laser photolysis of H_2CO and D_2CO at 3371 Å. They found that at the low pressure limit, the CO rate of production is more than 100 times slower than the fluorescence decay rate. They suggest that CO is not produced from the initially formed fluorescing state S_1 by light absorption but rather from an intermediate state I. The intermediate state I, either the $^3A_2$ or the vibrationally excited ground state, is formed from S_1 either by collisions or by a spontaneous decay process. The I state dissociates into $H_2 + CO$ to a small extent by a slow spontaneous process (>4 μsec) but to a large extent by collisions with each other or with NO and O_2 molecules. The quantum yield of CO production at 3371 Å is independent of formaldehyde pressure in the range 0.1 to 10 torr.

The yield of CO increases with the addition of NO [Houston and Moore (486), Tadasa et al. (956)]. Tadasa et al. (956) suggest the reaction

$$HCO + NO \rightarrow CO + HNO \qquad \text{(VII-54}a\text{)}$$

is responsible for the increase.

Further work is needed to obtain more information on quantum yields of two primary processes as a function of pressure and wavelength. It is also of interest to look into the vibrational excitation of H_2. Since the process of H_2 formation involves the simultaneous excitation of the C—H stretching and the H—C—H bending vibrations, the product H_2 must be highly vibrationally excited. The radical production process (VII-46) is expected to be faster than the molecular process (VII-47) if H + HCO is predissociated by way of a repulsive state as shown in Fig. II–11, p. 78.

VII–4.2. Photodissociation in the Vacuum Ultraviolet

The results of the photolysis at 1470 and 1236 Å indicate that two primary processes

$$H_2CO \xrightarrow{h\nu} H_2 + CO \qquad \phi_{55} \qquad \text{(VII-55)}$$

$$H_2CO \xrightarrow{h\nu} H + H + CO \qquad \phi_{56} \qquad \lambda < 2830\ \text{Å} \qquad \text{(VII-56)}$$

are equally important [Glicker and Stief (403)]. Quantum yields $\phi_{55} = \phi_{56} = 0.5$ have been obtained.

Formaldehyde has been detected recently in the interstellar medium by microwave spectroscopy (593). It is a combustion product of hydrocarbons. The photolysis of H_2CO by sunlight in the troposphere may produce HO_2 radicals by reactions such as

$$HCO + O_2 \rightarrow HO_2 + CO \qquad \text{(VII-57)}$$

$$H + O_2 + M \rightarrow HO_2 + M \qquad \text{(VII-58)}$$

The HO_2 radicals produced may oxidize NO to NO_2

$$HO_2 + NO \rightarrow OH + NO_2 \qquad \text{(VII-59)}$$

Thus, H_2CO may play a significant role for photochemical smog formation in polluted atmospheres [Calvert et al. (182); see Section (VIII-2), p. 335].

VII–5. DIIMIDE (N_2H_2)

Diimide is an unstable molecule with a typical half-life of several minutes at room temperature [Willis and Back (1048)]. $D_0(\text{H—N}_2\text{H}) = 3.35$ eV. The ground state has a trans structure and is $\tilde{X}\,^1A_g$ of C_{2h} symmetry.

The near ultraviolet absorption spectrum lies in the 3000 to 4200 Å region with many diffuse bands [Back et al. (52)]. Back et al. attribute the spectrum to ^1B_g–^1A_g transition forbidden by electric dipole. The absorption is weak with an absorption coefficient of 3.9 l mol^{-1} cm^{-1} (base 10) at 3650 Å.

A second absorption of N_2H_2 starts at about 1730 Å and consists of nine vibrational bands. Rotational lines are all diffuse [Trombetti (976)]. The photolysis in the near ultraviolet has been studied recently by Willis et al. (1050).

The products of the photolysis of N_2H_2–C_2H_4 mixtures at various wavelengths from 3100 to 4050 Å are C_2H_6 ($\Phi = 7$), N_2, and small amounts of C_4H_{10} ($\Phi = 0.05$).

The products of the pure N_2H_2 photolysis are N_2 and H_2 and small amounts of N_2H_4. The quantum yields of N_2H_2 disappearance are about 15 with and without C_2H_4. From the results Willis et al. conclude that the main primary process is

$$N_2H_2 \xrightarrow{h\nu} N_2H + H \text{ (or } N_2 + 2H) \qquad \text{(VII-59a)}$$

followed by (in the presence of C_2H_4) chain reactions

$$H + C_2H_4 \rightarrow C_2H_5 \qquad \text{(VII-59b)}$$

$$C_2H_5 + N_2H_2 \rightarrow C_2H_6 + N_2H \qquad \text{(VII-59c)}$$

$$N_2H \rightarrow N_2 + H \qquad \text{(VII-59d)}$$

In the absence of C_2H_4 the reaction sequence would be

$$N_2H_2 \xrightarrow{h\nu} N_2H + H \qquad \text{(VII-59e)}$$

$$H + N_2H_2 \rightarrow H_2 + N_2H \qquad \text{(VII-59f)}$$

$$N_2H \rightarrow N_2 + H \qquad \text{(VII-59g)}$$

VII–6. HYDROGEN PEROXIDE (H_2O_2)

The ground state is probably nonplanar (the point group C_2). $D_0(HO—OH) = 2.15 \pm 0.02$ eV, $D_0(HO_2—H) = 3.84 \pm 0.1$ eV. Absorption starts at about 3000 Å. The absorption coefficients in the near and vacuum ultraviolet are given in Fig. VII–5. The spectrum shows only a continuum for the region 2000 to 3000 Å.

VII–6.1. Photochemistry

Volman has reviewed the photolysis in the near ultraviolet (995). The primary process

$$H_2O_2 \xrightarrow{h\nu} 2OH \qquad \text{(VII-60)}$$

appears to be the main process at 2537 Å. The quantum yield of disappearance is 1.7 ± 0.4. The reaction products are water and oxygen only. The secondary reactions proposed are [Volman (993)].

$$OH + H_2O_2 \rightarrow H_2O + HO_2 \qquad k_{61} \qquad \text{(VII-61)}$$

$$HO_2 + HO_2 \rightarrow H_2O_2 + O_2 \qquad \text{(VII-62)}$$

The rate constant k_{61} is 8×10^{-13} cm^3 molec^{-1} sec^{-1} (1). The flash photolysis of H_2O_2 in the near ultraviolet has been performed by Greiner (422) who

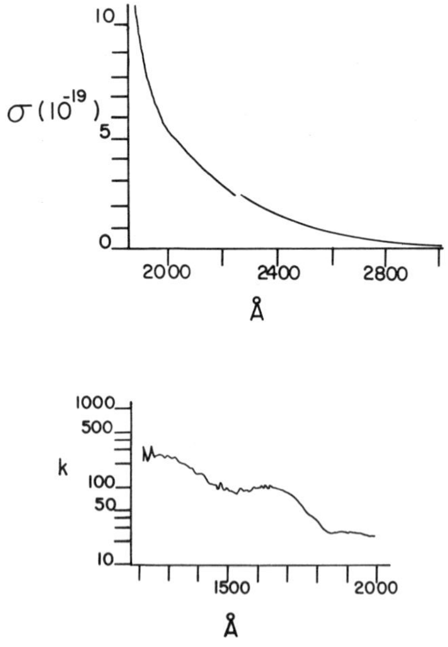

Fig. VII–5. Absorption coefficients of hydrogen peroxide in the near and vacuum ultraviolet regions. 2000 to 3000 Å region: σ is given in units of 10^{-19} cm^2 [Volman (995), p. 70. Originally from R. B. Holt, C. K. McLane, and O. Oldenberg, *J. Chem. Phys.* 16, 638 (1948) and H. C. Urey, L. H. Dawsey and F. O. Rice *J. Am. Chem. Soc.* 51, 1371 (1929). For more recent values see C. L. Lin, N. K. Rohatgi and W. B. DeMore, *Geophys. Res. Lett.* 5, 113 (1978).], base *e*, room temperature and 1200 to 2000 Å region: *k* is given in units of atm^{-1} cm^{-1}, base *e*, room temperature. From Schürgers and Welge (860), reprinted by permission of John Wiley & Sons and *Zeitschrift für Naturforschung*.

concludes that the main primary process is (VII-60) and that another primary process,

$$H_2O_2 \xrightarrow{h\nu} H_2O + O(^1D) \tag{VII-63}$$

is not more than 20% of process (VII-60).

The photolysis at 1236 Å appears to produce molecular hydrogen in the primary process [Stief and DeCarlo (930)].

$$H_2O_2 \xrightarrow{h\nu} H_2 + O_2 \text{ (or O + O)} \tag{VII-64}$$

with a quantum yield of 0.25. Another primary process proposed is

$$H_2O_2 \xrightarrow{h\nu} H + HO_2 \tag{VII-65}$$

with a quantum yield of about 0.25. At both 1470 and 2537 Å, primary process (VII-60) predominates. The rate constant of the reaction

$$H + H_2O_2 \rightarrow OH + H_2O \tag{VII-66}$$

has been measured recently by Klemm et al. (573), who provide a value of $5 \times 10^{-12} \exp(-1390/T)$ cm^3 molec^{-1} sec^{-1}. Meagher and Heicklen (689) have proposed another reaction path

$$H + H_2O_2 \rightarrow H_2 + HO_2 \tag{VII-67}$$

to be as equally important as (VII-66).

VII–7. ISOCYANIC ACID (HNCO); ISOTHIOCYANIC ACID (HNCS)

The ground state of HNCO is planar $\tilde{X}^1A'$ with an H—N—C angle of 128° and a linear NCO group. D_0(H—NCO) = 4.90 ± 0.01 eV, D_0(HN—CO) = 3.5 ± 0.1 eV. An absorption in the 2000 to 2800 Å region has been measured by Dixon and Kirby (287). The absorption spectrum is diffuse and shows long progressions, suggesting a bent NCO structure. Rotational analysis shows an NCO angle of 119° in the first excited state (287).

The absorption coefficients in the region 1200 to 2000 Å have been measured by Okabe (765) and those in the 2100 to 2500 Å region have been measured by Dixon and Kirby (287). They are shown in Figs. VII–6*a* and VII–6*b*.

The primary processes in the near ultraviolet photolysis are

$$HNCO \xrightarrow{h\nu} NH + CO \qquad \phi_{68} \tag{VII-68}$$

$$HNCO \xrightarrow{h\nu} H + NCO \qquad \phi_{69} \tag{VII-69}$$

Process (VII-68) becomes more important at shorter wavelengths [Bradley et al. (125)]. At 2062 Å $\phi_{68} = \phi_{69} = 0.5$ [Woolley and Back (1057)]. The

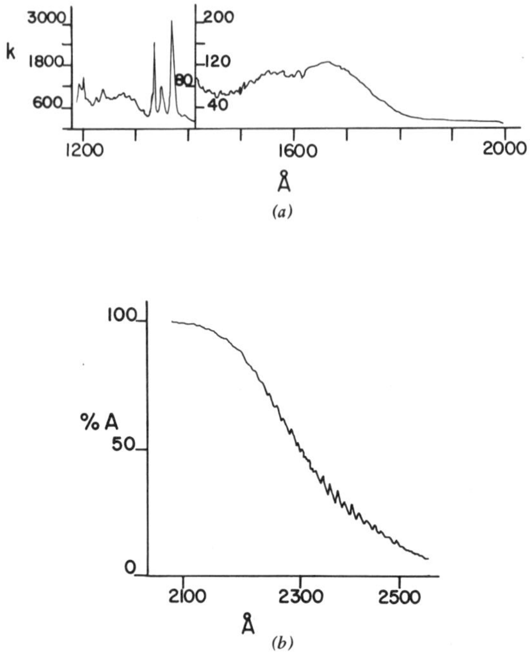

Fig. VII–6. **(*a*) The absorption coefficients of isocyanic acid in the vacuum ultraviolet. *k* is given in units of atm^{-1} cm^{-1} at room temperature, base *e*. From Okabe (765), reprinted by permission. Copyright 1970 by the American Institute of Physics. (*b*) Percent absorption of isocyanic acid in the near ultraviolet. Pressure is 100 torr and path length is 10 cm. From Dixon and Kirby (287), reprinted by permission of The Chemical Society.**

products of the 2062 Å photolysis (1057) are CO, N_2, and H_2 with respective quantum yields of about 1, 0.4, and 0.13.

The secondary reactions are complex since six free radicals, NH, NCO, H, NH_2CO, NH_2, and N, are involved. The reaction

$$H + HNCO \rightarrow H_2 + NCO \qquad \text{(VII-70)}$$

can be ruled out since the reaction is endothermic by 10 kcal mol^{-1}. The H atoms disappear by reactions such as

$$H + HNCO \rightarrow NH_2CO$$
$$H + NH_2CO \rightarrow NH_3 + CO$$
$$H + NH_2CO \rightarrow H_2 + HNCO$$

The near ultraviolet photolysis of HNCO in the presence of NO, O_2, and C_2H_4 has been studied by Back et al. (50, 51, 1057).

The $Hg(^3P_1)$ sensitized reaction of HNCO yields CO, N_2, and H_2. The primary process

$$Hg(^3P_1) + HNCO \rightarrow Hg + H + NCO \qquad \text{(VII-71)}$$

has been suggested by Friswell and Back (376).

In the vacuum ultraviolet photolysis of HNCO emissions from $NCO(A^2\Sigma)$ and $NH(A^3\Pi, c^1\Pi)$ have been observed [Okabe (765)]. In addition, weak fluorescence bands of $NCO(B^2\Pi)$ were found.

Two main primary processes associated with the production of emitting species are

$$HNCO \xrightarrow{h\nu} H + NCO(A^2\Sigma) \qquad \text{(VII-72)}$$

$$HNCO \xrightarrow{h\nu} NH(c^1\Pi) + CO \qquad \text{(VII-73)}$$

The $NCO(A^2\Sigma)$ radicals are highly excited in bending vibration, indicating that the upper state responsible for the production of $NCO(A^2\Sigma)$ is bent. The $NH(c^1\Pi)$ radicals are produced with high rotational excitation. The $NH(A^3\Pi)$ state is most likely produced from a secondary process such as

$$CO(a^3\Pi) + NH(X^3\Sigma) \rightarrow CO(X^1\Sigma) + NH(A^3\Pi)$$

and not from the primary process in accordance with the spin conservation rules.

The ground state of HNCS is $\tilde{X}^1A'$ of C_s symmetry (821a). The bond energy is unknown. A first weak diffuse absorption is in the 2100 to 2700 Å region with a maximum at 2450 Å followed by a second much stronger diffuse absorption in the 1900 to 2100 Å region with a maximum at 1970 Å [McDonald et al. (674a)]. The near ultraviolet flash photolysis of HNCS yields $S_2(X^3\Sigma)$, S_3, and NCS, but no NH radicals are detected [Boxall and Simons (123a)].

The primary processes suggested are

$$HNCS \rightarrow HNC + S(^3P)$$

$$HNCS \rightarrow H + NCS$$

followed by a rapid reaction

$$S(^3P) + HNCS \rightarrow S_2(X^3\Sigma^-) + HNC$$
$$k \geq 3 \times 10^{-11}\ \text{cm}^3\ \text{molec}^{-1}\ \text{sec}^{-1}$$

VII–8. FORMYL FLUORIDE (HCFO)

The ground state is planar $\tilde{X}^1A'$ belonging to the point group C_s. $D_0(H—CFO) = 4.5$ eV. An absorption in the 2000 to 2700 Å region shows vibrational structure corresponding to the $\tilde{A}^1A$–$\tilde{X}^1A'$ transition [Fischer (358)].

The results of the flash photolysis of HFCO above 1650 Å indicate the occurrence of two primary processes [Klimek and Berry (577)].

$$HFCO \xrightarrow{h\nu} HF + CO \qquad (VII\text{-}74)$$

$$HFCO \xrightarrow{h\nu} F + HCO \qquad (VII\text{-}75)$$

VII–9. NITROUS ACID (HNO_2)

The ground state nitrous acid exists in planar cis and trans forms of comparable stability, the trans form being lower in energy by about 0.5 kcal mol^{-1}. $D_0(HO—NO) = 2.09 \pm 0.03$ eV, $D_0(H—ONO) = 3.36 \pm 0.03$ eV. Nitrous acid in the gas phase is in equilibrium with NO, NO_2, and H_2O together with N_2O_3, N_2O_4, and a trace of HNO_3. At pressures above that in equilibrium with NO, NO_2, and H_2O, HNO_2 is relatively unstable. Hence, one cannot prepare pure samples at any desired pressure. The ground state structure of *cis*- and *trans*-nitrous acid has recently been determined by microwave spectroscopy [Cox et al. (242)].

Nitrous acid exhibits diffuse absorption bands in the region 3000 to 4000 Å. These bands are ascribed to the $^1A''–^1A'$ transition [King and Moule (567)]. The absorption cross sections in the region 2000 to 4000 Å have been measured by Cox and Derwent (249) and are given in Fig. VII–7.

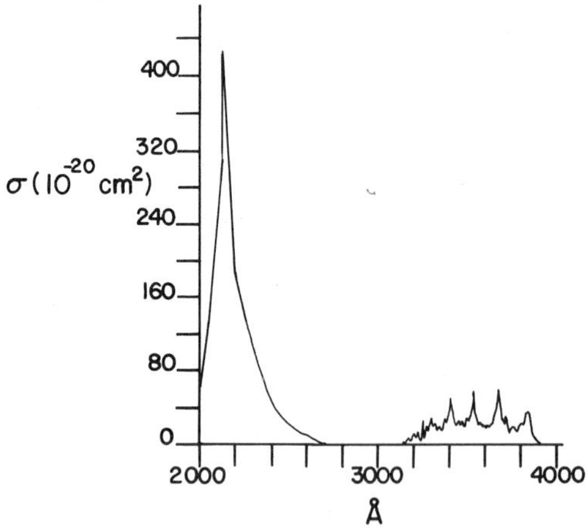

Fig. VII–7. The absorption cross section σ of nitrous acid in the region 2000 to 4000 Å. σ is given in units of 10^{-20} cm^2 $molec^{-1}$, base *e*, room temperature. From Cox and Derwent (249), reprinted by permission of Elsevier Sequoia, S. A.

VII–9.1. Photodissociation

Cox (245) has irradiated HNO_2–N_2–O_2 mixtures with light of wavelengths 3300 to 3800 Å. From the analysis of the products, NO and NO_2, he concludes that the main primary process is

$$HNO_2 \xrightarrow{h\nu} HO + NO \qquad \text{(VII-76)}$$

Another minor primary process

$$HNO_2 \xrightarrow{h\nu} H + NO_2 \qquad \text{(VII-77)}$$

may be operative to an extent of 10%.

Secondary reactions are

$$OH + NO \xrightarrow{M} HNO_2 \qquad \text{(VII-78)}$$

$$OH + HNO_2 \rightarrow H_2O + NO_2 \qquad k_{79} \qquad \text{(VII-79)}$$

$$OH + NO_2 \xrightarrow{M} HNO_3 \qquad \text{(VII-80)}$$

Cox et al. (248) have derived a rate constant $k_{79} = 6.6 \pm 0.3 \times 10^{-12}$ cm^3 molec^{-1} sec^{-1}.

Cox (246) also has studied the photolysis of HNO_2 in N_2–O_2 mixtures in the presence of CO and SO_2.

VII–9.2. Nitrous Acid in the Atmosphere

Cox (246) estimates the concentration of HNO_2 to be 10^9 molec cm^{-3} in the daytime natural troposphere. The photolysis of HNO_2 may be an important source of OH in the troposphere, since HNO_2 absorbs the sun's radiation above 3000 Å. The reactions of OH with hydrocarbons (either hydrogen abstraction from paraffins or addition to the double bond in olefins) in the troposphere are known to be the initial steps for photochemical smog formation [see Section VIII–2, p. 333].

VII–10. HYDRAZOIC ACID (HN_3)

The ground state HN_3 is planar with an H—N—N angle of 110° and an N—N—N angle of 180° (the point group C_s). D_0(H—N_3) = 4.18 eV, D_0(HN—N_2) = 0.47 ± 0.06 eV.

The absorption spectrum begins at 3000 Å. There are at least two absorption systems above 2000 Å, a weak one with a maximum near 2700 Å and a strong one near 2040 Å [Bonnemay and Verdier (120)]. The absorption coefficients in the ultraviolet have been measured by Beckman and Dickinson (89). The absorption coefficients in the region 1150 to 2100 Å have been measured by Okabe (763) and are shown in Fig. VII–8.

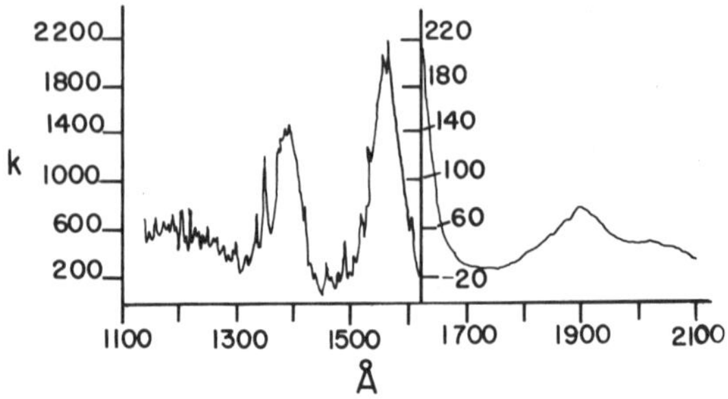

Fig. VII–8. Absorption coefficients of hydrazoic acid in the vacuum ultraviolet, 1100 to 2100 Å. *k* is given in units of atm^{-1} cm^{-1}, base *e*, room temperature. From Okabe (763), reprinted by permission. Copyright 1968 by the American Institute of Physics.

VII–10.1. Photodissociation

Near Ultraviolet Photolysis. The photolysis products at 1850 to 1990 Å lines are H_2, N_2, and NH_3 [Beckman and Dickinson (88)].

The flash photolysis of HN_3 above 2000 Å has revealed the main primary process to be

$$HN_3 \xrightarrow{h\nu} NH(a^1\Delta) + N_2 \qquad \text{(VII-81)}$$

in accordance with the spin conservation rules [Paur and Bair (800)]. Another primary process

$$HN_3 \xrightarrow{h\nu} H + N_3 \qquad \text{(VII-82)}$$

appears to be minor [Konar et al. (581)]. The $NH(^1\Delta)$ radicals rapidly react with HN_3,

$$NH(^1\Delta) + HN_3 \rightarrow NH_2 + N_3 \qquad k_{83} \qquad \text{(VII-83)}$$

with $k_{83} = 9.3 \times 10^{-11}$ cm^3 molec^{-1} sec^{-1}. Other secondary reactions are probably

$$2NH_2 \xrightarrow{M} N_2H_4$$
$$2NH_2 \rightarrow NH_3 + NH$$
$$NH_2 + HN_3 \rightarrow NH_3 + N_3$$
$$N_3 + HN_3 \rightarrow H + 3N_2$$
$$NH(^1\Delta) + HN_3 \rightarrow 2N_2 + H_2$$

The rate constant of a minor process,

$$H + HN_3 \rightarrow NH_2 + N_2 \qquad \text{(VII-84)}$$

has been measured to be $2.54 \times 10^{-11} \exp(-2300/T)$ cm^3 molec^{-1} sec^{-1} [LeBras and Combourieu (617)].

Vacuum Ultraviolet Photolysis. Welge (1031) has observed $NH(c^1\Pi)$ and $NH(A^3\Pi)$ emissions when HN_3 was irradiated by the Kr and Xe resonance lines. Since the direct production of NH $(A^3\Pi)$,

$$HN_3 \xrightarrow{h\nu} NH(A^3\Pi) + N_2 \qquad \text{(VII-85)}$$

is spin-forbidden, it is likely that $NH(A^3\Pi)$ is formed by a secondary process. Since the ratio of $NH(A^3\Pi)$ to $NH(c^1\Pi)$ increases with an increase of HN_3 pressure, Okabe (763) has concluded that $NH(A^3\Pi)$ is produced from a secondary process involving metastable N_2, such as

$$N_2(B^3\Pi) + HN_3 \rightarrow NH(A^3\Pi) + 2N_2 \qquad \text{(VII-86)}$$

The $NH(c^1\Pi)$ fluorescence excitation spectrum shows diffuse vibrational structure corresponding to the absorption spectrum below 1450 Å, while the spectrum is continuous above 1450 Å. The results indicate that the $NH(c^1\Pi)$ state may be formed from predissociation of the electronically excited HN_3 below the incident wavelength 1450 Å, while the $NH(c^1\Pi)$ is dissociated directly above 1450 Å.

VII–11. PHOSGENE ($OCCl_2$)

The ground state $OCCl_2$ $\tilde{X}^1A_1$ is planar and belongs to the point group C_{2v}. $D_0(Cl—COCl) = 3.3$ eV. The near ultraviolet absorption system in the region 2380 to 3050 Å probably belongs to the transition $^1A_2 \leftarrow {}^1A_1$ [Moule and Foo (721)]. LaPaglia and Duncan (611) have found several electronic transitions with vibrational structure in the region 1133 to 1545 Å.

The absorption coefficients in the near and vacuum ultraviolet regions have been measured by Moule and Foo (721) and by Okabe et al. (766) and are shown in Fig. VII–9.

The enthalpies required for the following reactions are (28)

$$OCCl_2 \rightarrow CO + Cl_2, \qquad \Delta H = 1.09 \pm 0.03 \text{ eV}$$

$$OCCl_2 \rightarrow CO + 2Cl, \qquad \Delta H = 3.56 \pm 0.03 \text{ eV}$$

VII–11.1. Photolysis

The photolysis in the near ultraviolet has been studied by Wijnen (1040) and recently by Heicklen (452). The processes may be expressed by

$$OCCl_2 \xrightarrow{h\nu} COCl + Cl \qquad \text{(VII-87)}$$

$$COCl \rightarrow CO + Cl \qquad \Delta H = 6 \text{ kcal mol}^{-1} \qquad \text{(VII-88)}$$

$$2Cl \xrightarrow{M} Cl_2 \qquad \text{(VII-89)}$$

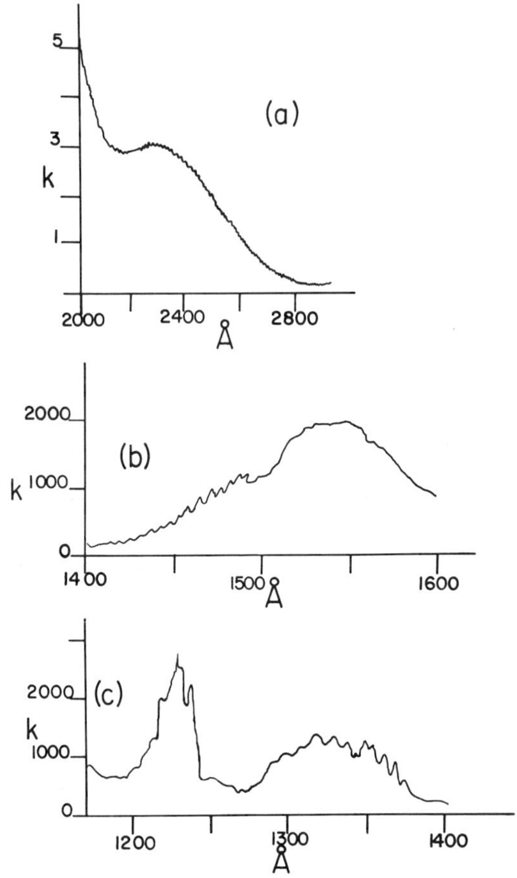

Fig. VII–9. Absorption coefficients of phosgene in the ultraviolet and in the vacuum ultraviolet. (*a*) Approximate values calculated from the data of Moule and Foo (721) and Heicklen (452). *k* is given in units of atm^{-1} cm^{-1}, base *e*, 25°C. (*b*) Absorption coefficients in the 1400 to 1600 Å region; From Okabe et al. (766). (*c*) Absorption coefficients in the 1150 to 1400 Å region; *k* is given in units of atm^{-1} cm^{-1}, base *e*, 25°C. From Okabe et al. (766). Reprinted by permission. Copyright 1971 by the American Institute of Physics.

The radicals, OCCl, formed in the primary process dissociate immediately into CO + Cl and do not appear to participate in the reactions (1040). The secondary reaction

$$Cl + OCCl_2 \rightarrow Cl_2 + OCCl \qquad \text{(VII-90)}$$

is endothermic by about 20 kcal mol^{-1}. Alkyl radicals also do not react with $OCCl_2$ at room temperature. The photolysis of $OCCl_2$ has been used as a source of Cl atoms (452, 1040).

The quantum yield Φ_{CO} is unity at 2537 Å [Okabe(774)]. The photolysis of $OCCl_2$ in the vacuum ultraviolet has resulted in the production of two electronically excited Cl_2 states, Cl_2^* at 7.21 and Cl_2^{**} at 7.93 eV [Okabe et al. (766)]

$$OCCl_2 \xrightarrow{h\nu} CO + Cl_2^* \quad \text{(VII-91)}$$

$$OCCl_2 \xrightarrow{h\nu} CO + Cl_2^{**} \quad \text{(VII-92)}$$

VII–12. THIOPHOSGENE ($SCCl_2$)

The ground state is planar $\tilde{X}^1A_1$ of C_{2v} symmetry. D_0(Cl—CSCl) = 2.75 ± 0.02 eV. Very weak absorption bands in the region 5300 to 7000 Å correspond to the $\tilde{a}^3A_2$–$\tilde{X}^1A_1$ transition [Moule and Subramaniam (722)]. Weak absorption bands with fine structure in the region 3900 to 5950 Å are ascribed to the $\tilde{A}^1A_2$–$\tilde{X}^1A_1$ transition by Brand et al. (127). Strong bands with diffuse vibrational structure in the region 2400 to 2970 Å have been assigned to the $\tilde{B}^1A_1$–$\tilde{X}^1A_1$ transition by Farnworth and King (343). The absorption coefficients in the visible and ultraviolet have been measured by Levine et al. (627) and are shown in Fig. VII–10.

VII–12.1. Photochemistry

Fluorescence from the $\tilde{A}^1A_2$ state has been observed by absorption of light wavelength 4550 Å and above [McDonald and Brus (674)]. The collision-free lifetime of the fluorescence is about 40 μsec. The breaking-off of the emission

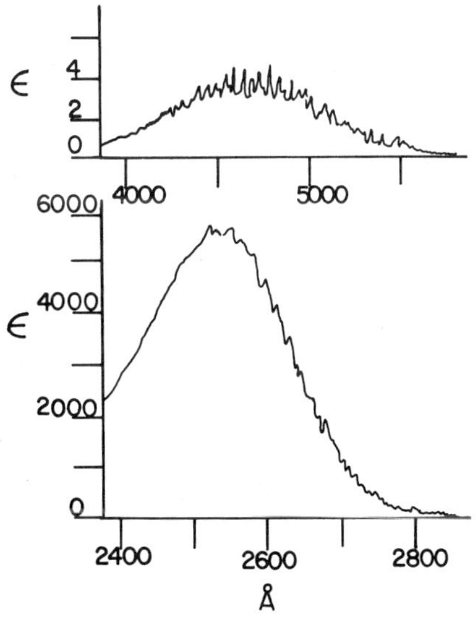

Fig. VII–10. The absorption coefficients of thiophosgene in the visible and ultraviolet regions. ε is given in units of l mol^{-1} cm^{-1}, base 10, room temperature. From Levine et al. (627), reprinted by permission of North-Holland Publishing Company.

bands has been observed below the excitation wavelength 4550 Å. The process has been ascribed to a predissociation,

$$Cl_2CS(^1A_2) \rightarrow Cl + ClCS \qquad \text{(VII-93)}$$

by Okabe (774). The quantum yield of Cl atom production decreases with an increase of pressure both at 3660 and at 4358 Å, indicating the occurrence of two competing processes, (VII-93) and

$$Cl_2CS(^1A_2) + Cl_2CS \rightarrow \text{product} \qquad \text{(VII-94)}$$

The lifetimes of $^1A_2$ at 3660 and 4358 Å are estimated to be about 6 and 55 nsec, respectively (774).

At 4658 and 4706 Å $^{37}Cl^{37}ClCS$ and $^{35}Cl^{35}ClCS$, respectively, are preferentially excited in mixtures of other chlorine isotopic species.

The preferentially excited isotopic species react with diethoxyethylene to form an addition product. Hence, it is possible to selectively reduce the concentration of a particular isotopic species in mixtures by choosing an appropriate exciting wavelength [Lamotte et al. (597)].

Fluorescence has been seen from the second singlet state $\tilde{B}^1A_1$ formed by absorption of light near 2800 Å corresponding to the 0–0 transition [Levine et al. (627), Oka et al (757)]. The quantum yield of fluorescence is high (0.5 to 1) [Oka et al. (758)].

At 2537 Å Okabe (774) has found that the quantum yield of Cl production is near unity and is independent of pressure in the range 0.4 to 80 torr, indicating immediate dissociation. The absorption spectrum of thiophosgene is extremely diffuse near 2537 Å, suggesting also a direct dissociation process.

VII–13. THIONYL CHLORIDE ($OSCl_2$)

The ground state of $OSCl_2$ is probably pyramidal, belonging to C_s symmetry [Martz and Lagemann (664)]. D_0(Cl—SOCl) is not known. The minimum energy for the reaction

$$Cl_2SO \rightarrow SO + 2Cl$$

is 4.70 ± 0.01 eV.

Absorption starts at about 2900 Å and is continuous in the ultraviolet region [Donovan et al. (306)]. The absorption coefficients in the region 1150 to 1350 Å have been measured by Okabe (768) and are shown in Fig. VII–11.

Photodissociation. Few photochemical studies of thionyl chloride have been made. In the near ultraviolet flash photolysis Donovan et al. (306) have found Cl and SO in absorption. Since the amounts of SO formed are much less than those of $OSCl_2$ decomposed, they have concluded the primary

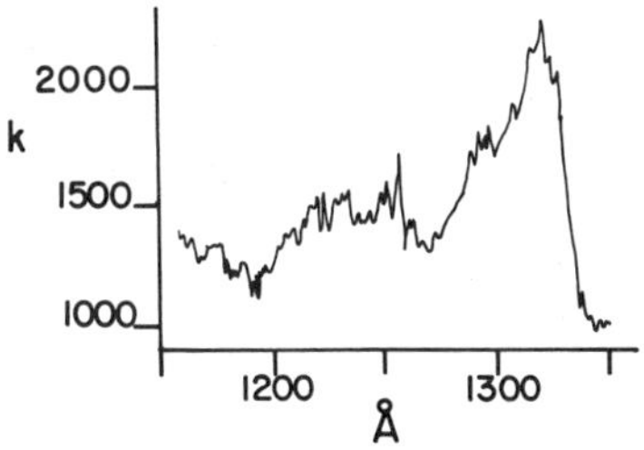

Fig. VII–11. Absorption coefficients of thionyl chloride in the region 1150 to 1350 Å. *k* is given in units of atm^{-1} cm^{-1}, base *e*, 25°C. From Okabe (768), reprinted by permission. Copyright 1972 by the American Institute of Physics.

process is

$$OSCl_2 \xrightarrow{h\nu} Cl + ClSO^{\dagger} \tag{VII-95}$$

followed by

$$ClSO^{\dagger} \rightarrow Cl + SO$$
$$ClSO^{\dagger} + M \rightarrow ClSO + M$$
$$2Cl + M \rightarrow Cl_2 + M$$
$$SO + Cl_2SO \rightarrow SO_2 + SCl_2$$
$$SO + ClSO \rightarrow SO_2 + SCl$$

where $ClSO^{\dagger}$ signifies vibrationally excited ClSO.

In the vacuum ultraviolet $SO(B^3\Sigma, A^3\Pi)$ states are formed below incident wavelength 1318 Å [Okabe (768)], indicating primary processes

$$Cl_2SO \xrightarrow{h\nu} SO(A^3\Pi) + 2Cl \tag{VII-96}$$
$$Cl_2SO \xrightarrow{h\nu} SO(B^3\Sigma) + 2Cl \tag{VII-97}$$

The $SO(A^3\Pi, B^3\Sigma)$ fluorescence excitation spectrum shows many diffuse features, indicating that the processes are predissociative.

VII–14. CYANOGEN (C_2N_2)

The ground state is $X^1\Sigma_g^+$ of $D_{\infty h}$ symmetry. D_0(NC—CN) = 5.58 ± 0.05 eV.

The first weak absorption bands are in the 2400 to 3020 Å region, corresponding to $a^3\Sigma_u^+ – X^1\Sigma_g^+$. The second absorption bands are in the 1820 to 2260 Å region, associated probably with the $A^1\Delta_u – X^1\Sigma_g^+$ transition. Two additional bands are in the 1450 to 1680 and 1250 to 1320 Å regions. The absorption coefficients in the region 1100 to 1700 Å have been measured by Connors et al. (233) and are shown in Fig. VII–12.

The photodissociation of C_2N_2 below 1410 Å yields the $CN(B^2\Sigma^+)$ state [Davis and Okabe (264)].

$$C_2N_2 \xrightarrow{h\nu} CN(X^2\Sigma^+) + CN(B^2\Sigma^+) \tag{VII-98}$$

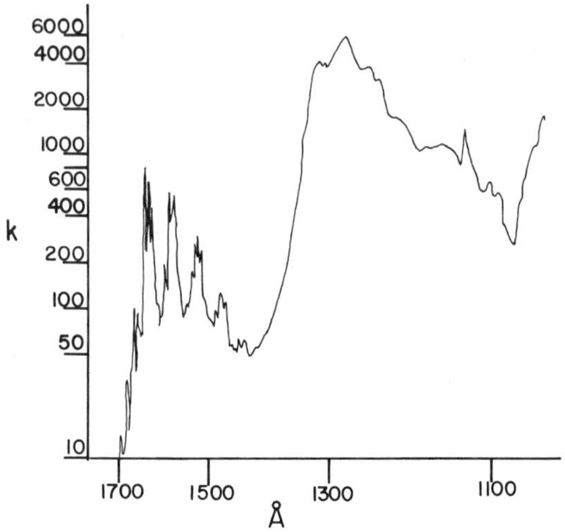

Fig. VII–12. The absorption coefficients of cyanogen in the region 1700 to 1050 Å. *k* is given in units of atm^{-1} cm^{-1}, base *e*, 0°C. From Connors et al. (233), reprinted by permission. Copyright 1974 by the American Institute of Physics.

At 1600 Å the primary process, according to Cody et al. (224), must be

$$C_2N_2 \xrightarrow{h\nu} CN(X^2\Sigma) + CN(A^2\Pi) \quad \text{(VII-99)}$$

Using a tunable laser as a probe they have observed that CN($X^2\Sigma$) radicals produced at this wavelength are vibrationally and rotationally excited. The rotational distribution follows the Boltzmann law, indicating that dissociation is not immediate but occurs after many vibrations of the electronically excited molecule. Thus, the distribution of the rotational population reflects the statistical nature of the dissociation processes. The distribution of the excess energy beyond that required to break the C—C bond is 54% in electronic, 20% in translational, 14% in vibrational, and 11% in rotational energies. See also p. 87.

West and Berry (1037) have observed laser emissions due to the transition CN($A^2\Pi \rightarrow X^2\Sigma$) in the vacuum ultraviolet flash photolysis of C_2N_2, HCN, ClCN, BrCN, and ICN. The CN($A^2\Pi$) radicals are produced within a low-loss optical cavity for effective laser action.

VII–15. SULFUR MONOCHLORIDE (S_2Cl_2)

The ground state structure is probably nonplanar (15). The bond energy estimated by Donovan et al. (100) is D_0(ClS—SCl) = 1.5 eV.

A strong continuum has been found in the near ultraviolet with a maximum at 2650 Å.

The near ultraviolet flash photolysis of S_2Cl_2 has been performed by Donovan et al. (100), who have found transient spectra of vibrationally excited $S_2(X^3\Sigma)$, as well as $S_2(a^1\Delta)$, $S(^3P)$, and probably SCl.

They have proposed the primary process

$$S_2Cl_2 \xrightarrow{h\nu} 2SCl \qquad \text{(VII-99a)}$$

followed by

$$SCl + S_2Cl_2 \rightarrow SCl_2 + S_2Cl \qquad \text{(VII-99b)}$$

$$SCl + S_2Cl \rightarrow SCl_2 + S_2\ (v'' \text{ up to } 8) \qquad \text{(VII-99c)}$$

VII–16. FOUR ATOM RADICALS

VII–16.1. Methyl (CH_3)

The ground state is $\tilde{X}^2A_2''$ (planar) in D_{3h} symmetry with an H—C—H angle of 120°; $D_0(\text{H—CH}_2) = 4.69 \pm 0.05$ eV.

Four absorption bands have been observed at 2160, 1503, 1497, and 1408 Å corresponding, respectively, to transitions $\tilde{B}^2A_1'$–$\tilde{X}^2A_2''$, $\tilde{C}^2E''$–$\tilde{X}^2A_2''$, $\tilde{D}^2A_1'$–$\tilde{X}^2A_2''$, and $\tilde{E}^2A_1'$–$\tilde{X}^2A_2''$ (16). Below 1400 Å several Rydberg transitions have been found. Most of the absorption bands are diffuse and apparently no fluorescence has been found. The oscillator strengths of the 2160, 1503, and 1497 Å bands are $1.2 \pm 0.2 \times 10^{-2}$, 5.1×10^{-2}, and 1.0×10^{-2}, respectively [Van den Berg et al. (988), Pilling et al. (809)].

The convenient sources of CH_3 radicals are the photolysis of azomethane

$$CH_3NNCH_3 \rightarrow 2CH_3 + N_2 \qquad \text{(VII-100)}$$

and the photolysis of dimethyl mercury

$$Hg(CH_3)_2 \rightarrow 2CH_3 + Hg \qquad \text{(VII-101)}$$

The reactions of CH_3 with various molecules have been extensively studied for many years and are reviewed by Steacie (26).

The combination rate

$$CH_3 + CH_3 \rightarrow C_2H_6 \qquad \text{(VII-102)}$$

has been measured recently and the rate constant is 5×10^{-11} cm^3 molec^{-1} sec^{-1} [James and Simons (524)]. The rate constant of the reaction

$$CH_3 + O \rightarrow CH_2O + H \qquad \text{(VII-103)}$$

is $1.23 \pm 0.25 \times 10^{-10}$ cm^3 molec^{-1} sec^{-1} [Washida and Bayes (1012)].

The combination rate of CH_3 with SO_2 has been measured by James et al. (522), who obtained a high-pressure value of 2.91×10^{-13} cm^3 molec^{-1} sec^{-1}

$$CH_3 + SO_2 \xrightarrow{M} CH_3SO_2 \qquad \text{(VII-104)}$$

The combination rates of CH_3 with NO and O_2 have been measured by Van den Bergh and Callear (989)

$$CH_3 + NO \xrightarrow{M} CH_3NO \qquad \text{(VII-105)}$$

$$CH_3 + O_2 \xrightarrow{M} CH_3O_2 \qquad \text{(VII-106)}$$

The high-pressure rate constant is 1.7×10^{-11} cm^3 molec^{-1} sec^{-1} for $CH_3 + NO$ and 1.8×10^{-12} cm^3 molec^{-1} sec^{-1} for $CH_3 + O_2$.

VII–16.2. Trifluoromethyl (CF_3), Trichloromethyl (CCl_3)

The ground state CF_3 is pyramidal (C_{3v} symmetry); $D_0(F—CF_2) = 3.75 \pm 0.1$ eV. Basco and Hathron (67) have observed the absorption bands of CF_3 in the region 1450 to 1650 Å. The CF_3 radicals are produced by the photolysis of hexafluoroacetone or hexafluoroazomethane

$$CF_3COCF_3 \rightarrow 2CF_3 + CO \qquad \text{(VII-107)}$$

$$CF_3N_2CF_3 \rightarrow 2CF_3 + N_2 \qquad \text{(VII-108)}$$

They have also obtained a combination rate constant of 5×10^{-12} cm^3 molec^{-1} sec^{-1} for CF_3

$$CF_3 + CF_3 \rightarrow C_2F_6 \qquad \text{(VII-109)}$$

The ground state CCl_3 is pyramidal (C_{3v} symmetry) (861). The absorption bands have apparently not been found. $D_0(Cl—CCl_2) = 2.9 \pm 0.3$ eV.

The source of CCl_3 radicals is the photolysis of hexachloroacetone

$$CCl_3COCCl_3 \rightarrow 2CCl_3 + CO \qquad \text{(VII-110)}$$

or the photolysis of bromotrichloromethane

$$CCl_3Br \rightarrow CCl_3 + Br \qquad \text{(VII-111)}$$

Tedder and Walton (964) report a combination rate constant of 5.3×10^{-11} cm^3 molec^{-1} sec^{-1} by a rotating-sector method

$$CCl_3 + CCl_3 \rightarrow C_2Cl_6 \qquad \text{(VII-112)}$$

VII–16.3. Nitrogen Trioxide (NO_3)

The ground state NO_3 is planar $\tilde{X}^2A_2'$ (D_{3h} symmetry). $D_0(O—NO_2) = 2.1 \pm 0.2$ eV. The absorption bands of NO_3 have been found in the region

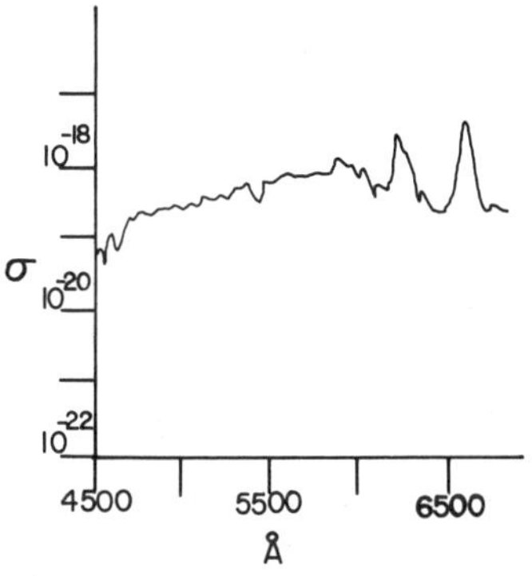

Fig. VII–13. Absorption cross sections of NO_3 in the region 4500 to 7000 Å. σ is in units of cm^2 $molec^{-1}$, base *e*, room temperature. From Johnston and Graham (541). Reproduced by permission of the National Research Council of Canada from the *Canadian Journal of Chemistry*, 52, 1415 (1974).

5000 to 6650 Å (16). Recently, Johnston and Graham (541) have measured the absorption cross sections in the region 4300 to 6800 Å (see Fig. VII–13). The NO_3 radicals are prepared by mixing NO_2 with excess O_3. The spectrum is characterized by diffuse bands. Energetically possible primary photochemical processes are

$$NO_3 \xrightarrow{h\nu} NO_2 + O \qquad < 5900\ \text{Å} \qquad \text{(VII-113)}$$

$$NO_3 \xrightarrow{h\nu'} NO + O_2 \qquad \Delta H = 3\ \text{kcal mol}^{-1} \qquad \text{(VII-114)}$$

By a molecular modulation technique. Johnston and Graham have found NO as a product of the photolysis above 6000 Å. The NO_3 radicals are formed as an intermediate of the HNO_3 photolysis.

VII–16.4. Sulfur Trioxide (SO_3)

The ground state SO_3 is planar (D_{3h}). $D_0(O—SO_2) = 3.55 \pm 0.01$ eV. The absorption spectrum starts at about 3100 Å and consists of weak diffuse bands superimposed on a continuum [Fajans and Goodeve (340)]. The absorption coefficient at 2200 Å is about 200 (l mol^{-1} cm^{-1}, base 10). The Raman band at 1067 cm^{-1} has been found by Skotnicki et al. (895), who performed an analysis of SO_3 in SO_2 by measuring band intensity at 1067 cm^{-1} (SO_3) relative to that at 1151 cm^{-1} (SO_2).

The photolysis of SO_3 in the near ultraviolet has been studied by Norrish and Oldershaw (746). They proposed that the primary process is either

$$SO_3 \xrightarrow{h\nu} SO + O_2 \qquad \text{(VII-115)}$$

or

$$SO_3 \xrightarrow{h\nu} SO_2 + O \qquad \text{(VII-116)}$$

The SO_3 radical may be formed in photooxidation processes of SO_2 in polluted atmospheres by

$$SO_2\ (\text{triplet}) + SO_2 \rightarrow SO_3 + SO \tag{VII-117}$$

$$SO_2\ (\text{triplet}) + O_2 \rightarrow SO_3 + O \tag{VII-118}$$

where SO_2 (triplet) is formed either directly by photon absorption in the range 3400 to 4000 Å or by intersystem crossing from SO_2 (singlet) formed by light absorption in the region 2900 to 3400 Å. See p. 248.

However, the detailed mechanism of photooxidation of SO_2 is still unknown [Sidebottom et al. (879)].

FIVE-ATOM MOLECULES

VII–17. METHANE (CH_4)

Ground state methane is $\tilde{X}\,^1A_1$ (tetrahedron) of T_d symmetry. $D_0(\text{H—CH}_3) = 4.48 \pm 0.01$ eV. The absorption spectrum is continuous in the region 1100 to 1600 Å, which is shown in Fig. VII–14.

Mount et al. (723a) have found recently that absorption coefficients above 1475 Å are approximately 200 times smaller than those reported by Watanabe et al. (31). The photolysis of CH_4 has been studied by many workers and has been reviewed by Ausloos and Lias (49).

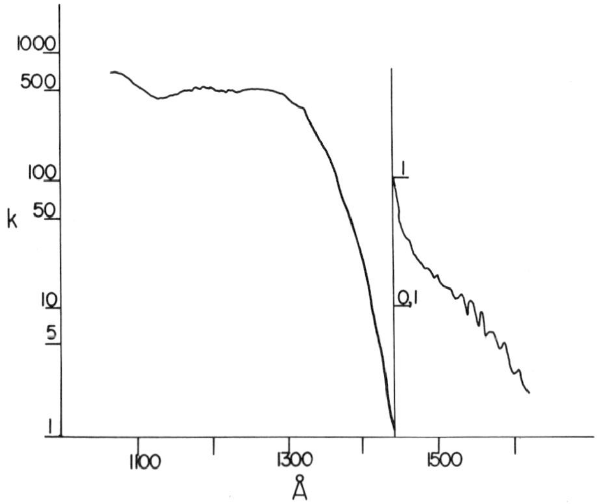

Fig. VII–14. Absorption coefficients of CH_4 in the region 1065 to 1610 Å. *k* is in units of atm^{-1} cm^{-1}, base *e*, 0°C. From Watanable et al. (31), reprinted by permission of the Air Force Geophysics Laboratory.

The primary processes may be represented by

$$CH_4 \xrightarrow{h\nu} CH_3 + H \qquad \text{(VII-119)}$$

$$CH_4 \xrightarrow{h\nu} CH_2 + H_2 \qquad \phi_{120} \qquad \text{(VII-120)}$$

$$CH_4 \xrightarrow{h\nu} CH + H_2 + H \qquad \phi_{121} \qquad \text{(VII-121)}$$

The relative importance of three processes is not well known. The quantum yield of (VII-120) is about 0.5 at 1236 Å [Laufer and McNesby (603)]. The quantum yield of (VII-121) is about 0.06 at 1236 Å [Rebbert and Ausloos (827)].

Ethylene is a major hydrocarbon product in the vacuum ultraviolet flash photolysis of methane (137). Braun et al. suggest that the following processes are responsible for ethylene formation

$$CH_2 + CH_2 \xrightarrow{M} C_2H_4 \qquad \text{(VII-122)}$$

$$CH + CH_4 \rightarrow C_2H_4 + H \qquad \text{(VII-123)}$$

From the isotopic analysis of ethylene produced from the photolysis of $CH_4 + CD_4$ mixtures, they conclude (VIII-123) is a dominant process for the ethylene production.

Braun et al. (139) have measured the rate constant of (VIII-123) and obtained a value of 2.5×10^{-12} cm^3 $molec^{-1}$ sec^{-1}. Other secondary reactions of importance are

$$CH_2 + CH_4 \rightarrow C_2H_6$$

$$2CH_3 \rightarrow C_2H_6$$

$$H + CH_3 \xrightarrow{M} CH_4$$

$$H + H \xrightarrow{M} H_2$$

$$2CH \xrightarrow{M} C_2H_2$$

where M is a third body.

VII–18. HALOGENATED METHANES

The photochemistry of halogenated methanes has been of great interest recently ever since it was recognized that chloromethanes in the stratosphere may release Cl atoms upon absorption of solar radiation and that Cl atoms so produced may catalytically decompose O_3.

Primary processes of halogenated methanes may be summarized as follows:

1. The near ultraviolet photolysis of halogenated methanes gives rise to one halogen atom [Takacs and Willard (956a)].

$$RX \xrightarrow{h\nu} R + X \qquad (X = Cl, Br, I) \qquad \text{(VII-124)}$$

2. If they contain more than one kind of halogen, the weakest C—X bond breaks and $D_0(\mathrm{R—F}) > D_0(\mathrm{R—H}) > D_0(\mathrm{R—Cl}) > D_0(\mathrm{R—Br}) > D_0(\mathrm{R—I})$ for R = CH_3. Hence, for example, CCl_3Br dissociates into CCl_3 + Br rather than CCl_2Br + Cl.
3. In the vacuum ultraviolet photolysis the simultaneous rupture of the two weakest bonds occurs concurrently with (VII-124). For example, CF_2Cl_2 dissociates into CF_2 + 2Cl and CF_2Cl + Cl, the former process becoming more important at shorter wavelengths.
4. The three-bond scission is a rare event even at the shortest wavelength used, 1470 Å, that is, $CFCl_3 \xrightarrow{h\nu} CF + Cl_2 + Cl$, is only a few percent of the total process.
5. Donovan and Husain (299) have detected $Br(^2P_{1/2})$ in the vacuum ultraviolet flash photolysis of $CHCl_2Br$ and CF_3Br but not from CH_3Br and CH_2Br_2.

VII–18.1. Methyl Chloride (CH_3Cl), Methyl Bromide (CH_3Br)

The ground state of CH_3X (X = Cl, Br) is $\tilde{X}\,^1A_1$ of C_{3v} symmetry. $D_0(\mathrm{Cl—CH_3}) = 3.57 \pm 0.01$ eV; $D_0(\mathrm{Br—CH_3}) = 2.97$ eV. The absorption spectrum of CH_3Cl in the region 1600 to 2000 Å is continuous with a maximum at 1730 Å. Below 1600 Å three diffuse transitions have been found. The absorption coefficients in the vacuum ultraviolet have been measured by Raymonda et al. (826) and Russell et al. (845). The absorption spectrum of CH_3Br is also continuous in the region 1800 to 2850 Å with a maximum at 2050 Å. Several transitions have been found in the vacuum ultraviolet (16). The absorption coefficients in the vacuum ultraviolet have been measured by Causley and Russell (199). The absorption coefficients of CH_3Cl in the region 1700 to 2300 Å and of CH_3Br in the region 1700 to 2700 Å have recently been measured by Robbins (837) and are shown in Fig. VII–15*a* and VII–15*b*, respectively.

Hubrich et al. (486*c*) also have measured the absorption cross sections of CH_3Cl in the 1600 to 2750 Å region at 298 and 208°K. Their results are in good agreement with those shown in Fig. VII–15*a* obtained by Robbins (837).

The primary process in the ultraviolet region must be

$$CH_3X \xrightarrow{h\nu} CH_3 + X \qquad (X = Cl, Br) \qquad \text{(VII-125)}$$

Ting and Weston (973) have studied reactions of CH_3 radicals produced by the photolysis of CH_3Br at 1849 Å. The CH_3 radicals thus formed have been found to carry an excess energy sufficient to overcome the activation energy for the H atom abstraction from CH_3Br or H_2. Only $Br(^2P_{3/2})$ has been detected in the vacuum ultraviolet photolysis of CH_3Br [Donovan and Husain (299)]. Very recent results by Shold and Rebbert (873a) indicate that CH_3Cl dissociates into CH_2Cl + H, CH_2 + HCl and CHCl + H_2 as well as (VII-125) at 1470 and 1236 Å.

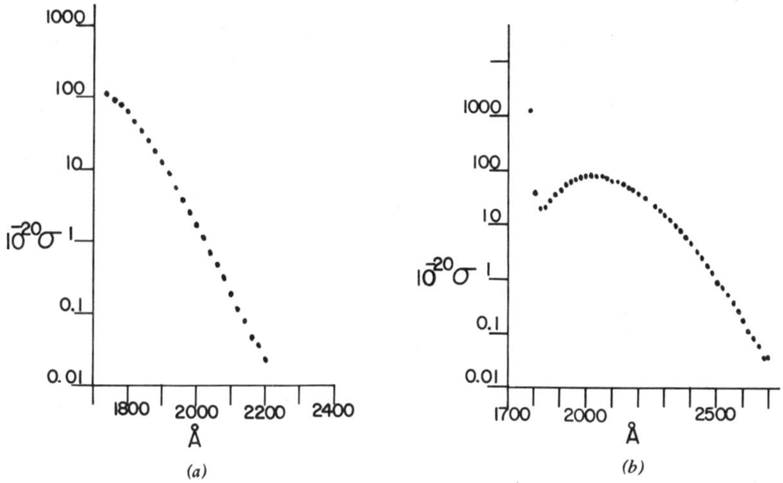

Fig. VII–15. (*a*) **Absorption cross sections of CH_3Cl in the near ultraviolet. σ in units of cm^2 $molec^{-1}$, base *e*, room temperature. From Robbins (837), reprinted by permission. Copyright by the American Geophysical Union.** (*b*) **Absorption cross sections of CH_3Br in the near ultraviolet. σ is in units of cm^2 $molec^{-1}$, base *e*, room temperature. From Robbins (837), reprinted by permission. Copyright by the American Geophysical Union.**

VII–18.2. Methyl Iodide (CH_3I), Trifluoroiodiodomethane (CF_3I)

Primary photochemical processes of CH_3I and CF_3I have been studied extensively in conjunction with the laser emission $I(^2P_{1/2}) \rightarrow I(^2P_{3/2}) + h\nu$ observed at 1.315 μm in the near ultraviolet flash photolysis. Over 90% of I atoms produced from CF_3I are in the $^2P_{1/2}$ state.

The ground state of CH_3I is $\tilde{X}^1A_1$ of C_{3v} symmetry. $D_0(I—CH_3) = 2.42$ eV. An absorption in the region 2000 to 3600 Å is continuous with a maximum at 2590 Å. Below 2000 Å several band systems are observed (16).

Photochemistry. Riley and Wilson (834) have measured the translational energy of I atoms produced from the photolysis of CH_3I at 2662 Å by a pulsed laser. They suggest the following two primary processes

$$CH_3I \xrightarrow{h\nu} CH_3 + I(^2P_{1/2}) \qquad \text{(VII-126)}$$

$$CH_3I \xrightarrow{h\nu} CH_3 + I(^2P_{3/2}) \qquad \text{(VII-127)}$$

The production of $I(^2P_{1/2})$ is 78% of the primary process. Palmer and Padrick (790) as well as Donohue and Wiesenfeld (294) have measured the fraction of $I(^2P_{1/2})$ produced in the flash photolysis of CH_3I. They have found 76 (Palmer and Padrick) and 90% (Donohue and Wiesenfeld) of I atoms produced are in the $^2P_{1/2}$ state.

Chou et al. (209) have suggested another minor process in the near ultraviolet photolysis,

$$CH_3I \xrightarrow{h\nu} CH_2 + HI \quad \text{(VII-128)}$$

as a result of the product analysis in the photolysis of mixtures of CH_3I and hydrocarbons. The excess energies beyond those required for processes (VII-126) and (VII-127) to occur are 30 and 51 kcal mol^{-1}, respectively, at the 2662 Å photolysis. The excess energy goes mainly into the translational energies of the fragments and only 12% of the available energy resides in CH_3 radicals (834). The CH_3 radicals with excess kinetic and internal energies ($CH_3^{\ddagger}$) are known to react with CH_3I to form CH_4

$$CH_3^{\ddagger} + CH_3I \rightarrow CH_4 + CH_2I \quad \text{(VII-129)}$$

while for thermal CH_3 radicals the reaction requires an activation energy of about 9 kcal mol^{-1}.

Rice and Truby (835) suggest that CH_3 radicals probably carry more vibrational than translational energy at wavelengths shorter than 2537 Å, since below 2537 Å photolysis deactivation rates of $CH_3^{\ddagger}$ by He, Ar, N_2, and CH_3I follow the order

$$He < Ar < N_2 < CH_3I$$

while at 2537 Å quenching rates by He, Ne, Ar, N_2, and CO_2 are about equal. The reactions of $I(^2P_{1/2})$ are

$$I(^2P_{1/2}) + CH_3I \rightarrow I_2 + CH_3 \qquad k_{130} \quad \text{(VII-130)}$$

$$I(^2P_{1/2}) + CH_3I \rightarrow I(^2P_{3/2}) + CH_3I \qquad k_{131} \quad \text{(VII-131)}$$

Palmer and Padrick (790) have determined rate constants $k_{130} < 5 \times 10^{-15}$ and $k_{131} = 2 \times 10^{-13}$ cm^3 $molec^{-1}$ sec^{-1}, that is, the deactivation process is more important than the chemical reaction. Mains and Lewis (659) have measured the quantum yield of methane production in low and high intensity photolysis of CH_3I in the near ultraviolet. The quantum yield of CH_4 is a function of pressure and ranges from 0.1 to 0.001 in the high intensity photolysis and from 0.02 to 0.05 in the low intensity photolysis.

The ground state of CF_3I is $\tilde{X}^1A_1$ of C_{3v} symmetry. The near ultraviolet absorption is continuous with a maximum at 2650 Å in the region 2480 to 2815 Å. Three transitions have been observed in the vacuum ultraviolet (16). $D_0(I—CF_3) = 2.31 \pm 0.05$ eV.

The photolysis of CF_3I in the near ultraviolet is represented as in the case of CH_3I by two primary processes

$$CF_3I \xrightarrow{h\nu} CF_3 + I(^2P_{1/2}) \quad \text{(VII-132)}$$

$$CF_3I \xrightarrow{h\nu} CF_3 + I(^2P_{3/2}) \quad \text{(VII-133)}$$

The fraction of $I(^2P_{1/2})$ production is 0.91 of the primary processes [Donohue and Wiesenfeld (295)]. The production of $I(^2P_{1/2})$ atoms in the near ultraviolet photolysis of CF_3I has been shown to occur by means of a mass spectrometer in conjunction with an inhomogeneous magnetic field [Talroze et al. (959)].

VII–18.3. Methylene Iodide (CH_2I_2), Iodoform (CHI_3), Chloroform ($CHCl_3$)

The ground states of CH_2I_2, CHI_3, and $CHCl_3$ are tetrahedral with $C_{2v}(CH_2I_2)$ and $C_{3v}(CHI_3, CHCl_3)$ symmetries. $D_0(I—CH_2I) = 2.1$ eV, $D_0(I—CHI_2) \simeq 2.0$ eV (377), $D_0(Cl—CHCl_2) \simeq 3.2$ eV.

The continuous absorption spectrum of CH_2I_2 starts at about 3600 Å and that of CHI_3 at about 3900 Å. The absorption coefficients of CH_2I_2 and CHI_3 in the near ultraviolet are given by Kawasaki et al. (560). The absorption coefficients of $CHCl_3$ in the vacuum ultraviolet have been measured by Russel et al. (845).

The photolyses of CH_2I_2 and CHI_3 in molecular beams have been investigated by Kawasaki et al. (560) using a broad-band polarized light source in conjunction with a mass spectrometer. The primary product of the photolysis in the near ultraviolet is the I atom. Hence, primary processes are

$$CH_2I_2 \xrightarrow{h\nu} CH_2I + I \qquad \text{(VII-134)}$$

$$CHI_3 \xrightarrow{h\nu} CHI_2 + I \qquad \text{(VII-135)}$$

The angular dependence of I atoms with respect to the direction of polarization suggests that the excited states of CH_2I_2 and CHI_3 are B_1 and E, respectively.

Kroger et al. (586) have measured the flight time of the CH_2I fragment dissociated from CH_2I_2 at 2660 Å by a pulsed polarized laser. They concluded that CH_2I radicals contain 80 to 90% of the available energy, that is, the energy beyond that required to break the $I—CH_2I$ bond, and that I atoms are probably in the ground state.

The fraction of the available energy residing in the CH_2I radicals is much larger than that in the CH_3 radicals dissociated from CH_3I, which is only 12% (834). Qualitatively, this difference in the energy partitioning can be understood from (II-23) based on the impulsive model (see p. 93).

$$\frac{E_{int}^{BC}}{E_{avl}} = 1 - \frac{\mu_{A-B}}{\mu_{A-BC}} \qquad \text{(VII-135a)}$$

where E_{int}^{BC} is the internal energy of the fragment BC, E_{avl} is the available energy, μ_{A-B} is the reduced mass of A and B, and μ_{A-BC} is the reduced mass of A and BC. From the equation the internal energies of the CH_2I and CH_3

radicals are 84% and 18%, respectively (A = I, B = carbon atom, BC = CH_2I and CH_3 for CH_2I_2 and CH_3I, respectively).

In addition, CH_2 radicals are also produced. The energetic consideration rules out the direct dissociation path at 2660 Å

$$CH_2I_2 \xrightarrow{h\nu} CH_2 + I_2 \quad \text{(VII-136)}$$

Instead, Kroger et al. suggest consecutive two-photon absorption process, that is, process (VII-134) followed by

$$CH_2I \xrightarrow{h\nu} CH_2 + I \quad \text{(VII-137)}$$

Style and Ward (953) and Dyne and Style (330) have found fluorescence from an electronically excited $I_2(I_2^*)$ in the vacuum ultraviolet photolysis of CH_2I_2, indicating a primary process

$$CH_2I_2 \xrightarrow{h\nu} CH_2 + I_2^* \quad \text{(VII-137a)}$$

The photolysis of $CHCl_3$ has been performed by Yu and Wijnen (1077) in the near ultraviolet in the presence of ethane. The results of the product analysis suggest the primary process

$$CHCl_3 \xrightarrow{h\nu} Cl + CHCl_2 \quad \text{(VII-138)}$$

VII–18.4. Trichlorofluoromethane ($CFCl_3$, Freon-11); Dichlorodifluoromethane (CF_2Cl_2, Freon-12); Dibromodifluoromethane (CF_2Br_2)

The ground state of $CFCl_3$ is tetrahedral (C_{3v} symmetry) and those of CF_2Cl_2 and CF_2Br_2 are tetrahedral (C_{2v} symmetry). $D_0(Cl—CCl_2F)$ is not known but it is probably about 3.3 eV, which is between $D_0(Cl—CCl_3) = 3.0$ eV and $D_0(Cl—CClF_2) = 3.50 \pm 0.1$ eV. $D_0(Br—CF_2Br)$ is unknown but is probably close to $D_0(Br—CH_2Br) = 2.9$ eV.

The absorption coefficients of $CFCl_3$ and CF_2Cl_2 in the near ultraviolet have been measured by Rowland and Molina (843) and Robbins et al. (836) and are shown in Fig. VII–16. The absorption coefficients of CF_2Br_2 in the region 2200 to 3100 Å have been measured by Walton (1002). The absorption coefficients of $CFCl_3$ and CF_2Cl_2 in the 1600 to 2750 Å region at 208 and 298°K have been measured by Hubrich et al. (486c) and in the 1900 to 2200 Å region from 212 to 257°K by Chou et al. (210a).

Photochemistry. Both $CFCl_3$ and CF_2Cl_2 are used as aerosol propellants and refrigerants in large quantities. They are chemically inert in the troposphere. However, when they diffuse into the stratosphere they are photodissociated by solar radiation to produce Cl atoms. The Cl atoms so formed would catalytically destroy O_3 in the stratosphere (see Section VIII–2, p. 350).

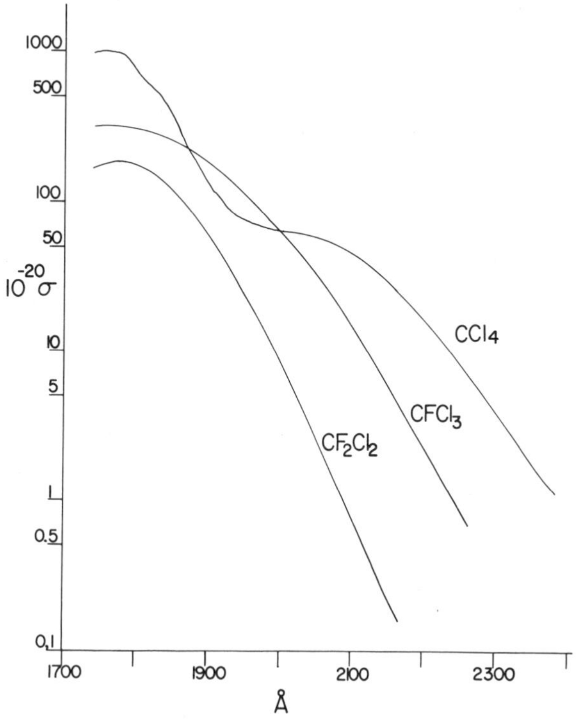

Fig. VII–16. Absorption cross sections of CF_2Cl_2, $CFCl_3$, and CCl_4 in the ultraviolet region. σ is given in units of cm^2 $molec^{-1}$, base *e*, room temperature. From Rowland and Molina (843) and Robbins et al. (836), reprinted by permission of the authors. Copyright by the American Geophysical Union.

The photolyses of $CFCl_3$ and CF_2Cl_2 have recently been studied by Milstein and Rowland (708) and Rebbert and Ausloos (828). The primary processes are

$$CFCl_3 \xrightarrow{h\nu} Cl + CFCl_2 \qquad \phi_{139} \qquad \text{(VII-139)}$$

$$CFCl_3 \xrightarrow{h\nu} 2Cl + CFCl \qquad \phi_{140} \qquad \text{(VII-140)}$$

$$CF_2Cl_2 \xrightarrow{h\nu} Cl + CF_2Cl \qquad \phi_{141} \qquad \text{(VII-141)}$$

$$CF_2Cl_2 \xrightarrow{h\nu} 2Cl + CF_2 \qquad \phi_{142} \qquad \text{(VII-142)}$$

Milstein and Rowland have measured the quantum yields of CF_2Cl_2 disappearance and of the production of CF_2Cl radicals in the photolysis of CF_2Cl_2–O_2 mixtures at 1849 Å and find near unit quantum yield for both.

The photolyses of $CFCl_3$ and CF_2Cl_2 in the presence of CH_4 or C_2H_6 have been studied at 2139, 1849, 1633, and 1470 Å (828).

From the product analysis, the following primary quantum yields are derived (282);

	2139 Å	1849 Å	1633 Å	1470 Å
$CFCl_3$				
ϕ_{139}	0.98	0.7	0.50	<0.1
ϕ_{140}	0.03	0.3	0.45	>0.87
CF_2Cl_2				
ϕ_{141}	0.91	0.65	0.56	<0.1
ϕ_{142}	0.07	0.34	0.40	>0.8

The production of one Cl atom is predominant at 2139 Å, while the rupture of the two C—Cl bonds becomes increasingly more important at shorter wavelengths.

The photolysis of CF_2Br_2 at 2650 Å has been performed by Walton (1002). A main product is $C_2F_4Br_2$, the quantum yield of which decreases with an increase of pressure or by the addition of CO_2. Walton suggests the following primary processes:

$$CF_2Br_2 \xrightarrow{h\nu} CF_2Br_2^* \quad \text{(VII-143)}$$

$$CF_2Br_2^* \rightarrow Br + CF_2Br \quad \text{(VII-144)}$$

$$CF_2Br_2^* \xrightarrow{M} CF_2Br_2 \quad \text{(VII-145)}$$

where $CF_2Br_2^*$ denotes an electronically excited molecule.

VII–18.5. Carbon Tetrachloride (CCl_4); Bromotrichloromethane (CCl_3Br)

The ground states of CCl_4 and CCl_3Br are tetrahedral with $T_d(CCl_4)$ and $C_{3v}(CCl_3Br)$ symmetries.

Near ultraviolet absorption of CCl_4 starts at about 2500 Å and is continuous. The absorption cross sections in the near ultraviolet have been measured by Rowland and Molina (843) and Robbins et al. (836). They are shown in Fig. VII–16. $D_0(Cl—CCl_3) = 3.04 \pm 0.1$ eV.

Photochemistry. Primary processes of CCl_4 may be given by

$$CCl_4 \xrightarrow{h\nu} CCl_3 + Cl \qquad \phi_{146} \quad \text{(VII-146)}$$

$$CCl_4 \xrightarrow{h\nu} CCl_2 + 2Cl \text{ (or } Cl_2) \qquad \phi_{147} \quad \text{(VII-147)}$$

$$CCl_4 \xrightarrow{h\nu} CCl + Cl_2 + Cl \qquad \phi_{148} \quad \text{(VII-148)}$$

Davis et al. (266) have measured the products in the photolysis of CCl_4–Br_2 mixtures at 2537, 1849, and 1470 Å. At 2537 Å only CCl_3Br is found, while at 1849 Å both CCl_3Br and CCl_2Br_2 are present. At 1470 Å CCl_3Br, CCl_2Br_2, and $CClBr_3$ are produced. From the results Davis et al. conclude that at 2537 Å (VII-146) is predominant while at shorter wavelengths (VII-147) and (VII-148) occur in conjunction with (VII-146).

Rebbert and Ausloos (829) have performed the photolysis of CCl_4 in the presence of HCl, HBr, and C_2H_6. From the product analysis they conclude that $\phi_{146} = 0.9$, $\phi_{147} = 0.05$ at 2139 Å; $\phi_{146} = 0.25$, $\phi_{147} = 0.76$ at 1633 Å; and $\phi_{146} = 0.04$, $\phi_{147} = 0.6$ at 1470 Å. The production of CCl appears unimportant even at 1470 Å. Roquitte and Wijnen (841) have investigated the photolysis of CCl_4 in the presence of ethane or ethylene in the near ultraviolet. The results are consistent with the occurrence of (VII-146) followed by the addition of Cl atoms to C_2H_4 or the abstraction of hydrogen from C_2H_6 to form $C_2H_5 + HCl$.

Jayanty et al. (529) have studied the photolysis of CCl_4 and O_2 or O_3 mixtures at 2139 Å. They have postulated an excited CCl_4 that dissociates into $CCl_2 + Cl_2$ at low pressures, while at high pressures the excited CCl_4 is quenched. In view of other studies and the continuous nature of the absorption spectrum of CCl_4 observed, it is unlikely that the excited state is formed at 2139 Å.

The products of the photolysis of CCl_3Br at 3650 Å are Br_2, CCl_4, CCl_2Br_2, and C_2Cl_6 [Sidebottom et al. (875)]. The quantum yields of C_2Cl_6 decrease with an increase of CF_4 or CO_2 pressure. $D_0(Br—CCl_3) = 2.4$ eV.

Primary processes proposed are

$$CCl_3Br \xrightarrow{h\nu} CCl_3Br^* \qquad \text{(VII-149)}$$

$$CCl_3Br^* \rightarrow CCl_3 + Br \qquad \text{(VII-150)}$$

$$CCl_3Br^* \xrightarrow{M} CCl_3Br \qquad \text{(VII-151)}$$

where CCl_3Br^* signifies an electronically excited molecule. The results suggest that more than one excited state may be involved.

VII–18.6. Dichlorofluoromethane ($CHFCl_2$); Chlorodifluoromethane (CHF_2Cl)

The absorption cross sections of $CHFCl_2$ and CHF_2Cl have been measured by Hubrich et al. (486c) in the 1600 to 2400 Å region at 208 and 298°K. They are shown in Fig. VII–17. The photolysis of $CHFCl_2$ has been studied by Rebbert et al. (829a) at 2139, 1633, and 1470 Å. At 2139 Å the main primary process is

$$CHFCl_2 \xrightarrow{h\nu} CHFCl + Cl$$

while at shorter wavelengths, two other primary processes

$$CHFCl_2 \xrightarrow{h\nu} CHF + 2Cl$$

$$CHFCl_2 \xrightarrow{h\nu} CF + HCl + Cl$$

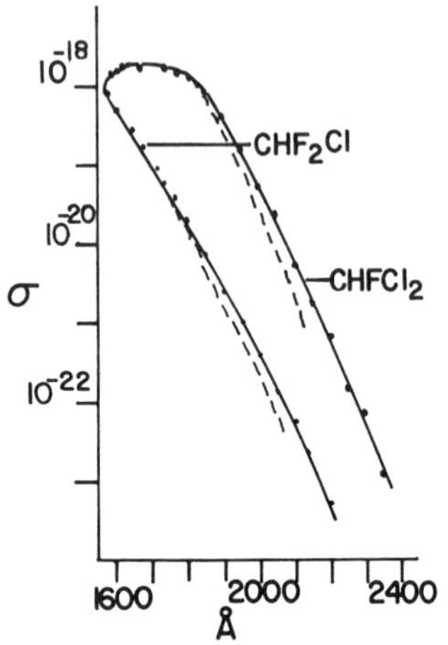

Fig. VII–17. The absorption cross sections of $CHFCl_2$ and CHF_2Cl in the 1600 to 2400 Å region. σ is in units of cm^2 $molec^{-1}$, base *e*, the solid line at 298°K, the dashed line at 208°K. The absorption cross sections decrease at 208°K for $\lambda > 1700$ Å. From Hubrich et al. (486c), reprinted by permission of Verlag Chemie.

become just as important as the CHFCl production process. The results indicate that at 2139 Å absorption is localized in the C—Cl bond, while at shorter wavelengths the C—H bond absorption occurs as well.

VII–19. DIAZOMETHANE (CH_2N_2), DIAZIRINE (CYCLIC CH_2N_2)

The ground state of CH_2N_2 is planar $\tilde{X}^1A_1$ of C_{2v} symmetry with a H—C—H angle of 127° and a C—N—N angle of 180°. $D_0(N_2—CH_2) = 1.81$ eV (605). A first absorption system is in the 3200 to 4750 Å region and is very diffuse. A second system is in the 2000 to 2650 Å region and is continuous with a maximum at 2175 Å. Several transitions have been found in the vacuum ultraviolet (16). The absorption coefficients, measured by Brinton and Volman (150) in the region 2500 to 5000 Å, are given in Fig. VII–18.

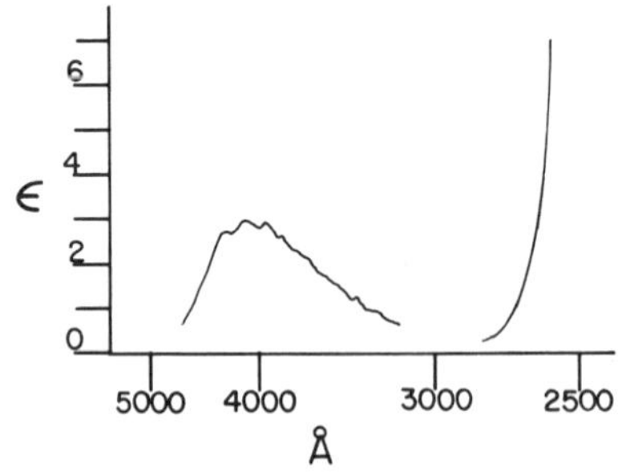

Fig. VII–18. Absorption coefficients ϵ of diazomethane in the 2500 to 5000 Å region. ϵ is in units of $l\ mol^{-1}\ cm^{-1}$, base 10, room temperature. From Brinton and Volman (150), reprinted by permission. Copyright 1951 by the American Institute of Physics.

The flash photolysis of CH_2N_2 in the near and vacuum ultraviolet results in the production of the singlet CH_2 [Herzberg and Johns (465), Herzberg (464), Braun et al. (143)].

$$CH_2N_2 \xrightarrow{h\nu} CH_2(\tilde{a}\,^1A_1) + N_2 \qquad \text{(VII-152)}$$

in agreement with the spin conservation rule.

At higher N_2 pressures, the triplet CH_2 concentration is increased, indicating that the singlet CH_2 radicals are deactivated to the triplet ground state [Braun et al. (143)].

$$CH_2(\tilde{a}\,^1A_1) \xrightarrow{N_2} CH_2(\tilde{X}\,^3B_1) \qquad \text{(VII-153)}$$

The photolysis of CH_2N_2 in the vacuum ultraviolet yields $CH(A^2\Delta)$ [Laufer and Okabe (605)].

$$CH_2N_2 \xrightarrow{h\nu} CH(A^2\Delta) + H + N_2 \qquad \text{(VII-154)}$$

The ground state of diazirine is probably $\tilde{X}\,^1A_1$ of C_{2v} symmetry. The enthalpy change of the reaction

$$\text{Cyclic-}CH_2N_2 \rightarrow CH_2(\tilde{X}\,^3B_1) + N_2 \qquad \text{(VII-155)}$$

is 1.38 eV (607).

The absorption spectrum, starting at 3230 Å, is diffuse and its rotational analysis is not possible. Several band progressions have been found. The 0–0 band is at 30,964 cm^{-1} [Merritt (697)].

The primary process in the near ultraviolet photolysis, in analogy with diazomethane photolysis, must be

$$\text{Cyclic-}CH_2N_2 \xrightarrow{h\nu} CH_2(\tilde{a}\,^1A_1) + N_2 \qquad \text{(VII-156)}$$

Diazirine, as well as diazomethane and ketene, has been used as a convenient source of CH_2 radicals. The photolysis of diazirine in the vacuum ultraviolet leads to

$$\text{Cyclic-}CH_2N_2 \xrightarrow{h\nu} CH(A^2\Delta) + H + N_2 \qquad \text{(VII-157)}$$

$$\text{Cyclic-}CH_2N_2 \xrightarrow{h\nu} N_2(A^3\Sigma_u^+) + CH_2(\tilde{X}\,^3B_1) \qquad \text{(VII-158)}$$

By measuring threshold photon energies required to initiate reactions (VII-157) and (VII-158), Laufer and Okabe (607) have obtained the heat of formation of diazirine.

VII–20. KETENE (CH_2CO)

The ground state of CH_2CO is planar $\tilde{X}\,^1A_1$ of C_{2v} symmetry with an H—C—H angle of 122.3°. $D_0(OC—CH_2) = 3.32 \pm 0.05$ eV. A first absorption system is in the region 2600 to 4000 Å with diffuse bands. A second

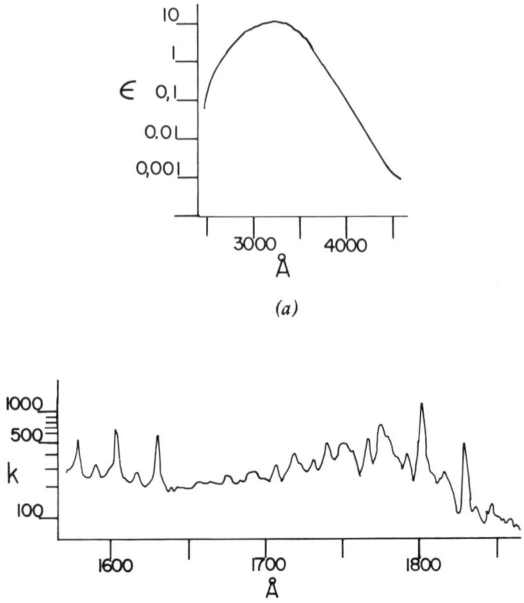

Fig. VII–19. (*a*) Absorption coefficients ε of ketene in the region 2500 to 4500 Å. ε is given in units of l mol^{-1} cm^{-1}, base 10, room temperature. Reprinted with permission from A. H. Laufer and R. A. Keller, *J. Am. Chem. Soc.* 93, 61 (1971). Copyright by the American Chemical Society. (*b*) Absorption coefficients of ketene in the region 1550 to 1850 Å. *k* is in units of atm^{-1} cm^{-1}, base *e*, room temperature. From Braun et al. (143), reprinted by permission. Copyright 1970 by the American Institute of Physics.

region of absorption is 1930 to 2130 Å with diffuse structure. Several sharp-banded systems have been found in the vacuum ultraviolet (16). The absorption coefficients in the region 2500 to 4500 Å have been measured by Laufer and Keller (606) and are shown in Fig. VII–19*a*. The absorption coefficients in the 1550 to 1850 Å region have been measured by Braun et al. (143) and are given in Fig. VII–19*b*.

VII–20.1. Photochemistry

The 2400 to 3700 Å Region. The photochemistry of ketene has been extensively studied for the past 30 years and was reviewed by Noyes et al. (750) in 1956 and by Calvert and Pitts in 1966 (4).

The relevant features of the photochemistry of ketene in the 2400 to 3700 Å region are given by Zabransky and Carr (1078) and Kelley and Hase (561).

Photochemical processes may be summarized as follows:

1. Below 3130 Å the quantum yield of CO is 2 and the ratio CO to C_2H_4 is about 2.2, indicating the following main processes

$$CH_2CO \xrightarrow{h\nu} CH_2 + CO \qquad \text{(VII-159)}$$

$$CH_2 + CH_2CO \rightarrow C_2H_4 + CO \qquad \text{(VII-160)}$$

   The primary quantum yield of dissociation is $\frac{1}{2}\Phi_{CO}$.
2. The primary yield of dissociation at the low-pressure limit is unity at 3130 Å, slightly less than unity, at 3340 Å, and about 0.04 at 3660 Å. The Φ_{CO} values decrease at higher pressures of ketene or added inert gas and the rate of decrease is much larger at longer wavelengths. The dissociation lifetimes are 0.3, 4, and 850 nsec, respectively, at 3130, 3340, and 3660 Å [Porter and Connelly (814)].
3. Fluorescence from ketene upon absorption of light in the near ultraviolet has not been observed. The quantum yield of fluorescence is less than 10^{-5}. Since the radiative lifetime calculated from the integrated absorption coefficient is 40 μsec, the lifetime of the excited state must be less than 0.4 nsec. Since the lifetime of the initially formed excited state is much shorter than the dissociative lifetimes, an excited state responsible for dissociation must be different from the one initially formed at 3340 and 3660 Å.
4. Although there is no direct spectroscopic evidence that CH_2 is formed in the near ultraviolet flash photolysis [Herzberg (464)], the results of the low-intensity photolysis indicate that CH_2 radicals are formed. Since the absorption coefficient of ketene is small, the concentrations of CH_2 formed in Herzberg's experiment would have been too low to be observed. The results of reactions of CH_2 with O_2 and butenes have led to the following conclusions [see Section VI–14.1 for reactions of methylene, p. 258]:

   a. The reaction products due to $^1CH_2(\tilde{a}\,^1A_1)$ with butenes are not suppressed by the addition of O_2, while those due to $^3CH_2(\tilde{X}\,^3B_1)$ are eliminated by O_2. Since the rate constant of the reaction

$$^3CH_2 + O_2 \rightarrow \text{products} \qquad k_{161} \qquad \text{(VII-161)}$$

   is $1.5 \pm 0.1 \times 10^{-12}$ cm^3 molec^{-1} sec^{-1} [Laufer and Bass (608)], rate constants of $^3CH_2$ with butenes must be much smaller than 10^{-12} cm^3 molec^{-1} sec^{-1}. On the other hand, rate constants of $^1CH_2$ with butenes must be comparable to that of $^1CH_2$ with O_2 ($\simeq 3 \times 10^{-11}$ cm^3 molec^{-1} sec^{-1}).

   b. The $^1CH_2$ adds to *cis* 2 butene to form mainly *cis*-1,2-dimethylcyclopropane, while the $^3CH_2$ gives both *cis*- and *trans*-1,2 dimethylcyclopropane. Hence, the product analysis with and without O_2 should be able to indicate the ratio of $^1CH_2$ and $^3CH_2$. The ratios of $^1CH_2$ to $^3CH_2$ produced in the photolysis of ketene are apparently wavelength dependent and the ratio increases (that is, a larger fraction of $^1CH_2$ is produced) as the incident wavelength decreases. At 3660 Å practically all methylenes are the triplet and at 3130 Å they are almost all $^1CH_2$ [Kelley and Hase (561)]. However, the ratios are dependent on the kind of methylene interceptor, reflecting the complex chemistry of reactions. For example, it is not clear how the initial $^1CH_2$ to $^3CH_2$ ratios are affected by ketene and by an added methylene interceptor. It is very likely that some of $^1CH_2$ initially produced is deactivated to $^3CH_2$ by collisions

$$^1CH_2 \xrightarrow{M} {}^3CH_2 \qquad \text{(VII-162)}$$

5. The following scheme has been proposed by Zabransky and Carr (1078) for the near ultraviolet photochemistry of ketene. The $^1A''$ state is formed initially by light absorption in the ultraviolet

$$CH_2CO(\tilde{X}^1A_1) \xrightarrow{h\nu} CH_2CO(^1A'') \qquad \text{(VII-163)}$$

The $^1A''$ state crosses over to another state, probably the ground state, $\tilde{X}^1A_1$, by internal conversion within 0.4 nsec

$$CH_2CO(^1A'') \rightarrow CH_2CO^\dagger(\tilde{X}^1A_1) \qquad \text{(VII-164)}$$

where $CH_2CO^\dagger$ denotes ground state ketene with excess vibrational energy. The fate of the $\tilde{X}^1A_1$ state with excess vibrational energy may be represented by the following three processes; the dissociation into into $^1CH_2 + CO$, the deactivation to the ground state and intersystem crossing to $^3A'$

$$CH_2CO^\dagger(\tilde{X}^1A_1) \rightarrow {}^1CH_2 + CO \qquad \text{(VII-165)}$$
$$CH_2CO^\dagger(\tilde{X}^1A_1) \xrightarrow{M} CH_2CO(\tilde{X}^1A_1) \qquad \text{(VII-166)}$$
$$CH_2CO^\dagger(\tilde{X}^1A_1) \rightarrow CH_2CO(^3A') \qquad \text{(VII-167)}$$

The triplet CH_2CO dissociates into $^3CH_2 + CO$

$$CH_2CO(^3A') \rightarrow {}^3CH_2 + CO \qquad \text{(VII-168)}$$

The observed dependence of the $^1CH_2$ to $^3CH_2$ ratio may be explained on the basis of a less energetically favorable process producing $^1CH_2 + CO$ (VII-165) as the wavelength is increased. The pressure dependence of Φ_{CO} is explained by the two competing processes (VII-165) and (VII-166). At 3660 Å, close to the dissociation limit, only a fraction of molecules dissociate even at the low pressure limit.

The rapid internal conversion process (VII-164) in comparison with fluorescence in ketene may be treated as a case of a so-called statistical limit in large molecules discussed extensively by Bixon and Jortner (110, 111). They conclude that in large molecules such as naphthalene and anthracene, the lifetime of the excited state is governed by the intramolecular relaxation, such as (VII-164), since the level density of $CH_2CO^{\dagger}(\tilde{X}^1A_1)$ at the point of interaction is very large. Once a crossover to the $CH_2CO^{\dagger}(\tilde{X}^1A_1)$ occurs, it will be deactivated to the ground state, since the process of returning to the $^1A''$ and emitting fluorescence would be much slower than the collisional deactivation or intramolecular relaxation processes.

Photochemistry below 2400 Å. The photolysis of ketene at 2139 Å yields CO and C_2H_4 as major products and H_2, C_2H_2, and C_2H_6 as minor products. The quantum yields of CO and C_2H_4 are 2 and 0.8, respectively [Kistiakowsky and Walter (572)].

They suggest the following processes:

$$CH_2CO \xrightarrow{h\nu} CH_2 + CO \quad \text{(VII-168a)}$$

$$CH_2 + CH_2CO \rightarrow C_2H_4^* + CO \quad \text{(VII-168b)}$$

$$C_2H_4^* \rightarrow C_2H_2 + H_2 \quad \text{(VII-168c)}$$

$$C_2H_4^* \xrightarrow{M} C_2H_4 \quad \text{(VII-168d)}$$

$$CH_2 + CH_2CO \rightarrow CH_3 + CHCO \quad \text{(VII-168e)}$$

$$2CH_3 \rightarrow C_2H_6 \quad \text{(VII-168f)}$$

$$2CHCO \rightarrow C_2H_2 + 2CO \quad \text{(VII-168g)}$$

$$CH_3 + CHCO \rightarrow C_2H_4 + CO \quad \text{(VII-168h)}$$

where $C_2H_4^*$ denotes vibrationally excited C_2H_4 and M represents a third body.

The results of the photolysis of ketene and *n*-butane mixtures suggest that the ratio of the singlet to the triplet methylene is 7 to 3 on the assumption that the singlet CH_2 only inserts into the C—H bond of butane while the triplet CH_2 only abstracts hydrogen from butane to form CH_3.

Laufer (604) has measured the isotopic composition of hydrogen formed (to an extent of several percent of CO) in the vacuum ultraviolet photolysis of 1:1 mixtures of CH_2CO and CD_2CO. He found the ratios $H_2/HD/D_2 = 54:13:33$ at 1470 Å, suggesting that the hydrogen-forming process is

$$CH_2CO \rightarrow H_2 + C_2O \quad \text{(VII-168i)}$$

rather than the decomposition of $C_2H_4^*$ proposed by Kistiakowsky and Walter.

Pilling and Robertson (809b) have recently measured end products CH_4, C_2H_2, C_2H_4, and C_2H_6 in the flash photolysis of CH_2CO–H_2 mixtures above 1600 Å in the presence of inert gases. The suggested mechanism is

$$CH_2CO \xrightarrow{h\nu} {}^1CH_2 + CO$$
$${}^1CH_2 \xrightarrow{M} {}^3CH_2$$
$${}^1CH_2 + H_2 \rightarrow CH_3 + H$$
$${}^3CH_2 + {}^3CH_2 \rightarrow C_2H_2 + H_2$$
$$CH_3 + CH_3 \rightarrow C_2H_6$$
$$CH_3 + H \xrightarrow{M} CH_4$$
$$H + CH_2CO \rightarrow CH_3 + CO$$
$${}^3CH_2 + CH_3 \rightarrow C_2H_4 + H$$
$${}^1CH_2 + CH_2CO \rightarrow C_2H_4 + CO$$

where ${}^1CH_2$ and ${}^3CH_2$ are ${}^1A_1$ and ${}^3B_1$ methylenes, respectively. The products from ${}^1CH_2$ are CH_4, C_2H_6, and C_2H_4, and a triplet methylene product is C_2H_2. The yields of CH_4 and C_2H_6, the singlet methylene products, decrease at first but remain constant even above 600 torr of total pressure and that of C_2H_2 is insensitive to a change in total pressure. The results are not in agreement with the suggested mechanism. Pilling and Robertson have proposed that the two kinds of ${}^1CH_2$, ${}^1B_1$ and ${}^1A_1$, are produced in the primary process in a ratio of 1:1, and ${}^1B_1$, 0.88 eV above ${}^1A_1$, is not quenched by inert gases. The ${}^1B_1$ methylene reacts with H_2 to form $CH_3 + H$, leading to the formation of CH_4 and C_2H_6 observed at high total pressures.

VII–21. FORMIC ACID (HCOOH)

The ground state is planar $\tilde{X}\,{}^1A'$ of C_s symmetry (H—C(=O)—O′—H′) with O—C—O′, H—C—O, and C—O′—H′ angles of 124.9, 124.1, and 106.3°, respectively (16, 737). D_0(HO—CHO) = 4.50 ± 0.1 eV

D_0(H—COOH) = 4.0 ± 0.2 eV; D_0(HCOO—H) = 4.6 ± 0.2 eV. The first absorption system in the 2250 to 2600 Å region is diffuse merging into continuous absorption below 2250 Å. The excited state is nonplanar [Ng and Bell (737)].

Below 1800 Å several Rydberg transitions have been found leading to an ionization potential of 11.329 ± 0.002 eV [Bell et al. (93)]. The absorption coefficients in the region 2000 to 2500 Å have been measured by McMillan [quoted in (4), p. 428].

The photolysis products in the near ultraviolet are H_2, CO, and CO_2. The addition of C_2H_4 or O_2 reduces the H_2 yield to 16% of that without the scavenger but the CO and CO_2 yields are only slightly reduced.

The primary processes are predominantly those forming radicals and (VII-170) may be the most likely process on energetic ground

$$HCOOH \xrightarrow{h\nu} OH + HCO \qquad \text{(VII-169)}$$

$$HCOOH \xrightarrow{h\nu} H + COOH \qquad \text{(VII-170)}$$

$$HCOOH \xrightarrow{h\nu} HCOO + H \qquad \text{(VII-171)}$$

The molecular elimination process

$$HCOOH \xrightarrow{h\nu} CO_2 + H_2 \qquad \text{(VII-172)}$$

would account for less than 5% of the primary processes [Gorden and Ausloos (412), Yankwich and Steigelmann (1064)].

The vacuum ultraviolet photolysis of formic acid yielded blue fluorescence, although the emitter was not identified [Style and Ward (953)].

VII–22. CYANOACETYLENE (C_2HCN)

The ground state of C_2HCN is linear $X^1\Sigma^+$. $D_0(NC—C_2H) = 6.21 \pm 0.04$ eV. Two absorption spectra have been identified in the near ultraviolet, one in the 2300 to 2715 Å region and the other in the 2100 to 2300 Å region. They correspond probably to the forbidden transitions $^1\Sigma^-–^1\Sigma^+$ or $^1\Delta–^1\Sigma^+$ [Connors et al. (233)]. The absorption spectra in the 1100 to 1650 Å region consist of many sharp bands. They are assigned to $^1\Pi–^1\Sigma^+$ (1610 Å), $^1\Sigma^+–^1\Sigma^+$ (1450 Å) and two Rydberg series leading to an ionization potential of 11.60 eV. The absorption coefficients in the region 1050 to 1650 Å have been measured by Connors et al. (233) and are shown in Fig. VII–20.

The photolysis of C_2HCN in the vacuum ultraviolet forms the $CN(B^2\Sigma)$ [Okabe and Dibeler (771)]

$$C_2HCN \xrightarrow{h\nu} C_2H + CN(B^2\Sigma) \qquad \text{(VII-173)}$$

The fluorescence excitation spectra closely follow the Rydberg bands, indicating that the dissociation occurs from the Rydberg states. The fluorescence yield is only 1% at 1216 Å.

Cyanoacetylene has recently been detected in outer space by Turner (982).

VII–23. NITRIC ACID (HNO_3)

The ground state structure is planar $\left(H—O'—\overset{\overset{\displaystyle O}{|}}{N}—O\right)$ with H—O′—N, O′—N—O, and O—N—O angles of 90, 115, and 130°, respectively (28). $D_0(HO—NO_2) = 2.07 \pm 0.02$ eV, $D_0(H—ONO_2) = 4.3 \pm 0.3$ eV, $D_0(HONO—O) = 3.11 \pm 0.01$ eV. The absorption coefficients in the ultra-

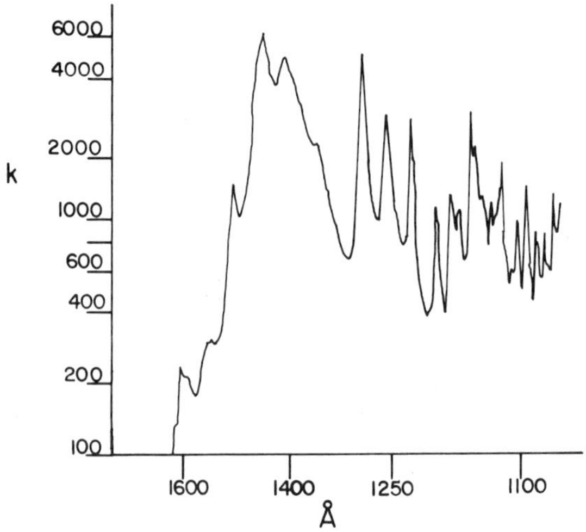

Fig. VII–20. Absorption coefficients of cyanoacetylene in the 1050 to 1650 Å region. k is in units of atm^{-1} cm^{-1}, base e, 0°C. From Connors et al. (233), reprinted by permission. Copyright 1974 by the American Institute of Physics.

violet (1800 to 3350 Å) have been measured by Johnston and Graham (539) and by Biaume (107) and are shown in Fig. VII–21*a*. The absorption is continuous and shows at least two different transitions.

The vacuum ultraviolet absorption coefficients have been measured by Beddard et al. (90) and are shown in Fig. VII–21*b*.

VII–23.1. Photodissociation

The photolysis of HNO_3 in the near ultraviolet has been studied by Berces et al. (96–98) and more recently by Johnston et al. (540).

The primary process appears mainly to be

$$HNO_3 \xrightarrow{h\nu} OH + NO_2 \qquad \text{(VII-174)}$$

with a quantum yield of unity (540) in the region 2000 to 3000 Å. This conclusion is based on the results that the quantum yield of NO_2 in the photolysis of HNO_3 and excess CO and O_2 mixtures is unity. Under these conditions OH radicals produced from (VII-174) react with CO to form CO_2 and H atoms

$$OH + CO \rightarrow CO_2 + H \qquad \text{(VII-175)}$$

and H atoms react with O_2 to form HO_2 radicals which recombine to form H_2O_2 and O_2

$$H + O_2 \xrightarrow{M} HO_2 \rightarrow \tfrac{1}{2}H_2O_2 + \tfrac{1}{2}O_2 \qquad \text{(VII-176)}$$

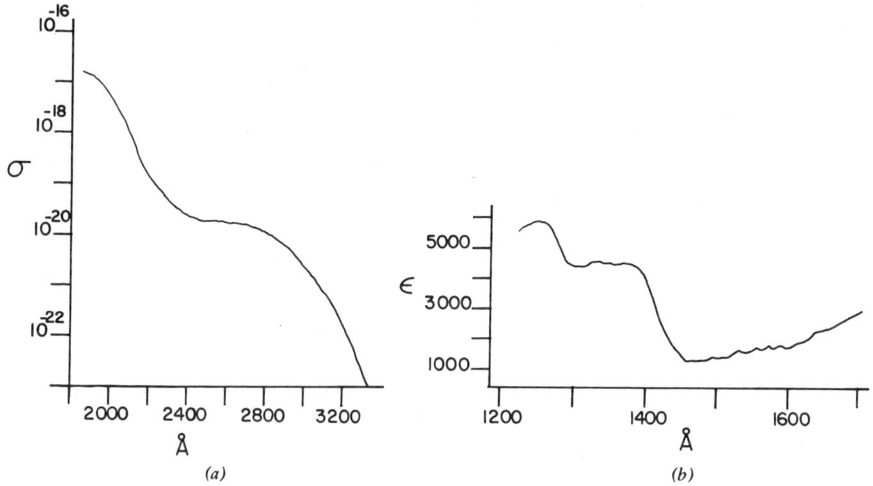

Fig. VII–21. (*a*) Absorption cross sections of HNO_3 in the region 1850 to 3350 Å. σ is in units of cm^2 $molec^{-1}$, base *e*, room temperature. From Biaume (107) and Johnston and Graham (539), reprinted by permission. Copyright 1973 by the American Institute of Physics. (*b*) Absorption coefficients of nitric acid in the region 1200 to 1700 Å. ϵ is in units of l mol^{-1} cm^{-1}, base 10, room temperature. From Beddard et al. (90), reprinted by permission of Elsevier Sequoia, S. A.

If the extent of photolysis is kept at less than 1% the photolysis of the product NO_2 may be neglected. Hence, from (VII-174) through (VII-176) the primary quantum yield of dissociation must be equal to $\Phi_{NO_2} = 1$.

The low quantum yields of NO_2 (~ 0.1) observed by Berces et al. in the photolysis of HNO_3 have been explained by Johnston et al. (540) on the basis of the following reactions

$$OH + HNO_3 \rightarrow H_2O + NO_3 \qquad k_{177} \qquad \text{(VII-177)}$$

$$NO_2 + NO_3 \overset{M}{\rightarrow} N_2O_5 \qquad \text{(VII-178)}$$

$$N_2O_5 + H_2O \overset{W}{\rightarrow} 2HNO_3 \qquad \text{(VII-179)}$$

where k_{177} is 1.5×10^{-13} cm^3 $molec^{-1}$ sec^{-1} (10), and *W* signifies a wall reaction. Another complication is the photolysis of NO_2 producing NO which reacts heterogeneously with nitric acid.

In the vacuum ultraviolet the fluorescence due to $OH(A^2\Sigma^+)$ radicals has been observed below the incident wavelength 1500 Å [Okabe, unpublished results]

$$HNO_3 \overset{h\nu}{\rightarrow} OH(A^2\Sigma^+) + NO_2 \qquad \text{(VII-180)}$$

Nitric Acid in the Atmosphere. Nitric acid is present in the stratosphere at a maximum concentration of about 2×10^{10} molec cm^{-3} at an altitude of 20 km [Williams et al. (1045)].

It is probably formed from the reaction

$$OH + NO_2 \xrightarrow{M} HNO_3 \qquad (VII\text{-}181)$$

where M signifies a third body. It may also be formed by N_2O_5 reacting with water in aqueous sulfuric acid droplets

$$N_2O_5 + H_2O \rightarrow 2HNO_3 \qquad (VII\text{-}182)$$

VII–24. CYANOGEN AZIDE (N_3CN)

The ground state structure is not known but probably is planar in analogy with hydrazoic acid and chlorine azide. $D_0(N_2\text{—}NCN) = 0.3 \pm 0.1$ eV, $D_0(N_3\text{—}CN) = 4.0 \pm 0.2$ eV. The absorption spectra of N_3CN in cyclohexane in the near ultraviolet have been observed in two regions with maxima at 2750 and 2200 Å, the latter about 20 times as strong as the former [Marsh and Hermes (662)].

The absorption coefficients in the vacuum ultraviolet have been measured by Okabe and Mele (764), and are shown in Fig. VII–22. The flash photolysis

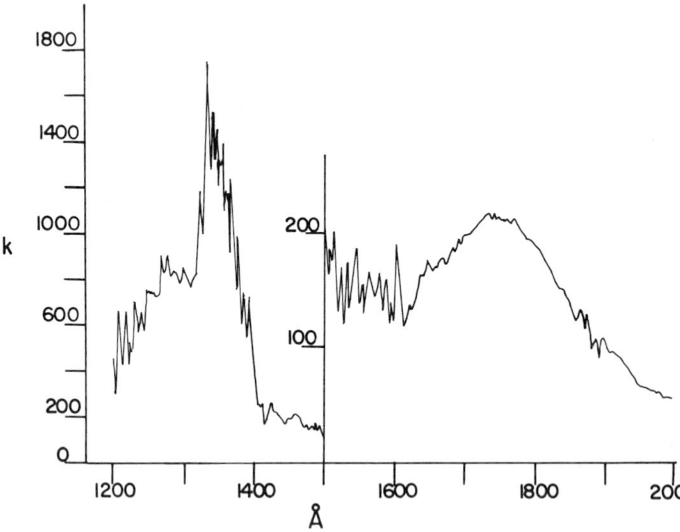

Fig. VII–22. The absorption coefficients of cyanogen azide in the vacuum ultraviolet. *k* is in units of atm^{-1} cm^{-1}, base *e*, 25°C. From Okabe and Mele (764), reprinted by permission. Copyright 1969 by the American Institute of Physics.

of N_3CN has first been studied by Pontrelli and Anastassiou (812). Kroto (588, 589) has observed transient absorption spectra of $N_3(^2\Sigma^+ \leftarrow {}^2\Pi)$, $CN(^2\Sigma \leftarrow {}^2\Sigma)$, $NCN(^3\Pi \leftarrow {}^3\Sigma_g^-)$, and $NCN(^1\Pi_u \leftarrow {}^1\Delta_g)$ in the near ultraviolet flash photolysis of N_3CN. Since the absorption bands of $NCN(^1\Delta_g)$ appear immediately after the flash and those of $NCN(X^3\Sigma_g^-)$ grow in intensity as the singlet NCN bands decay with time, Kroto concludes that a major primary process must be

$$N_3CN \xrightarrow{h\nu} N_2 + NCN(^1\Delta_g) \qquad \text{(VII-183)}$$

followed by collisional deactivation of $NCN(^1\Delta_g)$ to the ground state

$$NCN(^1\Delta_g) \xrightarrow{M} NCN(X^3\Sigma_g^-) \qquad \text{(VII-184)}$$

Another primary process

$$N_3CN \xrightarrow{h\nu} CN + N_3 \qquad \text{(VII-185)}$$

appears unimportant, since the absorption of $N_3(^2\Sigma \leftarrow {}^2\Pi)$ at 2719 Å is weak [Kroto et al. (590)].

In the low intensity photolysis of N_3CN in the vacuum ultraviolet Okabe and Mele (764) have found emissions originating from $CN(B^2\Sigma)$ and $NCN(A^3\Pi)$ indicating a primary process

$$N_3CN \xrightarrow{h\nu} N_3 + CN(B^2\Sigma) \qquad \text{(VII-186)}$$

The threshold energy of incident photons to produce $NCN(A^3\Pi)$ is only 6.5 eV which is insufficient to induce the spin-allowed process

$$N_3CN \xrightarrow{h\nu} NCN(A^3\Pi) + N_2(A^3\Sigma) \qquad \text{(VII-187)}$$

Okabe and Mele concluded that the most likely mechanism for the production of $NCN(A^3\Pi)$ must be the initial production of $N_2(A^3\Sigma)$

$$N_3CN \xrightarrow{h\nu} NCN(X^3\Sigma) + N_2(A^3\Sigma) \qquad \text{(VII-188)}$$

followed by a sensitized reaction by $N_2(A^3\Sigma)$ to produce $NCN(A^3\Pi)$

$$N_2(A^3\Sigma) + N_3CN \rightarrow NCN(A^3\Pi) + 2N_2 \qquad \text{(VII-189)}$$

VII–25. CARBON SUBOXIDE (C_3O_2)

The ground state of C_3O_2 is linear $(X^1\Sigma_g^-)$. $D_0(OC_2\text{—}CO) = 3.3 \pm 0.7$ eV. The main uncertainty in the bond energy is the heat of formation of C_2O [67 to 93 kcal mol^{-1}, see Willis and Bayes (1047)].

The near ultraviolet absorption spectrum consists of a weak continuum in the 2400 to 3300 Å region with a maximum around 2700 Å. A second absorption starts at 1900 Å and is very intense. The absorption coefficients in the region 2000 to 3000 Å have been measured by Bayes (74) and in the region 1100 to 1900 Å by Roebber et al. (566, 838, 839). They are shown in Figs. VII-23*a* and VII–23*b*. The required minimum energy for the production $2CO + C(^3P)$ is 6.00 ± 0.01 eV, corresponding to the incident wavelength 2066 Å.

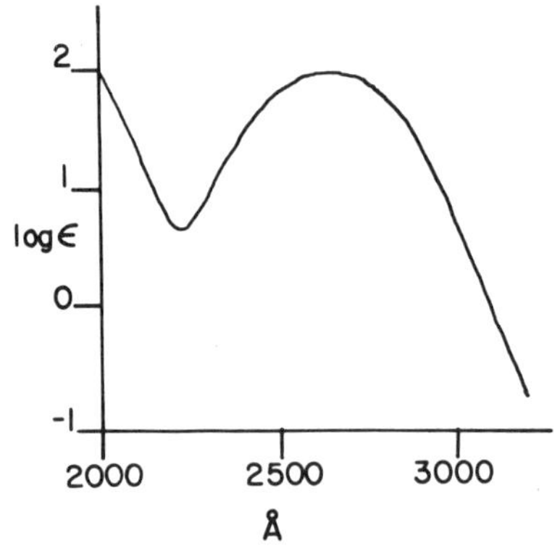

(a)

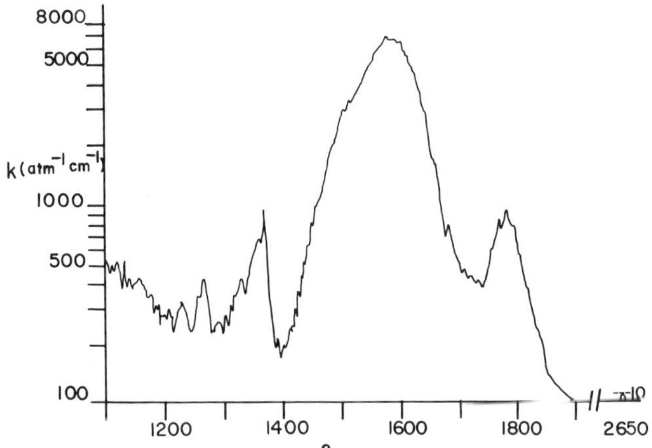

(b)

Fig. VII–23. (*a*) Absorption coefficients of carbon suboxide in the region 2000 to 3000 Å. ∈ is given in units of l mol^{-1} cm^{-1}, base 10, room temperature. Reprinted with permission from Bayes, *J. Am. Chem. Soc.*, 84, 4077 (1962). Copyright by the American Chemical Society. (*b*) Absorption coefficients of carbon suboxide in the region 1100 to 2700 Å. *k* is given in atm^{-1} cm^{-1}, base *e*, 0°C. See Ref. 838 for more extended and improved *k* values. From Kim and Roebber (566), reprinted by permission. Copyright 1966 by the American Institute of Physics.

VII–25.1. Photolysis of C_3O_2 in the Near Ultraviolet

The photolysis of carbon suboxide in the near ultraviolet has been studied by many workers. The formation of $C(^3P)$ atoms is not energetically possible above the incident wavelength, 2066 Å. Hence, the primary process must be

$$C_3O_2 \xrightarrow{h\nu} CO + C_2O \qquad \text{(VII-190)}$$

Devillers (284) and Devillers and Ramsay (285) have found a transient absorption spectrum in the 5000 to 9000 Å region in the near ultraviolet flash photolysis of C_3O_2. They have conclusively assigned the spectrum to the transition $A^3\Pi$–$X^3\Sigma^-$ of C_2O.

The infrared absorption of C_2O has been observed in the vacuum ultraviolet photolysis of matrix-isolated C_3O_2 at 4°K [Jacox et al. (520)]. On the other hand, no absorption due to a singlet C_2O has been reported.

The C_2O produced reacts with C_3O_2 to yield polymer and CO [Bayes (74), Forchioni and Willis (368)]

$$C_2O + C_3O_2 \rightarrow \text{polymer} + n\text{CO} \qquad (n = 1\text{–}3) \qquad \text{(VII-191)}$$

In the presence of ethylene C_3H_4 (methylacetylene and allene) and CO are produced.

$$C_2O + C_2H_4 \rightarrow C_3H_4 + CO \qquad \text{(VII-192)}$$

A major product, allene, is effectively eliminated by the addition of O_2 when C_3O_2 and ethylene mixtures are irradiated with light of wavelengths above 2900 Å, while at shorter wavelengths O_2 has little effect on the production of allene. From the results Bayes (76) has suggested the production of two kinds of C_2O radicals, a singlet and a triplet. The triplet $C_2O(X^3\Sigma)$ radicals are formed at longer wavelengths (3000 Å) and react with O_2 135 times as fast as with ethylene, while the singlet C_2O formed at 2500 Å is less reactive with O_2 than with C_2H_4.

The reactions of the triplet C_2O with O_2, NO (1045a), and olefins (1047) and those of the singlet C_2O with H_2 (368), fluoroethylenes (57), alkanes (57), and olefins (56) have been studied. In general the reactivities of the singlet with olefins are all comparable within a factor of 3, while those of the triplet with olefins are widely different [Williamson and Bayes (1046)]. See Table VI–9 for the reaction rates of singlet and triplet C_2O.

VII–25.2. Photolysis of C_3O_2 in the Vacuum Ultraviolet

The primary processes energetically possible below 2000 Å are, in addition to (VIII-190),

$$C_3O_2 \xrightarrow{h\nu} 2CO + C(^3P) \qquad < 2070\ \text{Å} \qquad \text{(VII-193)}$$

$$C_3O_2 \xrightarrow{h\nu} 2CO + C(^1D) \qquad < 1710\ \text{Å} \qquad \text{(VII-194)}$$

$$C_3O_2 \xrightarrow{h\nu} 2CO + C(^1S) \qquad < 1430\ \text{Å} \qquad \text{(VII-195)}$$

Braun et al. (141) have detected $C(^1S, ^1D, ^3P)$ and CO in the flash photolysis of C_3O_2 near 1600 Å.

The production of C_2O in the primary process appears unimportant, since two CO molecules were found for each C_3O_2 destroyed and since no absorption due to the $C_2O(X^3\Sigma^-)$ radical was found, indicating

$$C_3O_2 \xrightarrow{h\nu} 2CO + C \qquad \text{(VII-196)}$$

The production of C_2O in the primary process appears to be not more than 25%. The relative concentrations of $C(^3P)$, $C(^1D)$, and $C(^1S)$ are $4:1:<0.1$. Hence, (VII-193) appears to be a major primary process in contradiction to the spin conservation rules. The $C(^1S)$ may be formed by the photolysis of C_2O, since its concentration does not increase linearly with the flash intensity but shows a higher order dependence on the intensity. More detailed description of process (VII-196) may be the following. By absorption of light near 1600 Å the C_3O_2 molecule is excited to a Rydberg state (838)

$$C_3O_2 \xrightarrow{h\nu} C_3O_2^* \qquad \text{(VII-197)}$$

where $C_3O_2^*$ signifies the Rydberg state. Let us assume that the molecule $C_3O_2^*$ initially formed in the Rydberg state with an average energy of 7.85 eV internally converts to the ground state with high vibrational energy $C_3O_2^\dagger$ from which it dissociates into $C_2O + CO$ only after many vibrations so that statistical equilibrium is established among all available vibrational degrees of freedom in the molecule prior to dissociation. Then the available energy for fragments after dissociation is roughly the difference between the photon energy (7.85 eV) and the bond energy $D_0(OC_2—CO) = 3.3$ eV, that is, 4.55 eV. The fraction of the available energy that the fragment C_2O can carry as vibration is 4.55 eV times the ratio of the number of vibrational degrees of freedom for C_2O to C_3O_2. This ratio is 0.4 assuming both $C_3O_2^\dagger$ and C_2O are linear molecules. Hence, C_2O has only 1.8 eV in vibrational energy on the basis of a simple statistical model that all vibrational degrees of freedom are in equilibrium on the entire potential energy surface. Since the bond energy $D_0(C—CO)$ is about 2.2 eV, it is highly unlikely that all C_2O dissociates into $C(^3P) + CO$, which is found experimentally. Onc possible explanation of the results is the two-step photolysis, that is, process (VII-197) followed by

$$C_3O_2^* \rightarrow C_2O\ (\text{singlet}) + CO(X^1\Sigma) \qquad \text{(VII-198a)}$$

$$C_2O\ (\text{singlet}) \xrightarrow{h\nu} C(^3P) + CO \qquad \text{(VII-198b)}$$

Since the excited C_2O formed by light absorption is most likely a singlet it must cross over to a triplet repulsive C_2O for (VII-198b) to occur. The occurrence of a two-photon process is found by Herzberg and Johns (467) in the near ultraviolet flash photolysis of CH_2N_2. The absorption spectra due to CH are found with photon energies insufficient to break CH_2N_2 into $CH + H + N_2$ (6.17 eV is required for the process). Hence, the production of CH must be

$$CH_2N_2 \xrightarrow{h\nu} CH_2 + N_2$$
$$CH_2 \xrightarrow{h\nu} CH + H$$

The reactions of C atoms with C_3O_2 may produce C_2 and CO (978)

$$C(^3P) + C_3O_2 \rightarrow C_2 + 2CO \qquad \text{(VII-199)}$$
$$C(^1D) + C_3O_2 \rightarrow C_2 + 2CO \qquad \text{(VII-200)}$$

The C_2 molecules may then react further with C_3O_2

$$C_2 + C_3O_2 \rightarrow \text{polymer}$$

In the photolysis of C_3O_2–CH_4 mixtures at 1470 Å, CO, C_2H_2, and C_2H_4 are main products [Tschuikow-Roux et al. (978), Stief and DeCarlo (931)].

The results are explained by the following processes

$$C_3O_2 \xrightarrow{h\nu} C + 2CO \qquad \text{(VII-201)}$$
$$C + CH_4 \rightarrow C_2H_4^{\dagger} \qquad \text{(VII-202)}$$
$$C_2H_4^{\dagger} \xrightarrow{M} C_2H_4 \qquad \text{(VII-203)}$$
$$C_2H_4^{\dagger} \rightarrow C_2H_2 + H_2 \qquad \text{(VII-204)}$$

where $C_2H_4^{\dagger}$ signifies internally excited C_2H_4 and M is a third body.

The photolysis of C_3O_2 in the presence of CH_3F at 1470 Å has been studied by Tschuikow-Roux and Kodama (977).

VII–26. CHLORINE NITRATE ($ClONO_2$)

The ground state is either a planar [Miller et al. (705)] or a nonplanar structure with the ClO group perpendicular to the NO_2 group [Shamir et al. (867)]. $D_0(ClO—NO_2) = 1.15 \pm 0.01$ eV, $D_0(Cl—ONO_2) = 1.8 \pm 0.2$ eV. The absorption coefficients in the ultraviolet have been measured by Rowland et al. (844) and are shown in Fig. VII–24.

Chlorine nitrate may be formed in the stratosphere by the combination of ClO and NO_2

$$ClO + NO_2 \xrightarrow{M} ClONO_2 \qquad \text{(VII-205)}$$

where M signifies a third body. When $M = N_2$ the rate constant is 1.5×10^{-31} cm^6 molec^{-2} sec^{-1} at 298°K (844).

The primary process has not been well established. In analogy with the photolysis of HNO_3 and on the basis of the bond energy, the most likely

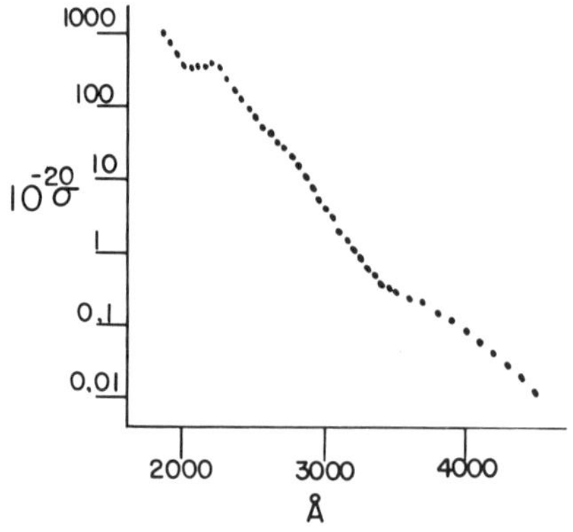

Fig. VII–24. Absorption cross sections of chlorine nitrate in the near ultraviolet; σ is in units of cm^2 $molec^{-1}$, base *e*, room temperature. Reprinted with permission from F. S. Rowland, J. E. Spencer, and M. J. Molina, *J. Phys. Chem.* 80, 2711 (1976). Copyright by the American Chemical Society.

process in the near ultraviolet photolysis is

$$ClONO_2 \xrightarrow{h\nu} ClO + NO_2 \qquad \text{(VII-206)}$$

Very recent results by Smith et al. (916a) indicate that the main primary process at 3025 Å is $ClONO_2 \xrightarrow{h\nu} ClONO + O$ rather than (VII-206).

The reaction of $O(^3P)$ atoms with $ClONO_2$ has recently been studied by Molina et al. (712):

$$O + ClONO_2 \rightarrow ClO + NO_3 \qquad \Delta H = -23 \text{ kcal mol}^{-1} \qquad \text{(VII-207)}$$

The rate constant is $3.4 \times 10^{-12} \exp(-840/T)$ cm^3 $molec^{-1}$ sec^{-1}.

chapter VIII

Various Topics Related to Photochemistry

Three topics related to photochemistry are treated in this chapter. Isotopic enrichment takes advantage of the monochromatic nature of a light source in exact coincidence with an absorption line of a desired isotopic species in mixtures of other species. The recent advancement of tunable lasers in the visible and ultraviolet regions has extended the possibility of isotopic enrichment not only in the atomic system, but also in the molecular system.

Photochemical air pollution of the earth's troposphere and stratosphere involves a series of complex reactions initiated by sunlight. Thanks to the large body of information accumulated in recent years, the main processes leading to the formation of photochemical smog are well understood, although the details of some reactions are still unknown.

The recent space probes have stimulated the laboratory experiments on the photochemistry of the constituent gases present in the atmospheres of Mars, Venus, and Jupiter. Based on these experiments and on the results of recent space probes a number of atmospheric models have been presented.

VIII-1. ISOTOPE ENRICHMENT

If the isotopic shift of a spectral line in an atom or in a molecule is more than the Doppler width, it is in principle possible to selectively excite a particular isotopic species from isotopic mixtures by monochromatic light of wavelength in coincidence with the absorption of the particular isotopic species. In a typical example, ^{202}Hg atoms in natural Hg vapor containing 204, 202, 201, 200, 199, and 198 isotopes are preferentially excited by the 2537 Å resonance line of ^{202}Hg atoms. It has recently been demonstrated that ^{235}U atoms are enriched in the photoionization processes of ^{235}U and ^{238}U mixtures by using tunable lasers. The isotopic enrichment of the carbon monoxide and hydrogen products has been demonstrated in the near ultraviolet photolysis of isotopic mixtures of formaldehyde using a tunable dye laser. Following the selective excitation of the desired isotopic species, the electronically excited isotopic species must be removed from the system by physical or chemical means. The efficiency of this second step is important in determining the final isotope enrichment yield. In many cases the efficiency is considerably less than unity.

Other isotopic species, such as boron, carbon, sulfur, and silicon, have been found to be enriched by illumination of isotopic mixtures of respective polyatomic molecules with intense focused CO_2 laser pulses at 10.6 μm (375, 438, 655).

VIII–1.1. The Atomic System

Enrichment of ^{202}Hg *Atoms.* As is discussed in Section I–6.2, under certain conditions the emission line profile of the Hg 2537 Å lamp may be represented by that of the Doppler line modified by self-absorption. Figure I–9 shows such a line profile. The width of the line is about 0.08 cm^{-1}. Figure VIII–1 shows the emission intensities of hyperfine components in natural mercury at 2537 Å. The separations of the ^{202}Hg component from adjacent ones, that is, from ^{199}Hg, ^{200}Hg, ^{201}Hg, and ^{204}Hg are more than 0.1 cm^{-1}, which is larger than the line width of ^{202}Hg. Hence, in principle, it is possible to excite only the ^{202}Hg isotope in natural mercury provided a lamp containing only the ^{202}Hg isotope is used as a light source.

Gunning and his coworkers (429a) have in fact succeeded in enriching the reaction product of ^{202}Hg preferentially excited with a lamp containing ^{202}Hg operated in a microwave discharge. Upon illumination of mixtures of natural mercury and HCl by the ^{202}Hg lamp, the maximum fractional isotopic abundance of ^{202}HgCl (calomel) obtained is 0.45 with intermittent

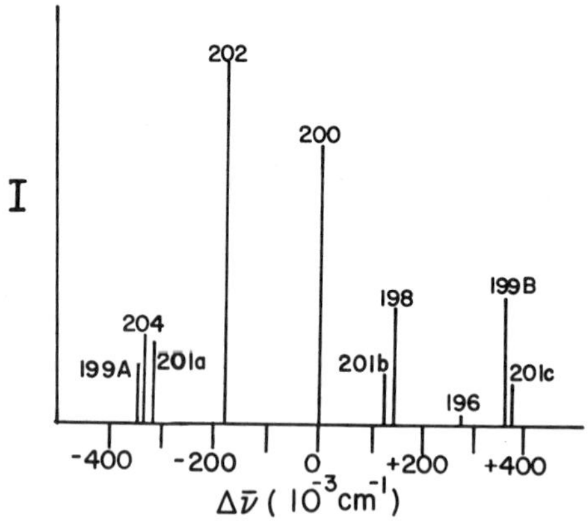

Fig. VIII–1. Relative emission intensities *I*, of hyperfine components of the 2537 Å line of natural mercury. There are 10 hyperfine lines, 5 from even mass isotopes, 2 from ^{199}Hg, and 3 from ^{201}Hg. From Gunning and Strausz (429a), reprinted by permission of John Wiley & Sons.

illumination in comparison with an original abundance of 0.296. The sequence of reactions to give this yield is

$$Hg \xrightarrow{h\nu} {}^{202}Hg^* \tag{VIII-1}$$

$$^{202}Hg^* + HCl \rightarrow {}^{202}HgCl + H \tag{VIII-2a}$$

$$^{202}Hg^* + HCl \rightarrow {}^{202}Hg + Cl + H \tag{VIII-2b}$$

$$Cl + Hg + M \rightarrow HgCl + M \tag{VIII-3}$$

$$H + HCl \rightarrow H_2 + Cl \tag{VIII-4}$$

$$2H + M \rightarrow H_2 + M \tag{VIII-5}$$

Reaction (VIII-2b) explains why the quantum yield of isotopically specific $^{202}HgCl$ formation is much less than unity, although process (VIII-1), the selective production of $^{202}Hg^*$ from the isotopic mixtures, is near unity. Reactions (VIII-3) and (VIII-4) increase the nonspecific formation of HgCl and consequently decrease enrichment. The addition of butadience is found to increase enrichment by scavenging H atoms

$$H + C_4H_6 \rightarrow C_4H_7 \rightarrow \text{polymer} \tag{VIII-6}$$

Isotopic enrichment has also been found by monoisotopic photosensitization for mixtures of natural mercury and alkyl chlorides and vinyl chloride by similar processes. Isotopic enrichment is dependent on such factors as lamp temperatures, flow rates, and substrate pressures. Enrichment increases with decreasing lamp temperature and increasing flow rate, since process (VIII-1) is more efficient at low temperatures and Cl atoms react with natural mercury containing higher fractions of ^{202}Hg in (VIII-3) at higher flow rates of HCl or under intermittent illumination. The intermittent illumination results in higher enrichment than the steady illumination.

Enrichment of ^{235}U *Atoms.* As much as 50% enrichment of ^{235}U (1:1 for $^{235}U/^{238}U$) out of the initial 1:140 $^{235}U/^{238}U$ mixtures has recently been achieved by Janes et al. (527) using a two-photon ionization process of uranium by lasers.

The process involves two steps: an atomic beam of uranium vapor produced by electron-beam evaporation is excited by light of wavelength 4266.266 Å from a pulsed laser and is subsequently ionized by light of wavelength 3609 Å from a second laser. The ions produced are detected by a mass spectrometer. To achieve selectivity for the excitation process the incident wavelength must coincide exactly with one of many absorption lines of ^{235}U atoms. The isotope shift of the absorption lines between ^{235}U and ^{238}U near 4266 Å is about 0.06 Å or about 0.32 cm^{-1}. Hence, the width of the laser line must be less than 0.32 cm^{-1}.

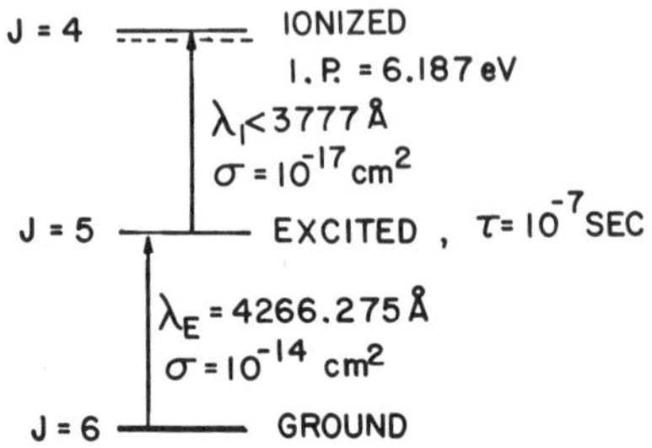

Fig. VIII–2. Energy level diagram for ^{235}U atoms. The first laser pulse at $\lambda_E = 4266.275$ Å preferentially excites ^{235}U atoms, which are subsequently ionized within 10^{-7} sec by the second laser pulse at $\lambda_I < 3777$ Å. σ signifies absorption cross section. From Janes et al. (527), reprinted by permission of the Institute of Electrical and Electronics Engineers, Inc.

Figure VIII–2 shows the principle of isotope enrichment by two-photon ionization of ^{235}U atoms. The excitation wavelength is 4266.275 ± 0.02 Å. A band width of 0.1 cm^{-1} is much narrower than an isotope shift of 0.32 cm^{-1}. Since the preferentially excited ^{235}U atoms decay in 10^{-7} sec the second laser source to ionize the excited atoms must be pulsed within 10^{-7} sec. The wavelength of the second laser must be shorter than 3777 Å, as the combined photon energy must exceed the ionization potential, 6.187 eV, of U atoms. If the first laser is set at 4266.325 Å in coincidence with an absorption line of ^{238}U atoms, an isotopic yield ratio of 3000:1 for $^{238}U/^{235}U$ is obtained in comparison with 140:1 for the same ratio in the starting material.

VIII–1.2. The Molecular System

Isotopic enrichment in the molecular system can in principle be achieved by a two-step process, namely, the selective excitation of a specific isotopic species by monochromatic light and the removal of the specific isotopic product from other isotopic species by physical or chemical means.

As in the atomic system the first step involves a transition to a discrete upper state by light absorption. The wavelength of exciting light is chosen to coincide with an absorption band of a specific isotope species. Absorption bands of other isotope species must be sufficiently separated from the exciting line so that only a desired isotopic species is excited.

The ultraviolet absorption spectrum of formaldehyde consists of many sharp discrete bands of Doppler width. The isotopic shifts due to C and O atoms are sometimes 5 to 10 cm^{-1} in the 3000 to 3100 Å region [see Moore (715)]. Hence, it is possible to selectively excite a specific carbon or oxygen isotopic species in mixtures of other isotopic species.

Clark et al. (217) have found that the photolysis of $H_2\,^{12}CO$–$H_2\,^{13}CO$ mixtures near 3032 Å has produced more ^{12}CO than was present in the original mixtures.

Enrichment of ^{12}CO can be explained on the basis of Fig. VIII–3, which shows the fluorescence excitation spectra of $H_2{}^{12}CO$ and $H_2{}^{13}CO$ near

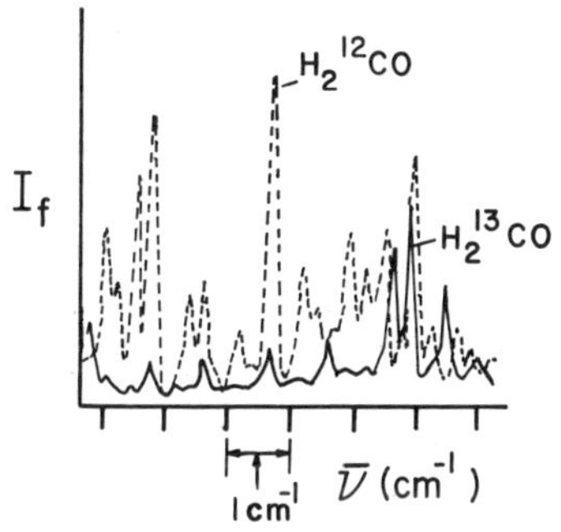

Fig. VIII–3. Fluorescence excitation spectrum of formaldehyde near 3032 Å (32,980 cm^{-1}). (---) $H_2{}^{12}CO$; (——) $H_2{}^{13}CO$. In this region $H_2{}^{12}CO$ is preferentially excited by laser. The width of the laser line is close to the Doppler widths of the absorption bands, $\simeq$ 0.1 cm^{-1}. From Clark et al. (217), reprinted by permission of North-Holland Publishing Company.

3032 Å. In this region absorption by $H_2{}^{12}CO$ is prominent. Hence, a laser line with a width of 0.1 cm^{-1} coincident with a peak of an $H_2{}^{12}CO$ absorption band should preferentially excite $H_2{}^{12}CO$ species in $H_2{}^{12}CO$–$H_2{}^{13}CO$ mixtures. The estimated selectivity by absorption is more than 40:1 for $H_2{}^{12}CO/H_2{}^{13}CO$ in a 1:1 mixture of $H_2{}^{12}CO$ and $H_2{}^{13}CO$. However, the actual ratio of ^{12}CO to ^{13}CO was 6.5:1. The proposed mechanism of the photolysis is

$$H_2CO \xrightarrow{h\nu} H_2{}^{12}CO^* \tag{VIII-7}$$

$$H_2{}^{12}CO^* \rightarrow H_2 + {}^{12}CO \tag{VIII-8}$$

$$H_2{}^{12}CO^* \rightarrow H + H^{12}CO \tag{VIII-9}$$

$$H_2{}^{12}CO^* \rightarrow H_2{}^{12}CO + h\nu' \tag{VIII-10}$$

$$H_2{}^{12}CO^* + H_2CO \rightarrow H_2CO + H_2CO \tag{VIII-11}$$

$$H_2{}^{12}CO^* + H_2CO \rightarrow H_2CO + H_2 + CO \text{ (or H + HCO)} \tag{VIII-12}$$

$$H_2{}^{12}CO^* + H_2CO \rightarrow H_2COH + HCO \tag{VIII-13}$$

$$H + H_2CO \rightarrow H_2 + HCO \tag{VIII-14}$$

$$2HCO \rightarrow H_2 + 2CO \tag{VIII-15}$$

where H_2CO^* signifies the electronically excited formaldehyde $^1A_2$.

Processes (VIII-9) and (VIII-11) to (VIII-15) reduce the ^{12}CO enrichment achieved by the initial act of light absorption. The lifetime of H_2CO^* becomes shorter at shorter wavelengths and, accordingly, dissociation processes (VIII-8) and (VIII-9) become more important than quenching processes (VIII-11) through (VIII-13). Hence, isotopic abundance at first sight could be improved at shorter wavelengths. However, this effect is counterbalanced by the increased occurrence of (VIII-9) followed by (VIII-14) and (VIII-15) and much more extensive overlapping of absorption bands of isotopic formaldehydes. Thus, no advantage is expected at the shorter wavelength photolysis.

A similar technique was used by Yeung and Moore (1075) for D_2 enrichment by the photolysis of H_2CO–D_2CO mixtures. Marling (660a) has succeeded in enriching carbon monoxide photoproducts $C^{18}O$, $C^{17}O$, and ^{13}CO by illuminating formaldehyde with a Ne ion laser at 3323.74 Å (for $C^{18}O$) and at 3323.71 Å (for $C^{17}O$, ^{13}CO) in exact coincidence with the absorption bands of corresponding formaldehyde isotope species. Deuterium photoproducts also are found enriched when a proper incident wavelength is chosen.

A somewhat different approach was made by Lamotte et al. (597) for photochemical enrichment of chlorine isotope species. They have used laser lines at 4657 and 4706 Å that coincide with vibrational bands of $SC^{37}Cl^{37}Cl$ and $SC^{35}Cl^{35}Cl$, respectively. The absorption of a laser line by isotopically mixed thiophosgene induces selective excitation of a particular isotopic species. The width of the laser line must be much narrower than isotope shifts of vibrational bands in thiophosgene ranging up to 7 cm^{-1} (2 Å). The excited isotopically pure thiophosgene is removed from the system by forming an addition product with added diethoxyethylene. After irradiation the abundance of the particular isotopic thiophosgene that is excited by the laser line has decreased, while those of other isotopic species have not changed. At these wavelengths photodissociation of thiophosgene does not take place [see Section VII–12, p. 291].

A similar technique has been used by Zare et al. (261, 643) for chlorine isotope separation. Isotopic mixtures of iodine monochloride ($I^{35}Cl$, $I^{37}Cl$) are irradiated in the presence of dibromoethylene by a laser line at 6053 Å which selectively excites $I^{37}Cl$. An adjacent vibrational band of $I^{35}Cl$ is about 15 Å away. The excited $I^{37}Cl$ reacts with added 1,2-dibromoethylene to form the product *trans*-ClHC=CHCl enriched in ^{37}Cl. At this wavelength no photodissociation of ICl takes place. See p. 191.

Koren et al. (582) have used an intense focused pulse of CO_2 laser to photodissociate HDCO (formaldehyde). The 944.18 cm^{-1} laser line nearly coincides with an absorption line of HDCO, however, there is no absorption band of H_2CO in the region of the laser line. Thus, the authors found an enrichment factor of 40 (the ratio of HD to H_2 after illumination to that in the original material).

VIII–2. PHOTOCHEMISTRY OF AIR POLLUTION

VIII–2.1. The Earth's Atmosphere

It has been known since 1900 that the earth's temperature varies with altitude in a complicated manner. It decreases first at a rate of about 6°C km^{-1} to a minimum of about 200°K at a height of about 10 to 15 km depending on latitude. This region of temperature decrease is called the troposphere. Above the troposphere the temperature starts to increase gradually with altitude to a maximum of about 280°K at about 50 km. This region of temperature increase is designated the stratosphere and the boundary between the troposphere and the stratosphere is called the tropopause. With increasing height above the stratosphere the temperature again decreases rapidly to a minimum of about 140°K. The region of the second temperature decrease is named the mesosphere and the region of the maximum temperature at about 50 km is called the stratopause. Figure VIII–4 shows the temperature profile of the earth's atmosphere with nomenclatures for various regions. The stratosphere is an inversion layer where the temperature increases with altitude. Since dense cold air is at the bottom of the layer, the stratosphere is stable against vertical mixing. This slow vertical mixing becomes important in determining the atmospheric distribution of minor constituents in the region where the photochemical equilibrium is extremely slow.

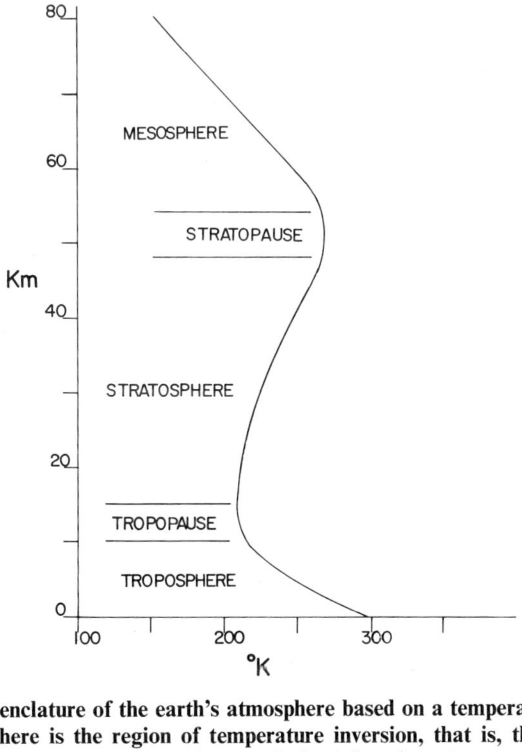

Fig. VIII–4. Nomenclature of the earth's atmosphere based on a temperature classification. The stratosphere is the region of temperature inversion, that is, the temperature increases with height and is stable against vertical mixing since dense cold air is at the bottom of the layer.

VIII–2.2. Atmospheric Air Pollution

Photochemical air pollution in the troposphere results from a complex interplay between sunlight and primary air pollutants emitted in ambient air that leads to the formation of ozone and other oxidizing and eye-irritating agents. On the other hand, pollutants injected into the stratosphere by such human activities as supersonic transports (SST's) and release of chlorofluoromethanes in air by their use as aerosol propellants and refrigerants may eventually reduce the protective layer of ozone from harsh solar ultraviolet radiation. Although the full impact of injected air pollutants in the stratosphere is not apparent at present, various model calculations show conclusively that the continuous future release of chlorofluoromethanes and NO_x (NO and NO_2) would result in substantial reduction of ozone in the stratosphere.

Photochemical air pollution in the troposphere was first recognized in the Los Angeles Basin in 1944 when crop damage occurred. Since then the nature of air pollutants and the mechanism of their formation have been studied extensively. Tropospheric air pollution has turned out to be a complex phenomenon arising from the interaction of sunlight with emission gases from automobiles and power plants. It involves hundreds of chemical reactions and hundreds of chemical species. Although the detailed mechanism is still unknown, at least the main reactions responsible for photochemical air pollution have been well established. The time history of air pollutants observed in simulated smog chamber experiments can be reproduced reasonably well by computer calculations based on known rate constants of various reactions.

VIII–2.3. Photochemical Air Pollution in the Troposphere

Photochemical air pollution in the troposphere is induced by the action of solar ultraviolet radiation upon mixtures of NO_x (NO and NO_2), SO_x (SO_2 and sulfates), and reactive hydrocarbons (mostly olefins) emitted in the atmosphere by automobiles and power plants.

Photochemical air pollution is characterized by the formation of a so-called "photochemical oxidant" and the reduction of visibility due to the simultaneous production of aerosol particles or particulates. This type of air pollution is commonly known as "photochemical smog."

Photochemical smog generally occurs at low relative humidities and high ambient temperatures with the aid of sunlight, while London type smog occurs at high relative humidities and low temperatures. Photochemical smog is oxidizing (mainly ozone) and London smog is reducing (mainly sulfur dioxide). The adverse effects of photochemical smog include eye irritation, plant damage, and reduced visibility. For detailed discussions of the subject the reader is referred to review articles by Berry and Lahman (101) and Finlayson and Pitts (357, 811) and books by Leighton (18) and Heicklen (12).

Figure VIII–5 shows the diurnal variations of NO_x and photochemical oxidant observed in Pasadena, California. A photochemical oxidant consists mainly of ozone and small amounts of other species, such as peroxyacetyl nitrate (PAN), capable of oxidizing aqueous iodide ions. The formation of a photochemical oxidant is commonly accompanied by the significant formation of an aerosol. Figure VIII–5 indicates a rapid conversion of NO to NO_2 prior to the buildup of oxidant.

Mechanism of Smog Formation. A mechanism initially proposed to explain the time history of air pollutants was the dissociation of NO_2 by solar radiation since other primary pollutants NO and hydrocarbons do

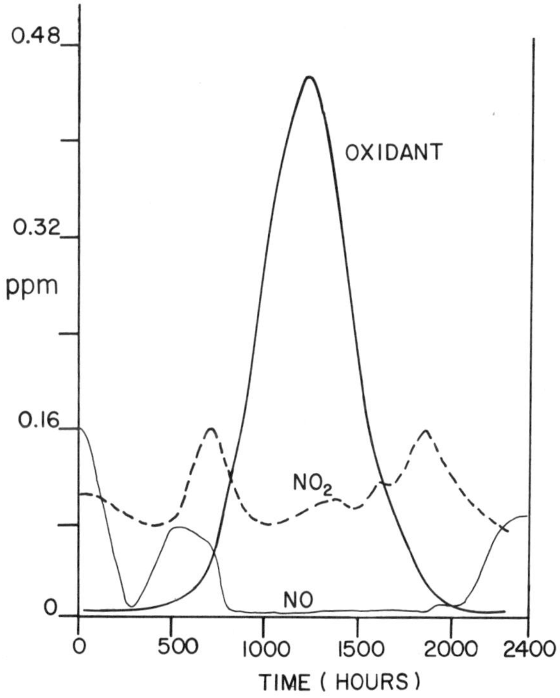

Fig. VIII–5. Diurnal variation of some pollutants in Pasadena, California on July 25, 1973. Concentrations of NO are small while those of photochemical oxidant are large. From Finlayson and Pitts (357), reprinted by permission. Copyright 1976 by the American Association for the Advancement of Science.

not absorb solar radiation of wavelengths above 2900 Å

$$NO_2 \xrightarrow{\lambda > 2900\,\text{Å}} NO + O(^3P) \qquad J_{NO_2} \qquad \text{(VIII-16)}$$

followed by

$$O + O_2 \xrightarrow{M} O_3 \qquad k_{17} \qquad \text{(VIII-17)}$$

$$NO + O_3 \rightarrow NO_2 + O_2 \qquad k_{18} \qquad \text{(VIII-18)}$$

$$(k_{18} = 1.5 \times 10^{-14}\ \text{cm}^3\ \text{molec}^{-1}\ \text{sec}^{-1}\ \text{at}\ 298°\text{K})$$

where M is air molecules and J_{NO_2} is the photodissociation coefficient of NO_2 defined in Section VIII–2.4. The spectral distribution of sunlight in the troposphere is shown in Fig. VIII–6 by the broken line. The active solar wavelengths are above 2900 Å.

Reaction (VIII-16) is brought about by absorption of sunlight in the region 2900 to 4300 Å [see Section VI–9]. From (VIII-16) to (VIII-18) we obtain the relationship

$$[O_3] = \frac{[NO_2]J_{NO_2}}{[NO]k_{18}} \qquad \text{(VIII-18a)}$$

where the bracket signifies concentrations. The level of ozone expected before sunrise is below 0.01 ppm using typical values of $[NO_2]/[NO] =$

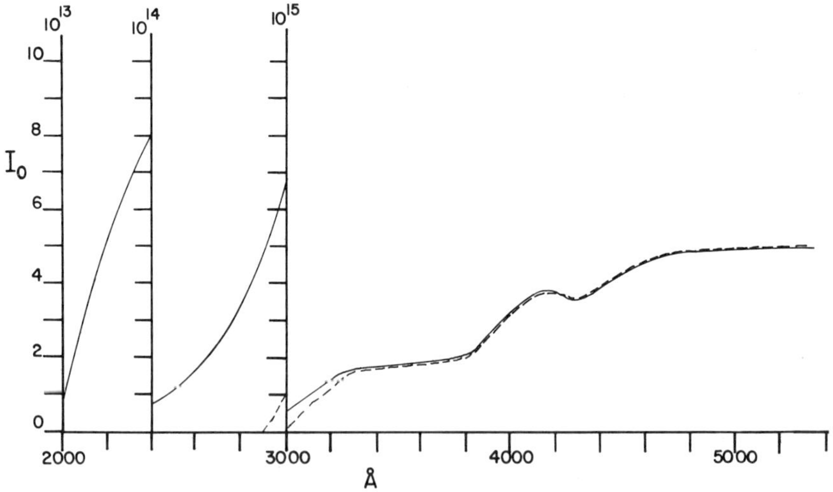

Fig. VIII–6. Mean intensities of solar radiation as a function of wavelength with bandwidths of 100 Å. The solar intensity I_0 above the atmosphere is given by the solid line in units of number of photons cm^{-2} sec^{-1} 100 Å^{-1}. The dashed line represents the solar flux in the troposphere. The radiation below 2900 Å is absorbed by the ozone layer. The curve is drawn using the data of Leighton (18), p. 29 and Nicolet (740).

0.3 and $J_{NO_2} = 8 \times 10^{-3}\ sec^{-1}$, $k_{18} = 1.5 \times 10^{-14}\ cm^3\ molec^{-1}\ sec^{-1}$ or $0.38\ ppm^{-1}\ sec^{-1}$ [Calvert and McQuigg (184)].

It is apparent from (VIII-18a) that a rapid conversion of NO to NO_2 is needed for the buildup of O_3 concentration levels. It was recognized that the conversion rate of NO to NO_2, observed in the real atmosphere, was several hundred times as fast as the rate of the reaction $2NO + O_2 \rightarrow 2NO_2$ in the range of NO concentrations of 0.05 to 0.5 ppm. Furthermore, in simulated smog chamber experiments it was found that the rate of butene consumption by illumination of a $NO–NO_2–H_2O$–butene mixture in air was much faster than that calculated on the basis of reactions of $O(^3P)$ atoms and O_3 with butene. Apparently a new mechanism based on radical chain reactions was needed to explain these observations. Two groups of scientists, Heicklen and coworkers and Weinstock and coworkers, were the first to propose independently the following sequence of reactions in 1970.

$$OH + CO \rightarrow CO_2 + H \qquad \text{(VIII-19)}$$
$$H + O_2 \xrightarrow{M} HO_2 \qquad \text{(VIII-20)}$$
$$HO_2 + NO \rightarrow OH + NO_2 \qquad \text{(VIII-21)}$$

The reactions of OH radicals with various hydrocarbons have recently been studied extensively. The reaction rates are in general fast (0.01 to 0.1 of the gas kinetic collision rate) and hence, it is reasonable to consider another similar chain mechanism involving aldehyde and hydrocarbon to convert NO to NO_2, for example,

$$HO + CH_2O \rightarrow H_2O + HCO \qquad \text{(VIII-19a)}$$
$$HCO + O_2 \rightarrow HO_2 + CO \qquad \text{(VIII-20a)}$$
$$HO_2 + NO \rightarrow OH + NO_2 \qquad \text{(VIII-21a)}$$

and in general

$$OH + RH \rightarrow H_2O + R \qquad \text{(VIII-22)}$$
$$R + O_2 \rightarrow RO_2 \qquad \text{(VIII-23)}$$
$$RO_2 + NO \rightarrow RO + NO_2 \qquad \text{(VIII-24)}$$
$$RO + O_2 \rightarrow R'CHO + HO_2 \qquad \text{(VIII-25)}$$
$$HO_2 + NO \rightarrow OH + NO_2 \qquad \text{(VIII-26)}$$

where RH signifies hydrocarbons. In summary, the following three reactions are most important in oxidizing NO to NO_2.

$$O_3 + NO \rightarrow NO_2 + O_2 \qquad \text{(VIII-18)}$$
$$HO_2 + NO \rightarrow OH + NO_2 \qquad \text{(VIII-21a)}$$
$$RO_2 + NO \rightarrow RO + NO_2 \qquad \text{(VIII-24)}$$

The reaction intermediates of OH with paraffins and olefins in air are not well known, but in general, one may write

$$OH + RH \text{ (paraffin, olefin)} \rightarrow \alpha HO_2 + \beta RO_2 \text{ (or } RCOO_2) + \text{other products}$$

where α and β are some unknown numbers.

Likewise, O_3 reacts with hydrocarbons to produce unknown numbers of HO_2 and RO_2 (or $RCOO_2$) [see below]. From the computer analysis of simulated smog formation involving the hypothetical illumination of NO–NO_2–H_2O–butene–aldehydes–CO–CH_4 mixtures in air, Calvert and McQuigg (184) estimate that HO_2 and RO_2 radicals, formed mainly by the addition of OH to butene, account for 70% of NO to NO_2 conversion. The HO_2 and RO_2 radicals formed from the photolysis of aldehydes and OH reactions with aldehydes are responsible for 25% of the conversion. Carbon monoxide is only 5% effective for the NO to NO_2 conversion. The effect of paraffins on the NO to NO_2 conversion rate is very small.

Calvert and McQuigg have also suggested that the rate of decay of *trans*-2-butene in the initial stage is mainly determined by the reaction of OH with the hydrocarbon. In the later stage of smog formation OH and O_3 attacks on the hydrocarbon must be equally important.

Sources of OH *and* HO_2 *Radicals.* It has been suggested recently that OH radicals are the most important intermediate in promoting the formation of a photochemical oxidant, the oxidation of NO, and the consumption of olefins in the atmosphere especially at the early stage of smog formation. The OH radicals are probably produced mainly from the photolysis of nitrous acid since it absorbs light of wavelengths below 4000 Å [see Section VII–9]. The bond energy $D_0(HO\text{–}NO) = 2.09$ eV corresponds to 5930 Å. Hence, the photolysis should occur in the region of solar wavelengths 3000 to 4000 Å

$$HONO \xrightarrow{h\nu} OH + NO \qquad \text{(VIII-27)}$$

Another minor source of OH radicals may be the photolysis of nitric acid since it absorbs light of wavelengths below 3300 Å [see Section VII–23].

$$HNO_3 \xrightarrow{h\nu} OH + NO_2 \qquad \text{(VIII-28)}$$

Nitrous and nitric acid are probably formed by the reactions

$$OH + NO \xrightarrow{M} HONO \qquad \text{(VIII-29)}$$

$$NO_2 + NO + H_2O \leftrightarrow 2HONO \qquad \text{(VIII-30)}$$

$$OH + NO_2 \xrightarrow{M} HNO_3 \qquad \text{(VIII-31)}$$

$$NO_2 + NO_3 + H_2O \rightarrow 2HNO_3 \qquad \text{(VIII-32)}$$

Both acids are detected in the atmosphere. Hydrogen peroxide dissociates into OH radicals by absorption of light below 3200 Å [see Section VII–6]

$$H_2O_2 \xrightarrow{h\nu} 2OH \tag{VIII-33}$$

and may contribute to OH. Its presence in air has been confirmed recently.

The OH radicals may also be produced by the photolysis of ozone

$$O_3 \xrightarrow{\lambda < 3190\ \text{Å}} O_2(^1\Delta) + O(^1D) \tag{VIII-34}$$

followed by

$$O(^1D) + H_2O \rightarrow 2OH \tag{VIII-35}$$

Major destruction routes of OH radicals are the addition to olefins, the H atom abstraction from olefins and aldehydes, and the reaction with CO. Another radical, hydroperoxyl (HO_2), has been considered as a major oxidizing agent for NO and to a lesser extent for hydrocarbons. The HO_2 radicals are probably formed by the photolysis of formaldehyde [see Section VII–4, p. 277]

$$HCHO \xrightarrow{\lambda < 3700\ \text{Å}} H + CHO \tag{VIII-36}$$

followed by

$$H + O_2 \xrightarrow{M} HO_2 \tag{VIII-37}$$

$$CHO + O_2 \rightarrow HO_2 + CO \tag{VIII-38}$$

and hydrogen atom abstraction from alkoxy radicals by O_2

$$RCH_2O + O_2 \rightarrow RCHO + HO_2 \tag{VIII-39}$$

Major loss processes of HO_2 are probably the oxidation of NO (VIII-21*a*), the reaction $2HO_2 \rightarrow H_2O_2 + O_2$, and the reactions with NO_2 and olefins.

Ambient Concentrations of O, OH, *and* HO_2. The O atoms are produced from the photolysis of NO_2 and are lost by the combination with O_2 to form O_3. The peak value of O atom concentration is calculated by Graedel et al. (415) to be 7.5×10^4 molec cm^{-3}. The OH radical peak concentration is estimated to be about 2×10^6 molec cm^{-3} by Graedel et al. (415) and by Calvert and McQuigg (184). On the other hand, the HO_2 radical peak value must be about 10^9 molec cm^{-3}, which is about 1000 times as large as OH radical concentration. The large difference of concentration is due to the much slower reaction rate of HO_2 with NO than that of OH with olefin.

The ambient concentration of OH radicals has recently been measured by Davis et al. (267) and by Perner et al. (805).

Davis et al. have used laser induced fluorescence of OH radicals excited at 2820.6 Å. The OH radical concentrations are in the range 10^6 to 10^7 molec cm^{-3}. Perner et al., on the other hand, used the absorption of OH at

3079.95 Å with a path length of 7.8 km. They obtained OH concentrations of 10^6 to 10^7 molec cm^{-3}.

Wang et al. (1058) have recently measured OH radical concentrations in a simulated smog chamber by the laser induced fluorescence of OH. The OH concentrations in the chamber range from 0.5 to 1.5×10^7 molec cm^{-3}. In view of the difficulties involved in the absolute determination of OH radicals at such low levels, the uncertainty must be larger than $\pm 50\%$. Table VIII–1A summarizes the ambient concentrations of reactive species and their rate constants with hydrocarbons and NO in polluted air.

Table VIII–1A Concentrations and Rate Constants of Reactive Species in Polluted Air

Reactive Species	Concentration in Ambient Air (molec cm^{-3})	Rate Constant[a] (cm^3 molec^{-1} sec^{-1}) Olefin	Paraffin	NO
O	10^{4} [b]	10^{-11} to 10^{-12}	10^{-14} to 10^{-15}	1×10^{-31} [e]
O_3	10^{11} to 10^{13}	10^{-17} to 10^{-18}	$<10^{-20}$	1.5×10^{-14}
OH	3×10^{6} [d], 2×10^{6} [f], $10^6 - 10^{7}$ [c]	10^{-10} to 10^{-12}	10^{-12} to 10^{-13}	3×10^{-30} [e]
HO_2	5×10^{9} [d], 5×10^{8} [f]	10^{-15} to 10^{-16}	10^{-20} to 10^{-22}	1.7×10^{-13}
CH_3O_2	[2×10^{7} [f]] [g]	[10^{-16}] [g]		[6×10^{-13}] [g]

[a] From Anderson (1).
[b] Estimated by Johnston (543).
[c] Davis et al. (267), Perner et al. (805), observed values.
[d] Peak values, estimated by Calvert and McQuigg (184).
[e] N_2 as a third body, units of cm^6 molec^{-2} sec^{-1}.
[f] Estimated by Graedel et al. (415), peak values.
[g] The brackets signify estimated values.

The comparison of the reaction rates of O_3, HO_2, and CH_3O_2 with olefin, paraffin, and NO reveals that the predominant reactions of these reactive species are the oxidations of NO [(VIII-18), (VIII-21a), and (VIII-24)]. The major destruction processes of olefin are the reactions with O_3 and with OH. (The rate of olefin destruction is proportional to the rate constant times the concentration of the active species.) The destruction process of olefins by HO_2 is less important and those by O atoms and CH_3O_2 radicals are also minor.

Reactions of OH *and* O_3 *with Hydrocarbons.* The initial reactions of OH with olefins are mainly the addition to the double bond and partially H atom abstraction. The final stable products are the corresponding aldehydes and ketones. For example, the reactions of OH with propylene produce C_2H_5CHO and CH_3COCH_3. However, the detailed mechanism of their formation is not known. In a polluted atmosphere the OH addition product to propylene may oxidize NO

$$OH + CH_3CHCH_2 \rightarrow CH_3CHCH_2OH$$
$$CH_3CHCH_2OH + O_2 \rightarrow CH_3CH(OO)CH_2OH$$
$$CH_3CH(OO)CH_2OH + NO \rightarrow NO_2 + CH_3CHOCH_2OH$$
$$CH_3CHOCH_2OH \rightarrow CH_3CHO + CH_2OH$$
$$CH_2OH + O_2 \rightarrow HO_2 + HCHO$$

The reactions of OH with paraffins and aldehydes proceed by H atom abstraction to produce alkyl (R) and carbonyl (RCO) radicals, respectively. In a polluted atmosphere R and RCO radicals react with O_2 to give RO_2 and $RC(O)O_2$, which further oxidize NO to NO_2

$$RO_2 + NO \rightarrow NO_2 + RO$$
$$RC(O)O_2 + NO \rightarrow NO_2 + RCO_2$$

$RC(O)O_2$ may combine with NO_2

$$RC(O)O_2 + NO_2 \rightarrow RC(O)OONO_2 \qquad (\text{PAN, if } R = CH_3) \qquad (\text{VIII-40})$$

Ozone reacts slowly with olefins. The reaction of O_3 with propylene is believed to yield the so-called zwitterions and aldehydes in liquid. In the gas phase zwitterions ($HC^+HOO^-, CH_3C^+HOO^-$) are probably biradicals

$$O_3 + CH_3CH{=}CH_2 \rightarrow H\dot{C}HOO\cdot + CH_3CHO$$
$$O_3 + CH_3CH{=}CH_2 \rightarrow CH_3\dot{C}HOO\cdot + HCHO$$

The HCHOO species has recently been identified as dioxirane, $H_2\overline{COO}$, by Lovas and Suenram (649a) in the low temperature reaction of ozone with olefins. Zwitterions react with O_2 to yield oxidizing radicals in polluted air

$$H\dot{C}HOO\cdot + O_2 \rightarrow OH + HC(O)O_2$$
$$CH_3\dot{C}HOO\cdot + O_2 \rightarrow OH + CH_3C(O)O_2$$
$$HC(O)O_2 + NO \rightarrow NO_2 + HC(O)O$$
$$CH_3C(O)O_2 + NO \rightarrow NO_2 + CH_3C(O)O$$
$$CH_3C(O)O \rightarrow CO_2 + CH_3$$
$$CH_3 + O_2 \rightarrow CH_3OO$$
$$CH_3OO + NO \rightarrow NO_2 + CH_3O$$

Various oxidation products, including PAN (peroxyacetyl nitrate), are formed by the radical termination reactions

$$CH_3C(O)O_2 + NO_2 \rightarrow CH_3C(O)O_2NO_2 \text{ (PAN)}$$
$$CH_3C(O)O_2 + HO_2 \rightarrow CH_3C(O)OOH + O_2$$
$$CH_3OO + HO_2 \rightarrow CH_3OOH + O_2$$
$$CH_3O + NO \rightarrow CH_3ONO$$
$$CH_3O + NO_2 \rightarrow CH_3ONO_2$$

Aerosol Formation. The following is the composition of aerosol particles associated with the formation of a photochemical oxidant:

1. Trace metals (Pb, Na, Mg, Al, V, and Zn)
2. Sulfates, water, nitrates and ammonium compounds
3. Organic nitrates, carboxylic acids and their esters, carbonyl compounds, and alcohols.

The formation of sulfates by photooxidation of SO_2 is slow in pure air (see Section VI–12). However, in the presence of hydrocarbons and NO_x the photooxidation of SO_2 becomes 50 to 100 times more rapid. Undoubtedly, the oxidation mechanism is complex, involving many radicals such as OH, HO_2, O, NO_3, RO_2, and RO, as well as O_3. Various homogeneous and heterogeneous processes are proposed to explain aerosol formation.

Calvert and McQuigg suggest that yet unknown radicals, such as OCH_2O or those derived from it, formed in the O_3–olefin–air mixtures may oxidize SO_2 in the homogeneous reaction. It is known that OH and HO_2 radicals combine rapidly with SO_2. The addition products may eventually be transformed into sulfuric acid, peroxysulfuric acid, sulfates, and nitrates in a polluted atmosphere probably in a liquid phase of aerosol particles, although the detailed steps are still unknown. Finlayson and Pitts (357) believe that the oxidation of aromatic compounds by such species as OH, HO_2, O_3, and $O(^3P)$ may also be significant for the formation of organic aerosol.

VIII–2.4. Air Pollution in the Stratosphere

In the preceding section it was indicated that sunlight is the primary initiator for the buildup of ozone and other oxidants as a result of a rapid conversion of NO to NO_2 by catalytic cycles involving HO_2 and RO_2 radicals.

Pollution in the stratosphere may induce the reduction of ozone without participation of sunlight in the case of NO_x injected directly into the stratosphere by SST's, while in the case of chlorofluoromethanes, their photodissociation by sunlight to produce Cl atoms is required for the reduction of ozone by a catalytic cycle involving Cl and ClO. The time scale required for the buildup of tropospheric air pollution is of the order of hours, while the full impact of stratospheric air pollution would manifest itself in years.

The pollutants that have been considered to be potentially effective in reducing ozone in the stratosphere are NO_x (NO, NO_2, HNO_3) and ClO_x (Cl, ClO, HCl). The reduction of the ozone layer leads to the increased penetration of solar ultraviolet radiation below 3100 Å, damaging biological systems. The adverse effects of the ultraviolet radiation in the range 2800 to 3200 Å include the possible increase in skin cancer and other yet unknown genetic damage to plants and plankton. Detailed discussions of the problem are given in several recent reviews by Johnston (542, 543), Turco and Whitten (981), Rowland and Molina (843), and Nicolet. (740)

Ozone Balance in the Natural Stratosphere. Typical temperatures and concentrations of air and ozone at various altitudes in the stratosphere are given in Table VIII–1B. The ozone concentration first increases with increasing height to a maximum at about 25 km and then decreases at higher altitudes, while the temperature increases with an increase of altitude to a maximum at about 45 km.

For the discussion of the formation and destruction of ozone in the stratosphere it is convenient to define the photodissociation coefficient generally denoted by J (in units of sec^{-1}). J is the probability of dissociation of a molecule per second by light absorption.

Photodissociation coefficient. The photodissociation coefficient may be defined as

$$J = \int_0^\infty \phi_{\bar{v}} I_{\bar{v}} \sigma_{\bar{v}} \, d\bar{v} \qquad \text{(VIII-40a)}$$

where $\phi_{\bar{v}}$ is the quantum yield of dissociation of the molecule at a wave number $\bar{v}$, $I_{\bar{v}}$ is the intensity of sunlight in quanta cm^{-2} sec^{-1} at a wave

Table VIII–1B. Typical Temperature and Concentration of Air and Ozone in the Stratosphere[a]

Altitude (km)	Temperature (°K)	Total Concentration (molec cm^{-3})	Ozone (molec cm^{-3})
15	211	3.9×10^{18}	1.0×10^{12}
20	219	1.9×10^{18}	2.9×10^{12}
25	227	7.7×10^{17}	3.2×10^{12}
30	235	3.6×10^{17}	2.9×10^{12}
35	252	1.7×10^{17}	2.0×10^{12}
40	268	8.1×10^{16}	1.0×10^{12}
45	274	4.3×10^{16}	3.2×10^{11}
50	274	2.3×10^{16}	1.0×10^{11}

[a] From Nicolet (739)

number $\bar{\nu}$ and at a given altitude, and $\sigma_{\bar{\nu}}$ is the absorption cross section in cm^2 $molec^{-1}$ of the molecule at a wave number $\bar{\nu}$. $I_{\bar{\nu}}$ depends on the solar zenith angle χ and the absorption by O_2 and O_3 in the atmosphere. When the sun is overhead χ is zero and at horizon $\chi = 90°$. $I_{\bar{\nu}}$ is larger for smaller χ, since the loss by absorption, scattering, particulate diffusion, and so forth is smaller. The absorption by O_3 ($\lambda < 3200$ Å) and O_2 ($\lambda < 2400$ Å) is most important in attenuating the sun's radiation in the stratosphere.

The photodissociation coefficients of O_2, H_2O, N_2O, CF_2Cl_2, and H_2O_2 calculated by various authors are given in Fig. VIII–7. Values of J are those for an overhead sun ($\chi = 0$) and are given at various altitudes. If a molecule disappears solely by photodissociation, its lifetime is given by J^{-1}. For

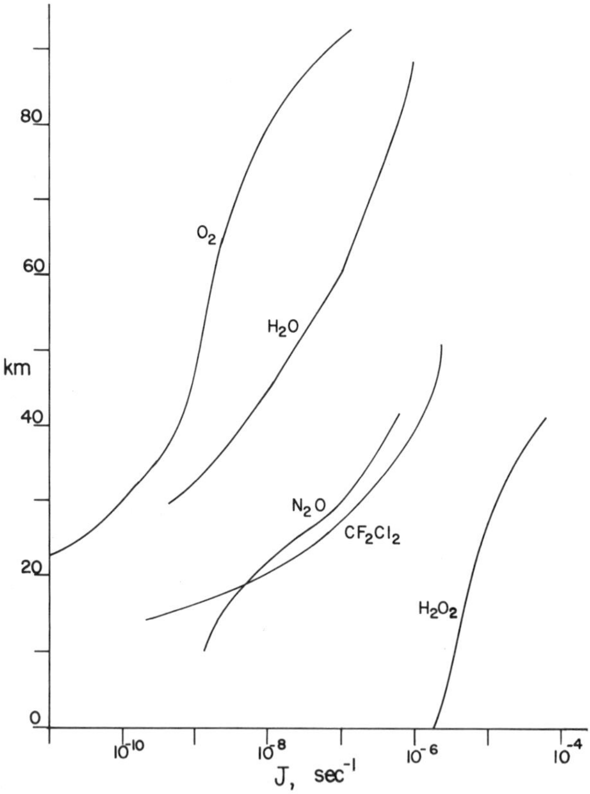

Fig. VIII–7. Photodissociation coefficients, J (sec^{-1}) of O_2, H_2O, N_2O, CF_2Cl_2, and H_2O_2 as a function of altitude (km) for an overhead sun. O_2, from Kockarts (579a), p. 174; H_2O, from Park (795); N_2O, from Nicolet (740); CF_2Cl_2, from Rowland and Molina (843); and H_2O_2; from Nicolet (739).

example, CF_2Cl_2 molecules, which are chemically inert in the troposphere, would diffuse into the stratosphere and would eventually be lost mainly by photodissociation. The photodissociation lifetime at 20 km is about 10^8 sec or 3 years. Since the lifetimes at other solar zenith angles are much longer, the average lifetime over all zenith angles would be about 66 years [Rowland and Molina (843)]. The production of O_3 by the photodissociation of O_2 is negligible below 20 km.

The photodissociation coefficients of O_3 and NO_2 are given in Fig. VIII–8. The photodissociation coefficient of NO_2 is nearly independent of altitude and is 8.6×10^{-3} and 9.2×10^{-3} sec^{-1}, respectively, at 20 and 50 km with $\chi = 45°$ [Shimazaki and Whitten (872b)]. Above 50 km O_3 dissociates into $O(^1D) + O(^1\Delta)$ with $J = 10^{-2}$ sec^{-1} predominantly by the absorption in the Hártley band, while below 50 km the contribution from the Chappuis (4000 to 10,000 Å) and Huggins bands (3100 to 3600 Å) increases successively with a decrease of altitude. The absorption in the Chappuis band results in the production of $O(^3P) + O_2(X^3\Sigma^-)$ and that in the Huggins band yields $O(^3P)$ and electronically excited $O_2(^1\Delta, {}^1\Sigma^+)$. As a result the fraction of $O(^1D)$ production decreases at lower altitudes as shown. The fraction of $O_2(^1\Delta)$ production in the lower stratosphere is not well established since O_3 may dissociate either into $O(^3P) + O_2(^1\Sigma^+)$ or $O(^3P) + O_2(^1\Delta)$.

The Chapman mechanism. The mechanism of ozone formation and destruction in the stratosphere was first formulated by Chapman (205) in 1930. He did not consider the effects of minor constituents and physical transport processes that have since been recognized as important factors to explain the discrepancy between the calculated results and the actual observation. According to his mechanism, ozone is formed by the photolysis

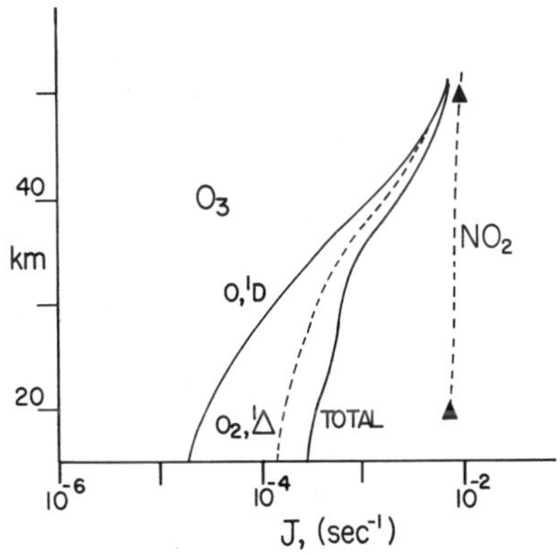

Fig. VIII–8. Photodissociation coefficients J of ozone and NO_2 in sec^{-1} as a function of altitude for an overhead sun and $\chi = 45°$, respectively. Above 50 km O_3 dissociates predominantly into $O(^1D) + O_2(^1\Delta)$ by photolysis in the Hartley band. Below 50 km the photolysis in the Chappuis and Huggins bands becomes progressively more important, producing $O(^3P)$, $O_2(X^3\Sigma^-)$, and $O_2(^1\Delta)$. The extent of $O_2(^1\Delta)$ production is uncertain below 50 km. From Nicolet (738) for O_3 and Shimazaki and Whitten (872b) for NO_2, reprinted by permission of Reidel and the American Geophysical Union.

of O_2 below 2420 Å of solar radiation followed by the three body combination of ground state O atoms with O_2 molecules to form O_3

$$O_2 \xrightarrow{\lambda < 2420\,\text{Å}} 2O \qquad J_{O_2} \qquad \text{(VIII-41)}$$

$$O + O_2 \overset{M}{\rightarrow} O_3 \qquad k_{42} \qquad \text{(VIII-42)}$$

$$k_{42} = 6 \times 10^{-34}\ \text{cm}^6\ \text{molec}^{-2}\ \text{sec}^{-1}$$

where J signifies the photodissociation coefficient. Ozone so formed is destructued by visible and ultraviolet sunlight and by O atoms

$$O_3 \overset{h\nu}{\rightarrow} O_2 + O \qquad J_{O_3} \qquad \text{(VIII-43)}$$

$$O_3 + O \rightarrow O_2 + O_2 \qquad k_{44} \qquad \text{(VIII-44)}$$

$$k_{44} = 1.9 \times 10^{-11} \exp(-2300/T)\ \text{cm}^3\ \text{molec}^{-1}\ \text{sec}^{-1}$$

Note that reaction (VIII-44) is a slow process with a high activation energy (4.6 kcal mol^{-1}). At 220°K (stratospheric temperature) k_{44} is only 5.5×10^{-16} cm^3 molec^{-1} sec^{-1}.

The combination of O atoms by a third body

$$O + O \overset{M}{\rightarrow} O_2$$

is much slower than (VIII-42) in the stratosphere and can be omitted in later discussion. The rate of formation of ozone decreases with a decrease of altitude because of the rapid attenuation of sunlight by O_2 below 2420 Å and is negligible below 20 km (see Fig. VIII–7). On the other hand, the photolysis of ozone takes place throughout the atmosphere by visible and ultraviolet sunlight with a near uniform rate. (See Fig. VIII–8.)

Equations (VIII-41) to (VIII-44) give the steady state concentrations of ozone molecules as a function of altitude as follows. The rate of ozone change, $d[O_3]/dt$, is

$$\frac{d[O_3]}{dt} = k_{42}[O][O_2][M] - J_{O_3}[O_3] - k_{44}[O][O_3] \qquad \text{(VIII-44a)}$$

where the concentration is denoted by a bracket, [M] is the concentration of air, and the rate of O atom production is

$$\frac{d[O]}{dt} = 2J_{O_2}[O_2] - k_{42}[O][O_2][M] + J_{O_3}[O_3] - k_{44}[O][O_3] \qquad \text{(VIII-44b)}$$

The steady state conditions may be assumed for O atoms since their concentration is extremely low

$$2J_{O_2}[O_2] + J_{O_3}[O_3] = k_{42}[O][O_2][M] + k_{44}[O][O_3] \qquad \text{(VIII-44c)}$$

From (VIII-44a) and (VIII-44c), we obtain

$$\frac{d[O_3]}{dt} = 2J_{O_2}[O_2] - 2k_{44}[O][O_3] \qquad \text{(VIII-44d)}$$

At an altitude of 30 km it can be seen (Figs. VIII–7 and VIII–8, Table VIII–1B) that

$$J_{O_3}[O_3] \gg J_{O_2}[O_2]$$

and

$$k_{42}[O][O_2][M] \gg k_{44}[O][O_3]$$

Hence from (VIII-44c)

$$J_{O_3}[O_3] \simeq k_{42}[O][O_2][M] \qquad \text{(VIII-44e)}$$

Combining (VIII-44d) and (VIII-44e), we obtain

$$\frac{d[O_3]}{dt} = 2J_{O_2}[O_2] - \frac{2k_{44}J_{O_3}[O_3]^2}{k_{42}[O_2][M]} \qquad \text{(VIII-44f)}$$

and at equilibrium the O_3 concentration is

$$[O_3]_{eq} = [O_2]\sqrt{k_{42}J_{O_2}[M]/k_{44}J_{O_3}} \qquad \text{(VIII-44g)}$$

To obtain the concentration profiles in an atmosphere, it is important to know the time required to attain the equilibrium concentration. If the equilibrium time scale is more than a year, physical transport processes become appreciable and a large departure from the equilibrium profile is expected. The equilibrium time scale τ_{eq} may be obtained from

$$\tau_{eq} = \frac{[O_3]_{eq}}{2J_{O_2}[O_2]} \qquad \text{(VIII-44h)}$$

From Table VIII–1B and J_{O_2} of Fig. VIII–7, $\tau_{eq}(O_3)$ is calculated as a function of altitude. This relationship is shown in Fig. VIII–9. The equilibrium time scale near the top of the stratosphere is about a day, while below 15 km $\tau_{eq}(O_3)$ is more than a year and downward physical transport processes of O_3 become important.

Physical transport processes and mixing ratio. The concentration profile of a minor constituent in an atmosphere is often expressed as a mixing ratio by volume or a mole fraction rather than the concentration by atmospheric modelers. Physical transport processes involve vertical and horizontal mixing by turbulence and molecular diffusion. The molecular diffusion process can be ignored in the stratosphere since it is important only above about 40 km.

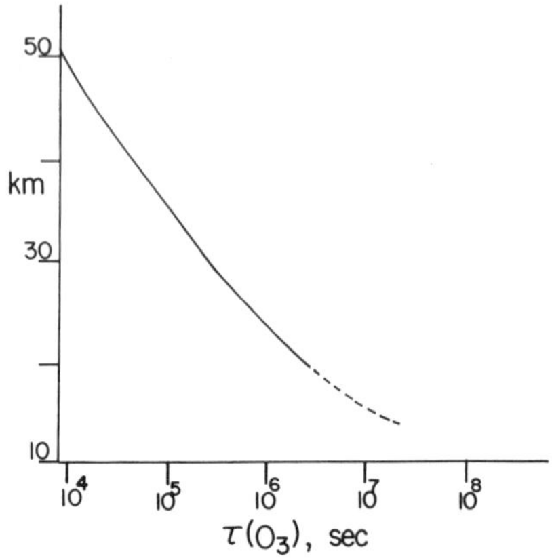

Fig. VIII–9. Equilibrium time scale in seconds for ozone in a pure oxygen atmosphere as a function of altitude (km) for an overhead sun. The time scale is a day at 50 km and more than a year below 15 km for daytime conditions.

The simplest approach to treat the physical transport in the stratosphere is one dimensional vertical mixing by eddy diffusion that is, by turbulent mixing.

In this case (VIII-44d) may be replaced by [see McElroy et al. (679)],

$$[M]\frac{\delta f_{O_3}}{\delta t} - \frac{\delta}{\delta z}\left([M]K_z\frac{\delta f_{O_3}}{\delta z}\right) = 2J_{O_2}[O_2] - 2k_{44}[O][O_3] \quad \text{(VIII-44i)}$$

assuming no thermal gradient, where $f_{O_3} = [O_3]/[M]$. K_z is a constant at a given height z and is called the eddy diffusion coefficient in units of $cm^2\ sec^{-1}$.

The quantity $[M]K_z\delta f_{O_3}/\delta z$ has units of molec $cm^{-2}\ sec^{-1}$ and signifies the flux of ozone through unit area (cm^2) per second. The eddy diffusion coefficient is a parameter adjusted to give the observed profile of the mixing ratio of a minor constituent in air. A typical value of K at the ground level is of the order of $10^5\ cm^2\ sec^{-1}$ and about 10^3 to $10^4\ cm^2\ sec^{-1}$ at the tropopause. With an eddy diffusion coefficient of $10^4\ cm^2\ sec^{-1}$, it will take about a year for a minor constituent to attain an equilibrium profile. (The time required to reach an equilibrium profile is approximately given by H^2/K, where H is the scale height defined by $H = RT/Mg$ and is typically 6.5 km (740). R is the gas constant, T is the absolute temperature, M is the molecular weight, and g is the acceleration due to gravity [see McEwan and Phillips (20), p. 67].)

In the real atmosphere horizontal motions along latitude and longitude must also be taken into consideration. Thus, the ozone concentration profile should show a significant derivation near the tropopause due to the downward transport of O_3 from the expected profile without vertical eddy diffusion.

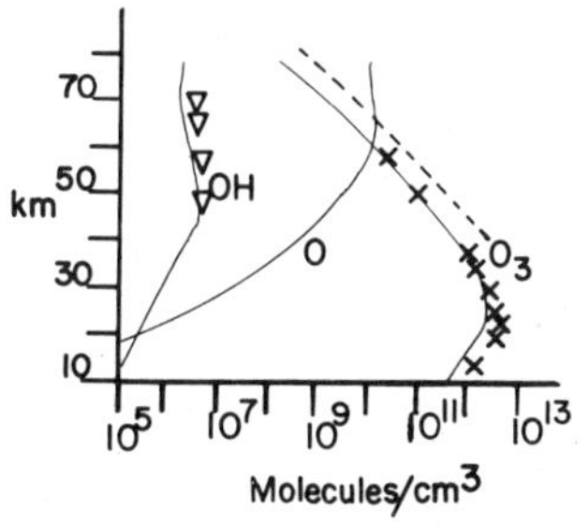

Fig. VIII–10. Concentrations of O, OH, and O_3 as a function of altitude in kilometers. Model calculations by Turco and Whitten (981); ($\triangle$ and $\times$) measured values; (---) results of the computation of ozone profile without NO_x and HO_x. From McElroy et al. (679), reprinted by permission of Pergamon Press and the American Meteorological Society.

Deviation from the Chapman mechanism. It was recognized by Nicolet (740) that the observed O_3 concentrations were much less than the calculated values even near the stratopause where the physical transport processes are not important (see Fig. VIII–10). He suggested that to explain the observed ozone concentration, the effective value of k_{44}, the rate constant for the destruction of O_3, must be much larger than that given in (VIII-44g) for a pure O_2–N_2 atmosphere.

Johnston, Crutzen, and others have also recognized that the natural ozone balance in the stratosphere cannot be explained on the basis of the Chapman mechanism and air motions. Johnston (542) has concluded that the calculated ozone destruction rate based on the Chapman reactions and air motions can explain only 20% of the natural destruction rate. About 80% of ozone produced by sunlight must be destroyed by a mechanism other than (VIII-43) and (VIII-44).

Minor species observed in the stratosphere are shown in Fig. VIII–11. Of these it is now believed that nitric oxide is the most effective agent to destroy ozone by a catalytic cycle

$$NO + O_3 \rightarrow NO_2 + O_2 \qquad k_{45} \qquad \text{(VIII-45)}$$

$$NO_2 + O \rightarrow NO + O_2 \qquad k_{46} \qquad \text{(VIII-46)}$$

$$\overline{O + O_3 \rightarrow 2O_2}$$

$$k_{46} = 9 \times 10^{-12} \text{ cm}^3 \text{ molec}^{-1} \text{ sec}^{-1}$$

Only a small fraction of nitrogen dioxide formed in (VIII-45) is destroyed by (VIII-46), and the major fraction of NO_2 is photodissociated into NO + O, regenerating ozone ($J_{NO_2} = 10^{-2}$ sec^{-1} and $k_{46}[O] = 10^{-4}$ sec^{-1})

$$NO_2 \xrightarrow{h\nu} NO + O \qquad J_{NO_2} \qquad \text{(VIII-47)}$$

$$O + O_2 \xrightarrow{M} O_3 \qquad k_{42} \qquad \text{(VIII-42)}$$

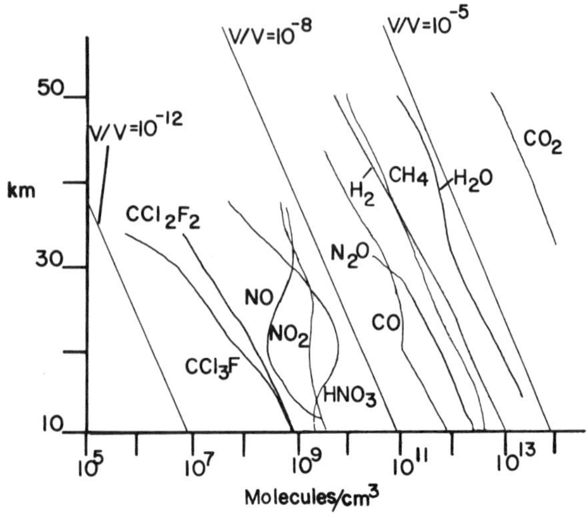

Fig. VIII–11. Observed concentrations of trace species in the stratosphere. From Johnston (543). Data on CCl_3F and CCl_2F_2 are from Heidt et al. (460). *V*/*V* is the volume mixing ratio. Reprinted by permission of Annual Reviews, Inc. and the American Geophysical Union.

Hence, (VIII-46) is the rate determining step of ozone destruction. From (VIII-45) to (VIII-47) we obtain the following steady state relation for NO and NO_2

$$k_{45}[O_3][NO] = J_{NO_2}[NO_2] + k_{46}[O][NO_2] \qquad \text{(VIII-47a)}$$

The rate of ozone change is

$$\frac{d[O_3]}{dt} = k_{42}[O][O_2][M] - J_{O_3}[O_3] - k_{44}[O][O_3] - k_{45}[NO][O_3] \qquad \text{(VIII-47b)}$$

Assuming as before the steady state conditions for O atoms

$$J_{NO_2}[NO_2] + 2J_{O_2}[O_2] + J_{O_3}[O_3] = k_{42}[O][O_2][M] + k_{44}[O][O_3] + k_{46}[O][NO_2] \qquad \text{(VIII-47c)}$$

From (VIII-47a) to (VIII-47c) we have for the rate of ozone change

$$\frac{d[O_3]}{dt} = 2J_{O_2}[O_2] - 2k_{44}[O][O_3] - 2k_{46}[O][NO_2] \qquad \text{(VIII-47d)}$$

At 20 km, $k_{46}/k_{44} = 16{,}600$ and $[O_3]/[NO_2] \simeq 1000$. Hencc, $k_{46}[O][NO_2]$ is more than 10 times as large as $k_{44}[O][O_3]$, that is, an atmosphere containing NO_2 is much more effective in destroying O_3 than one without NO_2.

Nitrogen dioxide is partially removed from the system by forming HNO_3 by combination with OH

$$NO_2 + OH \xrightarrow{M} HNO_3 \qquad \text{(VIII-48)}$$

A portion of HNO_3 is removed from the system by rainout when it diffuses to the troposphere and the remaining portion re-forms NO_2 by the photolysis

$$HNO_3 \xrightarrow{h\nu} OH + NO_2 \qquad \text{(VIII-49)}$$

The ratio of ozone destruction rate with and without NO_2 is sometimes called the catalytic ratio ρ,

$$\rho = 1 + \frac{k_{46}[NO_2]}{k_{44}[O_3]} \qquad \text{(VIII-50)}$$

Thus, a large fraction of ozone must be destroyed by NO_x. The source of NO in the natural stratosphere is probably the photolysis of N_2O (see Section VI–8) which is formed by bacteria in the soil followed by the reaction of $O(^1D)$ with N_2O [McElroy and McConnell (676)].

$$N_2O \xrightarrow{\lambda < 2300\ \text{Å}} N_2 + O(^1D) \qquad \text{(VIII-51)}$$

$$O(^1D) + N_2O \longrightarrow 2NO \qquad \text{(VIII-52)}$$

The destruction of ozone by another catalytic cycle (an HO_x cycle) is estimated to be about 10% of the NO_x cycle

$$HO + O_3 \rightarrow HO_2 + O_2 \qquad \text{(VIII-53)}$$

$$HO_2 + O \rightarrow HO + O_2 \qquad \text{(VIII-54)}$$

Photochemistry of the Polluted Stratosphere. The intensity of solar radiation reaching the stratosphere is attenuated by oxygen and ozone. Since O_2 is transparent to radiation of wavelengths above 1800 Å, while O_3 absorbs light weakly in the region 1900 to 2100 Å [see Figs. VI–12b and 12c], the effective wavelengths of solar radiation for photodissociation are 1800 to 2200 and above 2900 Å in the stratosphere (843).

Two cases have been considered as a possible threat to human health as a result of the partial destruction of ozone in the stratosphere: (1) injections of NO_x into the stratosphere by SST's (supersonic transports) and of HCl by the space shuttle and (2) release of chlorofluoromethanes into the troposphere.

Both will result in the possible partial destruction of the ozone layer, although the effect of NO_x injection into the O_3 layer will become apparent much sooner than the latter. The former involves the removal of O_3 by chain reactions (VIII-45) and (VIII-46), while the latter comprises a slow diffusion process of chlorofluoromethanes through the tropopause to the stratosphere, the photodissociation producing Cl atoms and the removal of ozone by a catalytic cycle involving Cl and ClO.

Injection of NO_x *by SST's.* Johnston (543) and others have calculated that 500 Boeing SST's would at least double the rate of injection of NO_x into the stratosphere over the natural rate of coming from the photolysis of N_2O.

If one assumes that the NO_2 concentration at 20 km becomes 3×10^{10} molec cm^{-3} [corresponding to 68 ppb (parts per billion by volume) of NO_x] as a result of 500 Boeing SST's, the rate of ozone destruction is $2k_{46}[O][NO_2] = 5.4 \times 10^5$ molec sec^{-1} from (VIII-47d) and $[O] = 10^6$ molec cm^{-3}. On the other hand, in the natural stratosphere the rate of ozone destruction is 9×10^3 molec sec^{-1} using $[NO_2] = 5 \times 10^8$ molec cm^{-3} in Fig. VIII–11. Since the ozone concentration at 20 km is about 3×10^{12} molec cm^{-3}, ozone would be destroyed in about 65 days in a polluted atmosphere in comparison with about 10 years in the natural stratosphere. The more detailed calculations including eddy diffusion (vertical mixing by turbulence), NO_x distributions, and ozone photochemistry indicate the possible reduction of ozone concentration by 2 to 60%, depending on the assumed rates of NO_x emissions at 20 km from SST fleets. The ozone recovery time is 10 to 15 years after the SST fleets are stopped.

Release of chlorofluoromethanes. Chlorofluoromethanes (CCl_3F Freon-11, CCl_2F_2 Freon-12) are released into air as a result of their use as aerosol propellants and refrigerants.

Molina and Rowland (711, 843) were the first to predict the possible destruction of the stratospheric ozone by a catalytic cycle involving Cl atoms released from the photolysis of chlorofluoromethanes by sunlight. Since chlorofluoromethanes are unreactive with atoms and radicals in the troposphere, they eventually reach the stratosphere where they are photodissociated into Cl atoms by solar radiation below 2300 Å.

$$CCl_3F \xrightarrow{\lambda < 2300\ \text{Å}} CCl_2F + Cl \qquad \text{(VIII-55)}$$
$$CCl_2F_2 \xrightarrow{\lambda < 2300\ \text{Å}} CClF_2 + Cl \qquad \text{(VIII-56)}$$

The Cl atoms formed start chain reactions to consume ozone

$$Cl + O_3 \rightarrow ClO + O_2 \qquad \text{(VIII-57)}$$
$$ClO + O \rightarrow Cl + O_2 \qquad k_{58} \qquad \text{(VIII-58)}$$
$$k_{58} = 5 \times 10^{-11}\ \text{cm}^3\ \text{molec}^{-1}\ \text{sec}^{-1}$$

The rate constant k_{58} is much larger than that for the $NO_2 + O$ reaction ($k_{46} = 9 \times 10^{-12}$ cm^3 molec^{-1} sec^{-1}). Hence, the catalytic effect by the

ClO–Cl cycle is even more pronounced than that by the NO_2–NO cycle, if $[NO_2] \approx [ClO]$.

Because of the slow diffusion process of chlorofluoromethanes through the tropopause into the stratosphere and a slow photodissociation process of chlorofluoromethanes by sunlight in the stratosphere the maximum reduction of O_3 is estimated to occur 10 years after the release of chlorofluoromethanes at ground level [Rowland and Molina (843)].

The ClO radicals produced also react with NO to form NO_2, which eventually regenerates ozone by the sequence discussed before [(VIII-47) and (VIII-42)]

$$ClO + NO \rightarrow Cl + NO_2 \qquad \text{(VIII-59)}$$

Hence, (VIII-58) is the rate-determining step for the ClO–Cl catalytic cycle. The chain-terminating step is the reaction with methane to form HCl

$$Cl + CH_4 \rightarrow HCl + CH_3 \qquad \text{(VIII-60)}$$

Hydrochloric acid would partially be removed by rainout after it diffuses to the troposphere, but it would also react with OH to regenerate Cl atoms

$$HCl + OH \rightarrow H_2O + Cl \qquad \text{(VIII-61)}$$

The chain processes (VIII-57) and (VIII-58) are similar to (VIII-45) and (VIII-46) for NO_x.

The ratio of the rate constants of the two rate-determining steps (VIII-58) and (VIII-46) ($ClO + O \rightarrow Cl + O_2, NO_2 + O \rightarrow NO + O_2$) is

$$\frac{k_{58}}{k_{46}} = 6 \qquad \text{(VIII-62)}$$

Nitrogen dioxide is about 20 to 50% of the total nitrogen oxides NO_x (NO, NO_2, HNO_3, N_2O_5), while ClO represents about 10 to 15% of the total chlorine species ClO_x (Cl, ClO, HCl) at 25 to 30 km. Hence, the rate of ozone removal by ClO_x is about equal to that by NO_x if the amounts of NO_x are equal to those of ClO_x. According to a calculation by Turco and Whitten (981), the reduction of ozone in the stratosphere in the year 2022 with a continuous use of chlorofluoromethanes at present levels would be 7%. Rowland and Molina (843) conclude that the ozone depletion level at present is about 1%, but it would increase up to 15 to 20% if the chlorofluoromethane injection were to continue indefinitely at the present rates. Even if release of chlorofluorocarbons were stopped after a large reduction of ozone were found, it would take 100 or more years for full recovery, since diffusion of chlorofluorocarbons to the stratosphere from the troposphere is a slow process. The only loss mechanism of chlorofluorocarbons is the photolysis in the stratosphere, production of HCl, diffusion back to the troposphere, and rainout.

VIII–3. PHOTOCHEMISTRY OF THE ATMOSPHERES OF OTHER PLANETS

The atmospheric composition of the terrestrial group (Mercury, Venus, Earth, and Mars) is radically different from that of the Jovian group (Jupiter, Saturn, Uranus, and Neptune). The former is oxidizing (CO_2, O_2) and the latter is reducing (H_2, CH_4, NH_3). The difference comes from the size and hence the strength of the gravitational field. The giant planets are capable of retaining light gases, He and H_2, which are abundant in the sun and in the other stars of our galaxy, while in the terrestrial group light gases must have escaped into space [Huntress (492)]. Among the terrestrial group, Earth has a unique atmospheric composition (N_2, O_2) because of the presence of oceans and biological activities. Venus has lost its water on account of its much higher temperature than that of Earth. Without water on the surface much of the CO_2 outgassed from the interior has remained in the atmosphere of Venus. Mars must have contained large quantities of water and hydrogen in its primordial atmosphere. Hydrogen has escaped into space because of the weak gravitational field of Mars. The water content in the Mars atmosphere at present is very low, however, because of the low atmospheric pressure and temperature. The Jovian planets have retained all the primitive stellar gases, since these gases cannot escape from the strong gravitational fields of the planets. Their atmospheric composition closely represents that of the cosmic abundance, that is, hydrogen and helium in a ratio of 10:1. In addition to hydrogen, methane and ammonia have been detected.

Tables VIII–2 through VIII–4 show the major and minor constituents detected in the atmospheres of Mars, Venus, and Jupiter.

VIII–3.1. Photochemistry of the Mars Atmosphere

Mars is almost free of clouds and the surface can be seen from the earth through a telescope. The results of the recent space probes (1073a) reveal that the surface temperature ranges from 188 to 243°K and the Martian poles are composed of substantial amounts of water ice, seasonally covered by CO_2 frost. The rusty-red color of the surface is caused by the presence of substantial amounts of iron oxides. The mean surface atmospheric pressure is 7.65 ± 0.1 mbar. The temperature profile of the Mars atmosphere is given in Fig. VIII–12.

The high quantum yield of photolysis of CO_2 suggests the rapid destruction of CO_2 and the formation of CO and O_2 by sunlight of wavelengths below about 2200 Å (see Section VI–5). According to an estimate by McElroy and McConnell (675), the column abundance of CO_2 in the atmosphere is 2×10^{23} molec cm^{-2}. With a dissociation rate of 2.5×10^{12} cm^{-2} sec^{-1}, the entire CO_2 may be destroyed in less than 10,000 years.

Table VIII–2. Chemical Composition of the Atmosphere[a] of Mars (20, 510, 788)

Major Component	Abundance
CO_2	80–100
Ar	<20, 1.5[b]
N_2	<6, 6[b]
Minor Component	**Mixing Ratio Relative to CO_2**
O_2	1.3×10^{-3} [c], 3×10^{-3} [b]
CO	0.9×10^{-3} [c]
H_2O	0.4 to 2×10^{-4}
O_3	$<0.2 \times 10^{-7}$
C_3O_2	$<0.2 \times 10^{-6}$
COS	$<0.6 \times 10^{-6}$
NO	$<0.7 \times 10^{-6}$
N_2O	$<1.8 \times 10^{-5}$
HCN	$<5 \times 10^{-5}$
C_2H_2	$<5 \times 10^{-5}$
C_2H_4, C_2H_6	$<6 \times 10^{-6}$

[a] Pressure, 7.3 mbars (surface) corresponding to a temperature of 241°K on the surface; from Ref. 742.
[b] Results by Viking space flight 1976; from Ref. 742.
[c] From Ref. 190.

Table VIII–3. Chemical Composition of the Atmosphere[a] of Venus (20, 403a, 510, 788)

Mojor Component	Relative Abundance
CO_2	97
H_2O	0.4 to 1
N_2	2
Minor Component	**Mixing Ratio Relative to CO_2**
CO	4.6×10^{-5}
HCl	6×10^{-7}
HF	1.5×10^{-9}
O_2	$<10^{-6}$
H_2O, above clouds	$\sim 10^{-4}$
H_2O, lower atmosphere	$5\text{–}10 \times 10^{-3}$
C_3O_2	10^{-5}
O_3	10^{-8}
NH_3	$<5 \times 10^{-8}$

[a] Pressure, 100 bars (surface), 100–200 mbars (cloud top). The surface temperature is about 750°K.

Table VIII–4. Chemical Composition of the Atmosphere[a] of Jupiter (735, 788, 943)

Major Component	Abundance (%)
He	~10
H_2	~90
Minor Component	**Mixing Ratio Relative to H_2**
CH_4	7×10^{-4}
NH_3	2×10^{-4} to 3×10^{-7}
C_2H_2	$<5 \times 10^{-7}$
C_2H_4	$<2 \times 10^{-5}$
C_2H_6	$<3 \times 10^{-5}$, 10^{-8} to 10^{-7}
PH_3	$<5 \times 10^{-7}$
H_2S	$<3 \times 10^{-6}$
HCN	$<6 \times 10^{-7}$
H_2O	$<6 \times 10^{-7}$
CO	10^{-9}[b]

[a] Pressure 2 bars (cloud top).
[b] Ref. 91.

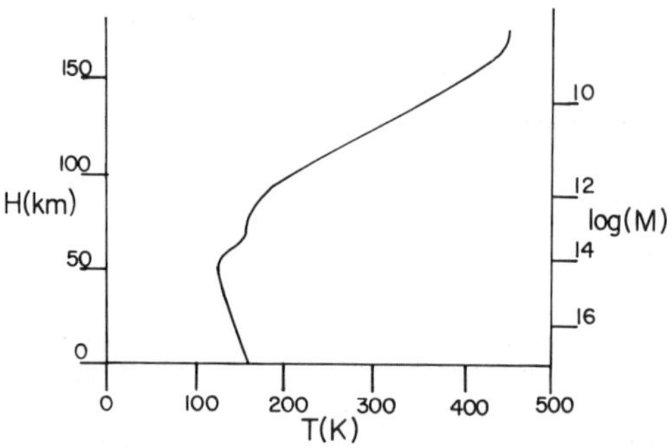

Fig. VIII–12. Temperature profile of the Mars atmosphere. The surface temperature ranges from 170 to 243°K and the mean atmospheric pressure is 7.65 mbar at the surface (1073a) (M) is the number of molecules per cm³. From McElroy (674c), reprinted by permission. Copyright by the American Geophysical Union.

$$CO_2 \xrightarrow{\lambda < 2200\ \text{Å}} CO + O \tag{VIII-63}$$

The O atoms produced combine to form O_2

$$O + O \overset{M}{\rightarrow} O_2 \tag{VIII-64a}$$

rather than CO_2

$$O + CO \overset{M}{\rightarrow} CO_2 \tag{VIII-64b}$$

since (VIII-64b) is 1000 times slower than (VIII-64a) (1).

From Table VIII–2 one can see that this is not the case. The mixing ratios of CO and O_2 are only about 0.1% of the amount of CO_2 that would be produced in only 2 years. Considerable efforts have been devoted to explain this unusual stability of CO_2 in Mars [see Hunten (491)]. Based on the abundant water vapor and HO_x (H, HO, HO_2) in the Martian atmosphere, McElroy et al. (675, 677) present a mechanism involving an HO_x cycle for the catalytic oxidation of CO to CO_2 similar to the one proposed for NO oxidation in the troposphere [see Section (VIII-2.3), p. 333].

$$CO + OH \rightarrow CO_2 + H \qquad k_{65} \tag{VIII-65}$$

$$H + O_2 \overset{M}{\rightarrow} HO_2 \qquad k_{66} \tag{VIII-66}$$

$$HO_2 + O \rightarrow OH + O_2 \qquad k_{67} \tag{VIII-67}$$

$$\text{net} \quad CO + O \rightarrow CO_2$$

The rate constant k_{65} is well established and is about 1.6×10^{-13} cm^3 molec^{-1} sec^{-1} at 298°K (1), k_{66} is 5×10^{-32} cm^6 molec^{-2} sec^{-1}, and k_{67} is estimated to be about 10^{-11} cm^3 molec^{-1} sec^{-1}.

The hydroxyl radicals are supplied by the photolysis of water

$$H_2O \overset{h\nu}{\rightarrow} H + OH \tag{VIII-68a}$$

or by the reaction of $O(^1D)$ from the photolysis of O_3 with H_2

$$O(^1D) + H_2 \rightarrow OH + H \tag{VIII-68b}$$

To explain the observed O atom concentrations, which are much lower than those expected from the photochemistry, large vertical mixing is assumed.

Parkinson and Hunten (798), on the other hand, have assumed the same HO_x catalytic cycle for CO oxidation but proposed another mechanism involving the H_2O_2 photolysis to explain the observed low abundances of O_2 and CO in the atmosphere:

$$HO_2 + HO_2 \rightarrow H_2O_2 + O_2 \tag{VIII-69}$$

$$H_2O_2 \overset{h\nu}{\rightarrow} OH + OH \tag{VIII-70}$$

$$CO + OH \rightarrow CO_2 + H \tag{VIII-71}$$

$$H + O_2 \overset{M}{\rightarrow} HO_2 \tag{VIII-72}$$

It is not known, however, whether sufficient amounts of H_2O_2 are present to support the hypothesis based on the H_2O_2 photolysis.

VIII–3.2. Photochemistry of the Venus Atmosphere

Venus is completely covered with dense clouds. The composition of the clouds has been the subject of much speculation for many years. It includes ice, carbon suboxide, sulfuric acid, hydrocarbons, mercuric chloride, ammonium chloride, and hydrated ferrous chloride. Recently, Young (1066a) has proposed, based on refractive index measurement, that the clouds are composed most probably of droplets of 75% H_2SO_4.

The temperature profile of the Venus atmosphere is shown in Fig. VIII–13. The surface temperature and pressure have recently been determined by space probes to be 747 ± 20°K and 88 ± 15 bars, respectively.

The atmospheric composition of Venus is similar to that of Mars (see Table VIII–3). Carbon dioxide is the main constituent. The CO mixing ratio is about 5×10^{-5}, but the O_2 mixing ratio is less than 10^{-6}. Minor constituents that are present in the Venus atmosphere but not in the Martian atmosphere are HCl and HF in mixing ratios of 6×10^{-7} and 1.5×10^{-9}, respectively.

Prinn (819) has proposed the ClO_x (Cl, ClO, ClO_2) cycle for the catalytic oxidation of CO to CO_2 in addition to the HO_x (H, OH, HO_2) cycle described in the preceding section [reactions (VIII-65) to (VIII-67)].

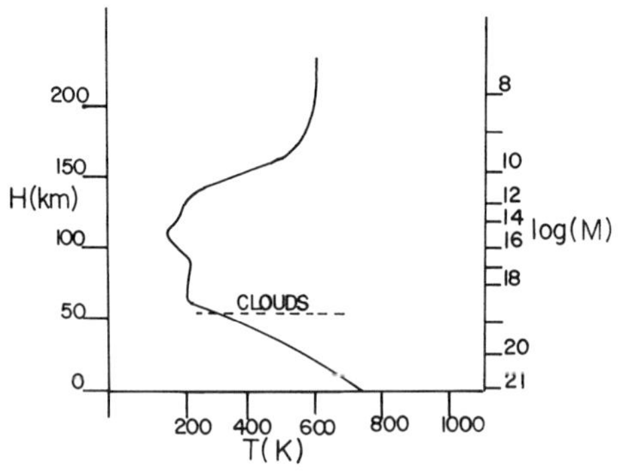

Fig. VIII–13. Temperature profile of the atmosphere of Venus. The surface corresponds to 6055 km from the center of Venus. (M) is the number of molecules per cm³. The surface pressure is 88 bars and the temperature is 750°K. Venus is covered with dense clouds (probably sulfuric acid droplets). After McEwan and Phillips (20), reprinted by permission of Edward Arnold Ltd.

$$HCl \xrightarrow{h\nu} H + Cl$$

$$Cl + O_2 \xrightarrow{M} ClOO \qquad \text{(VIII-73)}$$

$$ClOO + CO \rightarrow CO_2 + ClO \qquad \text{(VIII-74)}$$

$$ClO + CO \rightarrow Cl + CO_2 \qquad \text{(VIII-75)}$$

$$\text{net} \quad 2CO + O_2 \rightarrow 2CO_2$$

Because of the very low concentrations of O_2 observed in the Venus atmosphere, the proposed ClO_x cycle may be important in the Venus atmosphere. However, very little information is available on the rate constants involving ClO_x and the ClO_x cycle remains a hypothesis. Abundances of H_2O_2 and O_3 in the Venus atmosphere must be exceedingly small. McElroy et al. (678) considered the photolysis of HCl as another source of H atoms in addition to the photolysis of H_2O

$$HCl \xrightarrow{h\nu} H + Cl \qquad \text{(VIII-76)}$$

$$Cl + H_2 \rightarrow HCl + H \qquad \text{(VIII-77)}$$

The photochemistry of the Mars and Venus atmospheres may be better understood if the minor constituents, such as H_2O_2, O_3, and H_2, can be measured and if the rate constants involving HO_2 and ClOO radicals can be measured more accurately in the laboratory. McElroy et al. (678) believe that CO and H_2O are converted to CO_2 and H_2 in the hot region near the surface

$$CO + H_2O \rightarrow CO_2 + H_2$$

VIII–3.3. Photochemistry of the Jovian Atmosphere

Jupiter is a huge planet with a volume 1312 times that of Earth, but the average density is only 1.3 g cm^{-3}. It is covered by clouds marked with colored bands running parallel to the equator. The clouds also contain light and dark spots, including the famous Red Spot. Jupiter radiates more than twice as much heat as it absorbs from the sun, which indicates a convective interior all the way to the center. There is no evidence of a solid surface to which height can be referred. The temperature profile of the Jupiter atmosphere is shown in Fig. VIII–14 where the height above the tropopause is given [Hunten (490a, 490b)]. It appears that the boundary between stratosphere and mesosphere is not distinct with a uniform temperature of about 100°K. The observed cloud deck is probably ammonia ice.

The atmospheric composition of Jupiter is much different from those of Mars and Venus. It is similar to the primitive stellar atmospheres. The

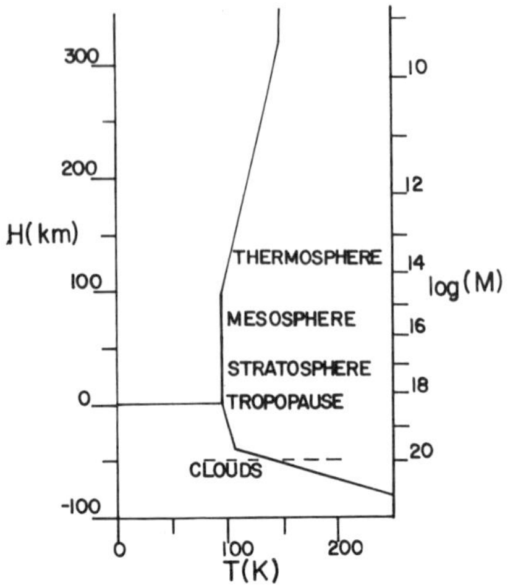

Fig. VIII–14. Proposed temperature profile of Jupiter's Atmosphere. The tropopause is chosen as height reference since there is no evidence of a solid surface. The temperature at the tropopause is 95.5°K and the number density is 2×10^{18} cm^{-3}. Contrary to the case of the upper atmosphere of earth, there appears to be no boundary between stratosphere and mesosphere. The observed cloud deck is believed to be solid ammonia. (M) signifies the number of molecules per cm^3. From Hunten (490b), reprinted by permission of the American Meteorological Society.

main constituents are H_2 and He and the minor constituents are CH_4 and NH_3. The composition is given in Table VIII–4.

Concentration profiles of H_2, He, CH_4, and NH_3 at 150°K and with the eddy diffusion coefficient $K = 3 \times 10^5$ cm^2 sec^{-1} have been calculated by Strobel (943) and are given in Fig. VIII–15. Because of the presence of CH_4 above the layer of NH_3, solar radiation of wavelengths only above about 1600 Å is effective in photodissociating NH_3.

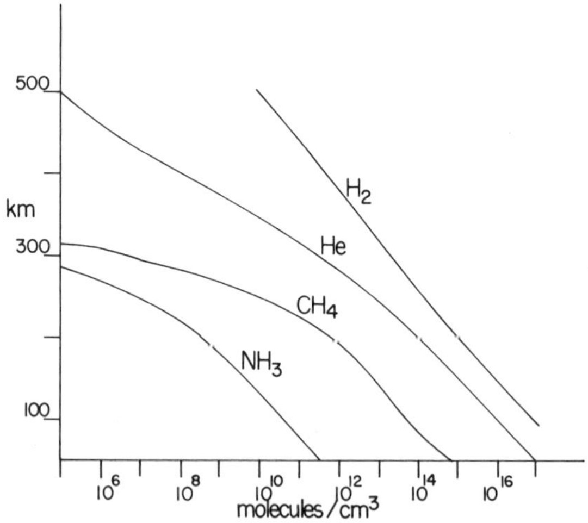

Fig. VIII–15. Concentration profiles of main constituents in the Jovian atmosphere. $T = 150$°K, $K = 3 \times 10^5$ cm^2 sec^{-1}. The height refers to the cloud top. From Strobel (943), reprinted by permission of Reidel.

Jupiter is much further away from the sun compared with the terrestrial planets and is colder with a mean temperature of about 100°K. Hence, any chemical reaction that requires an activation energy cannot occur in the Jovian atmosphere.

The stability of methane and ammonia against the photolysis by solar radiation has been discussed by McNesby (685) and Strobel (442, 944–946).

The photochemistry of CH_4 is predominantly the dissociation into $CH_3 + H$, $CH_2 + H_2$, and $CH + H + H_2$ by Lyman α radiation [see the methane photolysis Section VII–17, p. 298]

$$CH_4 \xrightarrow{h\nu} CH_3 + H \qquad \text{(VIII-78)}$$
$$CH_4 \xrightarrow{h\nu} CH_2 + H_2 \qquad \text{(VIII-79)}$$
$$CH_4 \xrightarrow{h\nu} CH + H + H_2 \qquad \text{(VIII-80)}$$

followed by

$$CH_2 + H_2 \rightarrow CH_3 + H \qquad \text{(VIII-81)}$$
$$CH_2 + CH_4 \rightarrow 2CH_3 \qquad \text{(VIII-82)}$$
$$CH + CH_4 \rightarrow C_2H_4 + H \qquad \text{(VIII-83)}$$
$$CH + H_2 \xrightarrow{M} CH_3 \qquad \text{(VIII-84)}$$
$$CH_3 + H \xrightarrow{M} CH_4 \qquad \text{(VIII-85)}$$
$$2CH_3 \xrightarrow{M} C_2H_6 \qquad \text{(VIII-86)}$$

Thus, the main photolysis products are C_2H_4 and C_2H_6. Acetylene, a minor observed product, may be formed from the photolysis of ethylene and ethane.

The stability of CH_4 against solar radiation has been explained on the basis of electronically excited CH_3 radicals (CH_3^*) formed by absorption of sunlight [McNesby (685)]. The CH_3^* radicals may be able to abstract H atoms from H_2, while the abstraction by ground state CH_3 radicals requires a high activation energy and hence does not occur.

$$CH_3^* + H_2 \rightarrow CH_4 + H \qquad \text{(VIII-87)}$$

Strobel (942), on the other hand, has proposed the sequence that higher hydrocarbons are transported downward into the hotter regions where they decompose thermally to produce methane, which in turn is transported upward to supplement methane lost by photolysis.

According to Strobel (945, 946) the upper atmosphere (> 100 km) photochemistry is dominated by the photolysis of methane. Only below 100 km the atmosphere contains sufficient ammonia to be photochemically important. The photochemically effective wavelengths for NH_3 photolysis are in the range from 1600 Å, the onset of CH_4 absorption, to 2300 Å, the onset of NH_3 absorption. The photolysis of NH_3 has already been discussed (see

Section VII–1) and may be summarized as follows

$$NH_3 \xrightarrow{\lambda > 1600 \text{ Å}} NH_2 + H \quad \text{(VIII-88)}$$

$$NH_2 + NH_2 \xrightarrow{M} N_2H_4 \quad \text{(VIII-89)}$$

$$NH_2 + H \xrightarrow{M} NH_3 \quad \text{(VIII-90)}$$

The combination of NH_2 to form N_2H_4 is a more favorable path than the reformation of NH_3, (VIII-90), and NH_3 is photochemically destructed. The N_2H_4 formed is partially converted into N_2 and H_2

$$N_2H_4 \xrightarrow{h\nu} N_2 + 2H_2 \quad \text{(VIII-91)}$$

$$H + N_2H_4 \rightarrow N_2H_3 + H_2 \quad \text{(VIII-92)}$$

$$H + N_2H_3 \rightarrow N_2 + 2H_2 \quad \text{(VIII-93)}$$

To explain the stability of ammonia, Strobel (945) further suggests a slow downward transport of N_2H_4 to the hotter dense regions of the deep atmosphere where it decomposes thermally to NH_2 radicals which react with H_2 to re-form NH_3.

On the other hand, McNesby (685) has proposed a mechanism to regenerate NH_3 by assuming the reaction of electronically excited $NH_2(NH_2^*)$ produced by absorption of sunlight in the visible region

$$NH_2\,(\tilde{X}^2B_1) \xrightarrow{h\nu} NH_2(\tilde{A}^2A_1) \quad \text{(VIII-94)}$$

The NH_2^* radicals produced may be capable of abstracting H atoms from H_2

$$NH_2^* + H_2 \rightarrow NH_3 + H \quad \text{(VIII-95)}$$

The probability of NH_2 disappearance by (VIII-89) is about 10^{-3} sec^{-1} using $k_{89} = 10^{-10}$ cm^3 molec^{-1} sec^{-1} and $[NH_2] = 10^7$ molec cm^{-3}, while that of $NH_2(\tilde{A}^2A_1)$ production (VIII-94) is 10^{-4} sec^{-1} on the basis of a solar intensity of 10^{16} photons cm^{-2} sec^{-1} and an absorption cross section of 10^{-20} cm^2 of NH_2 (625a) in the region 4300 to 9000 Å. Thus, (VIII-95) may not be significant in the lower stratosphere but may be important in the upper stratosphere. Strobel (945) believes that the NH_3 density profile in the stratosphere deviates significantly from a mixing ratio of 7.6×10^{-7} because of photochemical destruction and slow mixing.

appendix

Reference Tables

Table A–1. Fundamental Constants[a]

Quantity	Symbol	Value	Units: SI	Units: cgs
Velocity of light	c	2.99792458	10^8 m sec^{-1}	10^{10} cm sec^{-1}
Planck constant	h	6.626176	10^{-34} J sec	10^{-27} erg sec
Electronic charge	e	4.803242		10^{-10} esu
Electron rest mass	m_e	9.109534	10^{-31} kg	10^{-28} g
Rydberg constant	R_∞	1.097373177	10^7 m^{-1}	10^5 cm^{-1}
Bohr radius	a_0	5.2917706	10^{-11} m	10^{-9} cm
Avogadro number	N	6.022045	10^{23} mol^{-1}	10^{23} mol^{-1}
Boltzmann constant	k	1.380662	10^{-23} J K^{-1}	10^{-16} erg K^{-1}
Gas constant	R	8.31441	J mol^{-1} K^{-1}	10^7 erg mol^{-1} K^{-1}
One thermochemical calorie	cal	4.18400	Joules	10^7 ergs[b]
Standard volume of ideal gas	V_0	22.41383	10^{-3} m^3 mol^{-1}	10^3 cm^3 mol^{-1}
Loschmidt number	n_0	2.686754	10^{25} m^{-3}	10^{19} cm^{-3}

[a] From Ref. 6a.
[b] Ref. 28.

Table A–2. Energy Levels and Transition Probabilities of Some Atoms of Photochemical Interest

The data are taken mostly from Refs. 7, 21, 32, and 33 and are supplemented by recent individual papers as noted. The electronic energy E_0 above the ground state is given in electron volts and an odd term is designated by superscript o as $^2p^o$. The upper state is always given first for the transition. Only the electron involved in the transition is given in parenthesis and is designated by its principal and azimuthal quantum numbers. The symbols f and A mean, respectively, the oscillator strength and transition probability in sec^{-1} for the indicated transition, and $\tau = 1/A$ if only one transition is involved. For more than one transition $\tau = 1/\sum_i A_i$. The corresponding wavelength is given in angstrom units. The atoms are arranged according to the number of valence electrons.

State	E_0 (eV)	Transition	Wavelength (Å)	f	A (10^8 sec^{-1})	τ	Ref.
			Helium (He)[a]				
^1S	0						
		$^1P^o$–$^1S(2p$–$1s)$	584	0.2762	17.99		
$^1P^0_1$	21.217					0.56 nsec	
			Neon (Ne)[a]				
^1S	0						
		$^1P^o_1$–$^1S(3s$–$2p)$	736	0.162	6.64		
		$^3P^o_1$–$^1S(3s$–$2p)$	744	0.0118	0.476		
$^3P^o_1$	16.670					21.0 nsec	
$^1P^o_1$	16.848					1.51 nsec	
			Argon (Ar)[b]				
^1S	0						
		$^1P^o_1$–$^1S_0(4s'$–$3p)$	1,048	0.254	5.1	1.96 nsec	
		$^3P^o_1$–$^1S_0(4s$–$3p)$	1,067	0.061	1.19	8.40 nsec	
$^3P^o_1$	11.623					8.6 nsec	613
$^1P^o_1$	11.827					2.15 nsec	613

Table A.2. (*continued*)

State	E_0 (eV)	Transition	Wavelength (Å)	f	A (10^8 sec^{-1})	τ	Ref.
		Krypton (Kr)					
^1S	0						
		$^1P_1^o-^1S_0(5s'-4p)$	1,165	0.135	2.19		1041
				0.193			424
		$^3P_1^o-^1S_0(5s-4p)$	1,236	0.158	2.28		1041
				0.187	2.72		424
$^3P_1^o$	10.032					3.68 ± 0.11 nsec	
$^1P_1^o$	10.643					3.16 ± 0.15 nsec	424
		Xenon (Xe)					
^1S	0						
		$^1P_1^o-^1S_0(6s'-5p)$	1,295	0.270	3.57		1043
					2.58		424
		$^3P_1^o-^1S_0(6s-5p)$	1,470	0.260	2.67		1043
$^3P_1^o$	8.436					3.79 ± 0.12 nsec	
$^1P_1^o$	9.569					3.17 ± 0.19 nsec	41
						3.88 ± 0.10 nsec	424
		Hydrogen (H)[c]					
^2S	0						
		$^2P^o-^2S(3p-1s)$	1,026	7.91×10^{-2}	1.672	5.98 nsec	
		$^2P^o-^2S(2p-1s)$	1,216	0.4162	6.265	1.60 nsec	
$(2p)^2P^o$	10.198					1.60 nsec	213

			Sodium (Na)[b]				
^2S	0						
		$^2P^o_{3/2}$–$^2S(3p–3s)$	5,890	0.655	0.630	15.8 nsec	
				0.650			639
		$^2P^o_{1/2}$–$^2S(3p–3s)$	5,896	0.327	0.628	15.9 nsec	
				0.325			639
$^2P^o_{1/2}$	2.102					16.1 $\pm$ 0.3 nsec	639
$^2P^o_{3/2}$	2.104					16.1 $\pm$ 0.3 nsec	639
			Potassium (K)[b]				
^2S	0						
		$^2P^o_{3/2}$–$^2S(4p–4s)$	7,665	0.682	0.387	27.8 $\pm$ 0.5 nsec	
				0.639			639
		$^2P^o_{1/2}$–$^2S(4p–4s)$	7,699	0.339	0.382	27.8 $\pm$ 0.5 nsec	
				0.318			639
$^2P^o_{1/2}$	1.609						
$^2P^o_{3/2}$	1.617						
			Rubidium (Rb)				
^2S	0						
		$^2P^o_{3/2}$–$^2S(5p–5s)$	7,800	0.675		27.0 $\pm$ 0.5 nsec	639
		$^2P^o_{1/2}$–$^2S(5p–5s)$	7,948	0.335		28.1 $\pm$ 0.5 nsec	639
$^2P^o_{1/2}$	1.559						
$^2P^o_{3/2}$	1.588						

Table A.2. (*continued*)

State	E_0 (eV)	Transition	Wavelength (Å)	f	A (10^8 sec^{-1})	τ	Ref.
			Cesium (Cs)				
^2S	0						
		$^2P^o_{3/2}$–$^2S(6p–6s)$	8,521	0.732		30.5 ± 0.6 nsec	639
		$^2P^o_{1/2}$–$^2S(6p–6s)$	8,944	0.362		34.0 ± 0.6 nsec	639
$^2P^o_{3/2}$	1.454						
			Zinc (Zn)				
^1S	0						
		$^1P^o_1$–$^1S_0(4p–4s)$	2,139	1.46			653
		$^3P^o_1$–$^1S_0(4p–4s)$	3,076	0.0018			7
$^3P^o_1$	4.0295					20 μsec	653
$^1P^o_1$	5.7955					1.41 ± 0.04 nsec	653
			Cadmium (Cd)				
^1S	0						
		$^1P^o_1$–$^1S(5p–5s)$	2,288	1.20		1.98 nsec	21
		$^3P^o_1$–$^1S(5p–5s)$	3,261	0.0019		2.5 μsec	21
$^3P^o_0$	3.733						
$^3P^o_1$	3.800					2.39 ± 0.04 μsec	652
$^1P^o_1$	5.417					1.66 ± 0.5 nsec	652

Mercury (Hg)[d]

^1S	0						
		$^1P^o_1$–$^1S(6p–6s)$	1,849	1.18		1.31 nsec	654
		$^3P^o_1$–$^1S(6p–6s)$	2,537	0.0255		0.114 μsec	654
							21
				gf	**A (10^8 sec^{-1})**		
$^3P^o_0$	4.667						
		$^3S_1$–$^3P^o_0(8s–6p)$	2,753	0.12	1.1		
		$^3D_1$–$^3P^o_0(6d–6p)$	2,967	2.9	22		
		$^3S_1$–$^3P^o_0(7s–6p)$	4,047	8.8	36		
$^3P^o_1$	4.886						
		$^3S_1$–$^3P^o_1(8s–6p)$	2,894	0.23	1.9		
		$^3D_2$–$^3P^o_1(6d–6p)$	3,126	1.1	7.8		
		$^3D_1$–$^3P^o_1(6d–6p)$	3,131	0.92	6.2		
		$^1S_0$–$^3P^o_1(7s–6p)$	4,077	0.64	2.6		
		$^3S_1$–$^3P^o_1(7s–6p)$	4,358	24	86		
$^3P^o_2$	5.461						
		$^3D_3$–$^3P^o_2(7d–6p)$	3,021	0.54	4.0		
		$^3S_1$–$^3P^o_2(8s–6p)$	3,341	0.22	1.3		
		$^3D_3$–$^3P^o_2(6d–6p)$	3,650	13	64		
		$^3D_2$–$^3P^o_2(6d–6p)$	3,655	1.4	6.8		
		$^3D_1$–$^3P^o_2(6d–6p)$	3,663	0.37	1.8		
		$^1D_2$–$^3P^o_2(6d–6p)$	3,663	1.1	5.5		
		$^3S_1$–$^3P^o_2(7s–6p)$	5,460	38	86		
$^1P^o_1$	6.703						
		$^3D_2$–$^1P^o_1(6d–6p)$	5,770	4.3	8.6		
		$^1D_2$–$^1P^o_1(6d–6p)$	5,791	5.1	10		

Table A ,2. (*continued*)

State	E_0 (eV)	Transition	Wavelength (Å)	f	A (10^8 sec^{-1})	τ	Ref.
				f	A (sec^{-1})		
			Carbon (C)[a]				
^3P	0						
		$^3P^o$–$^3P(4s$–$2p)$	1,280	0.02	0.27×10^8		
		$^3P^o$–$^3P(2p$–$2s)$	1,329	0.039	0.49×10^8		
		$^3P^o$–$^3P(2p$–$2s)$	1,560	0.091	0.84×10^8		
		$^3P^o$–$^3P(3s$–$2p)$	1,657	0.17	1.4×10^8		
^1D	1.263					53 min	
		$^1F^o$–$^1D(3d$–$2p)$	1,463	0.093	2.1×10^8		
		$^1D^o$–$^1D(3d$–$2p)$	1,482	0.011	0.33×10^8		
		$^1P^o$–$^1D(3s$–$2p)$	1,931	0.082	2.4×10^8		
		^1D–$^3P(2p$–$2p)$	9,850		2.3×10^{-4}		
			9,823		7.8×10^{-5}		
^1S	2.683					2 sec	
		$^1P^o$–$^1S(3d$–$2p)$	1,752	0.12	0.87×10^8		
		$^1P^o$–$^1S(3s$–$2p)$	2,479	0.094	0.34×10^8		
		^1S–$^3P(2p$–$2p)$	4,622		0.0026		
		^1S–$^1D(2p$–$2p)$	8,727		0.50		
				gf	gA (10^8 sec^{-1})		
			Tin (Sn)[d]				
^3P	0	$^3P^o$–$^3P(6s$–$5p)$	2,863	0.65	5.3		
^1D	1.068	$^3F^o$–$^1D(5d$–$5p)$	2,851	1.3	11		
^1S	2.128	$^3D^o$–$^1S(6d$–$5p)$	2,914	1.2	9.5		
			Lead (Pb)[d]				
^3P	0	$^3P^o$–$^3P(7s$–$6p)$	2,833	0.22	1.8		

				f	A (sec $^{-1}$)	
			Nitrogen (N)[a]			
$^4S^o$	0					
		$^4P–^4S^o(2p–2s)$	1,135	0.13	2.3×10^8	
		$^4P–^4S^o(3s–2p)$	1,200	0.35	5.4×10^8	
$^2D^o$	2.38					12 hr
		$^2D–^2D^o(3s'–2p)$	1,243	0.11	4.6×10^8	
		$^2P–^2D^o(3s–2p)$	1,493	0.11	5.5×10^8	
		$^2D^o–^4S^o(2p–2p)$	5,199		1.6×10^{-5}	
			5,201		7×10^{-6}	
$^2P^o$	3.576					6 sec
		$^2D–^2P^o(3s'–2p)$	1,412	0.026	0.52×10^8	
		$^2P–^2P^o(3s–2p)$	1,744	0.091	2.0×10^8	
		$^2P^o–^4S^o(2p–2p)$	3,466		0.0087	
		$^2P^o–^2D^o(2p–2p)$	10,395		0.085	
			10,404		0.071	
				gf	gA (10^8 sec $^{-1}$)	
			Arsenic (As)[d]			
$^4S^o_{3/2}$	0					
		$^4P–^4S^o_{3/2}(5s–4p)$	1,972	0.29	5	
$^2D^o_{3/2}$	1.313					
		$^2P–^2D^o_{3/2}(5s–4p)$	2,350	2.1	26	
$^2D^o_{5/2}$	1.353					
		$^2P–^2D^o_{5/2}(5s–4p)$	2,288	1.2	15	
$^2P^o_{1/2}$	2.254					
		$^2P–^2P^o_{1/2}(5s–4p)$	2,860	4.1	33	
$^2P^o_{3/2}$	2.312					
		$^2P–^2P^o_{3/2}(5s–4p)$	2,780	6.9	60	

Table A, 2. (*continued*)

State	E_0 (eV)	Transition	Wavelength (Å)	f	A (10^8 sec^{-1})	τ	Ref.
				f	A (sec^{-1})	τ	
			Oxygen (O)[e]				
^3P	0						
		$^3S^o$–$^3P(3s$–$2p)$	1,302	0.031	2.1×10^8	2.4 nsec	856
				0.048		1.79 nsec	914
^1D	1.967					150 sec	
		$^1D^o$–$^1D(3s'$–$2p)$	1,152	0.090	4.5×10^8	1.9 nsec	380
				0.112		1.77 nsec	914,786
		^1D–$^3P(2p$–$2p)$	6,300		0.0051		
^1S	4.189					0.71 sec	
		$^1P^o$–$^1S(3s''$–$2p)$	1,218	0.13	2.0×10^8		
						0.9 nsec	914
		^1S–$^3P(2p$–$2p)$	2,972		0.067		
		^1S–$^1D(2p$–$2p)$	5,577		1.34		
			Sulfur (S)[b]				
^3P	0						
		$^3P^o$–$^3P(4s''$–$3p)$	1,296	0.12	4.8×10^8		
		$^3D^o$–$^3P(3d$–$3p)$	1,425	0.15	3.5×10^8	2.1 nsec	856
		$^3D^o$–$^3P(4s'$–$3p)$	1,474	0.075	1.6×10^8		
		$^3S^o$–$^3P(4s$–$3p)$	1,807	0.12	4.1×10^8	1.5 nsec	856
^1D	1.145					28 sec	
		$^1P^o$–$^1D(4s''$–$3p)$	1,448	0.13	6.9×10^8		
		$^1D^o$–$^1D(4s'$–$3p)$	1,667	0.24	5.8×10^8	1.5 nsec	856
		^1D–$^3P(3p$–$3p)$	10,820		0.0275		
			11,206		0.0880		

[illegible]	2.750					0.47 sec	
		$^1P^o$–$^1S(3p$–$3s)$	1,687	0.12	0.94×10^8	17 nsec	856
		$^1P^o$–$^1S(4s''$–$3p)$	1,782	0.22	1.5×10^8		
		^1S–$^3P(3p$–$3p)$	4,589		0.35		
		^1S–$^1D(3p$–$3p)$	7,725		1.78		
				f	**A (10^8 sec^{-1})**	**τ**	
			Fluorine (F)[a]				
$^2P^o_{3/2}$	0						
		^2P–$^2P^o_{3/2}(3s$–$2p)$	955				
		^4P–$^2P^o_{3/2}(3s$–$2p)$	974				
$^2P^o_{1/2}$	0.050					847 sec	
		^2P–$^2P^o_{1/2}(3s$–$2p)$	956				
			Chlorine (Cl)[e]				
$^2P^o_{3/2}$	0						
		^2P–$^2P^o_{3/2}(4s$–$3p)$	1,347	0.114	4.19		
$^2P^o_{1/2}$	0.109					81 sec	
		^2D–$^2P^o_{1/2}(4s'$–$3p)$	1,201	0.103	2.39		
		^2P–$^2P^o_{1/2}(4s$–$3p)$	1,351	0.088	3.23		
			Bromine (Br)				
$^2P^o_{3/2}$	0						
		^2P–$^2P^o_{3/2}(5s$–$4p)$	1,489		1.66		612
		^4P–$^2P^o_{3/2}(5s$–$4p)$	1,495		0.0212		612
		^4P–$^2P^o_{3/2}(5s$–$4p)$	1,541		1.28		612
		^4P–$^2P^o_{3/2}(5s$–$4p)$	1,577		0.0191		612
$^2P^o_{1/2}$	0.456					1.122 sec	387
		^2P–$^2P^o_{1/2}(5s$–$4p)$	1,532		1.95		612
		^4P–$^2P^o_{1/2}(5s$–$4p)$	1,582		0.082		612
		^4P–$^2P^o_{1/2}(5s$–$4p)$	1,634		0.10		612

Table A.2. (continued)

State	E_0 (eV)	Transition	Wavelength (Å)	f	A (10^8 sec^{-1})	τ	Ref.
			Iodine (I)				
$^2P^o_{3/2}$	0						
		^2P–$^2P^o_{3/2}(6s–5p)$	1,783		2.71		612
		^4P–$^2P^o_{3/2}(6s–5p)$	1,830		0.160		612
$^2P^o_{1/2}$	0.942					108 ± 10 msec	229
						0.127 sec	387
						0.02–0.045 sec	268
							495
		^2P–$^2P^o_{1/2}(6s–5p)$	1,799		2.11		612
		^4P–$^2P^o_{1/2}(6s–4p)$	1,844		0.0692		612
		^2P–$^2P^o_{1/2}(6s–5p)$	2,062		0.0296		612
		$^2P^o_{1/2}$–$^2P^o_{3/2}(5p–5p)$	13,152				

[a] From Ref. 32.
[b] From Ref. 33 unless otherwise noted.
[c] From Ref. 32 unless otherwise noted.
[d] From Ref. 7; g signifies the statistical weight.
[e] From Ref. 33.

Table A–3. Conversion Factors for Absorption Coefficients (9)

The absorption coefficient a is defined by the equation

$$I_t = I_0 e^{-abl} \quad \text{or} \quad I_t = I_0 10^{-abl}$$

where I_t and I_0 are the transmitted and incident light intensities, b denotes either pressure or concentration, and l is the path length in centimeters. Depending on whether b is given in pressure units (torr or atm) or in concentration units (mol dm^{-3}), a is designated as k or ϵ. If k (pressure^{-1} cm^{-1}) is used, it is necessary to specify the temperature to which the pressure is referred. At 25°C k is 9% less than that at 0°C.

Sometimes the absorption cross section σ, defined as $\sigma = k$ (atm^{-1} cm^{-1} at 0°C)$/n_0 = k/2.687 \times 10^{19}$ cm^2 $molec^{-1}$ is used instead of k or ϵ, where n_0 is the Loshmidt number. The absorption cross section is sometimes expressed in megabarns; 1 Mb = 10^{-18} cm^2. The absorption cross section is nearly temperature independent between 0°C and room temperature, but may change at much higher and lower temperatures.

To Convert From	Base	to	Base	Multiply By
k (atm, 298°K)$^{-1}$ cm^{-1}	e	σ (cm^2 $molec^{-1}$)	e	4.06×10^{-20}
k (atm, 298°K)$^{-1}$ cm^{-1}	e	k (atm, 273°K)$^{-1}$ cm^{-1}	e	1.09
k (atm, 298°K)$^{-1}$ cm^{-1}	e	ϵ (dm^3 mol^{-1} cm^{-1})	10	10.6
k (atm, 298°K)$^{-1}$ cm^{-1}	10	σ	e	9.35×10^{-20}
k (atm, 298°K)$^{-1}$ cm^{-1}	10	k (atm, 273°K)$^{-1}$ cm^{-1}	e	2.51
k (atm, 298°K)$^{-1}$ cm^{-1}	10	ϵ	10	24.4
k (mm Hg, 298°K)$^{-1}$ cm^{-1}	10	σ	e	7.11×10^{-17}
k (mm Hg, 298°K)$^{-1}$ cm^{-1}	10	k (atm, 273°K)$^{-1}$ cm^{-1}	e	1.91×10^{3}
k (mm Hg, 298°K)$^{-1}$ cm^{-1}	10	ϵ	10	1.86×10^{4}
k (atm, 273°K)$^{-1}$ cm^{-1}	e	σ	e	3.72×10^{-20}
k (atm, 273°K)$^{-1}$ cm^{-1}	e	ϵ	10	9.73
k (atm, 273°K)$^{-1}$ cm^{-1}	10	σ	e	8.57×10^{-20}
k (atm, 273°K)$^{-1}$ cm^{-1}	10	k (atm, 273°K)$^{-1}$ cm^{-1}	e	2.303
k (atm, 273°K)$^{-1}$ cm^{-1}	10	ϵ	10	22.4
ϵ (dm^3 mol^{-1} cm^{-1})	10	σ	e	3.82×10^{-21}
ϵ (dm^3 mol^{-1} cm^{-1})	10	k (atm, 273°K)$^{-1}$ cm^{-1}	e	0.103
σ (cm^2 $molec^{-1}$)	e	k (atm, 273°K)$^{-1}$ cm^{-1}	e	2.69×10^{19}
σ (cm^2 $molec^{-1}$)	e	ϵ	10	2.6×10^{20}

Table A–4. Conversion Factors for Second Order Rate Constants (9)

To Convert From	to	Multiply By
1 $cm^3\ mol^{-1}\ sec^{-1}$	$dm^3\ mol^{-1}\ sec^{-1}$	10^{-3}
1 $cm^3\ mol^{-1}\ sec^{-1}$	$cm^3\ molec^{-1}\ sec^{-1}$	0.166×10^{-23}
1 $cm^3\ mol^{-1}\ sec^{-1}$	$(mm\ Hg)^{-1}\ sec^{-1}$	$16.03 \times 10^{-6}\ K^{-1}$
1 $dm^3\ mol^{-1}\ sec^{-1}$	$cm^3\ mol^{-1}\ sec^{-1}$	10^{3}
1 $dm^3\ mol^{-1}\ sec^{-1}$	$cm^3\ molec^{-1}\ sec^{-1}$	0.166×10^{-20}
1 $dm^3\ mol^{-1}\ sec^{-1}$	$(mm\ Hg)^{-1}\ sec^{-1}$	$16.03 \times 10^{-3}\ K^{-1}$
1 $cm^3\ molec^{-1}\ sec^{-1}$	$cm^3\ mol^{-1}\ sec^{-1}$	6.023×10^{23}
1 $cm^3\ molec^{-1}\ sec^{-1}$	$dm^3\ mol^{-1}\ sec^{-1}$	6.023×10^{20}
1 $cm^3\ molec^{-1}\ sec^{-1}$	$(mm\ Hg)^{-1}\ sec^{-1}$	$96.53 \times 10^{17}\ K^{-1}$
1 $(mm\ Hg)^{-1}\ sec^{-1}$	$cm^3\ mol^{-1}\ sec^{-1}$	$62.40 \times 10^{3}\ K$
1 $(mm\ Hg)^{-1}\ sec^{-1}$	$dm^3\ mol^{-1}\ sec^{-1}$	$62.40\ K$
1 $(mm\ Hg)^{-1}\ sec^{-1}$	$cm^3\ molec^{-1}\ sec^{-1}$	$10.36 \times 10^{-20}\ K$
1 $atm^{-1}\ sec^{-1}$	$cm^3\ mol^{-1}\ sec^{-1}$	$82.10\ K$
1 $atm^{-1}\ sec^{-1}$	$dm^3\ mol^{-1}\ sec^{-1}$	$82.10 \times 10^{-3}\ K$
1 $atm^{-1}\ sec^{-1}$	$cm^3\ molec^{-1}\ sec^{-1}$	$13.63 \times 10^{-23}\ K$
1 $atm^{-1}\ sec^{-1}$	$(mm\ Hg)^{-1}\ sec^{-1}$	1.316×10^{-3}
1 $ppm^{-1}\ min^{-1}$	$cm^3\ mol^{-1}\ sec^{-1}$	4.11×10^{8} (300°K)[a]
1 $ppm^{-1}\ min^{-1}$	$dm^3\ mol^{-1}\ sec^{-1}$	4.11×10^{5} (300K)[a]
1 $ppm^{-1}\ min^{-1}$	$cm^3\ molec^{-1}\ sec^{-1}$	6.81×10^{-16} (300°K)[a]

[a] Calculated by the author.

Table A–5A. Conversion Factors for Third Order Rate Constants (9)

To Convert From	to	Multiply By
1 $cm^6\ mol^{-2}\ sec^{-1}$	$dm^6\ mol^{-2}\ sec^{-1}$	10^{-6}
1 $cm^6\ mol^{-2}\ sec^{-1}$	$cm^6\ molec^{-2}\ sec^{-1}$	2.76×10^{-48}
1 $cm^6\ mol^{-2}\ sec^{-1}$	$(mm\ Hg)^{-2}\ sec^{-1}$	$2.57 \times 10^{-10}\ K^{-2}$
1 $dm^6\ mol^{-2}\ sec^{-2}$	$cm^6\ mol^{-2}\ sec^{-1}$	10^{6}
1 $dm^6\ mol^{-2}\ sec^{-2}$	$cm^6\ molec^{-2}\ sec^{-1}$	2.76×10^{-42}
1 $dm^6\ mol^{-2}\ sec^{-2}$	$(mm\ Hg)^{-2}\ sec^{-1}$	$2.57 \times 10^{-4}\ K^{-2}$
1 $cm^6\ molec^{-2}\ sec^{-1}$	$cm^6\ mol^{-2}\ sec^{-1}$	36.28×10^{46}
1 $cm^6\ molec^{-2}\ sec^{-1}$	$dm^6\ mol^{-2}\ sec^{-1}$	36.28×10^{40}
1 $cm^6\ molec^{-2}\ sec^{-1}$	$(mm\ Hg)^{-2}\ sec^{-1}$	$93.18 \times 10^{36}\ K^{-2}$
1 $(mm\ Hg)^{-2}\ sec^{-1}$	$cm^6\ mol^{-2}\ sec^{-1}$	$38.94 \times 10^{8}\ K^{2}$
1 $(mm\ Hg)^{-2}\ sec^{-1}$	$dm^6\ mol^{-2}\ sec^{-1}$	$38.94 \times 10^{2}\ K^{2}$
1 $(mm\ Hg)^{-2}\ sec^{-1}$	$cm^6\ molec^{-2}\ sec^{-1}$	$1.07 \times 10^{-38}\ K^{2}$
1 $atm^{-2}\ sec^{-1}$	$cm^6\ mol^{-2}\ sec^{-1}$	$6.740 \times 10^{3}\ K^{2}$
1 $atm^{-2}\ sec^{-1}$	$dm^6\ mol^{-2}\ sec^{-1}$	$6.740 \times 10^{-3}\ K^{2}$
1 $atm^{-2}\ sec^{-1}$	$cm^6\ molec^{-2}\ sec^{-1}$	$1.86 \times 10^{-44}\ K^{2}$
1 $ppm^{-2}\ min^{-1}$	$cm^6\ mol^{-2}\ sec^{-1}$	1.01×10^{19} (300°K)[a]
1 $ppm^{-2}\ min^{-1}$	$dm^6\ mol^{-2}\ sec^{-1}$	1.01×10^{12} (300°K)[a]
1 $ppm^{-2}\ min^{-1}$	$cm^6\ molec^{-2}\ sec^{-1}$	2.79×10^{-29} (300°K)[a]

[a] Calculated by the author.

Table A–5B. Conversion from Pressure to Concentration Units

Unit	Equivalent
1 atm (0°C)	2.6867×10^{19} molec cm^{-3}
1 atm (0°C)	4.4615×10^{-5} mol cm^{-3}
1 atm (0°C)	4.4615×10^{-2} mol dm^{-3}
1 atm (0°C)	1.013250×10^{5} newton m^{-2} (Nm^{-2}, pascal)
1 atm (0°C)	760 torr (mm Hg, 0°C)
1 torr (mm Hg, 0°C)	3.5351×10^{16} molec cm^{-3}
1 torr (mm Hg, 0°C)	5.8704×10^{-8} mol cm^{-3}
1 torr (mm Hg, 0°C)	5.8704×10^{-5} mol dm^{-3}
1 torr (mm Hg, 0°C)	1.3332×10^{2} Nm^{-2}
1 Nm^{-2}	2.6515×10^{14} molec cm^{-3}
1 Nm^{-2}	4.4032×10^{-10} mol cm^{-3}
1 Nm^{-2}	4.4032×10^{-7} mol dm^{-3}
1 Nm^{-2}	7.5006 mtorr
1 bar	1.0000×10^{5} Nm^{-2}
1 bar	750.06 torr

Table A–6. Enthalpies of Formation of Atoms at 1 atm and 0°K in the Ideal Gas State

Atom	ΔHf_0° (kcal mol^{-1})	Ref.
H	51.634 ± 0.001	6
C	169.58 ± 0.45	28
N	112.5 ± 1	28
O	58.983 ± 0.024	6
F	18.36 ± 0.40	28
P	79.18 ± 0.05	28
S	65.75 ± 0.01	5
Cl	28.587 ± 0.002	5
Br	28.183 ± 0.029	6
I	25.613 ± 0.010	6

Table A–7. Enthalpies of Formation of Diatomic Radicals at 1 atm and 0°K in the Ideal Gas State

Radical	ΔHf_0^o (kcal mol^{-1})	Ref.[a]
CH	141.2 ± 0.1	
NH	82.6 ± 1.5[b]	666, 762
OH	9.290 ± 0.3	
PH	60.4 ± 8	
SH	34.4 ± 4	
C_2	198.2 ± 0.9	
CN	101 ± 1	264
CS	64.96 ± 0.4	769
CF	60.1 ± 2	
CCl	119.1 ± 5	
CBr	123 ± 15	
FO	22 ± 5	220, 628
ClO	24.211 ± 0.05	
BrO	32.0 ± 0.6	327
IO	43 ± 5	327
SO	1.64 ± 0.3	
	1.17 ± 0.03	5, 768
SCl	[15][c]	100

[a] Taken from Ref. 28 unless otherwise noted.
[b] From the threshold energy of $NH_3 \rightarrow NH(c^1\Pi) + H_2$ (762) and the energy difference of $NH(a^1\Delta - X^3\Sigma^-)$ (666).
[c] Estimated value.

Table A–8. Enthalpies of Formation of Triatomic Radicals at 1 atm and 0°K in the Ideal Gas State

Radical	ΔHf°_0 (kcal mol^{-1})	Ref.[a]
CH_2	92.25 ± 1.0	
NH_2	40.8 ± 3	
PH_2	33.1 ± 2	672
C_2H	127 ± 1	771
HCO	−3 ± 3 (Ref. 28), 10.3 ± 2 (Ref. 5), 3.7 ± 1.5[b]	
HCF	[29.9 ± 7][c]	
HCCl	[79.9 ± 10]	
N_2H	[61]	1049
HNO	24.5	
HO_2	6 ± 2	
HSO	≤14.9	859a
C_3	199.2 ± 0.4	999
C_2O	67.5 ± 15 (Ref. 28), 93 ± 5 (Ref. 1047)	
NCN	112.9 ± 5	5, 764
NCO	37 ± 3	765
FCO	[−41 ± 15]	
CF_2	−43.6 ± 1.5	
CCl_2	56.7 ± 5	
ClCO	−4	3
ClCS	43 ± 1	774
N_3	99.7 ± 5	5
NF_2	10.7 ± 2.0	
S_2O	−12.75 ± 0.25	100

[a] From Ref. 28 unless otherwise noted.
[b] Author's estimate from the breaking-off of H_2CO emission bands. See Section VII–4, p. 277.
[c] Estimated values are indicated by brackets.

Table A–9. Enthalpies of Formation of Four-atomic Radicals at 1 atm and 0°K in the Ideal Gas State

Radical	ΔHf°_0 (kcal mol^{-1})	Ref.
CH_3	35.62 ± 0.2	
CH_2Cl	30.0 (298°K)	270
CH_2Br	38.9 (298°K)	270
CH_2I	52.4 (298°K)	270
HCOO	-36 ± 4 (298°K)	3
COOH	-51 ± 3 (298°K)	3
$CHCl_2$	22[a]	
CHI_2	79.8 (298°K)	377
CF_3	-111.7 ± 1	
CF_2Cl	-64.3 ± 2 (298°K)	630
CF_2Br	?	
$CFCl_2$	-21[b]	
CCl_3	19.15 ± 2	
CBr_3	47	29
NO_3	18.5 ± 5	
SO_3	-93.22 ± 0.17	
Polyatomic Radical		
C_2H_3	59.6	647
CH_3CO	-5 ± 1 (298°K)	404
CH_3O	3.5	3

[a] Estimated from $D_0(Cl—CCl_3)$, $D_0(Cl—CH_2Cl)$, and $\Delta Hf^\circ_0(CHCl_3)$.
[b] Estimated from $D_0(Cl—CCl_3)$, $D_0(Cl—CF_2Cl)$, and $\Delta Hf^\circ_0(CFCl_3)$.

Table A–10. Enthalpies of Formation of Diatomic Molecules at 1 atm and 0°K in the Ideal Gas State[a]

Molecule	ΔHf°_0 (kcal mol^{-1})
HF	-65.13 ± 0.2
HCl	-22.019 ± 0.05
HBr	-6.84 ± 0.13
HI	6.82 ± 0.05
CO	-27.20 ± 0.04
NO	21.46 ± 0.04
S_2	30.80 ± 0.2
ClF	-12.12 ± 0.6
BrF	-12.1 ± 0.4
IF	-22.192 ± 0.9
Br_2	10.922 ± 0.030
BrCl	5.28 ± 0.30
I_2	15.66 ± 0.01
ICl	4.574 ± 0.025
IBr	11.91 ± 0.02

[a] From Ref. 28.

Table A–11. Enthalpies of Formation of Triatomic Molecules at 1 atm and 0°K in the Ideal Gas State[a]

Molecule	ΔHf^o_0 (kcal mol^{-1})
H_2O	−57.103
H_2S	−4.18 ± 0.15
HCN	32.39 ± 2
HFO	−22.8 ± 1[b]
CO_2	−93.965 ± 0.011
OCS	−33.11 ± 0.25
CS_2	27.79 ± 0.19
FCN	8.4 ± 1[c]
ClCN	32.8
BrCN	46.1 ± 1.5
ICN	54.1 ± 1.5
N_2O	20.43 ± 0.1
NO_2	8.59 ± 0.2
ONF	−15.1 ± 0.4
ONCl	12.83 ± 0.2
ONI	[27.6 ± 5][d]
O_3	34.8 ± 0.4
ClO_2	25.6 ± 1.5
Cl_2O	21.4 ± 0.6
F_2O	6.40 ± 0.38
SO_2	−70.341 ± 0.05

[a] From Ref. 28 unless otherwise noted.
[b] From Ref. 6.
[c] Calculated from D_0(F—CN) and enthalpies of formation of H and CN. See Ref. 264.
[d] Estimated value.

Table A–12. Enthalpies of Formation of Four-Atom Molecules at 1 atm and 0°K in the Ideal Gas State

Molecule	ΔHf^o_0 (kcal mol^{-1})	Ref.[a]
NH_3	−9.30 ± 0.1	
PH_3	7.0 ± 0.4	
C_2H_2	54.33 ± 0.19	
N_2H_2	36 ± 2	1049
H_2O_2	−31.025	
HCHO	−26.8 ± 1.5	
HN_3	71.72 ± 0.3	29
HNO_2, *cis*	−16.85 ± 0.32	
HNO_2, *trans*	−17.37 ± 0.32	
HNCO	−24 ± 3	765
FC_2H	[30][b]	773
ClC_2H	[52]	773
BrC_2H	64.2 ± 1.5	773
IC_2H	[77]	773
HCFO	[−90]	
C_2N_2	73.428 ± 0.43	
F_2CO	−152.0 ± 0.4	
ClFCO	−101 ± 8	
$OCCl_2$	−52.2 ± 0.8	
$SCCl_2$	7.92 ± 1	774
FNO_2	−24.6 ± 5	
$ClNO_2$	4.20 ± 0.4	
NF_3	−30.06 ± 0.3	
PF_3	−224.0 ± 0.9	
$OSCl_2$	−50.07	29
S_2Cl_2	−4.18	29

[a] From Ref. 28 unless otherwise noted.
[b] Estimated values are indicated by brackets.

Table A–13. Enthalpies of Formation of Five-Atom Molecules at 1 atm and 0°K in the Ideal Gas State

Molecule	ΔHf_0^o (kcal mol^{-1})	Ref.[a]
CH_4	-15.99 ± 0.08	
CH_3F	-54 ± 8	
CH_3Cl	-18.1 ± 0.5	5
CH_3Br	-4.72	29
CH_3I	5.38	29
CH_2N_2 (diazomethane)	≥ 51.3	605
CH_2N_2 (diazirine)	≥ 60.6	607
CH_2CO	-11.4 ± 0.4 (298°K)	751
HCOOH	-90.48 (298°K)	29
CH_2F_2	-105.9 ± 0.4	
CH_2FCl	[$-60.9 + 3$][b]	
CH_2Cl_2	-21.19 ± 0.3	
CH_2Br_2	-1	270
CH_2I_2	29.26	29
CHF_3	-164.9 ± 0.3	
CHF_2Cl	-113.6 ± 3	
$CHFCl_2$	[-66.36 ± 3]	
$CHCl_3$	-23.49 ± 0.3	
CHI_3	59.8 (298°K)	377
HNO_3	-29.76 ± 0.10	
C_2HCN	85 ± 1	771
N_3CN	108 ± 5	764
C_3O_2	-23.14 ± 0.44	
CF_4	-221.61 ± 0.3	
CF_3Cl	-168.0 ± 0.8	
CF_3Br	-152.2 ± 0.7	
CF_3I	-139.4 ± 0.8	
CF_2Cl_2	-116.5 ± 2	
$CFCl_3$	-68.24 ± 1.5	
CCl_4	-22.42 ± 0.5	
CCl_3Br	-8.81	29
CBr_4	38 (Ref. 377), 26.10 (Ref. 29)	
N_2O_3	19.80 ± 0.2 (298°K)	
$ClONO_2$	6.29 ± 0.08 (293°K)	578
SO_2F_2	-179.3 ± 2	5
SO_2Cl_2	-83.3 ± 0.5	5
Six-Atom Molecule		
N_2O_4	4.47 ± 0.4	
N_2O_5	2.7 ± 0.3 (298°K)	

[a] From Ref. 28 unless otherwise noted.
[b] Estimated values are indicated by brackets.

References

I. Books, Monographs, Tables

1. L. G. Anderson, "Atmospheric Chemical Kinetics Data Survey," *Rev. Geophys. Spaces Phys.* **14**, 151 (1976).

2. G. M. Barrow, *Introduction to Molecular Spectroscopy*, McGraw-Hill, New York, 1962.

2a. M. J. Beesley, *Lasers and Their Applications*, Halsted, New York, 1976.

3. S. W. Benson, *Thermochemical Kinetics; Methods for the Estimation of Thermochemical Data and Rate Parameters*, Wiley, New York, 1968.

4. J. C. Calvert and J. N. Pitts, Jr., *Photochemistry*, Wiley, New York, 1966.

5. M. W. Chase, J. L. Curnutt, A. T. Hu, H. Prophet, A. N. Syverud, and L. C. Walker, "JANAF Thermochemical Tables," 1974 Supplement, *J. Phys. Chem. Ref. Data 3*, 311 (1974).

6. M. W. Chase, J. L. Curnutt, H. Prophet, R. A. McDonald, and A. N. Syverud, JANAF Thermochemical Tables, 1975 Supplement, *J. Phys. Chem. Ref. Data* **4**, 1 (1975).

6a. E. R. Cohen and B. N. Taylor, The 1973 Least-Squares Adjustment of the Fundamental Constants, *J. Phys. Chem. Ref. Data* **2**, 663 (1973).

7. C. H. Corliss and W. R. Bozman, "Experimental Transition Probabilities for Spectral Lines of Seventy Elements," *Natl. Bur. Stand.* (*U.S.*) *Monogr.* **53** (1962).

8. A. G. Gaydon, *Dissociation Energies and Spectra of Diatomic Molecules*, Chapman and Hall, London, 1968.

9. R. F. Hampson, Ed., Survey of Photochemical and Rate Data for Twenty-Eight Reactions of Interest in Atmospheric Chemistry, *J. Phys. Chem. Ref. Data* **2**, 267–312 (1973).

10. R. F. Hampson and D. Garvin, Ed., "Chemical Kinetic and Photochemical Data for Modelling Atmospheric Chemistry," *Natl. Bur. Stand.* (*U.S.*) *Tech. Note* **866** (1975).

11. J. B. Hasted, *Physics of Atomic Collisions*, Butterworth, Washington, D.C. 1964.

12. J. Heicklen, *Atmospheric Chemistry*, Academic Press, New York, 1976.

13. G. Herzberg, *Atomic Spectra and Atomic Structure*, Dover Publications, New York, 1944.

14. G. Herzberg, *Molecular Spectra and Molecular Structure I, Spectra of Diatomic Molecules*, 2nd ed. Van Nostrand, Princeton, New Jersey, 1950.

15. G. Herzberg, *Molecular Spectra and Molecular Structure II, Infrared and Raman Spectra of Polyatomic Molecules*, Van Nostrand, New York, 1945.

16. G. Herzberg, *Molecular Spectra and Molecular Structure III, Electronic Spectra and Electronic Structure of Polyatomic Molecules*, Van Nostrand, Princeton, New Jersey, 1966.

17. R. D. Hudson, "Critical Review of Ultraviolet Photoabsorption Cross Sections for Molecules of Astrophysical and Aeronomic Interest," *Natl. Stand. Ref. Data Ser., Natl. Bur. Stand.* (U.S.) **38** (1971).

17a. W. Kirmse, *Carbene Chemistry*, 2nd ed., Academic Press, New York 1971.

17b. L. R. Koller, *Ultraviolet Radiation*, Wiley, New York, 1965.

18. P. A. Leighton, *Photochemistry of Air Pollution*, Academic Press, New York, 1961.

19. G. N. Lewis, M. Randall, K. S. Pitzer, and L. Brewer, *Thermodynamics*, McGraw-Hill, New York, 1961.

20. M. J. McEwan and L. F. Phillips, *Chemistry of the Atmosphere*, Halsted, New York, 1975.

21. A. C. G. Mitchell and M. W. Zemansky, *Resonance Radiation and Excited Atoms*, Cambridge Univ. Press, London, 1961.

22. W. A. Noyes and P. A. Leighton, *The Photochemistry of Gases*, Dover Publications, New York, 1966.

23. L. Pauling and E. B. Wilson, *Introduction to Quantum Mechanics with Applications to Chemistry*, McGraw-Hill, New York, 1935.

24. B. Rosen, "Spectroscopic Data Relative to Diatomic Molecules," *International Tables of Selected Constants*, Vol. 17, Pergamon Press, New York, 1970.

25. J. A. R. Samson, *Techniques of Vacuum Ultraviolet Spectroscopy*, Wiley, New York, 1967.

26. E. W. R. Steacie, *Atomic and Free Radical Reactions*, Reinhold, New York, 1954.

27. D. R. Stull, E. F. Westrum, Jr. and G. C. Sinke, *The Chemical Thermodynamics of Organic Compounds*, Wiley, New York, 1969.

28. D. R. Stull and H. Prophet, Project Directors, "JANAF Thermochemical Tables," 2nd ed., *Natl. Stand. Ref. Data Ser., Natl. Bur. Stand.* (U.S.) **37** (1971).

28a. A. F. Trotman-Dickenson and G. S. Milne, *Tables of Bimolecular Gas Reactions*, Natl. Stand. Ref. Data Ser-Nat. Bur. Stand. (U.S.), **9** (1967).

29. D. D. Wagman, W. H. Evans, V. B. Parker, I. Halow, S. M. Bailey and R. H. Schumm, "Selected Values of Chemical Thermodynamic Properties," *Natl. Bur. Stand. (U.S.) Tech. Note* **270–3** (1968).

30. L. Wallace, "Band-Head Wavelengths of C_2, CH, CN, CO, NH, NO, O_2, OH, and Their Ions," *Astrophys. J. Suppl.* **7**, 165 (1962).

31. K. Watanabe, M. Zelikoff, and E. C. Y. Inn, *Absorption Coefficients of Several Atmospheric Gases*, Air Force Cambridge Res. Center, Tech. Report No. **53-23** 1953.

32. W. L. Wiese, M. W. Smith and B. M. Glennon, *Atomic Transition Probabilities*, Natl. Stand. Ref. Data Ser. Nat. Bur. Stand. (U.S.) **4** (1966).

33. W. L. Wiese, M. W. Smith, and B. M. Miles, "Atomic Transition Probabilities," *Natl. Stand. Ref. Data Ser., Natl. Bur. Stand.* (U.S.) **22** (1969).

II. Individual Papers

34. K. Abe, F. Myers, T. K. McCubbin, Jr., and S. R. Polo, *J. Mol. Spectrosc.* **38**, 552 (1971).

35. K. Abe, *J. Mol. Spectrosc.* **48**, 395 (1973).

36. K. Abe, F. Myers, T. K. McCubbin, Jr., and S. R. Polo, *J. Mol. Spectrosc.* **50**, 413 (1974).
37. M. Ackerman and F. Biaume, *J. Mol. Spectrosc.* **35**, 73 (1970).
38. M. Ackerman, F. Biaume, and G. Kockarts, *Planet. Space Sci.* **18**, 1639 (1970).
39. F. Alberti, R. A. Ashby, and A. E. Douglas, *Can. J. Phys.* **46**, 337 (1968).
40. A. C. Allison, A. Dalgarno, and N. W. Pasachoff, *Planet. Space Sci.* **19**, 1463 (1973).
41. D. K. Anderson, *Phys. Rev.* **137**, A21 (1965).
42. J. G. Anderson, *Geophys. Res. Lett.* **3**, 165 (1976).
43. R. A. Anderson, J. Peacher, and D. M. Wilcox, *J. Chem. Phys.* **63**, 5287 (1975).
44. T. Aoki, T. Morikawa, and K. Sakurai, *J. Chem. Phys.* **59**, 1543 (1973).
45. S. J. Arnold, N. Finlayson, and E. A. Ogryzlo, *J. Chem. Phys.* **44**, 2529 (1966).
46. M. Arvis, *J. Chim. Phys.* **66**, 517 (1969).
47. M. N. R. Ashfold and J. P. Simons, *J. Chem. Soc. Faraday II* **73**, 858 (1977).
47a. M. N. R. Ashfold and J. P. Simons, *Chem. Phys. Lett.* **47**, 65 (1977).
48. R. Atkinson and K. H. Welge, *J. Chem. Phys.* **57**, 3689 (1972).
49. P. J. Ausloos and S. G. Lias, *Ann. Rev. Phys. Chem.* **22**, 85 (1971).
49a. P. Ausloos, R. E. Rebbert, and S. G. Lias, *J. Photochem.* **2**, 267 (1973/74).
50. R. A. Back, *J. Chem. Phys.* **40**, 3493 (1964).
51. R. A. Back and R. Ketcheson, *Can. J. Chem.* **46**, 531 (1968).
52. R. A. Back, C. Willis, and D. A. Ramsay, *Can. J. Chem.* **52**, 1006 (1974).
53. C. C. Badcock, H. W. Sidebottom, J. G. Calvert, G. W. Reinhardt, and E. K. Damon, *J. Am. Chem. Soc.*, **93**, 3115 (1971).
54. R. M. Badger, A. C. Wright, and R. F. Whitlock, *J. Chem. Phys.* **43**, 4345 (1965).
55. V. D. Baiamonte, L. G. Hartshorn, and E. J. Bair, *J. Chem. Phys.* **55**, 3617 (1971).
56. R. T. K. Baker, J. A. Kerr, and A. F. Trotman-Dickenson, *J. Chem. Soc.* **1966**, A975.
57. R. T. K. Baker, J. A. Kerr, and A. F. Trotman-Dickenson, *J. Chem. Soc.* **1967**, A1641.
57a. W. J. Balfour and A. E. Douglas, *Can. J. Phys.* **46**, 2277 (1968).
58. N. M. Ballash and D. A. Armstrong, *Spectrochim. Acta* **30A**, 941 (1974).
59. Y. B. Band and K. F. Freed, *J. Chem. Phys.* **63**, 3382 (1975).
60. J. R. Barker and J. V. Michael, *J. Opt. Soc. Am.* **58**, 1615 (1968).
61. A. J. Barnes, H. E. Hallam, and J. D. R. Howells, *J. Mol. Struct.* **23**, 463 (1974).
61a. R. H. Barnes, C. E. Moeller, J. F. Kircher, and C. M. Verber, *Appl. Opt.* **12**, 2531 (1973).
62. A. P. Baronavski, A. Hartford, Jr., and C. B. Moore, *J. Mol. Spectrosc.*, **60**, 111 (1976).
63. N. Basco and R. G. W. Norrish, *Proc. Roy. Soc.* (*Lond.*) **A268**, 291 (1962).
64. N. Basco, J. E. Nicholas, R. G. W. Norrish, and W. H. J. Vickers, *Proc. Roy. Soc.* (*Lond.*) **A272**, 147 (1963).
65. N. Basco and S. K. Dogra, *Proc. Roy. Soc.* (*Lond.*) **A323**, 29 (1971).
66. N. Basco and S. K. Dogra, *Proc. Roy. Soc.* (*Lond.*) **A323**, 401 (1971).
67. N. Basco and F. G. M. Hathorn, *Chem. Phys. Lett.* **8**, 291 (1971).

68. N. Basco and R. D. Morse, J. Mol. Spectrosc. **45**, 35 (1973).
69. N. Basco and R. D. Morse, *Chem. Phys. Lett.* **20**, 557 (1973).
70. N. Basco and R. D. Morse, *Proc. Roy. Soc.* (*Lond.*) **A336**, 495 (1974).
71. A. M. Bass, *Appl. Opt.* **5**, 1967 (1966).
71a. A. M. Bass and A. H. Laufer, *J. Photochem.* **2**, 465 (1973/74).
72. A. M. Bass, A. E. Ledford, Jr., and A. H. Laufer, *J. Res. Natl. Bur. Stand.* (*U.S.*) **80A**, 143 (1976).
73. D. L. Baulch and W. H. Breckenridge, *Trans. Faraday Soc.* **62**, 642 (1966).
74. K. D. Bayes, *J. Am. Chem. Soc.* **84**, 4077 (1962).
75. K. D. Bayes, K. H. Becker, and K. H. Welge, *Z. Naturforsch.* **17a**, 676 (1962).
76. K. D. Bayes, *J. Am. Chem. Soc.* **85**, 1730 (1963).
77. K. H. Becker and K. H. Welge, *Z. Naturforsch.* **18a**, 600 (1963).
78. K. H. Becker and K. H. Welge, *Z. Naturforsch.* **19a**, 1006 (1964).
79. K. H. Becker and K. H. Welge, *Z. Naturforsch.* **20a**, 1692 (1965).
80. K. H. Becker, W. Groth, and U. Schurath, *Chem. Phys. Lett.* **8**, 259 (1971).
81. K. H. Becker, D. Haaks, and M. Schürgers, *Z. Naturforsch.* **26a**, 1770 (1971).
82. K. H. Becker and D. Haaks, *J. Photochem.* **1**, 177 (1972/73).
83. K. H. Becker, E. H. Fink, P. Langen, and U. Schurath, *Z. Naturforsch.* **28a**, 1872 (1973).
84. K. H. Becker, D. Haaks, and T. Tatarczyk, *Chem. Phys. Lett.* **25**, 564 (1974).
85. K. H. Becker, G. Capelle, D. Haaks, and T. Tatarczyk, *Ber. Bunsenges. Phys. Chem.* **78**, 1157 (1974).
86. K. H. Becker, E. H. Fink, P. Langen, and U. Schurath, *J. Chem. Phys.* **60**, 4623 (1974).
87. H. D. Beckey, W. Groth, H. Okabe, and H. J. Rommel, *Z. Naturforsch.* **19a**, 1511 (1964).
88. A. O. Beckman and R. G. Dickinson, *J. Am. Chem. Soc.* **50**, 1870 (1928).
89. A. O. Beckman and R. G. Dickinson, *J. Am. Chem. Soc.* **52**, 124 (1930).
90. G. S. Beddard, D. J. Giachardi, and R. P. Wayne, *J. Photochem.* **3**, 321 (1974/75).
91. R. Beer, *Astrophys. J.* **200**, L167 (1975).
91a. J. A. Bell, *J. Phys. Chem.* **75**, 1537 (1971).
92. S. Bell, *J. Mol. Spectrosc.* **16**, 205 (1965).
93. S. Bell, T. L. Ng, and A. D. Walsh, *J. Chem. Soc. Faraday II* **71**, 393 (1975).
94. P. P. Bemand and M. A. A. Clyne, *J. Chem. Soc. Faraday II* **68**, 1758 (1972).
95. P. P. Bemand and M. A. A. Clyne, *J. Chem. Soc. Faraday II* **69**, 1643 (1973).
95a. W. Benesch, J. T Vanderslice, S. G. Tilford, and P. G. Wilkinson, *Astrophys. J.* **142**, 1227 (1965).
96. T. Berces and S. Förgeteg, *Trans. Faraday Soc.* **66**, 633 (1970).
97. T. Berces and S. Förgeteg, *Trans. Faraday Soc.* **66**, 640 (1970).
98. T. Berces, S. Förgeteg, and F. Marta, *Trans. Faraday Soc.* **66**, 648 (1970).
99. J. Berkowitz, *J. Chem. Phys.* **36**, 2533 (1962).
100. J. Berkowitz, J. H. D. Eland, and E. H. Appelman, *J. Chem. Phys.* **66**, 2183 (1977).
101. R. S. Berry and P. A. Lehman, *Ann. Rev. Phys. Chem.* **22**, 47 (1971).

102. J. M. Berthou, B. Pascat, H. Guenebaut, and D. A. Ramsay, *Can. J. Phys.* **50**, 2265 (1972).
103. G. W. Bethke, *J. Chem. Phys.* **31**, 662 (1959).
104. G. W. Bethke, *J. Chem. Phys.* **31**, 669 (1959).
105. M. J. Bevan and D. Husain, *J. Photochem.* **3**, 1 (1974/75).
106. K. D. Beyer and K. H. Welge, *J. Chem. Phys.* **51**, 5323 (1969).
107. F. Biaume, *J. Photochem.* **2**, 139 (1973/74).
108. D. Biedenkapp and E. J. Bair, *J. Chem. Phys.* **52**, 6119 (1970).
109. D. Biedenkapp, L. G. Hartshorn, and E. J. Bair, *Chem. Phys. Lett.* **5**, 379 (1970).
110. M. Bixon and J. Jortner, *J. Chem. Phys.* **50**, 3284 (1969).
111. M. Bixon and J. Jortner, *J. Chem. Phys.* **50**, 4061 (1969).
112. G. Black, T. G. Slanger, G. A. St. John, and R. A. Young, *J. Chem. Phys.* **51**, 116 (1969).
113. G. Black, R. L. Sharpless, T. G. Slanger, and D. C. Lorents, *J. Chem. Phys.* **62**, 4266 (1975).
114. G. Black, R. L. Sharpless, T. G. Slanger, and D. C. Lorents, *J. Chem. Phys.* **62**, 4274 (1975).
115. G. Black, R. L. Sharpless, and T. G. Slanger, *J. Chem. Phys.* **63**, 4546 (1975).
116. G. Black, R. L. Sharpless, and T. G. Slanger, *J. Chem. Phys.* **63**, 4551 (1975).
117. G. Black, R. L. Sharpless, and T. G. Slanger, *J. Photochem.* **5**, 435 (1976).
118. G. Black, R. L. Sharpless, and T. G. Slanger, *J. Chem. Phys.* **66**, 2113 (1977).
119. R. P. Blickensderfer, W. H. Breckenridge, and J. Simons, *J. Phys. Chem.* **80**, 653 (1976).
120. M. Bonnemay and E. T. Verdier, *J. Chim. Phys.* **41**, 113 (1944).
121. P. Borrell, P. Cashmore, and A. E. Platt, *J. Chem. Soc. A* **1968**, 3063.
122. W. L. Borst and E. C. Zipf, *Phys. Rev.* **A3**, 979 (1971).
122a. M. W. Bosnali and D. Perner, *Z. Naturforsch.* **26a**, 1768 (1971).
122b. J. W. Bottenheim and J. G. Calvert, *J. Phys. Chem.* **80**, 782 (1976).
123. M. J. Boxall, C. J. Chapman, and R. P. Wayne, *J. Photochem.* **4**, 435 (1975).
123a. C. R. Boxall and J. P. Simons, *J. Photochem.* **1**, 363 (1972/73).
124. D. J. Bradley, *Contemp. Phys.* **16**, 263 (1975).
125. J. N. Bradley, J. R. Gilbert, and P. Svejda, *Trans. Faraday Soc.* **64**, 911 (1968).
126. J. C. D. Brand and R. I. Reed, *J. Chem. Soc.* **1957**, 2386.
127. J. C. Brand, J. H. Callomon, D. C. Moule, J. Tyrrell, and T. H. Goodwin, *Trans. Faraday Soc.* **61**, 2365 (1965).
128. J. C. Brand, V. T. Jones, and C. diLauro, *J. Mol. Spectrosc.* **40**, 616 (1971).
129. J. C. Brand and K. Srikameswaran, *Chem. Phys. Lett.* **15**, 130 (1972).
130. J. C. Brand and R. Nanes, *J. Mol. Spectrosc.* **46**, 194 (1973).
131. J. C. Brand, D. R. Humphrey, A. E. Douglas, and I. Zanon, *Can. J. Phys.* **51**, 530 (1973).
132. J. C. Brand, J. L. Hardwick, R. J. Pirkle, and C. J. Seliskar, *Can. J. Phys.* **51**, 2184 (1973).
133. J. C. Brand, W. H. Chan, and J. L. Hardwick, *J. Mol. Spectrosc.* **56**, 309 (1975).

134. J. C. Brand, P. H. Chiu, A. R. Hoy, and H. D. Bist, *J. Mol. Spectrosc.* **60**, 43 (1976).

135. J. C. Brand, J. L. Hardwick, D. R. Humphrey, Y. Hamada, and A. J. Merer, *Can. J. Phys.* **54**, 186 (1976).

136. S. Braslavsky and J. Heicklen, *J. Photochem.*, **1**, 203 (1972/73).

137. W. Braun, K. H. Welge, and J. R. McNesby, *J. Chem. Phys.*, **45**, 2650 (1966).

138. W. Braun, A. M. Bass, and A. E. Ledford, Jr., *Appl. Opt.* **6**, 47 (1967).

139. W. Braun, J. R. McNesby, and A. M. Bass, *J. Chem. Phys.* **46**, 2071 (1967).

140. W. Braun and T. Carrington, *J. Quant. Spectrosc. Radiat. Transfer* **9**, 1133 (1969).

141. W. Braun, A. M. Bass, D. D. Davis, and J. D. Simmons, *Proc. Roy. Soc.* (*Lond.*) **A312**, 417 (1969).

142. W. Braun, C. Carlone, T. Carrington, G. V. Volkenburgh, and R. A. Young, *J. Chem. Phys.* **53**, 4244 (1970).

143. W. Braun, A. M. Bass, and M. Pilling, *J. Chem. Phys.* **52**, 5131 (1970).

144. W. H. Breckenridge and H. Taube, *J. Chem. Phys.* **53**, 1750 (1970).

145. W. H. Breckenridge and A. B. Callear, *Trans. Faraday Soc.* **67**, 2009 (1971).

145a. B. Brehm and H. Siegert, *Z. Angew. Phys.* **19**, 244 (1965).

146. W. R. Brennen, I. D. Gay, F. P. Glass, and H. Niki, *J. Chem. Phys.* **43**, 2569 (1965).

147. L. Brewer and J. Tellinghuisen, *J. Chem. Phys.* **56**, 3929 (1972).

148. W. B. Bridges, A. N. Chester, A. S. Halsted, and J. V. Parker, *Proc. IEEE* **59**, 724 (1971).

149. J. P. Briggs, R. B. Caton, and M. J. Smith, *Can. J. Chem.* **53**, 2133 (1975).

150. R. K. Brinton and D. H. Volman, *J. Chem. Phys.* **19**, 1394 (1951).

151. A. Brown and D. Husain, *J. Photochem.* **3**, 37 (1974/75).

152. A. Brown and D. Husain, *J. Less-Common Metals* **3**, 305 (1974/75).

153. M. Broyer, J. Vigue, and J. C. Lehmann, *J. Chem. Phys.* **63**, 5428 (1975).

154. C. F. Bruce and P. Hannaford, *Spectrochim. Acta* **26B**, 207 (1971).

155. L. E. Brus, *Chem. Phys. Lett.* **12**, 116 (1971).

156. L. E. Brus and J. R. McDonald, *Chem. Phys. Lett.* **21**, 283 (1973).

157. L. E. Brus and J. R. McDonald, *J. Chem. Phys.* **61**, 97 (1974).

158. J. J. Bufalini, *Environ. Sci. Tech.* **6**, 837 (1972).

159. G. E. Busch, R. T. Mahoney, R. I. Morse, and K. R. Wilson, *J. Chem. Phys.* **51**, 449 (1969).

160. G. E. Busch, R. T. Mahoney, R. I. Morse, and K. R. Wilson, *J. Chem. Phys.* **51**, 837 (1969).

161. G. E. Busch, J. F. Cornelius, R. T. Mahoney, R. I. Morse, D. W. Schlosser and K. R. Wilson, *Rev. Sci. Inst.* **41**, 1066 (1970).

162. G. E. Busch and K. R. Wilson, *J. Chem. Phys.* **56**, 3626 (1972).

163. G. E. Busch and K. R. Wilson, *J. Chem. Phys.* **56**, 3638 (1972).

164. G. E. Busch and K. R. Wilson, *J. Chem. Phys.* **56**, 3655 (1972).

165. P. Cadman and J. C. Polanyi, *J. Phys. Chem.* **72**, 3715 (1968).

166. A. B. Callear, *Proc. Roy. Soc.* (*Lond.*) **A276**, 401 (1963).

167. A. B. Callear and I. W. M. Smith, *Trans. Faraday Soc.* **59**, 1720 (1963).

168. A. B. Callear and I. W. M. Smith, *Trans. Faraday Soc.* **59**, 1735 (1963).

169. A. B. Callear and I. W. M. Smith, *Disc. Faraday Soc.* **37**, 96 (1964).
170. A. B. Callear and I. W. M. Smith, *Trans. Faraday Soc.* **61**, 2383 (1965).
171. A. B. Callear, M. J. Pilling, and I. W. M. Smith, *Trans. Faraday Soc.* **64**, 2296 (1968).
172. A. B. Callear and R. J. Oldman, *Trans. Faraday Soc.* **64**, 840 (1968).
173. A. B. Callear and J. McGurk, *Chem. Phys. Lett.* **7**, 491 (1970).
174. A. B. Callear and M. J. Pilling, *Trans. Faraday Soc.* **66**, 1886 (1970).
175. A. B. Callear and M. J. Pilling, *Trans. Faraday Soc.* **66**, 1618 (1970).
176. A. B. Callear and R. E. M. Hedges, *Trans. Faraday Soc.* **66**, 605 (1970).
177. A. B. Callear and R. E. M. Hedges, *Trans. Faraday Soc.* **66**, 615 (1970).
178. A. B. Callear and P. M. Wood, *Trans. Faraday Soc.* **67**, 3399 (1971).
179. A. B. Callear and P. M. Wood, *J. Chem. Soc. Faraday II* **68**, 302 (1972).
180. A. B. Callear and J. C. McGurk, *J. Chem. Soc. Faraday II* **69**, 97 (1972).
181. J. M. Calo and R. C. Axtmann, *J. Chem. Phys.* **54**, 1332 (1971).
182. J. G. Calvert, J. A. Kerr, K. L. Demerjian, and R. D. McQuigg, *Science* **175**, 751 (1972).
183. J. G. Calvert, *Chem. Phys. Lett.* **20**, 484 (1973).
184. J. G. Calvert and R. D. McQuigg, *Int. J. Chem. Kinet. Symp.* **1**, 113 (1975).
185. I. M. Campbell and B. A. Thrush, *J. Quant. Spectrosc. Radiat. Transfer* **8**, 1571 (1968).
186. G. A. Capelle, K. Sakurai, and H. P. Broida, *J. Chem. Phys.* **54**, 1728 (1971).
187. G. A. Capelle and H. P. Broida, *J. Chem. Phys.* **58**, 4212 (1973).
188. G. Cario and J. Franck, *Z. Phys.* **11**, 161 (1922).
189. N. P. Carleton and O. Oldenberg, *J. Chem. Phys.* **36**, 3460 (1962).
190. N. P. Carleton and W. A. Traub, *Science* **177**, 988 (1972).
191. T. Carrington, *J. Chem. Phys.* **41**, 2012 (1964).
192. P. K. Carroll, *Astrophys. J.* **129**, 794 (1959).
193. E. Castellano and H. J. Schumacher, *Z. Phys. Chem.* **NF34**, 198 (1962).
194. E. Castellano and H. J. Schumacher, *Z. Phys. Chem.* **NF65**, 62 (1969).
195. E. Castellano and H. J. Schumacher, *Z. Phys. Chem.* **NF76**, 258 (1971).
196. E. Castellano and H. J. Schumacher, *Chem. Phys. Lett.* **13**, 625 (1972).
197. R. B. Caton and A. B. F. Duncan, *J. Am. Chem. Soc.* **90**, 1945 (1968).
198. R. B. Caton and A. R. Gangadharan, *Can. J. Chem.* **52**, 2389 (1974).
199. G. C. Causeley and B. R. Russell, *J. Chem. Phys.* **62**, 848 (1975).
200. E. Cehelnik, C. W. Spicer, and J. Heicklen, *J. Am. Chem. Soc.* **93**, 5371 (1971).
201. E. Cehelnik, J. Heicklen, S. Braslavsky, L. Stockburger III, and E. Mathias, *J. Photochem.* **2**, 31 (1973/74).
202. G. A. Chamberlain and J. P. Simons, *Chem. Phys. Lett.* **32**, 355 (1975).
203. G. A. Chamberlain and J. P. Simons, *J. Chem. Soc. Faraday II* **71**, 402 (1975).
204. G. A. Chamberlain and J. P. Simons, *J. Chem. Soc. Faraday II* **71**, 2043 (1975).
205. S. Chapman, *Philos. Mag.* **10**, 369 (1930).
206. P. E. Charters, R. G. Macdonald, and J. C. Polanyi, *Appl. Opt.* **10**, 1747 (1971).
207. J. G. Chervenak and R. A. Anderson, *J. Opt. Soc. Am.* **61**, 952 (1971).
208. J. P. Chesick, *J. Chem. Educ.* **49**, 722 (1972).

209. C. C. Chou, P. Angelberger, and F. S. Rowland, *J. Phys. Chem.* **75**, 2536 (1971).

210. C. C. Chou, J. G. Lo, and F. S. Rowland, *J. Chem. Phys.* **60**, 1208 (1974).

210a. C. C. Chou, W. S. Smith, H. VeraRuiz, K. Moe, G. Grescentini, M. J. Molina, and F. S. Rowland, *J. Phys. Chem.* **81**, 286 (1977).

211. M. Y. Chu and J. S. Dahler, *Mol. Phys.* **27**, 1045 (1974).

212. K. Chung, J. C. Calvert, and J. N. Bottenheim, *Int. J. Chem. Kinet.* **7**, 161 (1975).

213. E. L. Chupp, L. W. Dotchin, and D. J. Pegg, *Phys. Rev.* **175**, 44 (1968).

213a. A. Chutjian, J. K. Link, and L. Brewer, *J. Chem. Phys.* **46**, 2666 (1967).

214. S. Cieslik and M. Nicolet, *Planet. Space Sci.* **21**, 925 (1973).

215. T. C. Clark and M. A. A. Clyne, *Trans. Faraday Soc.* **66**, 877 (1970).

216. I. D. Clark and J. F. Noxon, *J. Chem. Phys.* **57**, 1033 (1972).

217. J. H. Clark, Y. Haas, P. L. Houston, and C. B. Moore, *Chem. Phys. Lett.* **35**, 82 (1975).

218. R. D. Clear and K. R. Wilson, *J. Mol. Spectrosc.* **47**, 39 (1973).

219. R. D. Clear, S. J. Riley, and K. R. Wilson, *J. Chem. Phys.* **63**, 1340 (1975).

220. M. A. A. Clyne and R. T. Watson, *Chem. Phys. Lett.* **12**, 344 (1971).

221. M. A. A. Clyne and H. W. Cruse, *J. Chem. Soc. Faraday II* **68**, 1281 (1972).

222. M. A. A. Clyne and L. W. Townsend, *J. Chem. Soc. Faraday II* **70**, 1863 (1974).

223. M. A. A. Clyne, I. S. McDermid, and A. H. Curran, *J. Photochem.* **5**, 201 (1976).

224. R. J. Cody, M. J. Sabety-Dzvonik, and W. M. Jackson, *J. Chem. Phys.* **66**, 2145 (1977).

225. R. Colin, P. Goldfinger, and M. Jeunehomme, *Trans. Faraday Soc.* **60**, 306 (1964).

226. R. Colin, *Can. J. Chem.* **47**, 979 (1969).

227. R. J. Collins and D. Husain, *J. Chem. Soc. Faraday II* **69**, 145 (1973).

228. S. S. Collier, A. Morikawa, D. H. Slater, J. G. Calvert, G. Reinhardt, and E. Damon, *J. Am. Chem. Soc.* **92**, 217 (1970).

229. F. J. Comes and S. Pionteck, *Chem. Phys. Lett.* **42**, 558 (1976).

230. L. E. Compton, J. L. Gole, and R. M. Martin, *J. Phys. Chem.* **73**, 1158 (1969).

231. L. E. Compton, and R. M. Martin, *J. Phys. Chem.* **73**, 3474 (1969).

232. L. E. Compton and R. M. Martin, *J. Chem. Phys.* **52**, 1613 (1970).

233. R. E. Connors, J. L. Roebber, and K. Weiss, *J. Chem. Phys.* **60**, 5011 (1974).

234. G. R. Cook and P. H. Metzger, *J. Chem. Phys.* **41**, 321 (1964).

235. G. R. Cook, M. Ogawa, and R. W. Carlson, *J. Geophys. Res.* **78**, 1663 (1973).

236. T. J. Cook and D. H. Levy, *J. Chem. Phys.* **57**, 5059 (1972).

237. J. B. Coon and E. Ortiz, *J. Mol. Spectrosc.* **1**, 81 (1957).

238. C. D. Cooper and M. Lichtenstein, *Phys. Rev.* **109**, 2026 (1958).

239. D. M. Cooper and R. W. Nicholls, *J. Quant. Spectrosc. Radiat. Transfer* **15**, 139 (1975).

240. A. Corney and O. M. Williams, *J. Phys. Series B. Atom. Molec. Phys.* **5**, 686 (1972).

241. M. Cottin, J. Masanet, and C. Vermeil, *J. Chim. Phys.* **63**, 959 (1966).

242. A. P. Cox, A. H. Brittain, and D. J. Finnigan, *Trans. Faraday Soc.* **67**, 2179 (1971).

243. R. A. Cox and S. A. Penkett, *Atmos. Environ.* **4**, 425 (1970).

244. R. A. Cox, *J. Phys. Chem.* **76**, 814 (1972).

245. R. A. Cox, *J. Photochem.* **3**, 175 (1974).

246. R. A. Cox, *J. Photochem.* **3**, 291 (1974/75).

247. R. A. Cox and R. G. Derwent, *J. Photochem.* **4**, 139 (1975).

248. R. A. Cox, R. G. Derwent, and P. M. Holt, *J. Chem. Soc. Faraday II* **72**, 2031 (1976).

249. R. A. Cox and R. G. Derwent, *J. Photochem.* **6**, 23 (1976/77).

250. J. A. Coxon and D. A. Ramsay, *Can. J. Phys.* **54**, 1034 (1976).

251. C. L. Creel and J. Ross, *J. Chem. Phys.* **64**, 3560 (1976).

252. P. J. Crutzen, *Geophys. Res. Lett.* **1**, 205 (1974).

253. C. F. Cullis, D. J. Hucknall, and J. V. Shepherd, *Proc. Roy. Soc.* (*Lond.*) **A335**, 525 (1973).

254. D. L. Cunningham and K. C. Clark, *J. Chem. Phys.* **61**, 1118 (1974).

255. R. F. Curl, Jr., K. Abe, J. Bissinger, C. Bennett, and F. K. Tittel, *J. Mol. Spectrosc.* **48**, 72 (1973).

256. R. J. Cvetanović, *Prog. React. Kinet.* **2**, 41 (1964), Ed. G. Porter, Macmillan, New York, 1964.

257. R. J. Cvetanović, *J. Chem. Phys.* **43**, 1450 (1965).

258. E. E. Daby, J. S. Hitt, and G. J. Mains, *J. Phys. Chem.* **74**, 4204 (1970).

258a. J. Danon, S. V. Filseth, D. Feldmann, H. Zacharias, C. H. Dugan, and K. H. Welge 13th Informal Conference on Photochemistry, Clearwater Beach, Fl. Jan 4–7, 1978. *Chem. Phys.* **29**, 345 (1978).

259. B. deB. Darwent and R. Roberts, *Proc. Roy. Soc.* (*Lond.*) **A216**, 344 (1953).

260. B. deB. Darwent, R. L. Wadlinger, and M. J. Allard, *J. Phys. Chem.* **71**, 2346 (1967).

261. S. Datta, R. W. Anderson, and R. N. Zare, *J. Chem. Phys.* **63**, 5503 (1975).

262. J. A. Davidson, C. M. Sadowski, H. I. Schiff, G. E. Streit, and C. J. Howard, *J. Chem. Phys.* **64**, 57 (1976).

263. D. D. Davis and W. Braun, *Appl. Opt.* **7**, 2071 (1968).

264. D. D. Davis and H. Okabe, *J. Chem. Phys.* **49**, 5526 (1968).

265. D. D. Davis, W. Wong, and J. Lephardt, *Chem. Phys. Lett.* **22**, 273 (1973).

266. D. D. Davis, J. F. Schmidt, C. M. Neeley, and R. J. Hanrahan, *J. Phys. Chem.* **79**, 11 (1975).

267. D. D. Davis, W. Heaps, and T. McGee, *Geophys. Res. Lett.* **3**, 331 (1976).

268. J. J. Deakin, D. Husain, and J. R. Wiesenfeld, *Chem. Phys. Lett.* **10**, 146 (1971).

269. J. J. Deakin and D. Husain, *J. Chem. Soc. Faraday II* **68**, 41 (1972).

270. J. J. DeCorpo, D. A. Bafus, and J. L. Franklin, *J. Chem. Thermodyn.* **3**, 125 (1971).

271. R. A. DeGraff and J. G. Calvert, *J. Am. Chem. Soc.* **89**, 2247 (1967).

272. K. L. Demerjian, J. G. Calvert, and D. L. Thorsell, *Int. J. Chem. Kinet.* **6**, 829 (1974).

273. K. L. Demerjian and J. G. Calvert, *Int. J. Chem. Kinet.* **7**, 45 (1975).

274. W. DeMore and O. F. Raper, *J. Chem. Phys.* **37**, 2048 (1962).
275. W. DeMore and O. F. Raper, *J. Chem. Phys.* **46**, 2500 (1967).
276. W. DeMore, *J. Chem. Phys.* **47**, 2777 (1967).
277. W. DeMore and C. Dede, *J. Phys. Chem.* **74**, 2621 (1970).
278. W. DeMore and M. Mosesman, *J. Atmos. Sci.* **28**, 842 (1971).
279. W. DeMore and M. Patapoff, *J. Geophys. Res.* **77**, 6291 (1972).
280. W. DeMore and E. Tschuikow-Roux, *J. Phys. Chem.* **78**, 1447 (1974).
281. D. Demoulin and M. Jungen, *Theor. Chem. Acta* **34**, 1 (1974).
282. R. G. Derwent and B. A. Thrush, *Trans. Farad. Soc.* **67**, 2036 (1971).
283. M. deSorgo, A. J. Yarwood, O. P. Strausz, and H. E. Gunning, *Can. J. Chem.* **43**, 1886 (1965).
284. C. Devillers, *C. R. Acad. Sci.* **262**, 1485 (1966).
285. C. Devillers and D. A. Ramsay, *Can. J. Phys.* **49**, 2839 (1971).
285a. H. J. Dewey, *IEEE J. Quantum Electron.* **12**, 303 (1976).
286. R. L. deZafra, A. Marshall, and H. Metcalf, *Phys. Rev.* **A3**, 1557 (1971).
286a. G. DiLonardo and A. E. Douglas, *J. Chem. Phys.* **56**, 5185 (1972).
287. R. N. Dixon and G. H. Kirby, *Trans. Faraday Soc.* **64**, 2002 (1968).
288. R. N. Dixon and M. Halle, *Chem. Phys. Lett.* **22**, 450 (1973).
289. R. S. Dixon, *Radiat. Res. Rev.* **2**, 237 (1970).
290. R. W. Diesen, J. C. Wahr, and S. E. Adler, *J. Chem. Phys.* **50**, 3635 (1969).
291. R. W. Diesen, J. C. Wahr, and S. E. Adler, *J. Chem. Phys.* **55**, 2812 (1971).
292. M. C. Dodge and J. Heicklen, *Int. J. Chem. Kinet.* **3**, 269 (1971).
293. J. P. Doering and B. H. Mahan, *J. Chem. Phys.* **34**, 1617 (1961).
294. T. Donohue and J. R. Wiesenfeld, *Chem. Phys. Lett.* **33**, 176 (1975).
295. T. Donohue and J. R. Wiesenfeld, *J. Chem. Phys.* **63**, 3130 (1975).
296. R. J. Donovan and D. Husain, *Nature* **206**, 171 (1965).
297. R. J. Donovan and D. Husain, *Trans. Faraday Soc.* **62**, 1050 (1966).
298. R. J. Donovan and D. Husain, *Trans. Faraday Soc.* **62**, 2023 (1966).
299. R. J. Donovan and D. Husain, *Trans. Faraday Soc.* **62**, 2643 (1966).
300. R. J. Donovan and D. Husain, *Trans. Faraday Soc.* **62**, 2987 (1966).
301. R. J. Donovan, F. G. M. Hathorn, and D. Husain, *Trans. Faraday Soc.* **64**, 1228 (1968).
302. R. J. Donovan and D. Husain, *Trans. Faraday Soc.* **64**, 2325 (1968).
303. R. J. Donovan, F. G. M. Hathorn, and D. Husain, *Trans. Faraday Soc.* **64**, 3192 (1968).
304. R. J. Donovan, L. J. Kirsch, and D. Husain, *Nature* **222**, 1164 (1969).
305. R. J. Donovan, *Trans. Faraday Soc.* **65**, 1419 (1969).
306. R. J. Donovan, D. Husain, and P. T. Jackson, *Trans. Faraday Soc.* **65**, 2930 (1969).
307. R. J. Donovan, D. Husain, and C. D. Stevenson, *Trans. Faraday Soc.* **66**, 1 (1970).
308. R. J. Donovan, L. J. Kirsch, and D. Husain, *Trans. Faraday Soc.* **66**, 774 (1970).
309. R. J. Donovan, D. Husain, and L. J. Kirsch, *Trans. Faraday Soc.* **66**, 2551 (1970).
310. R. J. Donovan and D. Husain, *Chem. Rev.* **70**, 489 (1970).

311. R. J. Donovan, L. J. Kirsch, and D. Husain, *Chem. Phys. Lett.* **7**, 453 (1970).
312. R. J. Donovan, D. Husain, and L. J. Kirsch, *Chem. Phys. Lett.* **6**, 488 (1970).
313. R. J. Donovan, D. Husain, and L. J. Kirsch, *Trans. Faraday Soc.* **67**, 375 (1971).
314. R. J. Donovan and P. J. Robertson, *Spec. Lett.* **5**, 361 (1972).
315. R. J. Donovan and J. Konstantatos, *J. Photochem.* **1**, 75 (1972/73).
316. R. J. Donovan, K. Kaufmann, and J. Wolfrum, *Nature* **262**, 204 (1976).
317. F. H. Dorer and S. N. Johnson, *J. Phys. Chem.* **75**, 3651 (1971).
318. A. E. Douglas and J. M. Hollas, *Can. J. Phys.* **39**, 479 (1961).
319. A. E. Douglas, *Discuss. Faraday Soc.* **35**, 158 (1963).
320. A. E. Douglas and E. R. V. Milton, *J. Chem. Phys.* **41**, 357 (1964).
321. A. E. Douglas and I. Zanon, *Can. J. Phys.* **42**, 627 (1964).
322. A. E. Douglas and K. P. Huber, *Can. J. Phys.* **43**, 74 (1965).
323. A. E. Douglas and W. J. Jones, *Can. J. Phys.* **43**, 2216 (1965).
324. A. E. Douglas, *J. Chem. Phys.* **45**, 1007 (1966).
325. J. N. Driscoll and P. Warneck, *J. Phys. Chem.* **72**, 3736 (1968).
326. O. J. Dunn, S. V. Filseth, and R. A. Young, *J. Chem. Phys.* **59**, 2892 (1973).
327. R. A. Durie and D. A. Ramsay, *Can. J. Phys.* **36**, 35 (1958).
328. H. U. Dütsch, *Pure Appl. Geophys.* **106–108**, 1362 (1973).
329. P. J. Dyne and D. W. G. Style, *Discuss. Faraday Soc.* **2**, 159 (1947).
330. P. J. Dyne and D. W. G. Style, *J. Chem. Soc.* **1952**, 2122.
331. M. Dzvonik, S. Yang, and R. Bersohn, *J. Chem. Phys.* **61**, 4408 (1974).
332. B. L. Earl, R. R. Herm, S. M. Lin, and C. A. Mims, *J. Chem. Phys.* **56**, 867 (1972).
333. B. L. Earl and R. R. Herm, *J. Chem. Phys.* **60**, 4568 (1974).
333a. T. W. Eder and R. W. Carr, Jr., *J. Phys. Chem.* **73**, 2074 (1969).
334. F. H. C. Edgecombe, R. G. W. Norrish, and B. A. Thrush, *Proc. Roy. Soc.* (*Lond.*) **A243**, 24 (1957).
335. R. Engleman, Jr., *J. Am. Chem. Soc.* **87**, 4193 (1965).
336. R. Engleman, Jr. *J. Photochem.* **1**, 317 (1972/73).
337. K. Evans and S. A. Rice, *Chem. Phys. Lett.* **14**, 8 (1972).
338. K. Evans, D. Heller, S. A. Rice, and R. Scheps, *J. Chem. Soc. Faraday II* **69**, 856 (1973).
339. E. T. Fairchild, *Appl. Opt.* **12**, 2240 (1973).
340. E. Fajans and C. F. Goodeve, *Trans. Faraday Soc.* **32**, 511 (1936).
341. A. J. D. Farmer, W. Fabian, B. R. Lewis, K. H. Lohan, and G. N. Haddad, *J. Quant. Spectrosc. Radiat. Transfer* **8**, 1739 (1968).
342. A. J. D. Farmer, V. Hasson, and R. W. Nicholls, *J. Quant. Spectrosc. Radiat. Transfer* **12**, 635 (1972).
343. E. R. Farnworth and G. W. King, *J. Mol. Spectrosc.* **46**, 419 (1973).
344. R. A. Fass, *J. Phys. Chem.* **74**, 984 (1970).
345. A. M. Fatta, E. Mathias, J. Heicklen, L. Stockburger III, and S. Braslavsky, *J. Photochem.* **2**, 119 (1973/74).
346. W. Felder, W. Morrow, and R. A. Young, *J. Geophys. Res.* **75**, 7311 (1970).
347. B. M. Ferro and B. G. Reuben, *Trans. Faraday Soc.* **67**, 2847 (1971).
348. S. V. Filseth and K. H. Welge, *J. Chem. Phys.* **51**, 839 (1969).

349. S. V. Filseth, F. Stuhl, and K. H. Welge, *J. Chem. Phys.* **52**, 239 (1970).
350. S. V. Filseth, A. Zia, and K. H. Welge, *J. Chem. Phys.* **52**, 5502 (1970).
351. F. D. Findlay and D. R. Snelling, *J. Chem. Phys.* **54**, 2750 (1971).
352. E. H. Fink and K. H. Welge, *Z. Naturforsch.* **19a**, 1193 (1964).
353. E. H. Fink and K. H. Welge, *J. Chem. Phys.* **46**, 4315 (1967).
353a. P. Fink and C. F. Goodeve, *Proc. Roy. Soc.* (*Lond.*) **A163**, 592 (1937).
354. W. H. Fink, *J. Chem. Phys.* **49**, 5054 (1968).
355. W. H. Fink, *J. Chem. Phys.* **54**, 2911 (1971).
356. W. Finkelnburg, H. J. Schumacher, and G. Stieger, *Z. Phys. Chem.* **B15**, 127 (1931).
357. B. J. Finlayson and J. N. Pitts, Jr., *Science* **192**, 111 (1976).
358. G. Fischer, *J. Mol. Spectrosc.* **29**, 37 (1969).
359. E. R. Fischer and G. K. Smith, *Appl. Opt.* **10**, 1803 (1971).
360. E. R. Fisher and E. Bauer, *J. Chem. Phys.* **57**, 1966 (1972).
361. R. V. Fitzsimmons and E. J. Bair, *J. Chem. Phys.* **40**, 451 (1964).
362. I. S. Fletcher and D. Husain, *Chem. Phys. Lett.* **39**, 163 (1976).
363. R. A. Fletcher and G. Pilcher, *Trans. Faraday Soc.* **66**, 794 (1970).
364. A. L. Flores and B. deB. Darwent, *J. Phys. Chem.* **73**, 2203 (1969).
365. D. Florida and S. A. Rice, *Chem. Phys. Lett.* **33**, 207 (1975).
366. P. J. Flory and H. L. Johnston, *J. Am. Chem. Soc.* **57**, 2641 (1935).
367. P. D. Foo and K. K. Innes, *Chem. Phys. Lett.* **22**, 439 (1973).
368. A. Forchioni and C. Willis, *J. Phys. Chem.* **72**, 3105 (1968).
369. H. W. Ford and S. Jaffe, *J. Chem. Phys.* **38**, 2935 (1963).
370. C. J. Fortin, D. R. Snelling, and A. Tardif, *Can. J. Chem.* **50**, 2747 (1972).
371. P. Fowles, M. deSorgo, A. J. Yarwood, O. P. Strausz, and H. E. Gunning, *J. Am. Chem. Soc.* **89**, 1352 (1967).
372. T. C. Frankiewicz and R. S. Berry, *J. Chem. Phys.* **58**, 1787 (1973).
372a. T. C. Frankiewicz and R. S. Berry, *Environ. Sci. Technol.* **6**, 365 (1972).
373. C. G. Freeman, M. J. McEwan, R. F. C. Claridge, and L. F. Phillips, *Trans. Faraday Soc.* **66**, 2974 (1970).
374. C. G. Freeman and L. F. Phillips, *Chem. Phys. Lett.* **20**, 96 (1973).
375. S. M. Freund and J. J. Ritter, *Chem. Phys. Lett.* **32**, 255 (1975).
376. N. J. Friswell and R. A. Back, *Can. J. Chem.* **46**, 527 (1968).
377. S. Furuyama, D. M. Golden, and S. W. Benson, *J. Am. Chem. Soc.* **91**, 7564 (1969).
378. H. Gaedtke, H. Hippler and J. Troe, *Chem. Phys. Lett.* **16**, 177 (1972).
379. H. Gaedtke and J. Troe, *Ber. Bunsenges. Phys. Chem.* **79**, 184 (1975).
380. M. Gaillard and J. E. Hesser, *Astrophys. J.* **152**, 695 (1968).
381. T. D. Gaily, *J. Opt. Soc. Am.* **59**, 536 (1969).
382. A. R. Gallo and K. K. Innes, *J. Mol. Spectrosc.* **54**, 472 (1975).
383. R. G. Gann and J. Dubrin, *J. Chem. Phys.* **47**, 1867 (1967).
384. R. A. Gangi and L. Burnelle, *J. Chem. Phys.* **55**, 843 (1971).
385. R. A. Gangi and L. Burnelle, *J. Chem. Phys.* **55**, 851 (1971).
385a. R. A. Gangi and R. F. W. Bader, *J. Chem. Phys.* **55**, 5369 (1971).
386. P. J. Gardner, *Chem. Phys. Lett.* **4**, 167 (1969).
387. R. H. Garstang, *J. Res. Natl. Bur. Stand.* (*U.S.*) **68A**, 61 (1964).

388. J. A. Gelbwachs, M. Birnbaum, A. W. Tucker, and C. L. Fincher, *Opto-Electronics* **4**, 155 (1972).

389. B. Gelernt, S. V. Filseth, and T. Carrington, *Chem. Phys. Lett.* **36**, 238 (1975).

390. K. R. German, *J. Chem. Phys.* **63**, 5252 (1975).

391. J. A. Ghormley, R. L. Ellsworth, and C. J. Hochanadel, *J. Phys. Chem.* **77**, 1341 (1973).

392. D. J. Giachardi and R. P. Wayne, *Proc. Roy. Soc.* (*Lond.*) **A330**, 131 (1972).

393. G. E. Gibson and N. S. Bayliss, *Phys. Rev.* **44**, 188 (1933).

394. A. Gilles, J. Masanet, and C. Vermeil, *Chem. Phys. Lett.* **25**, 346 (1974).

395. G. D. Gillispie, A. U. Khan, A. C. Wahl, R. P. Hosteny, and M. Krauss, *J. Chem. Phys.* **63**, 3425 (1975).

396. H. M. Gillespie and R. J. Donovan, *Chem. Phys. Lett.* **37**, 468 (1976).

397. R. Gilpin and K. H. Welge, *J. Chem. Phys.* **55**, 975 (1971).

398. R. Gilpin, H. I. Schiff, and K. H. Welge, *J. Chem. Phys.* **55**, 1087 (1971).

398a. D. P. Gilra, *J. Chem. Phys.* **63**, 2263 (1975).

399. L. C. Glasgow and P. Potzinger, *J. Phys. Chem.* **76**, 138 (1972).

400. L. G. Glasgow and J. E. Willard, *J. Phys. Chem.* **74**, 4290 (1970).

401. W. S. Gleason and R. Pertel, *Rev. Sci. Inst.* **42**, 1638 (1971).

402. S. D. Gleditsch and J. V. Michael, *J. Phys. Chem.* **79**, 409 (1975).

403. S. Glicker and L. J. Stief, *J. Chem. Phys.* **54**, 2852 (1971).

403a. K. A. Goettel and J. S. Lewis, *J. Atmos. Sci.* **31**, 828 (1974).

404. D. M. Golden and S. W. Benson, *Chem. Rev.* **69**, 125 (1969).

405. A. Goldman, D. G. Murcray, F. H. Murcray, and W. J. Williams, *J. Opt. Soc. Am.* **63**, 843 (1973).

406. C. S. Goldman, R. I. Greenberg, and J. Heicklen, *Int. J. Chem. Kinet.* **3**, 501 (1971).

407. D. Golomb, K. Watanabe, and F. F. Marmo, *J. Chem. Phys.* **36**, 958 (1962).

408. M. A. Gonzalez, G. Karl, and P. J. S. Watson, *J. Chem. Phys.* **57**, 4054 (1972).

409. C. F. Goodeve and J. I. Wallace, *Trans. Faraday Soc.* **26**, 254 (1930).

410. C. F. Goodeve and N. O. Stein, *Trans. Faraday Soc.* **27**, 393 (1931).

411. C. F. Goodeve and S. Katz, *Proc. Roy. Soc.* (*Lond.*) **A172**, 432 (1939).

412. R. Gorden, Jr. and P. Ausloos, *J. Phys. Chem.* **65**, 1033 (1961).

413. R. Gorden, Jr. R. E. Rebbert, and P. Ausloos, *Natl. Bur. Stand.* (*U.S.*) *Tech. Note* **496** (1969).

414. T. A. Gover and H. G. Bryant, Jr., *J. Phys. Chem.* **70**, 2070 (1966).

415. T. E. Graedel, L. A. Farrow, and T. A. Weber, *Atmos. Environ.* **10**, 1095 (1976).

416. W. R. M. Graham, K. I. Dismuke, and W. Weltner, Jr., *J. Chem. Phys.* **60**, 3817 (1974).

417. A. Granzow, M. Z. Hoffman, N. N. Lichtin, and S. K. Wason, *J. Phys. Chem.* **72**, 3741 (1968).

418. A. Granzow, M. Z. Hoffman, and N. N. Lichtin, *J. Phys. Chem.* **73**, 4289 (1969).

419. R. I. Greenberg and J. Heicklen, *Int. J. Chem. Kinet.* **2**, 185 (1970).

420. R. I. Greenberg and J. Heicklen, *Int. J. Chem. Kinet.* **4**, 417 (1972).

421. K. F. Greenough and A. B. F. Duncan, *J. Am. Chem. Soc.* **83**, 555 (1961).
422. N. R. Greiner, *J. Chem. Phys.* **45**, 99 (1966).
423. N. R. Greiner, *J. Chem. Phys.* **47**, 4373 (1967).
424. P. M. Griffin and J. W. Hutcherson, *J. Opt. Soc. Am.* **59**, 1607 (1969).
425. M. Griggs, *J. Chem. Phys.* **49**, 857 (1968).
426. W. E. Groth and H. Schierholz, *Planet. Space Sci.* **1**, 333 (1959).
427. W. E. Groth, W. Pessara and H. J. Rommel, *Z. Phys. Chem.* **NF 32**, 192 (1962).
428. W. E. Groth, H. Okabe, and H. J. Rommel, *Z. Naturforsch.* **19a**, 507 (1964).
429. W. E. Groth, U. Schurath, and R. N. Schindler, *J. Phys. Chem.* **72**, 3914 (1968).
429a. H. E. Gunning and O. P. Strausz, in *Advances in Photochemistry*, Vol. 1, W. A. Noyes, Jr., G. S. Hammond, and J. N. Pitts, Jr., Eds., Interscience, New York, 1963, p. 209.
430. H. E. Gunning and O. P. Strausz, in *Advances in Photochemistry*, Vol. 4, W. A. Noyes, Jr., G. S. Hammond, and J. N. Pitts, Jr., Eds., Interscience, New York, 1966, p. 143.
431. T. C. Hall, Jr., and F. E. Blacet, *J. Chem. Phys.* **20**, 1745 (1952).
432. J. B. Halpern, G. Hancock, M. Lenzi, and K. H. Welge, *J. Chem. Phys.* **63**, 4808 (1975).
433. D. Hakala, P. Harteck, and R. R. Reeves, *J. Phys. Chem.* **78**, 1583 (1974).
434. Y. Hamada and A. J. Merer, *Can. J. Phys.* **52**, 1443 (1974).
435. Y. Hamada and A. J. Merer, *Can. J. Phys.* **53**, 2555 (1975).
436. R. F. Hampson, Jr., and H. Okabe, *J. Chem. Phys.* **52**, 1930 (1970).
437. G. Hancock, W. Lange, M. Lenzi, and K. H. Welge, *Chem. Phys. Lett.* **33**, 168 (1975).
438. G. Hancock, J. D. Campbell, and K. H. Welge, *Opt. Commun.* **16**, 177 (1976).
439. M. H. Hanes and E. J. Bair, *J. Chem. Phys.* **38**, 672 (1963).
440. I. Hansen, K. Höinghaus, C. Zetzsch, and F. Stuhl, *Chem. Phys. Lett.* **42**, 370 (1976).
441. H. G. Hanson, *J. Chem. Phys.* **23**, 1391 (1955).
442. J. L. Hardwick and J. C. D. Brand, *Chem. Phys. Lett.* **21**, 458 (1973).
443. P. Harteck, R. R. Reeves, Jr., and B. A. Thompson, *Z. Naturforsch.* **19a**, 2 (1964).
444. V. Hasson, G. R. Hebert, and R. W. Nicholls, *J. Phys. Series B. Atom. Mol. Phys.* **3**, 1188 (1970).
445. V. Hasson and R. W. Nicholls, *J. Phys. Series B. Atom. Mol. Phys.* **4**, 1769 (1971).
446. V. Hasson and R. W. Nicholls, *J. Phys. Series B. Atom. Mol. Phys.* **4**, 1778 (1971).
447. V. Hasson and R. W. Nicholls, *J. Phys. Series B. Atom. Mol. Phys.* **4**, 1789 (1971).
448. F. G. M. Hathorn and D. Hussain, *Trans. Faraday Soc.* **65**, 2678 (1969).
449. P. J. Hay and W. A. Goddard III, *Chem. Phys. Lett.* **14**, 46 (1972).
450. P. J. Hay, T. H. Dunning, Jr., and W. A. Goddard III, *Chem. Phys. Lett.* **23**, 457 (1973).

451. J. Heicklen, *J. Am. Chem. Soc.* **85**, 3562 (1963).

452. J. Heicklen, *J. Am. Chem. Soc.* **87**, 445 (1965).

453. J. Heicklen, *J. Phys. Chem.* **70**, 2456 (1966).

454. J. Heicklen and N. Cohen, in *Advances in Photochemistry*, Vol. 5, W. A. Noyes, Jr., G. S. Hammond, and J. N. Pitts, Jr., Eds., Interscience, New York, 1968, p. 157.

455. R. F. Heidner III, D. Husain, and J. R. Wiesenfeld, *Chem. Phys. Lett.* **16**, 530 (1972).

456. R. F. Heidner and D. Husain, *Nature Phys. Sci.* **241**, 10 (1973).

457. R. F. Heidner and D. Husain, *Int. J. Chem. Kinet.* **5**, 819 (1973).

458. R. F. Heidner and D. Husain, *J. Chem. Soc. Faraday II* **69**, 927 (1973).

459. F. E. Heidrich, K. R. Wilson, and D. Rapp, *J. Chem. Phys.* **54**, 3885 (1971).

460. L. E. Heidt, R. Lueb, W. Pollock, and D. H. Ehhalt, *Geophys. Res. Lett.* **2**, 445 (1975).

461. J. Heimerl, *J. Geophys. Res.* **75**, 5574 (1970).

461a. C. Hellner and R. A. Keller, *J. Air Poll. Control Assoc.* **22**, 959 (1972).

462. G. Herzberg, A. Lagerqvist and E. Miescher, *Can. J. Phys.* **34**, 622 (1956).

463. G. Herzberg and K. K. Innes, *Can. J. Phys.* **35**, 842 (1957).

464. G. Herzberg, *Proc. Roy. Soc.* (*Lond.*) **A262**, 291 (1961).

465. G. Herzberg and J. W. C. Johns, *Proc. Roy. Soc.* (*Lond.*) **A295**, 107 (1966).

466. J. E. Hesser, *J. Chem. Phys.* **48**, 2518 (1968).

467. G. Herzberg and J. W. C. Johns, *Astrophys. J.* **158**, 399 (1969).

468. G. Herzberg, *J. Mol. Spectrosc.* **33**, 147 (1970).

469. G. Herzberg and J. W. C. Johns, *J. Chem. Phys.* **54**, 2276 (1971).

470. J. E. Hesser and B. L. Lutz, *Astrophys. J.* **159**, 703 (1970).

471. T. Hikida, N. Washida, S. Nakajima, S. Yagi, T. Ichimura, and Y. Mori, *J. Chem. Phys.* **63**, 5470 (1975).

471a. T. Hikida, S. Nakajima, T. Ichimura, and Y. Mori, *J. Chem. Phys.* **65**, 1317 (1976).

472. I. H. Hillier and V. R. Saunders, *Mol. Phys.* **22**, 193 (1971).

473. J. Hinze, G. C. Lie, and B. Liu, *Astrophys. J.* **196**, 621 (1975).

474. C. J. Hochanadel, J. A. Ghormley, and P. J. Ogren, *J. Chem. Phys.* **56**, 4426 (1972).

475. R. T. Hodgson, *J. Chem. Phys.* **55**, 5378 (1971).

476. R. T. Hodgson and R. W. Dreyfus, *Phys. Lett.* **38A**, 213 (1972).

477. P. Hogan and D. D. Davis, *J. Chem. Phys.* **62**, 4574 (1975).

478. K. E. Holdy, L. C. Klotz, and K. R. Wilson, *J. Chem. Phys.* **52**, 4588 (1970).

479. G. W. Holleman and J. I. Steinfeld, *Chem. Phys. Lett.* **12**, 431 (1971).

480. J. L. Holmes and P. Rodgers, *Trans. Faraday Soc.* **64**, 2348 (1968).

481. H. Horiguchi and S. Tsuchiya, *Bull. Chem. Soc. Jap.* **44**, 1213 (1971).

482. A. Horowitz and J. G. Calvert, *Int. J. Chem. Kinet.* **4**, 175 (1972).

483. A. Horowitz and J. G. Calvert, *Int. J. Chem. Kinet.* **4**, 191 (1972).

484. A. Horowitz and J. G. Calvert, *Int. J. Chem. Kinet.* **5**, 243 (1973).

485. J. A. Horsley and W. H. Fink, *J. Chem. Phys.* **50**, 750 (1969).

486. P. L. Houston and C. B. Moore, *J. Chem. Phys.* **65**, 757 (1976).

486a. C. J. Howard and K. M. Evenson, *Geophys. Res. Lett.* **4**, 437 (1977).

486b. D. K. Hsu and W. H. Smith, *J. Chem. Phys.* **66**, 1835 (1977).

486c. C. Hubrich, C. Zetzsch, and F. Stuhl, *Ber. Bunsen. Ges. Phys. Chem.* **81**, 437 (1977).

487. R. D. Hudson and V. L. Carter, *J. Geophys. Res.* **74**, 393 (1969).

488. R. D. Hudson and V. L. Carter, *Can. J. Chem.* **47**, 1840 (1969).

489. B. J. Huebert and R. M. Martin, *J. Phys. Chem.* **72**, 3046 (1968).

489a. W. M. Hughes, J. Shannon, and R. Hunter, *Appl. Phys. Lett.* **24**, 488 (1974).

490. M. H. Hui and S. A. Rice, *Chem. Phys. Lett.* **17**, 474 (1972).

490a. D. M. Hunten, *Space Sci. Rev.* **12**, 539 (1971).

490b. D. M. Hunten, *J. Atmos. Sci.* **26**, 826 (1969).

491. D. M. Hunten, *Rev. Geophys. Space Phys.* **12**, 529 (1974).

492. W. T. Huntress, Jr., *J. Chem. Educ.* **53**, 204 (1976).

493. H. E. Hunziker and H. R. Wendt, *J. Chem. Phys.* **60**, 4622 (1974).

494. G. S. Hurst, E. B. Wagner and M. G. Payne, *J. Chem. Phys.* **61**, 3680 (1974).

495. D. Husain and T. R. Wiesenfeld, *Nature* **213**, 1227 (1967).

496. D. Husain and T. R. Wiesenfeld, *Trans. Faraday Soc.* **63**, 1349 (1967).

497. D. Husain and L. J. Kirsch, *Chem. Phys. Lett.* **8**, 543 (1971).

498. D. Husain and L. J. Kirsch, *Trans. Faraday Soc.* **67**, 2025 (1971).

499. D. Husain and L. J. Kirsch, *Trans. Faraday Soc.* **67**, 2886 (1971).

500. D. Husain and L. J. Kirsch, *Trans. Faraday Soc.* **67**, 3166 (1971).

501. D. Husain and J. G. F. Littler, *Chem. Phys. Lett.* **16**, 145 (1972).

502. D. Husain, J. G. F. Littler, and J. R. Wiesenfeld, *Faraday Discuss. Chem. Soc.* **53**, 201 (1972).

503. D. Husain and J. G. F. Littler, *J. Chem. Soc. Faraday II* **68**, 2110 (1972).

504. D. Husain and J. G. F. Littler, *J. Chem. Soc. Faraday II* **69**, 842 (1973).

505. D. Husain and L. J. Kirsch, *J. Photochem.* **2**, 297 (1973/74).

506. D. Husain, S. K. Mitra, and A. N. Young, *J. Chem. Soc. Faraday II* **70**, 1721 (1974).

507. J. W. Hutcherson and P. M. Griffin, *J. Opt. Soc. Am.* **63**, 338 (1973).

508. A. J. Illies and G. A. Takacs, *J. Photochem.* **6**, 35 (1976/77).

509. R. E. Imhof and F. H. Read, *Chem. Phys. Lett.* **11**, 326 (1971).

510. A. P. Ingersoll and C. B. Leovy, *Ann. Rev. Astron. Astrophys.* **9**, 147 (1971).

511. E. C. Y. Inn, K. Watanabe, and M. Zelikoff, *J. Chem. Phys.* **21**, 1648 (1953).

512. E. C. Y. Inn and J. M. Heimerl, *J. Atmos. Sci.* **28**, 838 (1971).

513. E. C. Y. Inn, *J. Geophys. Res.* **77**, 1991 (1972).

514. T. Ishiwata, H. Akimoto, and I. Tanaka, *Chem. Phys. Lett.* **21**, 322 (1973).

515. T. P. J. Izod and R. P. Wayne, *Proc. Roy. Soc. (Lond.)* **A308**, 81 (1968).

516. T. P. J. Izod and R. P. Wayne, *Nature* **217**, 947 (1968).

517. G. E. Jackson and J. G. Calvert, *J. Am. Chem. Soc.* **93**, 2593 (1971).

518. W. M. Jackson, *J. Chem. Phys.* **61**, 4177 (1974).

519. W. M. Jackson and R. J. Cody, *J. Chem. Phys.* **61**, 4183 (1974).

520. M. E. Jacox, D. E. Milligan, N. G. Moll, and W. E. Thompson, *J. Chem. Phys.* **43**, 3734 (1965).

520a. M. E. Jacox and D. E. Milligan, *J. Chem. Phys.* **54**, 919 (1971).

521. R. L. Jaffe, D. M. Hayes, and K. Morokuma, *J. Chem. Phys.* **60**, 5108 (1974).

522. F. C. James, J. A. Kerr, and J. P. Simons, *J. Chem. Soc. Faraday I* **69**, 2124 (1973).

523. F. C. James, J. A. Kerr, and J. P. Simons, *Chem. Phys. Lett.* **25**, 431 (1974).
524. F. C. James and J. P. Simons, *Int. J. Chem. Kinet.* **4**, 887 (1974).
525. T. C. James, *J. Mol. Spectrosc.* **40**, 545 (1971).
526. T. C. James, *J. Chem. Phys.* **55**, 4118 (1971).
527. G. S. Janes, I. Itzkan, C. T. Pike, R. H. Levy, and L. Levin, *IEEE J. Quantum Electron.* **12**, 111 (1976).
528. D. K. Jardine, N. M. Ballash, and D. A. Armstrong, *Can. J. Chem.* **51**, 656 (1973).
529. R. K. M. Jayanty, R. Simonaitis, and J. Heicklen, *J. Photochem.* **4**, 203 (1975).
530. R. K. M. Jayanty, R. Simonaitis, and J. Heicklen, *J. Phys. Chem.* **80**, 433 (1976).
531. M. Jeunehomme and A. B. F. Duncan, *J. Chem. Phys.* **41**, 1692 (1964).
532. M. Jeunehomme, *J. Chem. Phys.* **42**, 4086 (1965).
533. J. W. C. Johns, *Can. J. Phys.* **41**, 209 (1963).
534. J. W. C. Johns, S. H. Priddle, and D. A. Ramsay, *Disc. Faraday Soc.* **35**, 90 (1963).
535. A. W. Johnson and R. G. Fowler, *J. Chem. Phys.* **53**, 65 (1970).
536. P. D. Johnson, *J. Opt. Soc. Am.* **61**, 1451 (1971).
537. H. S. Johnston, *Science* **173**, 517 (1971).
538. H. S. Johnston, *Search* **3**, 276 (1972).
539. H. S. Johnston and R. Graham, *J. Phys. Chem.* **77**, 62 (1973).
540. H. S. Johnston, S. G. Chang, and G. Whitten, *J. Phys. Chem.* **78**, 1 (1974).
541. H. S. Johnston and R. Graham, *Can J. Chem.* **52**, 1415 (1974).
542. H. S. Johnston, *Rev. Geophys. Space Phys.* **13**, 637 (1975).
543. H. S. Johnston, *Ann. Rev. Phys. Chem.* **26**, 315 (1975).
544. H. S. Johnston and G. S. Selwyn, *Geophys. Res. Lett.* **2**, 549 (1975).
545. I. T. N. Jones, U. B. Kaczmar, and R. P. Wayne, *Proc. Roy. Soc.* (*Lond.*) **A316**, 431 (1970).
546. I. T. N. Jones and R. P. Wayne, *Proc. Roy. Soc.* (*Lond.*) A**319**, 273 (1970).
547. I. T. N. Jones and R. P. Wayne, *Proc. Roy. Soc.* (*Lond.*) A**321**, 409 (1971).
548. I. T. N. Jones and K. D. Bayes, *Chem. Phys. Lett.* **11**, 163 (1971).
549. I. T. N. Jones and K. D. Bayes, *J. Chem. Phys.* **59**, 3119 (1973).
550. I. T. N. Jones and K. D. Bayes, *J. Chem. Phys.* **59**, 4836 (1973).
551. P. R. Jones and H. Taube, *J. Phys. Chem.* **75**, 2991 (1971).
552. D. L. Judge and L. C. Lee, *J. Chem. Phys.* **58**, 104 (1973).
553. Ch. Jungen, D. N. Malm, and A. J. Merer, *Chem. Phys. Lett.* **16**, 302 (1972).
554. Ch. Jungen, D. N. Malm, and A. J. Merer, *Can. J. Phys.* **51**, 1471 (1973).
555. O. Kajimoto and R. J. Cvetanović, *Chem. Phys. Lett.* **37**, 533 (1976).
556. O. Kajimoto and R. J. Cvetanović, *J. Chem. Phys.* **64**, 1005 (1976).
557. G. Karl, P. Kruus, and J. C. Polanyi, *J. Chem. Phys.* **46**, 224 (1967).
558. D. Katakis and H. Taube, *J. Chem. Phys.* **36**, 416 (1962).
559. M. Kawasaki, Y. Hirata, and I. Tanaka, *J. Chem. Phys.* **59**, 648 (1973).
560. M. Kawasaki, S. J. Lee, and R. Bersohn, *J. Chem. Phys.* **63**, 809 (1975).
561. P. M. Kelley and W. L. Hase, *Chem. Phys. Lett.* **35**, 57 (1975).
562. L. F. Keyser, S. Z. Levine, and F. Kaufman, *J. Chem. Phys.* **54**, 355 (1971).
563. H. Kijewski and J. Troe, *Helv. Chim. Acta* **55**, 205 (1972).

564. T. T. Kikuchi and W. G. Daffron, *Appl. Opt.* **8**, 1738 (1969).
565. T. T. Kikuchi, *Appl. Opt.* **10**, 1288 (1971).
566. H. H. Kim and J. L. Roebber, *J. Chem. Phys.* **44**, 1709 (1966).
567. G. W. King and D. Moule, *Can. J. Chem.* **40**, 2057 (1962).
568. G. W. King and A. W. Richardson, *J. Mol. Spectrosc.* **21**, 339 (1966).
569. G. W. King and A. W. Richardson, *J. Mol. Spectrosc.* **21**, 353 (1966).
570. G. B. Kistiakowsky, *J. Am. Chem. Soc.* **52**, 102 (1930).
571. G. B. Kistiakowsky and J. C. Sternberg, *J. Chem. Phys.* **21**, 2218 (1953).
572. G. B. Kistiakowsky and T. A. Walter, *J. Phys. Chem.* **72**, 3952 (1968).
573. R. B. Klemm, W. A. Payne, and L. J. Stief, Int. *J. Chem. Kinet. Symp.* **1**, 61 (1975).
574. R. B. Klemm, S. Glicker, and L. J. Stief, *Chem. Phys. Lett.* **33**, 512 (1975).
575. D. Kley and K. H. Welge, *Z. Naturforsch.* **20a**, 124 (1965).
576. D. Kley and K. H. Welge, *J. Chem. Phys.* **49**, 2870 (1968).
577. D. E. Klimek and M. J. Berry, *Chem. Phys. Lett.* **20**, 141 (1973).
578. H. D. Knauth, H. Martin, and W. Stockmann, *Z. Naturforsch.* **29a**, 200 (1974).
579. A. R. Knudson and J. E. Kupperian, Jr., *J. Opt. Soc. Amer.* **47**, 440 (1957).
579a. G. Kockarts, in Mesopheric Model and Related Experiments, G. Fiocco, Ed., D. Reidel, Dordrecht, Holland, 1971, p. 160.
580. S. Koda, P. A. Hackett, and R. A. Back, *Chem. Phys. Lett.* **28**, 532 (1974).
581. R. S. Konar, S. Matsumoto, and B. deB Darwent, *Trans. Faraday Soc.* **67**, 1698 (1971).
582. G. Koren, U. P. Oppenheim, D. Tal, M. Okon, and R. Weil, *Appl. Phys. Lett.* **29**, 40 (1976).
583. I. Koyano and I. Tanaka, *J. Chem. Phys.* **40**, 895 (1964).
584. I. Koyano, T. S. Wauchop, and K. H. Welge, *J. Chem. Phys.* **63**, 110 (1975).
585. D. C. Krezenski, R. Simonaitis, and J. Heicklen, *Planet. Space Sci.* **19**, 1701 (1971).
586. P. M. Kroger, P. C. Demou, and S. J. Riley, *J. Chem. Phys.* **65**, 1823 (1976).
587. M. Kroll, *J. Chem. Phys.* **63**, 319 (1975).
588. H. W. Kroto, *J. Chem. Phys.* **44**, 831 (1966).
589. H. W. Kroto, *Can. J. Phys.* **45**, 1439 (1967).
590. H. W. Kroto, T. F. Morgan, and H. H. Sheena, *Trans. Faraday Soc.* **66**, 2237 (1970).
591. S. Kuis, R. Simonaitis, and J. Heicklen, *J. Geophys. Res.* **80**, 1328 (1975)
592. M. J. Kurylo, N. C. Peterson, and W Braun, *J. Chem. Phys.* **54**, 943 (1971).
593. M. Kutner and P. Thaddeus, *Astrophys. J.* **168**, L67 (1971).
594. C. Lalo and C. Vermeil, *J. Photochem.* **1**, 321 (1972/73).
595. C. Lalo and C. Vermeil, *J. Photochem.* **3**, 441 (1974/75).
596. C. Lambert and G. H. Kimbell, *Can. J. Chem.* **51**, 2601 (1973).
597. M. Lamotte, H. J. Dewey, R. A. Keller, and J. J. Ritter, *Chem. Phys. Lett.* **30**, 165 (1975).
598. A. L. Lane and A. Kuppermann, *Rev. Sci. Inst.* **39**, 126 (1968).
599. R. B. Langford and G. A. Oldershaw, *J. Chem. Soc. Faraday I* **68**, 1550 (1972).
600. R. B. Langford and G. A. Oldershaw, *J. Chem. Soc. Faraday I* **69**, 1389 (1973).

601. A. H. Laufer and J. R. McNesby, *Can. J. Chem.* **43**, 3487 (1965).
602. A. H. Laufer, J. A. Pirog, and J. R. McNesby, *J. Opt. Soc. Am.* **55**, 64 (1965).
602a. A. H. Laufer and J. R. McNesby, *J. Chem. Phys.* **42**, 3329 (1965).
603. A. H. Laufer and J. R. McNesby, *J. Chem. Phys.* **49**, 2272 (1968).
604. A. H. Laufer, *J. Phys. Chem.* **73**, 959 (1969).
605. A. H. Laufer and H. Okabe, *J. Am. Chem. Soc.* **93**, 4137 (1971).
606. A. H. Laufer and R. A. Keller, *J. Am. Chem. Soc.* **93**, 61 (1971).
607. A. H. Laufer and H. Okabe, *J. Phys. Chem.* **76**, 3504 (1972).
608. A. H. Laufer and A. M. Bass, *J. Phys. Chem.* **78**, 1344 (1974).
608a. A. H. Laufer and A. M. Bass, *Chem. Phys. Lett.* **46**, 151 (1977).
609. K. F. Langley and W. D. McGrath, *Planet. Space Sci.* **19**, 413 (1971).
610. K. F. Langley and W. D. McGrath, *Planet. Space Sci.* **19**, 416 (1971).
611. S. R. LaPaglia and A. B. F. Duncan, *J. Chem. Phys.* **34**, 125 (1961).
611a. C. W. Larson and H. E. O'Neal, *J. Phys. Chem.* **70**, 2475 (1966).
612. G. M. Lawrence, *Astrophys. J.* **148**, 261 (1967).
613. G. M. Lawrence, *Phys. Rev.* **175**, 40 (1968).
614. G. M. Lawrence, *Chem. Phys. Lett.* **9**, 575 (1971).
615. G. M. Lawrence, *J. Chem. Phys.* **56**, 3435 (1972).
616. G. M. Lawrence, *J. Chem. Phys.* **57**, 5616 (1972).
616a. G. M. Lawrence and S. C. Seitel, *J. Quant. Spectrosc. Radiat. Transfer* **13**, 713 (1973).
617. G. LeBras and J. Combourieu, *Int. J. Chem. Kinet.* **5**, 559 (1973).
618. F. J. LeBlanc, *J. Chem. Phys.* **48**, 1841 (1968).
619. E. K. C. Lee and W. M. Uselman, *Faraday Discuss. Chem. Soc.* **53**, 125 (1972).
620. J. H. Lee, J. V. Michael, W. A. Payne, D. A. Whytock, and L. J. Stief, *J. Chem. Phys.* **65**, 3280 (1976).
621. J. Lee and A. D. Walsh, *Trans. Faraday Soc.* **55**, 1281 (1959).
622. L. C. Lee and D. L. Judge, *Can. J. Phys.* **51**, 378 (1973).
623. P. H. Lee, H. P. Broida, W. Braun, and J. T. Herron, *J. Photochem.* **2** 165 (1973/74).
623a. P. S. T. Lee, R. L. Russel, and F. S. Rowland, *Chem. Commun.* **1970** 18.
624. A. G. Leiga and H. A. Taylor, *J. Chem. Phys.* **42**, 2107 (1965).
625. M. Lenzi and H. Okabe, *Ber. Bunsenges. Phys. Chem.* **72**, 168 (1968).
625a. M. Lenzi, J. R. McNesby, A. Mele, and C. N. Xuan, *J. Chem. Phys.* **57**, 319 (1972).
626. R. J. LeRoy and R. B. Bernstein, *J. Mol. Spectrosc.* **37**, 109 (1971).
627. S. Z. Levine, A. R. Knight, and R. P. Steer, *Chem. Phys. Lett.* **29**, 73 (1974).
628. D. H. Levy, *J. Chem. Phys.* **56**, 1415 (1972).
629. R. S. Lewis, K. Y. Tang, and E. K. C. Lee, *J. Chem. Phys.* **65**, 2910 (1976).
630. L. M. Leyland, J. R. Majer, and J. C. Robb, *Trans. Faraday Soc.* **66**, 898 (1970).
631. W. Lichten, *J. Chem. Phys.* **26**, 306 (1957).
632. P. L. Lijnse, *Chem. Phys. Lett.* **18**, 73 (1973).
633. P. L. Lijnse and Cj. vanderMaas, *J. Quant. Spectrosc. Radiat. Transfer* **13**, 741 (1973).
634. R. L. Lilly, R. E. Rebbert, and P. Ausloos, *J. Photochem.* **2**, 49 (1973/74).

635. C. L. Lin and F. Kaufman, *J. Chem. Phys.* **55**, 3760 (1971).
636. C. L. Lin and W. B. DeMore, *J. Phys. Chem.* **77**, 863 (1973).
637. C. L. Lin and W. B. DeMore, *J. Photochem.* **2**, 161 (1973/74).
637a. M. C. Lin, *J. Phys. Chem.* **77**, 2726 (1973).
637b. M. C. Lin, *J. Chem. Phys.* **61**, 1835 (1974).
638. J. H. Ling and K. R. Wilson, *J. Chem. Phys.* **63**, 101 (1975).
639. J. K. Link, *J. Opt. Soc. Am.* **56**, 1195 (1966).
640. E. Lissi and J. Heicklen, *J. Photochem.* **1**, 39 (1972/73).
641. H. S. Liszt and J. E. Hesser, *Astrophys. J.* **159**, 1101 (1970).
642. D. J. Little, A. Dalgleish, and R. J. Donovan, *Faraday Discuss. Chem. Soc.* **53**, 211 (1972).
643. D. D. S. Liu, S. Datta, and R. N. Zare, *J. Am. Chem. Soc.* **97**, 2557 (1975).
644. G. Liuti, S. Dondes, and P. Harteck, *J. Chem. Phys.* **44**, 4051 (1966).
645. M. Loewenstein, J. Heimerl, and E. C. Y. Inn, *Rev. Sci. Inst.* **41**, 1908 (1970).
646. G. London, R. Gilpin, H. I. Schiff, and K. H. Welge, *J. Chem. Phys.* **54**, 4512 (1971).
647. F. P. Lossing, *Can. J. Chem.* **49**, 357 (1971).
648. W. Lotmar, *Z. Phys.* **83**, 765 (1933).
649. L. F. Loucks and R. J. Cvetanović, *J. Chem. Phys.* **56**, 321 (1972).
649a. F. J. Lovas and R. D. Suenram, *Chem. Phys. Lett.* **51**, 453 (1977).
650. C. K. Luk and R. Bersohn, *J. Chem. Phys.* **58**, 2153 (1973).
651. M. Luria and J. Heicklen, *Can. J. Chem.* **52**, 3451 (1974).
652. A. Lurio and R. Novick, *Phys. Rev.* **134**, A608 (1964).
653. A. Lurio and R. L. deZafra, *Phys. Rev.* **134**, A1198 (1964).
654. A. Lurio, *Phys. Rev.* **140**, A1505 (1965).
655. J. L. Lyman and S. D. Rockwood, *J. Appl. Phys.* **47**, 595 (1976).
656. J. Y. MacDonald, *J. Chem. Soc.* **1928**, 1.
657. J. J. Magenheimer and R. B. Timmons, *J. Chem. Phys.* **52**, 2790 (1970).
658. B. H. Mahan and R. Mandal, *J. Chem. Phys.* **37**, 207 (1962).
659. G. J. Mains and D. Lewis, *J. Phys. Chem.* **74**, 1694 (1970).
660. K. A. Mantei and E. J. Bair, *J. Chem. Phys.* **49**, 3248 (1968).
660a. J. Marling, *J. Chem. Phys.* **66**, 4200 (1977).
661. F. F. Marmo, *J. Opt. Soc. Am.* **43**, 1186 (1953).
662. F. D. Marsh and M. E. Hermes, *J. Am. Chem. Soc.* **86**, 4506 (1964).
663. R. M. Martin and J. E. Willard, *J. Chem. Phys.* **40**, 2999 (1964).
664. D. E. Martz and R. T. Lagemann, *J. Chem. Phys.* **22**, 1193 (1954).
665. J. Masanet and C. Vermeil, *J. Chim. Phys.* **66**, 1249 (1969).
666. J. Masanet, A. Gilles, and C. Vermeil, *J. Photochem.* **3**, 417 (1974/75).
667. J. Masanet and C. Vermeil, *J. Chim. Phys.* **71**, 820 (1975).
668. C. W. Mathews, *Can. J. Phys.* **45**, 2355 (1967).
669. C. G. Matland, *Phys. Rev.* **92**, 637 (1953).
670. F. M. Matsunaga and K. Watanabe, *J. Chem. Phys.* **46**, 4457 (1967).
671. K. B. McAfee and R. S. Hozack, *J. Chem. Phys.* **64**, 2491 (1976).
672. T. McAllister and F. P. Lossing, *J. Phys. Chem.* **73**, 2996 (1969).
673. J. W. McConkey and J. A. Kernahan, *Planet. Space Sci.* **17**, 1297 (1969).
674. J. R. McDonald and L. E. Brus, *Chem. Phys. Lett.* **16**, 587 (1972).

674a. J. R. McDonald, U. M. Scherr, and S. P. McGlynn, *J. Chem. Phys.* **51**, 1723 (1969).

674b. J. R. McDonald, R. G. Miller, and A. P. Baronavski, *private communication.*

674c. M. B. McElroy, *J. Geophys. Res.* **74**, 29 (1969).

675. M. B. McElroy and J. C. McConnell, *J. Atmos. Sci.* **28**, 879 (1971).

676. M. B. McElroy and J. C. McConnell, *J. Atmos. Sci.* **28**, 1095 (1971).

677. M. B. McElroy and T. M. Donahue, *Science* **177**, 986 (1972).

678. M. B. McElroy, N. D. Sze, and Y. L. Yung, *J. Atmos. Sci.* **30**, 1437 (1973).

679. M. B. McElroy, S. C. Wofsy, J. E. Penner, and T. C. McConnell, *J. Atmos. Sci.* **31**, 287 (1974).

680. M. J. McEwan, G. M. Lawrence, and H. M. Poland, *J. Chem. Phys.* **61**, 2857 (1974).

681. J. J. McGee and J. Heicklen, *J. Chem. Phys.* **41**, 2974 (1964).

682. R. J. McNeal and G. R. Cook, *J. Chem. Phys.* **47**, 5385 (1967).

683. J. R. McNesby, I. Tanaka, and H. Okabe, *J. Chem. Phys.* **36**, 605 (1962).

684. J. R. McNesby and H. Okabe, in *Advances in Photochemistry*, Vol. 3, W. A. Noyes, G. S. Hammond and J. N. Pitts, Jr., Eds., Interscience, New York, 1964, p. 157.

685. J. R. McNesby, *J. Atmos. Sci.* **26**, 594 (1969).

685a. J. R. McNesby, W. Braun, and J. Ball, in *Creation and Detection of the Excited State*, A. A. Lamola, Ed., Dekker, New York, 1971, p. 503.

686. R. D. McQuigg and J. G. Calvert, *J. Am. Chem. Soc.* **91**, 1590 (1969).

687. G. M. Meaburn and D. Perner, *Nature* **212**, 1042 (1966).

688. G. M. Meaburn and S. Gordon, *J. Phys. Chem.* **72**, 1592 (1968).

689. J. F. Meagher and J. Heicklen, *J. Photochem.* **3**, 455 (1974/75).

690. H. Meinel, *Z. Naturforsch.* **30a**, 323 (1975).

691. L. G. Meira, Jr., *J. Geophys. Res.* **76**, 202 (1971).

692. A. Mele and H. Okabe, *J. Chem. Phys.* **51**, 4798 (1969).

693. L. A. Melton and W. Klamperer, *Planet. Space Sci.* **20**, 157 (1972).

694. L. A. Melton and W. Kamperer, *J. Chem. Phys.* **59**, 1099 (1973).

695. J. E. Mentall and E. P. Gentieu, *J. Chem. Phys.* **52**, 5641 (1970).

696. J. E. Mentall and E. P. Gentieu, *J. Chem. Phys.* **55**, 5471 (1971).

697. J. A. Merritt, *Can. J. Phys.* **40**, 1683 (1962).

698. H. D. Mettee, *J. Chem. Phys.* **49**, 1784 (1968).

699. K. A. Meyer and D. R. Crosley, *J. Chem. Phys.* **59**, 1933 (1973).

700. K. A. Meyer and D. R. Crosley, *J. Chem. Phys.* **59**, 3153 (1973).

701. J. V. Michael and R. E. Weston, Jr., *J. Chem. Phys.* **45**, 3632 (1966).

702. J. V. Michael and C. Yeh, *J. Chem. Phys.* **53**, 59 (1970).

703. J. V. Michael and G. N. Suess, *J. Phys. Chem.* **78**, 482 (1974).

704. K. J. Miller, S. R. Mielczarek, and M. Krauss, *J. Chem. Phys.* **51**, 26 (1969).

705. R. H. Miller, D. L. Bernitt, and I. C. Hisatsune, *Spectrochim. Acta* **23A**, 223 (1967).

706. R. G. Miller and E. K. C. Lee, *Chem. Phys. Lett.* **27**, 475 (1974).

707. R. G. Miller and E. K. C. Lee, *Chem. Phys. Lett.* **33**, 104 (1975).

708. R. Milstein and F. S. Rowland, *J. Phys. Chem.* **79**, 669 (1975).

709. R. C. Mitchell and J. P. Simons, *Discuss. Faraday Soc.* **44**, 208 (1967).

710. H. Mizutani, H. Mikuni, and M. Takahasi, *Chem. Lett.* **1972**, 573.

711. M. J. Molina and F. S. Rowland, *Nature*, **249**, 810 (1974).

712. L. T. Molina, J. E. Spencer, and M. J. Molina, *Chem. Phys. Lett.* **45**, 158 (1977).

713. N. C. Moll, D. R. Clutter, and W. E. Thompson, *J. Chem. Phys.* **45**, 4469 (1966).

714. C. B. Moore, *Ann. Rev. Phys. Chem.* **22**, 387 (1971).

715. C. B. Moore, *Acc. Chem. Res.* **6**, 323 (1973).

716. J. H. Moore, Jr., and D. W. Robinson, *J. Chem. Phys.* **48**, 4870 (1968).

717. G. K. Moortgat and P. Warneck, *Z. Naturforsch.* **30a**, 835 (1975).

718. F. A. Morse and F. Kaufman, *J. Chem. Phys.* **42**, 1785 (1965).

719. P. D. Morten, C. G. Freeman, M. J. McEwan, R. F. C. Claridge, and L. F. Phillips, *Chem. Phys. Lett.* **16**, 148 (1972).

720. P. D. Morten, C. G. Freeman, R. F. C. Claridge, and L. F. Phillips, *J. Photochem.* **3**, 285 (1974/75).

721. D. C. Moule and P. D. Foo, *J. Chem. Phys.* **55**, 1262 (1971).

722. D. C. Moule and C. R. Subramaniam, *J. Mol. Spectrosc.* **48**, 336 (1973).

723. D. C. Moule and A. D. Walsh, *Chem. Rev.* **75**, 67 (1975).

723a. G. H. Mount, E. S. Warden, and H. W. Moos, *Astrophys. J.* **214**, L47 (1977).

724. S. Mukamel and J. Jortner, *J. Chem. Phys.* **60**, 4760 (1974).

725. R. S. Mulliken, *J. Chem. Phys.* **8**, 382 (1940).

726. J. N. Murrell and J. M. Taylor, *Mol. Phys.* **16**, 609 (1969).

727. J. A. Myer and J. A. R. Samson, *J. Chem. Phys.* **52**, 266 (1970).

728. J. A. Myer and J. A. R. Samson, *J. Chem. Phys.* **52**, 716 (1970).

729. G. H. Myers, D. M. Silver, and F. Kaufman, *J. Chem. Phys.* **44**, 718 (1966).

730. R. S. Nakata, K. Watanabe, and F. M. Matsunaga, *Sci. Light* **14**, 54 (1965).

731. T. Nakayama, M. Y. Kitamura, and K. Watanabe, *J. Chem. Phys.* **30**, 1180 (1959).

731a. T. Nakayama and K. Watanabe, *J. Chem. Phys.* **40**, 558 (1964).

732. G. E. Nelson and R. F. Borkman, *J. Chem. Phys.* **63**, 208 (1975).

733. D. Neuberger and A. B. F. Duncan, *J. Chem. Phys.* **22**, 1693 (1954).

734. H. Neuimin and A. Terenin, *Acta Physicochim. URSS* **5**, 465 (1936).

735. R. L. Newburn, Jr. and S. Gulkis, *Space Sci. Rev.* **3**, 179 (1973).

736. R. H. Newman, C. G. Freeman, M. J. McEwan, R. E. C. Claridge, and L. F. Phillips, *Trans. Faraday Soc.* **66**, 2827 (1970).

737. T. N. Ng and S. Bell, *J. Mol. Spectrosc.* **50**, 166 (1974).

738. M. Nicholet, in Mesospheric Model and Related Experiments, G. Fiocco, Ed., D. Reidel, Dordrecht, Holland 1971, p. 1

739. M. Nicholet, *Planet. Space Sci.* **20**, 1671 (1972).

739a. M. Nicolet and W. Peetermans, *Ann. Geophys.* **28**, 751 (1972).

740. M. Nicholet, *Rev. Geophys. Space Phys.* **13**, 593 (1975).

741. J. R. Nielsen, V. Thornton, and E. B. Dale, *Rev. Mod. Phys.* **16**, 307 (1944).

742. A. O. Nier, W. B. Hanson, A. Seiff, M. B. McElroy, N. W. Spencer, R. J. Duckett, T. C. D. Knight, and W. S. Cook, *Science* **193**, 786 (1976).

743. H. Niki, E. E. Daby, and B. Weinstock, in *Photochemical Smog and Ozone Reactions*, R. F. Gould, Ed., American Chemical Society, Washington, D.C., 1972, p. 16.

744. J. F. Noxon, *Space Sci. Rev.* **8**, 92 (1968).
745. J. F. Noxon, *J. Chem. Phys.* **52**, 1852 (1970).
746. R. G. W. Norrish and G. A. Oldershaw, *Proc. Roy. Soc.* (*Lond.*) **A249**, 498 (1959).
747. R. G. W. Norrish and G. A. Oldershaw, *Proc. Roy. Soc.* (*Lond.*) **A262**, 1 (1961).
748. R. G. W. Norrish and R. P. Wayne, *Proc. Roy. Soc.* (*Lond.*) **A288**, 200 (1965).
749. R. G. W. Norrish and R. P. Wayne, *Proc. Roy. Soc.* (*Lond.*) **A288**, 361 (1965).
750. W. A. Noyes, Jr., G. B. Porter, and J. E. Jolley, *Chem. Rev.* **56**, 49 (1956).
751. R. L. Nuttall, A. H. Laufer, and M. V. Kilday, *J. Chem. Thermodyn.* **3**, 167 (1971).
752. R. J. O'Brien and G. H. Myers, *J. Chem. Phys.* **53**, 3832 (1970).
753. M. Ogawa, *J. Geophys. Res. Space Phys.* **73**, 6759 (1968).
754. M. Ogawa, *J. Chem. Phys.* **53**, 3754 (1970).
755. M. Ogawa, *J. Chem. Phys.* **54**, 2550 (1971).
756. J. F. Ogilvie, *J. Mol. Struct.* **31**, 407 (1976).
757. T. Oka, A. R. Knight, and R. P. Steer, *J. Chem. Phys.* **63**, 2414 (1975).
758. T. Oka, A. R. Knight, and R. P. Steer, *J. Chem. Phys.* **66**, 699 (1977).
759. H. Okabe and J. R. McNesby, *J. Chem. Phys.* **34**, 668 (1961).
760. H. Okabe, *J. Opt. Soc. Am.* **54**, 478 (1964).
761. H. Okabe, *J. Chem. Phys.* **47**, 101 (1967).
762. H. Okabe and M. Lenzi, *J. Chem. Phys.* **47**, 5241 (1967).
763. H. Okabe, *J. Chem. Phys.* **49**, 2726 (1968).
764. H. Okabe and A. Mele, *J. Chem. Phys.* **51**, 2100 (1969).
765. H. Okabe, *J. Chem. Phys.* **53**, 3507 (1970).
766. H. Okabe, A. H. Laufer, and J. J. Ball, *J. Chem. Phys.* **55**, 373 (1971).
767. H. Okabe, *J. Am. Chem. Soc.* **93**, 7095 (1971).
768. H. Okabe, *J. Chem. Phys.* **56**, 3378 (1972).
769. H. Okabe, *J. Chem. Phys.* **56**, 4381 (1972).
770. H. Okabe, P. L. Splitstone, and J. J. Ball, *J. Air Pollution Control Assoc.* **23**, 514 (1973).
771. H. Okabe and V. H. Dibeler, *J. Chem. Phys.* **59**, 2430 (1973).
772. H. Okabe, in *Chemical Spectroscopy and Photochemistry in the Vacuum Ultraviolet*, C. Sandorfy, P. J. Ausloos, and M. B. Robin, Ed., Reidel, Boston, 1974, p. 513.
773. H. Okabe, *J. Chem. Phys.* **62**, 2782 (1975).
774. H. Okabe, *J. Chem. Phys.* **66**, 2058 (1977).
775. S. Okuda, T. N. Rao, D. H. Slater, and J. G. Calvert, *J. Phys. Chem.* **73**, 4412 (1969).
776. G. A. Oldershaw and D. A. Porter, *J. Chem. Soc. Faraday I* **68**, 709 (1972).
777. G. A. Oldershaw, D. A. Porter, and A. Smith, *J. Chem. Soc. Faraday I* **68**, 2218 (1972).
778. R. J. Oldman, R. K. Sander, and K. R. Wilson, *J. Chem. Phys.* **54**, 4127 (1971).
779. R. J. Oldman, R. K. Sander, and K. R. Wilson, *J. Chem. Phys.* **63**, 4252 (1975).
780. R. C. Ormerod, T. R. Powers, and T. L. Rose, *J. Chem. Phys.* **60**, 5109 (1974).
781. M. H. Ornstein and V. E. Derr, *J. Opt. Soc. Am.* **66**, 233 (1967).

782. H. A. Ory, *J. Chem. Phys.* **40**, 562 (1964).
783. K. R. Osborn, C. C. McDonald, and H. E. Gunning, *J. Chem. Phys.* **26**, 124 (1957).
784. T. L. Osif and J. Heicklen, *J. Phys. Chem.* **80**, 1526 (1976).
785. K. Otsuka and J. G. Calvert, *J. Am. Chem. Soc.* **93**, 2581 (1971).
786. W. R. Ott, *Phys. Rev.* **A4**, 245 (1971).
787. R. Overend, G. Paraskevopoulos, J. R. Crawford, and H. A. Wiebe, *Can. J. Chem.* **53**, 1915 (1975).
788. T. Owen and C. Sagan, *Icarus* **16**, 557 (1972).
789. F. Paech, R. Schmiedl, and W. Demtröder, *J. Chem. Phys.* **63**, 4369 (1975).
790. R. E. Palmer and T. D. Padrick, *J. Chem. Phys.* **64**, 2051 (1976).
791. G. Paraskevopoulos and R. J. Cvetanović, *J. Am. Chem. Soc.* **91**, 7572 (1969).
792. G. Paraskevopoulos and R. J. Cvetanović, *J. Chem. Phys.* **52**, 5821 (1970).
793. G. Paraskevopoulos and R. J. Cvetanović, *Chem. Phys. Lett.* **9**, 603 (1971).
794. G. Paraskevopoulos, V. B. Symonds, and R. J. Cvetanović, *Can. J. Chem.* **50**, 1838 (1972).
795. J. H. Park, *J. Atmos. Sci.* **31**, 1893 (1974).
796. C. A. Parker, *Proc. Roy. Soc.* (*Lond.*) A**220**, 104 (1953).
797. D. A. Parkes, L. F. Keyser, and F. Kaufman, *Astrophys. J.* **149**, 217 (1967).
798. T. D. Parkinson and D. M. Hunten, *J. Atmos. Sci.* **29**, 1380 (1972).
798a. C. M. Pathak and H. B. Palmer, *J. Mol. Spectrosc.* **32**, 157 (1969).
799. T. T. Paukert and H. S. Johnston, *J. Chem. Phys.* **56**, 2824 (1972).
800. R. J. Paur and E. J. Bair, *Int. J. Chem. Kinet.* **8**, 139 (1976).
801. W. A. Payne, L. J. Stief, and D. D. Davis, *J. Am. Chem. Soc.* **95**, 7614 (1973).
802. W. A. Payne and L. J. Stief, *J. Chem. Phys.* **64**, 1150 (1976).
803. R. D. Penzhorn and B. deB. Darwent, *J. Phys. Chem.* **72**, 1639 (1968).
804. M. L. Perlman and G. K. Rollefson, *J. Chem. Phys.* **9**, 362 (1941).
805. D. Perner, D. H. Ehhalt, H. W. Pätz, U. Platt, E. P. Röth, and A. Volz, *Geophys. Res. Lett.* **3**, 466 (1976).
806. A. Pery-Thorne and F. P. Banfield, *J. Phys. Series B. Atom. Mol. Phys.* **3**, 1011 (1970).
807. L. F. Phillips, *J. Photochem.* **5**, 277 (1976).
808. R. F. Phillips and D. S. Sethi, *J. Chem. Phys.* **61**, 5473 (1974).
809. M. J. Pilling, A. M. Bass, and W. Braun, *Chem. Phys. Lett.* **9**, 147 (1971).
809a. M. J. Pilling and J. A. Robertson, *Chem. Phys. Lett.* **33**, 336 (1975).
809b. M. J. Pilling and J. A. Robertson, *J. Chem. Soc. Faraday I* **73**, 968 (1977).
810. J. N. Pitts, Jr., J. H. Sharp, and S. I. Chan, *J. Chem. Phys.* **42**, 3655 (1964).
811. J. N. Pitts, Jr., and B. J. Finlayson, *Angew Chem. Int. Ed.* **14**, 1 (1975).
812. G. J. Pontrelli and A. G. Anastassiou, *J. Chem. Phys.* **42**, 3735 (1965).
813. G. Porter, *Discuss. Faraday Soc.* **9**, 60 (1950).
814. G. B. Porter and B. T. Connelly, *J. Chem. Phys.* **33**, 81 (1960).
815. K. F. Preston and R. J. Cvetanović, *J. Chem. Phys.* **45**, 2888 (1966).
816. K. F. Preston and R. F. Barr, *J. Chem. Phys.* **54**, 3347 (1971).
817. K. F. Preston and R. J. Cvetanović, in *Comprehensive Chemical Kinetics*, Vol. 4, C. H. Bamford and C. F. H. Tipper, Eds., Elsevier, New York, 1972, p. 47.
818. W. C. Price and D. M. Simpson, *Proc. Roy. Soc.* (*Lond.*) A**165**, 272 (1938).

819. R. G. Prinn, in *Physics and Chemistry of Upper Atmospheres*, B. M. McCormac, Ed., Reidel, Dordrecht, Holland, 1973, p. 335.
820. M. Quack and J. Troe, *Ber. Bunsenges. Phys. Chem.* **79**, 469 (1975).
821. L. M. Quick and R. J. Cvetanović, *Can. J. Chem.* **49**, 2193 (1971).
821a. J. W. Rabalais, J. M. McDonald, V. Scherr, and S. P. McGlynn, *Chem. Rev.* **71**, 73 (1971).
822. J. S. Randhawa, *Pure Appl. Geophys.* **106–108**, 1490 (1973).
823. T. N. Rao, S. S. Collier, and J. G. Calvert, *J. Am. Chem. Soc.* **91**, 1609 (1969).
824. T. N. Rao, S. S. Collier, and J. G. Calvert, *J. Am. Chem. Soc.* **91**, 1616 (1969).
825. T. N. Rao and J. G. Calvert, *J. Phys. Chem.* **74**, 681 (1970).
826. J. W. Raymonda, L. O. Edwards, and B. R. Russel, *J. Am. Chem. Soc.* **96**, 1708 (1974).
827. R. E. Rebbert and P. Ausloos, *J. Photochem.* **1**, 171 (1972/73).
828. R. E. Rebbert and P. Ausloos, *J. Photochem.* **4**, 419 (1975).
829. R. E. Rebbert and P. Ausloos, *J. Photochem.* **6**, 265 (1976/77).
829a. R. E. Rebbert, S. G. Lias, and P. Ausloos, *J. Photochem.*, **8**, 17 (1978).
830. J. M. Ricks and R. F. Barrow, *Can. J. Phys.* **47**, 2423 (1969).
831. R. K. Ritchie, A. D. Walsh, and P. A. Warsop, *Proc. Roy. Soc.* (*Lond.*) A**266**, 257 (1962).
832. R. K. Ritchie and A. D. Walsh, *Proc. Roy. Soc.* (*Lond.*) A**267**, 395 (1962).
833. B. A. Ridley, J. A. Dovenport, L. J. Stief, and K. H. Welge, *J. Chem. Phys.* **57**, 520 (1972).
834. S. J. Riley and K. R. Wilson, *Faraday Discuss. Chem. Soc.* **53**, 132 (1972).
835. J. K. Rice and F. K. Truby, *Chem. Phys. Lett.* **19**, 440 (1973).
836. D. E. Robbins, L. T. Rose, and W. R. Boykin, Johnson Space Center Internal Note JSC-09937, 1975.
837. D. E. Robbins, *Geophys. Res. Lett.* **3**, 213 (1976).
838. J. L. Roebber, J. C. Larrabee, and R. E. Huffman, *J. Chem. Phys.* **46**, 4594 (1967).
839. J. L. Roebber, *J. Chem. Phys.* **54**, 4001 (1971).
840. J. Romand and B. Vodar, *C. R. Acad. Sci.* **226**, 238 (1948).
841. B. C. Roquitte and M. H. J. Wijnen, *J. Am. Chem. Soc.* **85**, 2053 (1963).
842. J. Rostas, D. Cossart, and J. R. Bastien, *Can. J. Phys.* **52**, 1274 (1974).
843. F. S. Rowland and M. J. Molina, *Rev. Geophys. Space Phys.* **13**, 1 (1975).
844. F. S. Rowland, J. E. Spencer, and M. J. Molina, *J. Phys. Chem.* **80**, 2711 (1976).
844a. F. S. Rowland, P. S. T. Lee, D. C. Montague, and R. L. Russell, *Faraday Discuss. Chem. Soc.* **53**, 111 (1972).
845. B. R. Russel, L. O. Edwards, and J. W. Raymonda, *J. Am. Chem. Soc.* **95**, 2129 (1973).
846. R. S. Sach, *Int. J. Radiat. Phys. Chem.* **3**, 45 (1971).
847. P. B. Sackett and J. T. Yardley, *Chem. Phys. Lett.* **9**, 612 (1971).
848. P. B. Sackett and J. T. Yardley, *J. Chem. Phys.* **57**, 152 (1972).
849. E. Safary, J. Romand, and B. Vodar, *J. Chem. Phys.* **19**, 379 (1951).
850. K. Sakurai and G. Kapelle, *J. Chem. Phys.* **53**, 3764 (1970).
851. K. Sakurai, J. Clark, and H. P. Broida, *J. Chem. Phys.* **54**, 1217 (1971).

852. K. Sakurai, G. Kapelle, and H. P. Broida, *J. Chem. Phys.* **54**, 1220 (1971).
853. K. Sakurai, G. Kapelle, and H. P. Broida, *J. Chem. Phys.* **54**, 1412 (1971).
854. D. R. Salahub and R. A. Boshi, *Chem. Phys. Lett.* **16**, 320 (1972).
855. D. Salomon and A. A. Scala, *J. Chem. Phys.* **62**, 1469 (1975).
856. B. D. Savage and G. M. Lawrence, *Astrophys. J.* **146**, 940 (1966).
857. H. I. Schiff, *Ann. Geophys.* **28**, 67 (1972).
858. H. J. Schumacher and R. V. Townend, *Z. Phys. Chem.* **B20**, 375 (1933).
859. U. Schurath, P. Tiedemann, and R. N. Schindler, *J. Phys. Chem.* **73**, 457 (1969).
859a. U. Schurath, M. Weber, and K. H. Becker, *J. Chem. Phys.* **67**, 110 (1977)
860. M. Schürgers and K. H. Welge, *Z. Naturforsch.* **23a**, 1508 (1968).
861. C. J. H. Schutte, *J. Chem. Phys.* **51**, 4678 (1969).
862. S. E. Schwartz and H. S. Johnston, *J. Chem. Phys.* **51**, 1286 (1969).
863. S. E. Schwartz and G. I. Senum, *Chem. Phys. Lett.* **32**, 569 (1975).
864. P. M. Scott, K. F. Preston, R. J. Andersen, and L. M. Quick, *Can. J. Chem.* **49**, 1808 (1971).
865. D. S. Sethi and H. A. Taylor, *J. Chem. Phys.* **48**, 533 (1968).
866. D. S. Sethi and H. A. Taylor, *J. Chem. Phys.* **49**, 3669 (1968).
867. J. Shamir, D. Yellin, and H. H. Claassen, *Isr. J. Chem.* **12**, 1015 (1974).
868. D. E. Shemansky, *J. Chem. Phys.* **51**, 689 (1969).
869. D. E. Shemansky and N. P. Carleton, *J. Chem. Phys.* **51**, 682 (1969).
870. D. E. Shemansky, *J. Chem. Phys.* **51**, 5487 (1969).
871. D. E. Shemansky, *J. Chem. Phys.* **56**, 1582 (1972).
872. S. Shida, Z. Kuri, and T. Furuoya, *J. Chem. Phys.* **28**, 131 (1958).
872a. S. Shih, S. D. Peyerimhoff, and R. J. Buenker, *J. Mol. Spectrosc.* **64**, 167 (1977).
872b. T. Shimazaki and R. C. Whitten, *Rev. Geophys. Space Phys.* **14**, 1 (1976).
873. K. Shimokoshi, *Chem. Phys. Lett.* **24**, 425 (1974).
873a. D. M. Shold and R. E. Rebbert, *J. Photochem.* (in preparation).
874. R. V. Shukla, S. K. Jain, S. K. Gupta, and A. N. Srivastava, *J. Chem. Phys.* **52**, 2744 (1970).
875. H. W. Sidebottom, J. M. Tedder, and C. Walton, *Trans. Faraday Soc.* **65**, 755 (1969).
876. H. W. Sidebottom, C. C. Badcock, J. G. Calvert, G. E. Reinhardt, B. R. Rabe and E. K. Damon, *J. Am. Chem. Soc.* **93**, 2587 (1971).
877. H. W. Sidebottom, C. C. Badcock, J. G. Calvert, B. R. Rabe, and E. K. Damon, *J. Am. Chem. Soc.* **93**, 3121 (1971).
878. H. W. Sidebottom, K. Otsuka, A. Horowitz, J. G. Calvert, B. R. Rabe, and E. K. Damon, *Chem. Phys. Lett.* **13**, 337 (1972).
879. H. W. Sidebottom, C. C. Badcock, G. E. Jackson, J. G. Calvert, G. W. Reinhardt, and E. K. Damon, *Environ. Sci. Technol.* **6**, 72 (1972).
880. K. S. Sidhu, I. G. Csizmadia, O. P. Strausz, and H. E. Gunning, *J. Am. Chem. Soc.* **88**, 2412 (1966).
881. L. W. Sieck, *J. Chem. Phys.* **50**, 1748 (1969).
882. S. J. Silvers and C. L. Chiu, *J. Chem. Phys.* **56**, 5663 (1972).
883. R. Simonaitis, E. Lissi, and J. Heicklen, *J. Geophys. Res.* **77**, 4248 (1972).

884. R. Simonaitis, R. I. Greenberg, and J. Heicklen, *Int. J. Chem. Kinet. 4*, 497 (1972).

885. R. Simonaitis and J. Heicklen, *J. Photochem.* **1**, 181 (1972/73).

886. R. Simonaitis and J. Heicklen, *J. Phys. Chem.* **77**, 1096 (1973).

887. R. Simonaitis, S. Braslavsky, J. Heicklen, and M. Nicolet, *Chem. Phys. Lett.* **19**, 601 (1973).

888. R. Simonaitis and J. Heicklen, *Int. J. Chem. Kinet.* **5**, 231 (1973).

889. R. Simonaitis and J. Heicklen, *J. Photochem.* **2**, 309 (1973/74).

890. R. Simonaitis and J. Heicklen, *J. Phys. Chem.* **78**, 653 (1974).

891. J. P. Simons and A. J. Yarwood, *Nature* **192**, 943 (1961).

892. J. P. Simons and A. J. Yarwood, *Trans. Faraday Soc.* **59**, 90 (1963).

893. J. P. Simons and P. W. Tasker, *Mol. Phys.* **26**, 1267 (1973).

894. J. P. Simons and P. W. Tasker, *Mol. Phys.* **27**, 1691 (1974).

894a. J. P. Simons, "Gas Kinetics and Energy Transfer, *Chem. Soc. Specialist Period, Rep.*, **2**, 53 (1977).

895. P. A. Skotnicki, A. G. Hopkins, and C. W. Brown, *Anal. Chem.* **45**, 2291 (1973).

896. P. A. Skotnicki, A. G. Hopkins, and C. W. Brown, *J. Phys. Chem.* **79**, 2450 (1975).

897. T. G. Slanger, *J. Chem. Phys.* **45**, 4127 (1966).

898. T. G. Slanger, *J. Chem. Phys.* **48**, 586 (1968).

899. T. G. Slanger and G. Black, *J. Chem. Phys.* **51**, 4534 (1969).

900. T. G. Slanger and G. Black, *Chem. Phys. Lett.* **4**, 558 (1970).

901. T. G. Slanger and G. Black, *J. Chem. Phys.* **54**, 1889 (1971).

902. T. G. Slanger and G. Black, *J. Chem. Phys.* **58**, 194 (1973).

903. T. G. Slanger and G. Black, *J. Chem. Phys.* **58**, 3121 (1973).

904. T. G. Slanger and G. Black, *J. Chem. Phys.* **60**, 468 (1974).

905. T. G. Slanger, R. L. Sharpless, and G. Black, *J. Chem. Phys.* **61**, 5022 (1974).

906. T. G. Slanger and G. Black, *J. Photochem.* **4**, 329 (1975).

907. T. G. Slanger and G. Black, *J. Chem. Phys.* **63**, 969 (1975).

908. T. G. Slanger and G. Black, *J. Chem. Phys.* **64**, 4442 (1976).

909. R. E. Smalley, L. Wharton, and D. H. Levy, *J. Chem. Phys.* **63**, 4977 (1975).

910. A. J. Smith, R. E. Imhof, and F. H. Read, *J. Phys. Series B. Atom. Mol. Phys.* **6**, 1333 (1973).

911. W. H. Smith, *J. Quant. Spectrosc. Radiat. Transfer* **9**, 1191 (1969).

912. W. H. Smith, *J. Chem. Phys.* **51**, 520 (1969).

913. W. H. Smith, *J. Chem. Phys.* **53**, 792 (1970).

914. W. H. Smith, J. Bromander, L. J. Curtis, H. G. Berry, and R. Buchta, *Astrophys. J.* **165**, 217 (1971).

915. W. H. Smith and G. Stella, *J. Chem. Phys.* **63**, 2395 (1975).

916. W. H. Smith, J. Brzozowski, and P. Erman, *J. Chem. Phys.* **64**, 4628 (1976).

916a. W. S. Smith, C. C. Chou, and F. S. Rowland, *Geophys. Res. Lett.*, **4**, 517 (1977)

917. D. R. Snelling, V. D. Baiamonte, and E. J. Bair, *J. Chem. Phys.* **44**, 4137 (1966).

918. D. R. Snelling and M. Gauthier, *Chem. Phys. Lett.* **9**, 254 (1971).

919. D. R. Snelling, *Can. J. Chem.* **52**, 257 (1974).

920. R. Solarz and D. H. Levy, *J. Chem. Phys.* **60**, 842 (1974).

921. T. Solomon, C. Jonah, P. Chandra, and R. Bersohn, *J. Chem. Phys.* **55**, 1908 (1971).

922. H. P. Sperling and S. Toby, *Can. J. Chem.* **51**, 471 (1973).

922a. D. H. Stedman, H. Alvord, and A. Baker-Blocker, *J. Phys. Chem.* **78**, 1248 (1974).

923. R. P. Steer, R. A. Ackerman, and J. N. Pitts, Jr., *J. Chem. Phys.* **51**, 843 (1969).

924. O. Stern and M. Volmer, *Phys. Z.* **20**, 183 (1919).

925. R. K. Steunenberg and R. C. Vogel, *J. Am. Chem. Soc.* **78**, 901 (1956).

926. C. G. Stevens, M. W. Swagel, R. Wallace, and R. N. Zare, *Chem. Phys. Lett.* **18**, 465 (1973).

927. L. J. Stief and R. J. Mataloni, *Appl. Opt.* **4**, 1674 (1965).

928. L. J. Stief, V. J. DeCarlo, and R. J. Mataloni, *J. Chem. Phys.* **42**, 3113 (1965).

929. L. J. Stief, *J. Chem. Phys.* **44**, 277 (1966).

930. L. J. Stief and V. J. DeCarlo, *J. Chem. Phys.* **50**, 1234 (1969).

931. L. J. Stief and V. J. DeCarlo, *J. Amer. Chem. Soc.* **91**, 839 (1969).

932. L. J. Stief, V. J. DeCarlo, and W. A. Payne, *J. Chem. Phys.* **51**, 3336 (1969).

933. L. J. Stief, W. A. Payne, and V. J. DeCarlo, *J. Chem. Phys.* **53**, 475 (1970).

934. L. J. Stief, *Nature* **237**, 29 (1972).

935. L. J. Stief, W. A. Payne, and R. B. Klemm, *J. Chem. Phys.* **62**, 4000 (1975).

936. L. Stockburger III, S. Braslavsky, and J. Heicklen, *J. Photochem.* **2**, 15 (1973/74).

937. E. J. Stone, G. M. Lawrence, and C. E. Fairchild, *J. Chem. Phys.* **65**, 5083 (1976).

938. O. P. Strausz, J. M. Campbell, H. S. Sandhu, and H. E. Gunning, *J. Am. Chem. Soc.* **95**, 732 (1973).

939. S. J. Strickler and R. A. Berg, *J. Chem. Phys.* **37**, 814 (1962).

940. S. J. Strickler and D. B. Howell, *J. Chem. Phys.* **49**, 1947 (1968).

941. S. J. Strickler, J. P. Vikesland, and H. D. Bier, *J. Chem. Phys.* **60**, 664 (1974).

942. D. F. Strobel, *J. Atmos. Sci.* **26**, 906 (1969).

943. D. F. Strobel, in *Physics and Chemistry of Upper Atmospheres*, B. M.. McCormac, Ed., Reidel, Dordrecht, Holland, 1973, p. 345.

944. D. F. Strobel, *J. Atmos. Sci.* **30**, 489 (1973).

945. D. F. Strobel, *J. Atmos. Sci.* **30**, 1205 (1973).

946. D. F. Strobel, *Rev. Geophys. Space Phys.* **13**, 372 (1975).

947. F. Stuhl and K. H. Welge, *Z. Naturforsch.* **18a**, 900 (1963).

948. F. Stuhl and K. H. Welge, *J. Chem. Phys.* **47**, 332 (1967).

949. F. Stuhl and K. H. Welge, *Can. J. Chem.* **47**, 1870 (1969).

950. F. Stuhl and H. Niki, *Chem. Phys. Lett.* **7**, 197 (1970).

951. F. Stuhl and H. Niki, *J. Chem. Phys.* **57**, 3671 (1972).

952. G. P. Sturm, Jr. and J. M. White, *J. Chem. Phys.* **50**, 5035 (1969).

953. D. W. G. Style and J. C. Ward, *J. Chem. Soc.* **1952**, 2125.

953a. F. Su, J. W. Bottenheim, D. L. Thorsell, J. G. Calvert, and E. K. Damon, *Chem. Phys. Lett.* **49**, 305 (1977).

954. J. O. Sullivan and P. Warneck, *J. Chem. Phys.* **46**, 953 (1967).

955. R. A. Sutherland and R. A. Anderson, *J. Chem. Phys.* **58**, 1226 (1973).
956. K. Tadasa, N. Imai, and T. Inaba, *Bull. Chem. Soc. Jap.* **49**, 1758 (1976).
956a. G. A. Takacs and J. E. Willards, *J. Phys. Chem.* **81**, 1343 (1977).
957. S. Takita, Y. Mori, and I. Tanaka, *J. Phys. Chem.* **72**, 4360 (1968).
958. S. Takita, Y. Mori, and I. Tanaka, *J. Phys. Chem.* **73**, 2929 (1969).
959. V. L. Talroze, M. N. Larichev, I. O. Leipunskii, and I. I. Morozov, *J. Chem. Phys.* **64**, 3138 (1976).
960. I. Tanaka and J. R. McNesby, *J. Chem. Phys.* **36**, 3170 (1962).
961. Y. Tanaka, E. C. Y. Inn, and K. Watanabe, *J. Chem. Phys.* **21**, 1651 (1953).
962. A. M. Tarr, O. P. Strausz, and H. E. Gunning, *Trans. Faraday Soc.* **61**, 1946 (1965).
963. G. W. Taylor and D. W. Setser, *J. Chem. Phys.* **58**, 4840 (1973).
964. J. M. Tedder and J. C. Walton, *Chem. Commun.* **1966**, 140.
965. T. B. Tellinghuisen, C. A. Winkler, S. W. Bennett, and L. F. Phillips, *J. Phys. Chem.* **75**, 3499 (1971).
966. S. G. Thomas, Jr., and W. A. Guillory, *J. Phys. Chem.* **77**, 2469 (1973).
967. B. A. Thompson, P. Harteck, and R. R. Reeves, Jr., *J. Geophys. Res.* **68**, 6431 (1963).
968. B. A. Thompson, R. R. Reeves, Jr., and P. Harteck, *J. Phys. Chem.* **69**, 3964 (1965).
969. R. Thomson and P. A. Warsop, *Trans. Faraday Soc.* **65**, 2806 (1969).
970. R. Thomson and P. A. Warsop, *Trans. Faraday Soc.* **66**, 1871 (1970).
971. W. B. Tiffany, *J. Chem. Phys.* **48**, 3019 (1968).
972. S. G. Tilford, P. G. Wilkinson, and J. T. Vanderslice, *Astrophys. J.* **141**, 427 (1965).
973. C. T. Ting and R. E. Weston, Jr., *J. Phys. Chem.* **77**, 2257 (1973).
974. S. Toby and G. O. Pritchard, *J. Phys. Chem.* **75**, 1326 (1971).
975. E. Treiber, J. Gierer, J. Rehnström, and K. E. Almin, *Acta Chem. Scand.* **11**, 752 (1957).
976. A. Trombetti, *Can. J. Phys.* **46**, 1005 (1968).
977. E. Tschuikow-Roux and S. Kodama, *J. Chem. Phys.* **50**, 5297 (1969).
978. E. Tschuikow-Roux, Y. Inel, S. Kodama, and A. W. Kirk, *J. Chem. Phys.* **56**, 3238 (1972).
979. M. Tsukada and S. Shida, *Bull. Chem. Soc. Jap.* **43**, 3621 (1970).
980. S. Tsurubuchi, *Chem. Phys.* **10**, 335 (1975).
980a. K. D. Tucker, M. L. Kutner, and P. Thaddeus, *Astrophys. J.* **193**, L115 (1974).
981. R. P. Turco and R. C. Whitten, *Atmos. Environ.* **9**, 1045 (1975).
982. B. E. Turner, *Astrophys. J.* **163**, L35 (1971).
983. A. Y. M. Ung and H. I. Schiff, *Can. J. Chem.* **44**, 1981 (1966).
984. A. Y. M. Ung, *Chem. Phys. Lett.* **28**, 603 (1974).
985. W. M. Uselman and E. K. C. Lee, *Chem. Phys. Lett.* **30**, 212 (1975).
986. W. M. Uselman and E. K. C. Lee, *J. Chem. Phys.* **64**, 3457 (1976).
987. W. M. Uselman and E. K. C. Lee, *J. Chem. Phys.* **65**, 1948 (1976).
988. H. E. Van den Bergh, A. B. Callear, and R. J. Norstrom, *Chem. Phys. Lett.* **4**, 101 (1969).

989. H. E. Van den Bergh, and A. B. Callear, *Trans. Faraday Soc.* **67**, 2017 (1971).

990. H. E. Van den Bergh, *Chem. Phys. Lett.* **43**, 201 (1976).

991. C. Vermeil, *Isr. J. Chem.* **8**, 147 (1970).

992. G. Van Volkenburgh, T. Carrington, and R. A. Young, *J. Chem. Phys.* **59**, 6035 (1973).

993. D. H. Volman, *J. Chem. Phys.* **17**, 947 (1949).

994. D. H. Volman, *J. Chem. Phys.* **24**, 122 (1956).

995. D. H. Volman, in *Advances in Photochemistry*, Vol. 1, W. A. Noyes, Jr., G. S. Hammond and J. N. Pitts, Jr., Eds., Interscience, New York, 1963, p. 43.

996. G. von Bunau and R. N. Schindler, *J. Chem. Phys.* **44**, 420 (1966).

997. G. von Bunau and R. N. Schindler, *Ber. Bunsenges. Phys. Chem.* **72**, 142 (1968).

998. G. von Ellerrieder, E. Castellano, and H. J. Schumacher, *Chem. Phys. Lett.* **9**, 152 (1971).

999. F. M. Wachi and D. E. Gilmartin, *High Temp. Sci.* **4**, 423 (1972).

1000. L. Wallace and D. M. Hunten, *J. Geophys. Res.* **73**, 4813 (1968).

1001. R. Walsh and S. W. Benson, *J. Am. Chem. Soc.* **88**, 4570 (1966).

1002. J. C. Walton, *J. Chem. Soc. Faraday I* **68**, 1559 (1972).

1003. F. B. Wampler, A. Horowitz, and J. G. Calvert, *J. Amer. Chem. Soc.* **94**, 5523 (1972).

1004. F. B. Wampler, J. G. Calvert, and E. K. Damon, *Int. J. Chem. Kinet.* **5**, 107 (1973).

1005. F. B. Wampler, K. Otsuka, J. G. Calvert, and E. K. Damon, *Int. J. Chem. Kinet.* **5**, 669 (1973).

1006. C. C. Wang and L. I. Davis, Jr., *Phys. Rev. Lett.* **32**, 349 (1974).

1007. D. Q. Wark and D. M. Mercer, *Appl. Opt.* **4**, 839 (1965).

1008. P. Warneck, *Appl. Opt.* **1**, 721 (1962).

1008a. P. Warneck, F. F. Marmo, and J. O. Sullivan, *J. Chem. Phys.* **40**, 1132 (1964).

1009. P. Warneck, *J. Opt. Soc. Am.* **56**, 408 (1966).

1010. N. Washida, H. Akimoto, and I. Tanaka, *Appl. Opt.* **9**, 1711 (1970).

1011. N. Washida, Y. Mori, and I. Tanaka, *J. Chem. Phys.* *54*, 1119 (1971).

1012. N. Washida and K. D. Bayes, *Chem. Phys. Lett.* **23**, 373 (1973).

1013. N. Washida, R. I. Martinez, and K. D. Bayes, *Z. Naturforsch.* *29a*, 251 (1974).

1013a. E. Wasserman, W. A. Yager, and V. J. Kuck, *Chem. Phys. Lett.* **7**, 409 (1970).

1014. K. Watanabe, E. C. Y. Inn, and M. Zelikoff, *J. Chem. Phys.* **21**, 1026 (1953).

1015. K. Watanabe and E. C. Y. Inn, *J. Opt. Soc. Am.* **43**, 32 (1953).

1016. K. Watanabe and M. Zelikoff, *J. Opt. Soc. Am.* **43**, 753 (1953).

1017. K. Watanabe, *J. Chem. Phys.* **22**, 1564 (1954).

1018. K. Watanabe and A. S. Jursa, *J. Chem. Phys.* **41**, 1650 (1964).

1019. K. Watanabe and S. P. Sood, *Sci. Light* **14**, 36 (1965)

1020. K. Watanabe, F. M. Matsunaga, and H. Sakai, *Appl. Opt.* **6**, 391 (1967).

1021. T. S. Wauchop and L. F. Phillips, *J. Chem. Phys.* **47**, 4281 (1967).

1022. T. S. Wauchop and L. F. Phillips, *J. Chem. Phys.* **51**, 1167 (1969).

1023. T. S. Wauchop, M. J. McEwan, and L. F. Phillips, *J. Chem. Phys.* **51**, 4227 (1969).

1024. T. S. Wauchop and H. P. Broida, *J. Geophys. Res.* **76**, 21 (1971).

1024a. R. W. Waynant, J. D. Shipman, Jr., R. C. Elton, and A. W. Ali, *Appl. Phys. Lett.* **17**, 383 (1970).

1024b. R. W. Waynant *Appl. Phys. Lett.*, **30** 235 (1977).

1025. R. P. Wayne, *Nature* **203**, 516 (1964).

1026. H. Webster III and E. J. Bair, *J. Chem. Phys.* **53**, 4532 (1970).

1027. H. Webster III and E. J. Bair, *J. Chem. Phys.* **57**, 3802 (1972).

1028. E. Weissberger, W. H. Breckenridge, and H. Taube, *J. Chem. Phys.* **47**, 1764 (1967).

1029. K. H. Welge, *J. Chem. Phys.* **45**, 166 (1966).

1030. K. H. Welge, *J. Chem. Phys.* **45**, 1113 (1966).

1031. K. H. Welge, *J. Chem. Phys.* **45**, 4373 (1966).

1032. K. H. Welge, J. Warnner, F. Stuhl, and A. Heindrichs, *Rev. Sci. Inst.* **38**, 1728 (1967).

1033. K. H. Welge and F. Stuhl, *J. Chem. Phys.* **46**, 2440 (1967).

1034. K. H. Welge, S. V. Filseth, and J. Davenport, *J. Chem. Phys.* **53**, 502 (1970).

1035. K. H. Welge, H. Zia, E. Vietzke, and S. V. Filseth, *Chem. Phys. Lett.* **10**, 13 (1971).

1036. K. H. Welge and R. Gilpin, *J. Chem. Phys.* **54**, 4224 (1971).

1037. G. A. West and M. J. Berry, *J. Chem. Phys.* **61**, 4700 (1974).

1038. J. M. White, R. L. Johnson, Jr., and D. Bacon, *J. Chem. Phys.* **52**, 5212 (1970).

1039. J. M. White and R. L. Johnson, Jr., *J. Chem. Phys.* **56**, 3787 (1972).

1039a. A. G. Whittaker and P. Kintner, *Rev. Sci. Inst.* **38**, 1743 (1967).

1040. W. H. J. Wijnen, *J. Am. Chem. Soc.* **83**, 3014 (1961).

1041. P. G. Wilkinson, *J. Quant. Spectrosc. Radiat. Transfer* **5**, 503 (1965).

1042. P. G. Wilkinson and E. T. Byram, *Appl. Opt.* **4**, 581 (1965).

1043. P. G. Wilkinson, *J. Quant. Spectrosc. Radiat. Transfer* **6**, 823 (1966).

1044. D. Wilcox, R. Anderson, and J. Peacher, *J. Opt. Soc. Am.* **65**, 1368 (1975).

1045. W. J. Williams, J. N. Brooks, D. G. Murcray, F. H. Murcray, P. M. Fried, and J. A. Weinman, *J. Atmos. Sci.* **29**, 1375 (1972).

1045a. D. G. Williamson and K. D. Bayes, *J. Am. Chem. Soc.* **89**, 3390 (1967).

1046. D. G. Williamson and K. D. Bayes, *J. Am. Chem. Soc.* **90**, 1957 (1968).

1047. C. Willis and K. D. Bayes, *J. Phys. Chem.* **71**, 3367 (1967).

1048. C. Willis and R. A. Back, *Can. J. Chem.* **51**, 3605 (1973).

1049. C. Willis, F. P. Lossing, and R. A. Back, *Can. J. Chem.* **54**, 1 (1976).

1050. C. Willis, R. A. Back, and J. M. Parsons, *J. Photochem.* **6**, 253 (1976/77).

1051. D. E. Wilson and D. A. Armstrong, *Radiat. Res. Rev.* **2**, 297 (1970).

1052. N. W. Winter, C. F. Bender, and W. A. Goddard III, *Chem. Phys. Lett.* **20**, 489 (1973).

1053. N. W. Winter, *Chem. Phys. Lett.* **33**, 300 (1975).

1054. K. L. Wray and S. S. Fried, *J. Quant. Spec. Radiat. Transfer* **11**, 1171 (1971).

1055 T. J. Wright, W. S. Spates, and S. J. Davis, *J. Chem. Phys.* **66**, 1566 (1977).

1056. G. O. Wood and J. M. White, *J. Chem. Phys.* **52**, 2613 (1970).

1056a. B. W. Woodward, V. J. Ehlers, and W. C. Lineberger, *Rev. Sci. Inst.* **44**, 882 (1973).

1057. W. D. Woolley and R. A. Back, *Can. J. Chem.* **46**, 295 (1968).

1058. C. H. Wu, C. C. Wang, S. M. Japar, L. I. Davis, Jr., M. Hanabusa, D. Killinger, H. Niki, and B. Weinstock, *Int. J. Chem. Kinet.* **8**, 765 (1976).
1059. S. Yamamoto, K. Tanaka and S. Sato. *Bull. Chem. Soc. Jap.* **48**, 2172 (1975).
1060. H. Yamazaki and R. J. Cvetanović *J. Chem. Phys.* **40**, 582 (1964).
1061. J. Y. Yang and F. M. Servedio, *J. Chem. Phys.* **47**, 4817 (1967).
1062. J. Y. Yang and F. M. Servedio, *Can. J. Chem.* **46**, 338 (1968).
1063. K. Yang, J. D. Paden, and C. L. Hassell, *J. Chem. Phys.* **47**, 3824 (1967).
1064. P. E. Yankwich and E. F. Steigelmann, *J. Phys. Chem.* **67**, 757 (1963).
1065. M. Yoshida and I. Tanaka, *J. Chem. Phys.* **44**, 494 (1966).
1066. K. Yoshino and Y. Tanaka, *J. Chem. Phys.* **48**, 4859 (1968).
1066a. A. T. Young, *Icarus* **18**, 564 (1973).
1067. R. A. Young and A. Y. M. Ung, *J. Chem. Phys.* **44**, 3038 (1966).
1068. R. A. Young and G. Black, *J. Chem. Phys.* **47**, 2311 (1967).
1069. R. A. Young and G. A. St. John, *J. Chem. Phys.* **48**, 898 (1968).
1070. R. A. Young, G. Black, and T. G. Slanger, *J. Chem. Phys.* **48**, 2067 (1968).
1071. R. A. Young, G. Black, and T. G. Slanger, *J. Chem. Phys.* **49**, 4769 (1968).
1072. R. A. Young, G. Black, and T. G. Slanger, *J. Chem. Phys.* **50**, 303 (1969).
1073. R. A. Young, G. Black, and T. G. Slanger, *J. Chem. Phys.* **50**, 309 (1969).
1073a. R. S. Young, *Am. Sci.* **64**, 620 (1976).
1074. E. S. Yeung and C. B. Moore, *J. Am. Chem. Soc.* **93**, 2059 (1971).
1075. E. S. Yeung and C. B. Moore, *Appl. Phys. Lett.* **21**, 109 (1972).
1076. E. S. Yeung and C. B. Moore, *J. Chem. Phys.* **58**, 3988 (1973).
1077. W. H. S. Yu and M. H. J. Wijnen, *J. Chem. Phys.* **52**, 2736 (1970).
1078. V. Zabransky and R. W. Carr, Jr., *J. Phys. Chem.* **79**, 1618 (1975).
1079. M. Zelikoff, K. Watanabe, and E. C. Y. Inn, *J. Chem. Phys.* **21**, 1643 (1953).
1080. M. Zelikoff and L. M. Aschenbrand, *J. Chem. Phys.* **22**, 1680 (1954).
1081. M. Zelikoff and L. M. Aschenbrand, *J. Chem. Phys.* **22**, 1685 (1954).
1082. M. Zelikoff and L. M. Aschenbrand, *J. Chem. Phys.* **24**, 1034 (1956).
1083. M. W. Zemansky, *Phys. Rev.* **36**, 919 (1930).
1084. C. Zetzsch and F. Stuhl, *Chem. Phys. Lett.* **33**, 375 (1975).
1084a. C. Zetzsch and F. Stuhl, *Ber. Bunsenges. Phys. Chem.* **80**, 1354 (1976).
1085. E. C. Zipf, *Can. J. Chem.* **47**, 1863 (1969).

Index

Greek letters used as symbols are arranged in alphabetical order according to their English names and are placed at the beginning of each corresponding section. *Italic* page numbers refer to the main discussions of the subjects. Letters F and T after page numbers denote figures and tables.

All molecules discussed in the book are given in chemical formulae. Their order follows that of the Hill indexing system as used by Chemical Abstracts. The order of atomic symbols in a chemical formula is alphabetical except for carbon containing molecules for which C comes first followed by H if H is present, and the remaining symbols are arranged in alphabetical order. The molecules with one carbon precede those with two carbon atoms and those with one hydrogen are placed before those with two hydrogen atoms. The following series illustrates the rules; Ar, BCl_3, CBr_2Cl_2, CCl_2O (phosgene), CCl_4, $CHBrCl_2$, CH_2Cl_2, CH_3Cl, CH_4, C_2HCl, C_2H_2, C_2H_4, Cl_2O, HNO_3, H_3N (ammonia), NO, O_2S (sulfur dioxide), O_3.

A

C

D

I

O

W

X

Z